PRECISION MEASUREMENT
IN THE
METAL WORKING INDUSTRY

Precision Measurement in the Metal Working Industry

PREPARED BY THE DEPARTMENT OF EDUCATION OF
INTERNATIONAL BUSINESS MACHINES CORPORATION

Syracuse University Press

Revised 1952
Second printing 1955
Third printing 1957
Fourth printing 1962
Fifth printing 1965
Sixth printing 1966
Seventh printing 1969
Eighth printing 1978, *first paperback binding*
Ninth printing 1982, *second paperback binding*
Tenth printing 1985, *third paperback binding*
Eleventh printing 1989, *fourth paperback binding*

International Business Machines Corporation. Educational Dept.

Precision measurement in the metal working industry/prepared by the Department of Education of International Business Machines Corporation. — Paperback ed. — Syracuse, N.Y.: Syracuse University Press, c1978.

365 p.: ill.; 26 cm.

Published in 1939-1941 under title: Measuring instruments.
Includes index.
ISBN 0-8156-2194-9 (pbk.)

1. Measuring instruments. 2. Metal-work. I. Title.
TJ1313.I58 1978 671′.028 78-104814
MARC

Library of Congress 78[8806]

Manufactured in the United States of America

PRECISION MEASUREMENT IN THE METAL WORKING INDUSTRY, prepared by the International Business Machines Corporation as a text to train workers for the war effort, first appeared in 1939, in two volumes. Reissued in 1943, it became the first book to bear the imprint of the newly founded Syracuse University Press. A one-volume revised edition, published in 1952, has been continuously in demand since then. Now available in paper binding, this edition is dedicated, with appreciation, to the founder of Syracuse University Press: Chancellor Emeritus William Pearson Tolley.

THE PUBLISHER

PREFACE

MODERN PRODUCTION depends on interchangeability and uniform accuracy of parts. In America a century ago, Eli Whitney first demonstrated the value of parts that could be interchanged; more recently American industry proved both for all time by pioneering the mass production of automobiles and, later, a great variety of other products--from home appliances to computing machines.

Today, in countless laboratories and factories, the search goes on for answers to the problems of production: better materials, more practical design, improved machines, better ways of making and assembling parts. Many of these answers are provided by precision measurement--through accuracy in production, high standards of inspection, new and improved instruments, uniform measuring practices. Here is a fast-growing, changing, and increasingly significant field.

This book deals with the fundamentals of precision measurement. In the early 1940's "Precision Measurement in the Metal Working Industry" was published in two volumes, based on previous informal publications bearing similar names. As the IBM Department of Education continued to work on problems of training in inspection and the use of new instruments and methods, technological advances called for revised and additional instructional material. Much of that information became the basis for this completely revised edition in a new single-volume format.

The new edition remains a practical description of actual shop experience. All pertinent material in former editions has been retained; new text has been added to bring the information up to date; new illustrations show important changes in instruments. The result is a modern version of a book that has long been a leader in its field.

IBM engineering and manufacturing personnel have collaborated actively in the preparation of this material. Manufacturers of precision instruments, representatives of standards organizations, and others--named in the Acknowledgements section--have contributed information, counsel, and illustrations. Usefulness of the book is enhanced by the experience and constructive criticism of both military and civilian instructors who have used the previous editions as a working textbook in government arsenals, Army, Navy,and Air Force schools, industrial training classes, and technological and vocational classes.

Syracuse, New York
September 17, 1952

ACKNOWLEDGEMENTS

MANY persons and organizations, as well as various members of the IBM Corporation, have made significant contributions to the revision of this edition.

To the National Bureau of Standards and to the American Standards Association, both of which reviewed the original text and made further helpful suggestions for the new edition, go the continued thanks of IBM and the publisher. Special thanks, too, are due the American Society of Mechanical Engineers for permission to quote in considerable detail from various bulletins.

The IBM Department of Education expresses appreciation to the following business firms for the many contributions of information and illustrations they have made to the original work and to this edition:

Acme Industrial Company
American Instrument Company
The B.C. Ames Company
Ames Precision Machine Works
Ansco Division of General Aniline and Film Corporation
Baldwin-Lima-Hamilton Corp.
Bausch & Lomb Optical Company
Brown & Sharpe Manufacturing Company
The Brush Development Company
Cadillac Gage Company
Canadian Radium & Uranium Corp.
Coats Machine Tool Company
The Comtor Company
The DoAll Company
Eastman Kodak Company
The Elkro Company
Engineers Specialties Division, The Universal Engraving and Colorplate Company, Incorporated
Federal Products Corporation
Ford Motor Company
The Gaertner Scientific Corporation
General Electric Company
Hanson-Whitney Company, Division of Whitney Chain Company
Hardinge Brothers, Incorporated
Hunter Spring Company
Johnson Gage Company
Jones & Lamson Machine Company
The Lufkin Rule Company
Magnaflux Corporation
Micrometrical Manufacturing Company
Newage International, Incorporated
Tinius Olsen Testing Machine Company
Pratt & Whitney Division, Niles-Bement-Pond Company
Robbins Engineering Company
George Scherr Company, Incorporated
The Sheffield Corporation
The Shore Instrument & Manufacturing Company, Incorporated
Sperry Products, Incorporated
Standard Gage Company
The L. S. Starrett Company
The Taft-Peirce Manufacturing Company
Thorndike Agency
The Van Keuren Company
The Velsey Company
Vinco Corporation
Wilson Mechanical Instrument Division, American Chain and Cable Company

CONTENTS

Chapter I

HISTORY AND BASIC PRINCIPLES

THE term "precision measurement" is applied to the field of measurement beyond the scope of non-precision line-graduated measuring instruments, such as the rule and scale. It refers to the art of reproducing and controlling dimensions expressed in thousandths of an inch or smaller. The instruments used to accomplish this purpose are known as precision measuring instruments.

In an essay published by the Ford Motor Company, "He Measured in Millionths," James Sweinhart tells how man has again and again devised means outside his physical self to help him extend his native abilities: the sound; the plumb; the prism and lens; the square, the angle and the triangle; the telescope. "But always, eventually," he continues, "their limit of use was reached. Always came a question of extension or refinement. Always, confronting, arose the inescapable problem of smaller and more accurate measurement. In many different branches of his explorations and investigations there came into being an unwritten but undeviating law that man's knowledge advances to the degree that he can measure precisely."

In the world's research and engineering laboratories this never-ending struggle to solve and measure the vast and mysterious phenomena of nature goes on. The achievements of these men of science read almost as fiction.

BRIEF HISTORY OF LINEAR MEASUREMENT

Evidences of linear measurement can be definitely traced back to the third century B.C., but there is reason to believe a system of measurement existed prior to this time. For example, the great pyramid of Khufu at Gizeh in Egypt was built about 4750 B.C. Although this edifice covers an area of 13 acres, the mean error in length of the sides of the base is about 6/10 of an inch, and the angles are not in error more than 12 seconds from a perfect square. The workmanship of these Egyptian artisans is remarkable even today. They finished the immense blocks of stone so accurately that the pyramids could be constructed without mortar. These men not only had an amazing amount of skill, but they also must have had an accurate system of linear and angular measurement.

The earliest known unit of measure is called the cubit. It was based on the length of the forearm from the elbow to the tip of the middle finger. Of several standard cubits discovered, two were outstanding: the Olympic cubit which averaged 18.24 inches, and the Royal Egyptian cubit which averaged 20.62 inches.

At later dates, various units, sub-units, and names were developed. Among these were:

Span -- One-half of the Olympic cubit, or about 9 inches.

Palm -- One-sixth of the Olympic cubit, or about 3 inches.

Digit -- One twenty-fourth of the Olympic cubit or about 3/4 of an inch.

Subsequently the cubit became known as a foot and was the equivalent of 12 thumbnail breadths or "unciae," which in Anglo-Saxon countries later became inches.

In the 12th century, the yard was determined by Britain as being the distance from the point of the nose of King Henry I to the end of his thumb. In 1324 the inch was defined as the combined length of three barleycorns set end to end. In 1558 the length of a certain bronze bar was decreed to be the standard yard.

An attempt was made in the 16th century to standardize the foot, which up to that time ranged in length from 9-3/4 inches to 19 inches. The order defined the foot in terms of the rod as follows: "On a certain Sunday as they come out of church, 16 men shall stand in line with the left feet touching, one behind the other. This distance shall be the legal rod and one-sixteenth of it shall be the foot."

In 1824 the yard of 1558 (known as Queen Elizabeth's yard) was superseded by "Bird's standard yard," but it was lost by fire in 1834.

Nearly all of these efforts to set up standards or units of measure involved measurements that were not stable or constant, parts of the body such as feet, arms, hands, or fingers. Not one was considered absolute nor could any one be referred to as a really basic, unvarying standard.

The Weights and Measures Act of 1878 defined the British Standard Yard in use today. It was declared to be the distance at 62.00° F. between two fine lines engraved on gold studs in a bronze bar. This bar was cast in 1844.

In France, scientists determined to establish a unit of measure that had a permanent relationship with something more concrete. They carefully computed the earth's diameter and circumference and then divided the meridian quadrant into ten million parts. To the unit of length produced, the designation "meter" was given. It became the foundation for what is now known as the metric system of linear measure. Later the computation was discovered to be inaccurate, and in 1927 the meter was redefined and is no longer associated with the size of the earth. The most recent definition of the meter, adopted in 1927, is as follows:

"The unit of length is the meter, defined by the distance at the temperature of melting ice between the centers of two lines traced on the plantinum-iridium bar deposited at the International Bureau of Weights and Measures, and declared prototype of the meter by the First General Conference on Weights and Measures, this bar being subjected to normal atmospheric pressure and supported by two rollers at least one centimeter in diameter situated in the same horizontal plane and at a distance of 572 millimeters from each other."

As an alternative or provisional definition which refers to some natural standard, the length of the meter is defined in terms of the wave length of light, as 1,553,164.13 wave lengths of the red light emitted by a cadmium vapor lamp excited under specified conditions. The relative accuracy of the value of the meter in terms of light waves is one part in ten million.

The International Bureau of Weights and Measures at the Pavillon de Breteuil, near Severs, France, has been declared international property by the French government and generally is regarded by the principal nations as possessing the basic standard of linear measure.

METRIC SYSTEM OF LINEAR MEASURE

The metric system of linear measure is based on the meter as the unit. The meter is subdivided into tenths, hundredths, and thousandths, and similarly is increased in multiples of ten, as shown in Table I.

TABLE I

10 millimeters (mm) equal 1 centimeter (cm)
10 centimeters (cm) equal 1 decimeter (dm)
10 decimeters (dm) equal 1 meter (m)
10 meters (m) equal 1 dekameter (dkm)
10 dekameters (dkm) equal 1 hektometer (hm)
10 hektometers (hm) equal 1 kilometer (km)

The following prefixes are used with the metric system of weights and measures:

milli	1/1000	(one-thousandth)
centi	1/100	(one-hundredth)
deci	1/10	(one-tenth)
deka	10	(ten)
hekto	100	(hundred)
kilo	1000	(thousand)

The metric system is almost universally used in Europe (except Great Britain), and considerably in the United States, especially in scientific laboratories and some types of manufacturing. The ball bearing industry, for example, lists all sizes of bearings in even millimeters, whereas the corresponding American figures run to four and five decimals. Manufacturers building machines and instruments largely for export to foreign countries occasionally use the metric system to facilitate assembly, servicing, and replacement of standard parts.

BRITISH IMPERIAL SYSTEM OF LINEAR MEASURE

The British imperial system differs from the metric in the size of the unit and in the manner in which it is subdivided. The names applied to the various multiples of the unit are those handed down from the Egyptians, Greeks, and Anglo-Saxons. The standard measuring length in the British imperial system is the yard, and the smallest unit identified by name is the inch. The present legal equivalent of the British yard is approximately 0.9143992 of a meter, a value which makes 1 inch equal to 2.54 cm, correct to two parts in a million. The system includes the following units:

12 inches equal 1 foot
3 feet equal 1 yard
5-1/2 yards equal 1 rod
40 rods equal 1 furlong
8 furlongs equal 1 mile

AMERICAN STANDARD OF LINEAR MEASURE

In 1856 the British government presented the United States with two bars, each of them one yard in length. These bars were regarded for many years as the best representation of the yard in this country, but they are no longer officially considered as having anything except historical significance.

In 1866 Congress legalized the use of the metric system. As a result of its participation in the metric convention, the United States government in 1889 was presented with a platinum-iridium meter which is an exact prototype of the International Standard Meter owned by France. This meter is called Meter No. 27, and since 1893 has been used as the standard in this country for all length measurements. It is the only standard of length authorized in the United States.

By law, the meter is declared equal to 39.37 inches, which makes the American legal yard 3600/3937 of a meter and the United States inch equal to 2.540005 + cm.

The following tables give the relationships between the American and metric systems, which permit conversion from one to the other. In calculating the figures, 1 inch is considered equal to 2.54 cm. This is an approximation to 2 parts in a million, as previously stated, making most of the relationships approximate:

1 millimeter equals .03937 inches exactly, or approximately 1/25 inch.
1 centimeter equals exactly .3937 inches.
1 meter equals 3.28083 feet or 1.09361 yards (approximately).
1 kilometer equals .62137 miles (approximately).

and conversely:

1 inch equals 25.4 millimeters or 2.54 centimeters (approximately).
1 foot equals .3048 meters (approximately).
1 yard equals .9144 meters (approximately).
1 mile equals 1.60934 kilometers.

The subdivisions or units of measure in the two systems have no uniform relation, and the only use for them is in the conversion from one system to the other.

INTERCHANGEABILITY OF PARTS

Eli Whitney, while known as the inventor of the cotton gin, is better remembered for his work as the forerunner of our modern mass production system, the manufacture of interchangeable parts.

Early craftsmanship failed during the 18th century when large armies demanded muskets. Each gun was individually made, each part fitted to each musket. Whitney recognized the deficiencies of such a system and decided to build a gun factory in which machinery would be used wherever possible to produce parts that were interchangeable. In order to produce parts usable on any gun he manufactured, he established measurement standards. By strict adherence to the gage standards set up, any part of a Whitney musket would fit any other Whitney musket, but the same muskets manufactured by others from identical drawings and dimensions were radically different, because of a lack of common standards.

Prior to World War I, interchangeable parts manufacture was confined largely to the products of a single plant. The sudden demand for war material and the inadequacy of the equipment in any one place for producing a complete unit made it necessary to delegate the manufacture of a machine to a number of plants. A great many difficulties arose from the fact that there were no universal standards of length and the parts from one plant would not fit those made in another.

The National Bureau of Standards undertook to calibrate and check all master industrial gages used in the manufacture of war material. Their efforts led to the subsequent use of gage blocks certified by the Bureau of Standards in controlling the accuracy of measuring instruments and gages in large industrial plants.

Rapid industrial development through mass production of popular-priced, high-quality machines required more exacting methods of manufacture and finer instruments to control them. The recognition of the value of precision gage blocks grew until today they are used by nearly every industrial concern making a product whose quality depends on the accuracy of its parts.

As faster and better-performing machines are developed, parts more accurate than ever before must be produced in great quantities. This means further use of precision gage blocks and further development in instruments capable of controlling smaller and smaller dimensional tolerances.

As plants continue to decentralize, as so many of the large automobile companies have done, universal standards of measure become increasingly important. In the assembly of the automobile, the component parts may come from hundreds of different points scattered over the country. To have these parts fit means that tolerances must be maintained and that all contributors to the final assembly must operate from a universal standard.

Products in the possession of customers must have interchangeable working parts throughout so that they may be properly serviced. This is possible only by using precision measurements based on universal standards.

STANDARD MEASURING TEMPERATURES

Until a few years ago there was no single method of converting the American inch accurately into the metric values of other countries. The standard measuring temperature in the United States and Canada was 68°F., and at this temperature one inch equals 25.40005 millimeters. The standard British equivalent was 25.39998 millimeters at 62°F., while the meter was defined at a measuring temperature of 0°C. (or 32°F.). Similar variations also existed among other countries. To overcome difficulties resulting from these different national standards, a group of large industrial concerns and standards organizations undertook the creation of a universal standard for conversion between the British imperial system and the French metric system. Many obstacles had to be overcome, including international prejudice or preference, but the largest of all was the revision of established standards and recalibrating industrial gages. Despite the difficulties involved, industrial nations of the world recognized the wisdom of having a universal standard based on a universal conversion factor between the imperial and metric systems, and a universal measuring temperature.

In 1929, at one of the sessions of the International Commission of Weights and Measures, the measuring temperature of 20°C., which is equivalent to 68°F., was established as standard for the metric system. In 1930, the British Standards Association adopted 68°F. as standard for the British imperial system and also adopted the conversion factor of 25.4 millimeters for one inch. In 1932, the American Standards Association submitted the standard measuring temperature (68°F.) and the fixed conversion factor (2.54), and they were approved by the American industries. However, this conversion factor is not yet legal so far as the United States government is concerned.

During the latter part of 1934, the Deutscher

Normenausshuss (German Standards Association) announced that it, too, had adopted these standards. Sweden, Finland, Denmark, Japan, and several other countries have cooperated with the International Standards Association (I. S. A.) and likewise have adopted the 2.54 conversion factor and the 68°F. or 20°C. measuring temperature.

Now that an international standard has been established, 2.54 is a fixed conversion factor and not a long irrational number having different values in different countries. The universally used measuring temperature is 68° F. or 20°C. and the centimeter and the inch are each the same all over the world.

AMERICAN STANDARDS ASSOCIATION

The American Standards Association is a clearing house for voluntary development of standards for all interested persons. It was organized in 1918 by five engineering societies as the American Engineering Standards Committee. Its purpose was to serve as a coordinating body, eliminating duplication and overlapping of their standardization activities.

As the work of the committee grew and its activities extended into new industrial fields, reorganization became necessary along broader, more flexible lines to meet the growing need and demand for adequate national standards. In 1928, the name was changed to American Standards Association. Today the membership of the Association includes some 100 national trade associations, technical societies and consumer organizations, and more than 2000 companies. Membership is open to any industrial, commercial, technical, or governmental group interested in standardization work.

The methods used by the Association have been developed through more than 32 years of experience in dealing with difficult inter-group problems. Based as they are on the assent of all groups having a substantial interest in the completed work, American standards are today becoming a positive force in American business. Eleven hundred standards and safety codes have thus far been approved.

The American Standards Association is the United States member of the International Organization for Standardization (I.S.O.) through which the national standardizing bodies of 30 countries carry on their general cooperative activities. Through this and other means, the A.S.A. makes available to American industry direct and authoritative contact with standardization developments in other countries.

The broad range of projects undertaken by the A.S.A. includes dimensional standards to allow for interchangeability of supplies or to secure the interworking of parts or interrelated apparatus; specifications for materials and methods of test; definitions of technical terms used in industry, industrial safety codes to make possible uniform requirements in safety devices for machines and other equipment in the fields of both public and industrial safety, industrial health codes for the prevention of occupational diseases, development of a national building code; specifications for consumer goods sold in retail trade.

NATIONAL BUREAU OF STANDARDS

Established by act of Congress in 1901, the National Bureau of Standards is the principal agency of the federal government for fundamental research in physics, mathematics, chemistry, and engineering. It has custody of the national standards of physical measurement, in terms of which all working standards in research laboratories and industry are calibrated, and carries on research leading to improvement in such standards and measurement methods. In addition to basic and applied research, the bureau determines physical constants and properties of materials and develops improved methods for testing materials and equipment.

The Bureau is authorized to make tests and calibrations for American industry, on a fee basis, when devices or materials must be checked with the Bureau's standards or when sufficient accuracy cannot be obtained elsewhere than at the Bureau. The Bureau assists industries in maintaining accurate standards of linear measurement by certifying master gage blocks to the American standard inch when this service is requested. General information on the testing program and the fees required is given in National Bureau of Standards Circular 483,

"Testing by the National Bureau of Standards," available from the Superintendent of Documents, U.S. Government Printing Office, Washington 25, D.C.

An important phase of the Bureau's work consists in cooperation with technical and trade associations, both in this country and abroad, on problems of concern to the Government and the nation, particularly those relating to the determination and establishment of scientific quantities and standards. In this way, organizations such as the American Society for Testing Materials and the American Standards Association are assisted in the development of specifications and industrial standards. As a large part of the research and testing has direct bearing on technical requirements for safe working and living conditions, the Bureau provides a central source of information to which federal, state, and municipal authorities, as well as industrial and trade associations, can turn when dealing with problems of safety or with building or plumbing codes. The Bureau also plays an important part in the development and establishment of federal specifications; these specifications insure quality and economy in federal purchase while providing an equal opportunity to all suppliers to compete for federal purchases.

The result of much of the Bureau's research and development are of direct interest to industry. These contributions are made available to the scientific and engineering world through publication in the "Journal of Research" of the National Bureau of Standards, and the National Bureau of Standards' "Technical News Bulletin," both of which are available from the Superintendent of Documents on a subscription basis.

The National Bureau of Standards now employs over 3,000 people, of whom approximately three-fourths are technically trained. Scientific and technical activities are carried out in 15 divisions concerned with electricity, optics and metrology, heat and power, atomic and radiation physics, chemistry, mechanics, organic and fibrous materials, metallurgy, mineral products, building technology, applied mathematics, electronics, ordnance development, and so on.

INDUSTRIAL STANDARDS OF LINEAR MEASURE

Manufacturing industries throughout the world have been instrumental in the establishment of basic standards of measurement. Through control of the accuracy of measuring instruments used to check vital dimensions of machine parts, manufacturers are able to guarantee high standards of quality and performance. By establishment of linear standards of measure and development of ways to bring them down to the bench, interchangeable part manufacture has been made possible. Universal standards enable manufacturers to secure standard parts, such as screws, nuts, and pins, made by many different concerns and to know those parts will fit into their proper places. These developments enable a manufacturer to order by blueprint, from a vendor in a distant city, parts to his specifications and receive them ready for use.

Modern production methods of manufacture require the assembly of machines with a minimum of fitting. Back of this mass production of interchangeable parts must be a set of universal standards or master gages with which the instruments employed to control accuracy may be calibrated. These instruments have gone through a long process of development. In 1851, the vernier caliper, the most accurate instrument of its time, was invented. It was capable of measuring to one thousandth of an inch. In 1867, the Système Palmer appeared; this became the micrometer caliper. In 1896, the metric gage block, the first of the precision gage blocks, was developed.

During the last 35 years greater accomplishments have been made. Measurements in terms of millionths of an inch are now possible by machines and instruments. Many modern industrial plants have master sets of gage blocks with a guaranteed accuracy of two and a half millionths of an inch (.0000025). Such measurement is difficult for the layman to visualize, but it is illustrated in Figure 1; one millionth of an inch has the same relationship to one inch as 1/16 of an inch has to one mile.

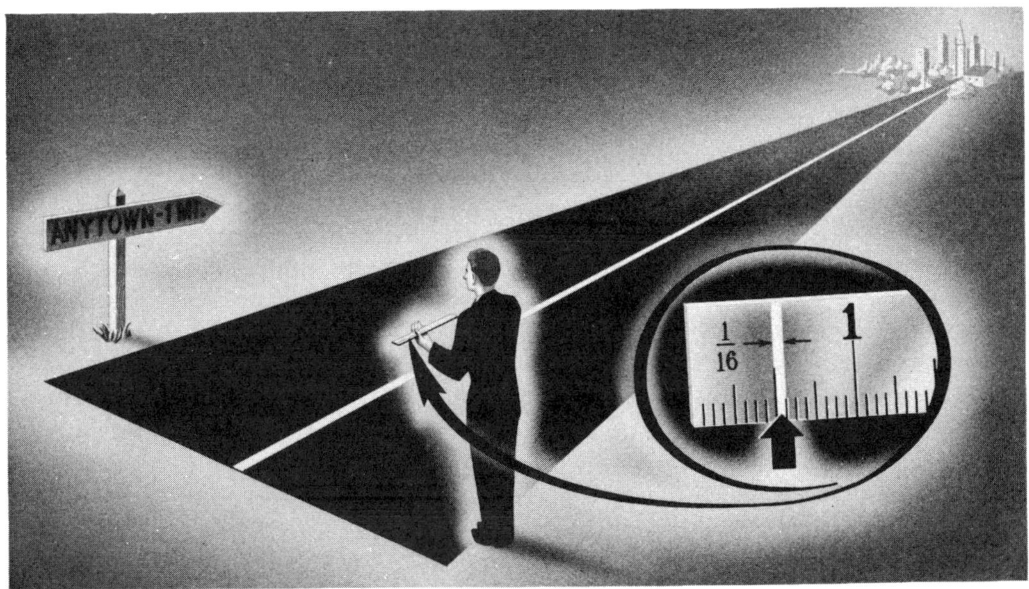

Figure 1

INTERCHANGEABILITY AND SELECTIVE ASSEMBLY

Interchangeability of parts is essential to mass production and in the large majority of manufactured products this objective is easily obtained with the standard gages commercially available.

However, when the successful operation or the required performance of a product depends upon holding dimensions closer than is desirable from the standpoint of production, it is customary to resort to selective assembly. Selective assembly means classifying by size both of the parts to be joined, and selecting mating parts which will subsequently be assembled into the machine as a unit. For example, wrist pins are graded into a series of sizes to be fitted selectively to pistons, which are similarly graded. Ball bearings are another example of parts which must be assembled selectively.

Following is a quotation of the American Standards Association on the subject of interchangeability: "Applied to manufactured material, the result sought is sufficient uniformity in size and contour to adapt the material without further fitting to the requirements of the industries. The fundamental principle involved in interchangeable manufacture requires that a system of standardization and classification of fits shall establish a clearly defined line at which interference between parts begins."

MEASUREMENT TERMS

Consider now the meaning of some of the terms which will be needed throughout this book. Anyone familiar with production processes knows that the parts of a machine cannot be made to an absolute dimension; all have some variation. In the majority of industrial applications, the mechanic or the inspector is not so interested in the exact value of a dimension as in its variation from the basic size. The allowable variation is determined by the function and design of the part, and is expressed in terms of limits between which the dimensions may vary.

DIMENSION. The measure of width, height, depth, or length in units of length of any object, as shown in Figure 2. A unit of length, as distinguished from a standard of length, is a measure in space without considering any physical conditions, such as temperature and pressure, while a standard of length is the physical representation of a defined unit of length under definite physical conditions.

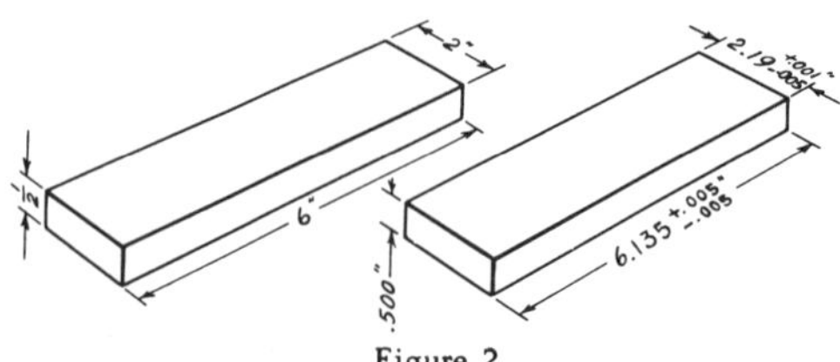

Figure 2

STANDARD SIZE. "A series of recognized or accepted sizes corresponding to various subdivisions of a recognized unit of length such as the yard or the meter. These are usually expressed in inches or in millimeters; sometimes by arbitrary numbers or letters" (American Standards Association). For example, cold-rolled bar stock may be obtained in standard sizes 1/64 inch, 1/32 inch, 3/64 inch and 1/16 inch, but if a bar between these sizes is desired, it is considered an odd size.

BASIC DIMENSION is the exact theoretical size from which all limiting variations are made. To illustrate: on a blueprint a dimension is given as 2.251 inch ± .001. The flat dimension (2.251) stripped of the allowable variation is the basic dimension.

Basic Dimension:

2.251" + .001 (plus tolerance)
- .001 (minus tolerance)

Upper Limit: 2.252

Lower Limit: 2.250

LIMITS are the maximum and minimum dimensions obtained by applying the tolerances to the basic dimensions and are the extreme dimensions beyond which the work cannot extend.

MEAN DIMENSION. The mean dimension is the average of the sum of the high and low limits. For example:

Basic Dimension:

.250" + .002
- .004

Upper Limit: .252"

Lower Limit: .246"

Mean Dimension: .249"

TOLERANCE. The permissible variation in the size of a part. The practice of showing the amount of permissible variation above and below the basic size is practically universal, because the tolerance on a dimension is the most pertinent information required in making the part. The meaning of the term "tolerance" is illustrated by the fact that dimensions such as $1.000''\,^{+.000}_{-.004}$, $1.000'' \pm .002$, $1.000''\,^{+.001}_{-.003}$, and $1.000''\,^{+.004}_{-.000}$, all have a tolerance of .004.

If the permissible variation is both plus and minus, it is referred to as a bilateral tolerance. If the tolerance is in one direction only, plus or minus, it is referred to as unilateral.

ALLOWANCE. "An intentional difference in the dimensions of mating parts or the minimum clearance space which is intended between mating parts. It represents the condition of the tightest permissible fit, or the largest internal member mated with the smallest external member. It is to provide for different classes of fit" (American Standards Association). The terms "tolerance" and "allowance" are often considered as having the same meaning. They are not synonyms, however, and have two entirely different meanings. Tolerance is the total permissible variation in one basic dimension on a single part, whereas allowance is the difference between the two tightest dimensions on mating parts.

To illustrate allowance, consider that a shaft is dimensioned .999" (+ .000, .001). The shaft is to be assembled to a gear, which has a hole in the hub dimensioned 1.000" (+ .001, - .000). The tightest condition possible between these parts occurs when the shaft is largest and the hole is smallest. The largest shaft is .999, the smallest hole is 1.000 inch, and so the minimum clearance or allowance is .001.

If the shaft diameter was 1.001" (+ .0000, - .0005), and the hole diameter 1.000" (+ .000, - .0005), it would be necessary to force the shaft into the hole because the hole is smaller than the shaft. It would be possible for the shaft and hole to have all the following combinations:

	(1)	(2)	(3)	(4)
Shaft	1.001	1.0005	1.001	1.0005
Hole	.9995	.9995	1.000	1.000
Negative Clearance	-.0015	-.0010	-.001	-.0005

In this case the allowance is always negative, and since the American Standards Association defines the allowances as the tightest fit, the allowance is -.0015.

If the dimensions of the shaft and hole were reversed, the clearances above would be positive, and the allowance would be + .0005, the condition representing the tightest fit. See Figure 3.

CODE OF TOLERANCE

In setting up a system to control the accuracy to which a product is to be manufactured, tolerances are given on the principal dimensions of all parts, and a code of tolerances is established to control the accuracy of the measuring instruments and gages used in the shop.

In general, the allowable error in a measuring instrument or gage is a small fraction of the tolerance on the dimension it is to control. The following excerpts from a typical code of tolerances illustrates the type of standards set up in a manufacturing industry for measuring instruments.

EXCERPTS, TYPICAL CODE OF TOLERANCES

GAGE BLOCKS. The master set of precision gage blocks shall be used only as a reference and shall be considered the basic standards of linear measure for the plant.

The master set of precision gage blocks must check a total error of not more than .000002 per inch of length. The working sets of precision gage blocks may have a total error of .000005 to as much as .000040 per inch of length, depending on the degree of accuracy required by the job.

SURFACE PLATES. Surface plates shall be flat, smooth, and true. The error in flatness of an 18 x 24 inch surface plate shall not exceed .001 of an inch. That is, no appreciable area of the working surface shall differ from the remainder of the surface by more than .001 inch.

OUTSIDE MICROMETER CALIPERS. The faces of the anvil and spindle must be flat and parallel with each other. The measuring error must not exceed .0002 of an inch at any point in the entire range of the graduations. This applies to all ordinary micrometer calipers whether equipped with a vernier for reading ten-thousandths or not.

VERNIER CALIPERS. The total error must not exceed .002 inch up to 18 inches of length and not exceed .001 of an inch for additional 12-inch lengths.

VERNIER HEIGHT GAGES. The total error must not exceed .002 inch up to 18 inches of length and not exceed .001 inch for additional 12-inch lengths. The blade must be square with the base within .005 inch in its entire length and the measuring surface of the sliding jaw must be parallel with the base within .005 inch.

PLUG, RING, AND LIMIT GAGES. In general, the error in size due to wear should not exceed .0002 inch. When gages are used to check parts having either very close or very wide limits, the allowable limits of error in the gage should be adjusted accordingly.

SOLID SQUARES. Solid squares must not have a total error in the 90-degree angle of

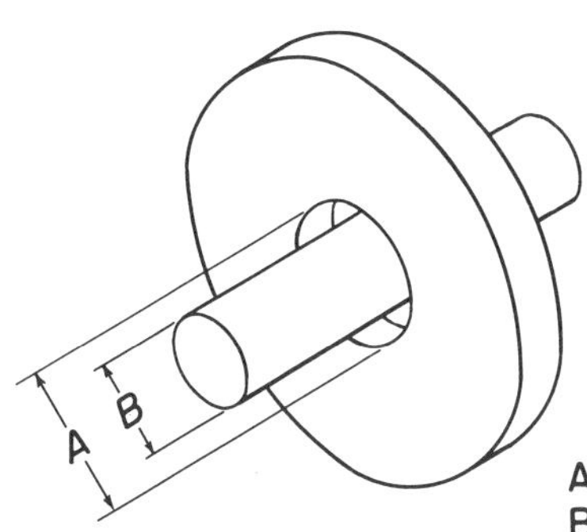

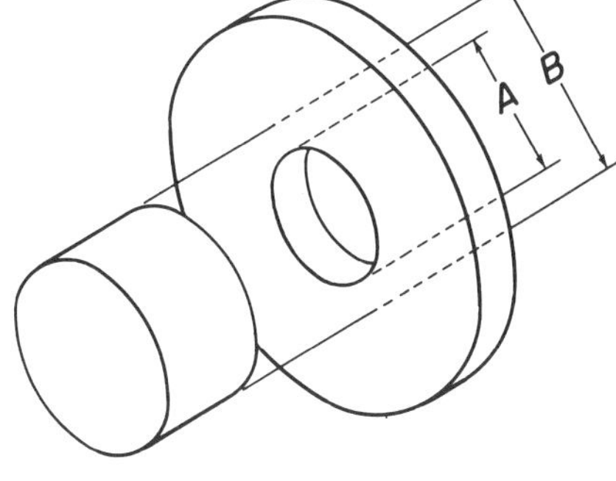

A — B = Clearance
B — A = Interference

Figure 3

more than .0005 inch in any six inches of length of blade.

CLASSIFICATION OF FITS

In order that terms describing various fits used in the assembly of parts may be universally understood to mean the same thing, the American Standards Association has classified fits into eight classes. The following classes are universally accepted as the standard:

CLASS 1 (Loose Fit - Large Allowance): This fit provides for considerable freedom and embraces certain fits where accuracy is not essential (on a one-inch size, an allowance of .003 inch).

CLASS 2 (Free Fit - Liberal Allowance): This is for running fits 600 rpm or over and journal pressures of 600 pounds per square inch (on a one-inch size, an allowance of .0014 inch).

CLASS 3 (Medium Fit - Medium Allowance): This is for running fits under 600 rpm and with journal pressures less than 600 pounds per square inch; also for the more accurate machine tool and automotive parts (on a one-inch size, an allowance of .0009 inch).

CLASS 4 (Snug Fit - Zero Allowance): This is the closest fit which can be assembled by hand, and necessitates work of considerable precision. It should be used where no perceptible shake is permissible and where moving parts are not intended to move freely under load (on any size, an allowance of .000 inch).

CLASS 5 (Wringing Fit - Zero to Negative Allowance): This is also known as a "tunking" fit, and it is practically metal to metal. Assembly is usually selective and not interchangeable (on a one-inch size, an allowance of minus .0004 inch).

CLASS 6 (Tight Fit - Slight Negative Allowance): Light pressure is required to assemble these fits, and the parts are more or less permanently assembled, such as the fixed ends of studs for gears, pulleys, and rocker arms. These fits are used for drive fits in thin sections or extremely long fits in thin sections and also for shrink fits on very light sections. They are used in automotive, ordnance, and general machine manufacturing (on a one-inch size, an allowance of minus .0009 inch).

CLASS 7 (Medium Force Fit - Negative Allowance): Considerable pressure is required to assemble these fits, and the parts are considered permanently assembled. These fits are used in fastening locomotive wheels, car wheels, armatures of dynamos and motors, and crank discs to their axles or shafts. They are also used for shrink fits on medium sections or long fits. These fits are the tightest which are recommended for cast iron holes or external members as they stress cast iron to its elastic limit (on a one-inch size, an allowance of minus .0011 inch).

CLASS 8 (Heavy Force and Shrink Fit - Considerable Negative Allowance): These fits are used for steel holes where the metal can be highly stressed without exceeding its elastic limit. These fits are used where heavy force fits are practical, as on locomotive wheel tires or heavy crank discs of large engines (on a one-inch size, an allowance of minus .0016 inch).

SUBDIVISION IN THE UNITS OF LINEAR MEASUREMENT

In the subdivision of units of linear measure there are two methods in general use, the fractional method and the decimal method.

The fractional method is developed from the natural tendency to subdivide by halving. Thus, the first subdivision creates halves; the second, quarters; the third, eighths; the fourth, sixteenths. On this basis, the scales used in a shop are calibrated in divisions which are 1/2, 1/4, 1/8, 1/16, or even 1/32 or 1/64 of the unit of measure. Divisions of 1/64 of an inch are about as fine as can be read easily without the aid of a magnifying glass. The dimensions on a blueprint which are given in terms of a fraction are known as scale dimensions, and are generally used to indicate the details on a part which do not require the same accuracy in size and location as other more vital details.

Fractional dimensions smaller than 1/64, such as 1/128 or 1/356, become cumbersome, and make it increasingly difficult to perform even simple calculations with them. For more accurate work, the decimal method of subdividing the unit of linear measure generally is used. The decimal method subdivides the unit of linear measure into tenths, hundredths, thousandths, ten-thousandths, hundred-thousandths, and millionths. Inasmuch as 1/64 inch is about the limit of measurement with the naked eye, practically all instruments calibrated to measure in terms of decimals are equipped with mechanical, electrical, or optical aids designed to magnify the otherwise imperceptible subdivisions of the inch.

With the decimal method, any fraction of an inch may be expressed in a form that is simple and convenient to use in calculations. For example, while it is rather difficult to add 1/328 and 1/36, it is quite a simple matter to add the decimal equivalents:

$$\begin{array}{r} .00305 \\ \underline{.02778} \\ .03083 \end{array}$$

The decimal form (.003) is also simpler and easier to understand than the fraction (1/328).

DECIMAL EQUIVALENTS

For purposes of reading an instrument, performing certain calculations, or interpreting specifications in shop work, conversion from fractions to decimals or vice versa is frequently desirable. For this purpose, a table is given in Figure 4. Those persons who do inspection work should commit to memory the more commonly used fractions and their decimal equivalents.

DIMENSIONING

The dimensions on a drawing may be in the form of fractions or decimals. Usually, in the case of fractional dimensions, only the basic dimension is given; that is, the flat dimension without tolerance. However, the tolerance should be explained either by a note or by specific job instructions, such as "the tolerance on fractional dimensions shall be plus or minus 1/64 inch unless otherwise specified." Thus, if a dimension is given as 7/16 inch, everyone working on the job understands that it may vary from 27/64 to 29/64 inch or the basic

1/64	0.015 625	11/32 . .	0.343 75	43/64	0.671 875
1/32 . .	0.031 25	23/64	0.359 375	11/16	0.687 5
3/64	0.046 875	3/8	0.375	45/64	0.703 125
1/16	0.062 5	25/64	0.390 625	23/32 . .	0.718 75
5/64	0.078 125	13/32 . .	0.406 25	47/64	0.734 375
3/32 . .	0.093 75	27/64	0.421 875	3/4	0.750
7/64	0.109 375	7/16	0.437 5	49/64	0.765 625
1/8	0.125	29/64	0.453 125	25/32 . .	0.781 25
9/64	0.140 625	15/32 . .	0.468 75	51/64	0.796 875
5/32 . .	0.156 25	31/64	0.484 375	13/16	0.812 5
11/64	0.171 875	1/2	0.500	53/64	0.828 125
3/16	0.187 5	33/64	0.515 625	27/32 . .	0.843 75
13/64	0.203 125	17/32 . .	0.531 25	55/64	0.859 375
7/32 . .	0.218 75	35/64	0.546 875	7/8	0.875
15/64	0.234 375	9/16	0.562 5	57/64	0.890 625
1/4	0.250	37/64	0.578 125	29/32 . .	0.906 25
17/64	0.265 625	19/32 . .	0.593 75	59/64	0.921 875
9/32 . .	0.281 25	39/64	0.609 375	15/16	0.937 5
19/64	0.296 875	5/8	0.625	61/64	0.953 125
5/16	0.312 5	41/64	0.640 625	31/32 . .	0.968 75
21/64	0.328 125	21/32 . .	0.656 25	63/64	0.984 375

Figure 4. Decimal Equivalents of Fractions of an Inch

dimension plus or minus the given tolerance and still pass inspection.

Decimal dimensions usually are given with a tolerance, as, for example:

$$2.537 \begin{matrix} +.002 \\ -.001 \end{matrix}.$$

This means that while 2.537 is the basic dimension, it may vary from 2.536 to 2.539 and still be within the limits of accuracy specified. The tolerance in this case is .003.

In the following chapters are described the many measuring instruments employed by manufacturing industries to control the accuracy of manufacture. Because of the number of concerns making measuring instruments, it will not be possible to illustrate the complete line of each manufacturer. So far as possible, each illustration will represent a certain type of measuring instrument, and the discussions of each of these types will, wherever possible, be made sufficiently general to include those made by all manufacturers.

QUESTIONS AND PROBLEMS

1. What was the earliest known unit of measurement? About how long was this unit and upon what was it based?
2. What was the natural basis upon which the first meter was established, and why is it no longer defined in the same terms?
3. What is the basic standard of length we use today; upon what was it based?
4. In what country is the International Bureau of Weights and Measures located?
5. What is the international standard conversion factor from the British Imperial System to the Metric System?
6. What is the International standard measuring temperature?
7. What is the function of the American Standards Association? The National Bureau of Standards?
8. Define the following:
   1. Tolerance 3. Basic Dimension
   2. Limits 4. Allowance
   5. Mean Dimension
9. If a dimension on a print is given as .432 + .0005 - .0001, what is the tolerance; the basic size; the low limit?
10. A bearing is dimensioned 1.347 + .002 - .000. The shaft which runs in this bearing is dimensioned 1.345 + .001 - .002. What is the allowance? What class of fit would this be?
11. Convert 527 millimeters to inches.
12. Convert 2.75 inches to millimeters.
13. A milling machine operator, in checking the piece in the machine, notes that 3/64 inch must be removed on the next cut. The calibrated hand wheel for making the adjustment is graduated in thousandths of an inch. How many divisions must the operator turn the wheel to take off 3/64 inch? Give your answer to the nearest division.
14. What is meant by a negative allowance? Give an example.
15. Convert the following decimals into fractions, accurate to the nearest 64th of an inch: .965, .3906, .225, .1385, and .703125.
16. Convert the following fractions to decimals: 3/8, 9/16, 31/32, 13/32, 1/2, and 11/32.
17. What is meant by "standard" size? Give an example.
18. Why are manufacturers interested in a universal standard of linear measure?
19. In most industrial plants, what is used as the standard of linear measure?
20. The hub of a gear is made separately from the gear blank. The tenon of the hub which fits in the blank is dimensioned .728 + .0005 - .0000. The hole in the gear blank is dimensioned .728 + .0005 - .0003. What is the allowance? What class of fit is it?
21. An inspector in checking the diameter of shafts notes that they range in diameter from .997 inch to 1.001 inch. From the blueprint of the part, he notes that the dimension of the shaft is given as 1.000 + .000 -.004. Should the lot of shafts be O.K.'d? Why?
22. What is meant by selective assembly? Name two parts that are selectively assembled.

Chapter II

LINE MEASUREMENT

THE short history of precision measurement in Chapter I shows that the advancement of measuring technology in industry today is the result of a long period of development. This development is still going on. As more accurate, simpler, faster, sturdier instruments of longer life and wider application are invented and marketed, they will take the places of those now being used. As new products are manufactured and new manufacturing processes developed, new instruments capable of controlling the accuracy and quality of the work also will appear.

Standard measuring instruments may be classified in two general groups:

1. Non-precision line-graduated measuring instruments.
2. Precision measuring instruments.

The term "line-graduated" is here applied to the simpler and more common types of instruments such as rules, calipers, surface gages and depth gages, for which accuracy of measurement is largely dependent on the user's ability to line up and read the graduations on a scale. Non-precision line-graduated measuring instruments of this type are used where accuracy in terms of thousandths, or tenths of thousandths of an inch, is not required.

The term "precision" is applied to all instruments used when accuracy of measurement is largely dependent upon the sensitivity of the instrument and the accuracy to which it is made. Precision instruments are used for measurements expressed in thousandths and tenths of thousandths of an inch or smaller units. A precision instrument usually has a mechanical, electrical or optical means of magnifying small units of length so that a thousandth of an inch is read more easily than 1/64 of an inch on a line-graduated scale.

With non-precision line-graduated measuring instruments such factors as temperature and measuring pressure are relatively unimportant; however, they must in many cases be considered in precision measurement. Although there is a definite dividing line between these two types of instruments, actually some instruments have features that might place them in either class.

SELECTION OF THE PROPER INSTRUMENT

The selection of the proper instrument is based on a number of considerations. Three main features, however, apply to all instruments, and determine their selection. These are the physical characteristics of the part, the dimensions to be measured, and the accuracy required.

Of course, if these were the only three features, there would be fewer instruments on the market. Convenience, adaptability, speed, and reliability are other considerations which often play major roles in the selection of the proper instrument. The physical characteristics of the part and the dimensions to be measured, will be covered in the subsequent discussion of the applications of each type of instrument.

No instrument is capable of controlling the accuracy of measurements to a degree finer than the finest graduation on its scale. This means that if a scale whose finest graduation is 1/32 inch is used to check a dimension, this dimension must be in terms of units not smaller than 32nds. If a dimension of 9/64, for example, is to be read with a 1/32 scale, it will fall between graduations. It will only be estimated, not controlled.

Some experienced mechanics scoff at any suggestion that they cannot read to two or three thousandths on a scale graduated in 64ths or hundredths, and they will demonstrate. There are a great many skilled mechanics and tool makers who have worked with tools and instruments so long that their eyes are practically graduated in thousandths. Still, it is an estimate when anyone attempts to read from a scale graduated in 64ths in any unit less than 64ths. Furthermore, it is not necessary because now many instruments are available to use in reading accurately in thousandths. In general, where an instrument is used in the shop to control the accuracy of a dimension on a part it should never be necessary to read part of a scale division.

The instruments described in this book have been classified partially by the limit of accuracy they are capable of controlling and partially by their application in the shop.

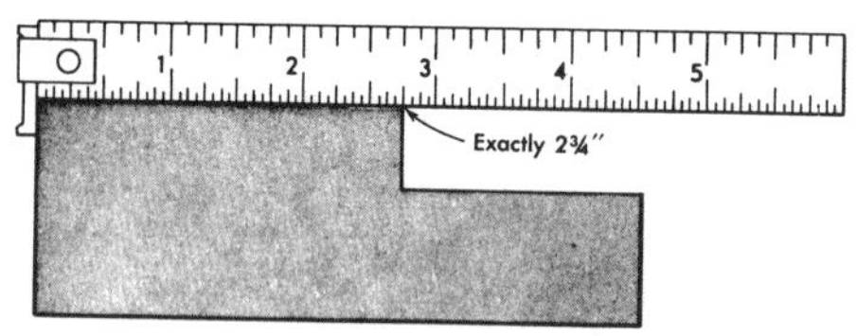

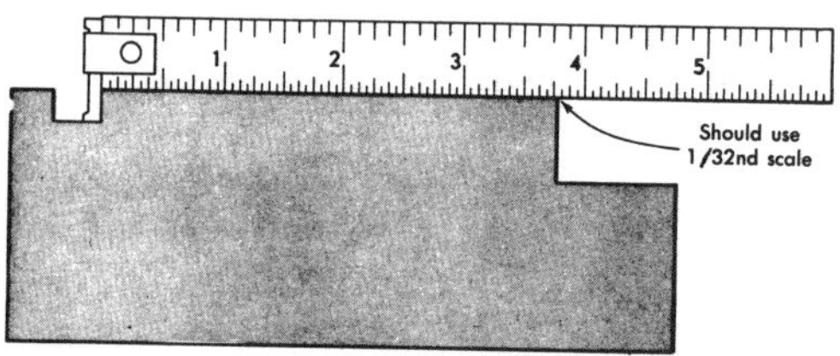

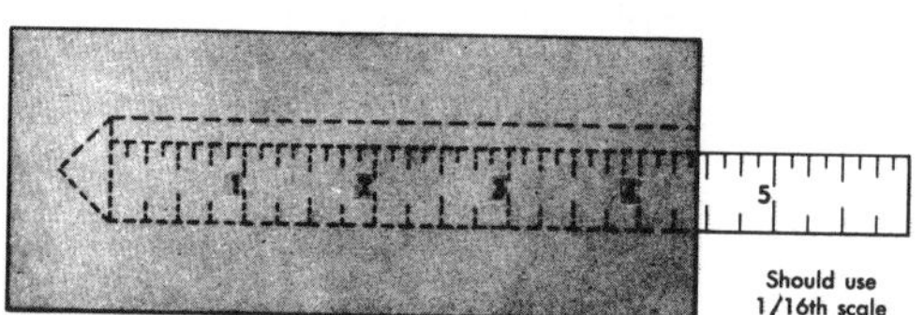

Figure 5. Use of Appropriate Graduations

LINE MEASUREMENT

Non-precision line-graduated measuring instruments such as rules, scales and graduated squares, are read by a comparison of the etched lines on a scale with an edge or surface. Most scale dimensions are measured by reading the scale with the naked eye, although a magnifying glass is frequently used by inspectors who are checking parts continuously.

When a scale is used to check a dimension, the proper graduated scale should be used to control the reading of the dimension as shown in Figure 5. If it is not possible to read a dimension on a 16th scale, a 1/32 scale should be used; and if it is impossible to read a dimension to a 32nd, the 1/64 scale should be used.

STEEL RULE OR SCALE

The rule is the oldest and most commonly used measuring instrument in the shop. The machinist's rule is a scale scribed on a strip of steel. However, to satisfy the requirements of a great variety of applications and the personal tastes of shop men, a number of different types of rules are commercially available.

Steel rules vary in length from a fraction of an inch up to four feet or more, but in machine shops the six-inch pocket rule is the most commonly used. Some of the more common types are shown in Figure 6.

There are also several standard systems of graduations. In the English system rules are obtainable graduated in:

(1) 10ths, 20ths, 50ths, and 100ths.

(2) 12ths, 24ths, and 48ths.

(3) 14ths and 28ths.

(4) 16ths, 32nds, and 64ths.

For uses involving the metric system, rules graduated in millimeters and one-half millimeters are obtainable. In the United States, rules graduated in 16ths, 32nds, and 64ths of an inch are the most commonly used.

Several features of rules which make them easier to use and more adaptable are:

1. Four Scales: Some rules have four scales, two on each side (one graduated in 32nds and the other in 64ths), with the scales on the reverse side running in the opposite direction. This

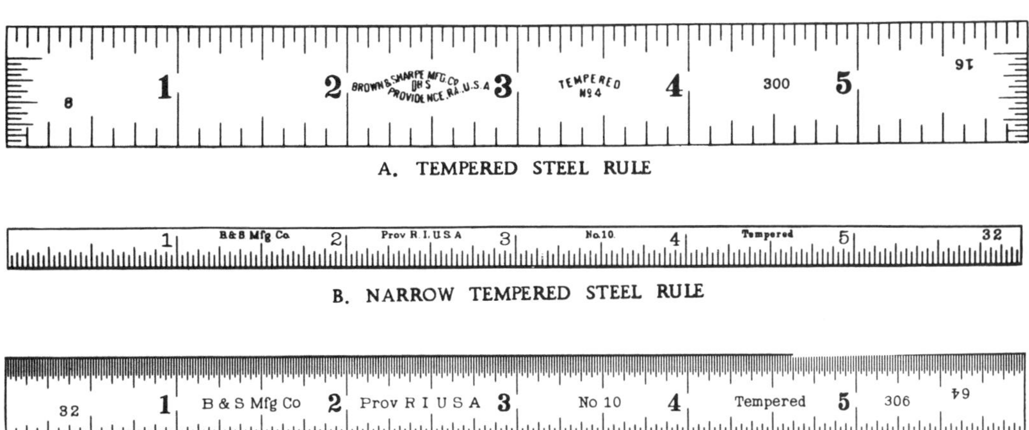

A. TEMPERED STEEL RULE

B. NARROW TEMPERED STEEL RULE

C. FLEXIBLE STEEL RULE

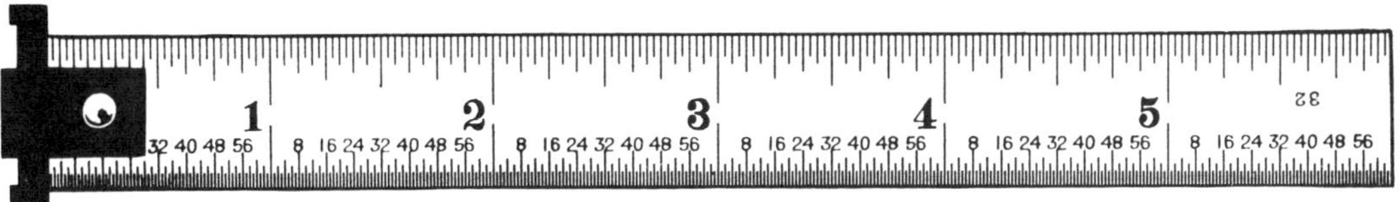

D. TEMPERED HOOK RULE

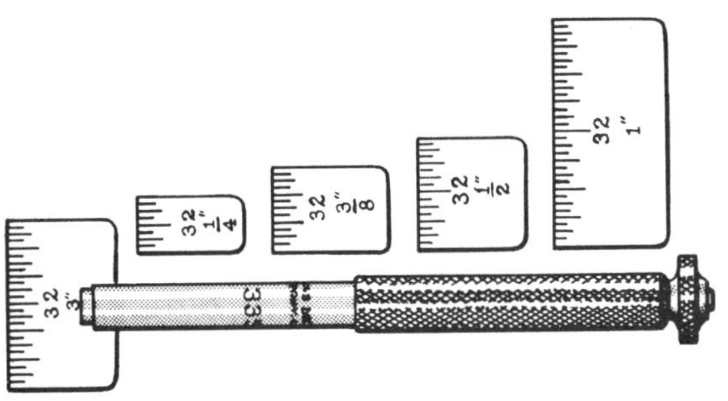

E. STEEL RULES WITH HOLDER

F. FLEXIBLE STEEL FILLET RULE

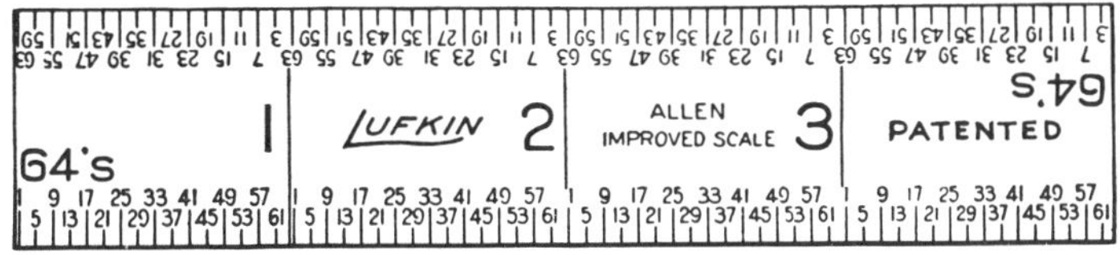

G. IMPROVED SCALE

Figure 6. Types of Rules

feature, a coarse and fine scale on each side, makes the scale easier to read and the rule more convenient to use.

2. Combination of English and Metric Measures: Some rules have both an inch scale and a millimeter scale which makes the rule adaptable to work involving both systems of measure.

3. Numbered Graduations: Some rules have numbered graduations which make the scale easier to read and reduces the possibility of error. See Figure 6D.

The rule shown in Figure 6G will measure to 64ths, yet its closest graduations are 1/32 apart, and all odd 64ths lines are 1/16 apart; all of these being numbered, it is easy to read to 64ths. The side shown bears the odd 64ths only. One edge carries odd 64ths, commencing with 1 and numbered 1, 5, 9, 13, etc. The other edge bears the remaining odd 64ths, such as 3, 7, 11, and 15. The reverse side is marked in the standard way, one edge 16ths, the other edge 32nds.

4. Flexible Rule: Some rules are made of thin, tempered spring steel which permits them to be bent over a rounded surface. See Figure 6C.

5. End Scale: Some rules are equipped with a scale across the end which facilitates measurement in restricted places. See Figure 6A.

6. Sliding Hook: Some rules are equipped with a sliding hook (Figure 6D) which facilitates measuring from a shoulder, particularly if the end of the rule is hidden so that it cannot be lined up with the shoulder. The sliding hook is also convenient in setting calipers and dividers.

7. Holder: Special short rules (Figure 6E) for measuring in a recess or in a restricted place are provided with a holder.

8. Fillet Rule: For the purpose of spanning fillets and corner fills which are frequently in the way when measuring flanges or shoulders, a special rule known as a fillet rule is available. See Figure 6F.

9. Attachments: The attachments used with rules adapt them to special types of work or increase their range of application. Key seat clamps, as shown in Figure 7, make the scale

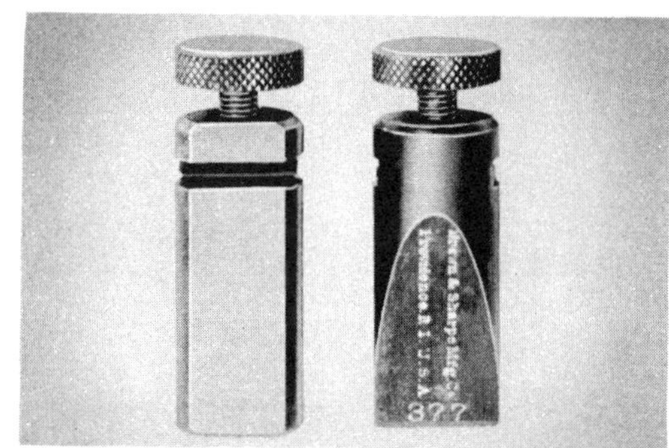

Figure 7. Key-Seat Clamps

adaptable to the more or less special job of laying out key-ways and scribing parallel lines on circular pieces. Rule clamps increase the range of application of a rule, making it possible to fasten two or more together and thus obtain a longer rule. Figure 8 illustrates a rule clamp. Figure 8A illustrates a right-angle rule clamp to be used with combination square blades and heads as shown in Figure 25C.

Figure 9 shows a standard six inch pocket rule used to check a dimension on a part. Note that in using the scale, the mechanic is holding

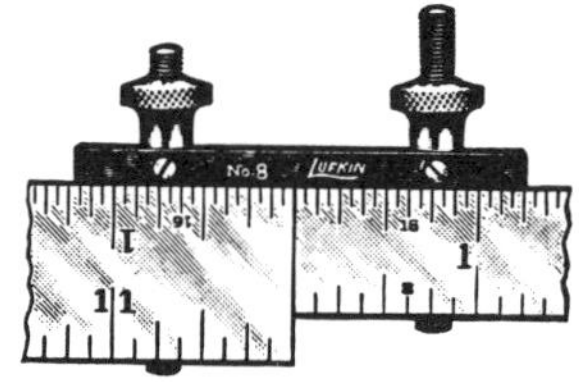

Rule Clamp

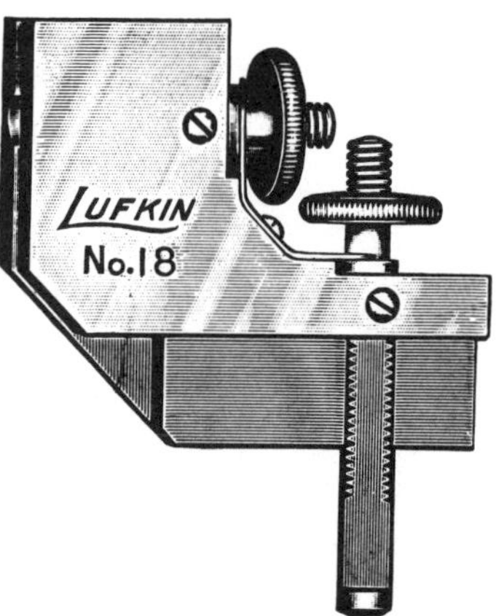

Right-Angle Rule Clamp

Figure 8. Rule Clamps

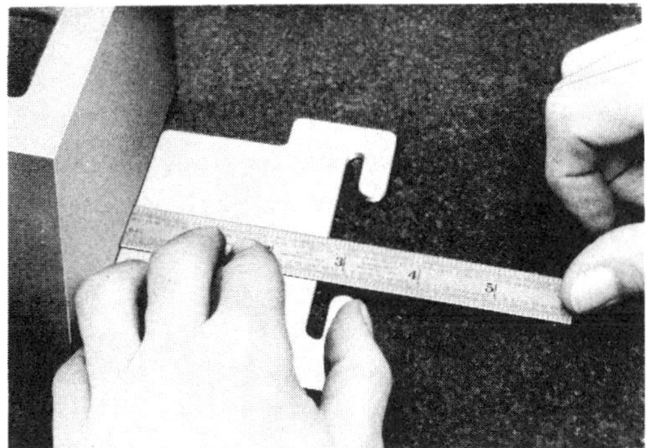

Figure 9. Measuring Dimension (Length)

the part and the rule firmly against an angle block. The end of the scale thus is lined up exactly with the surface from which the measurement is to be taken. This is considered good practice because the mechanic is able to hold the rule firmly against the block and devote his entire attention to reading the scale correctly.

Figure 10 illustrates the use of a scale in checking a dimension on a part from a surface plate. The surface plate in this case serves as a common base and lines up the scale with the part surface from which the measurement is taken. Figure 11 shows how a steel rule with holder may be applied to a measurement in a recess inaccessible to the ordinary rule.

Figure 12 shows an application of the narrow

Figure 10. Measuring Dimension (Height)

Figure 11. Measuring a Recess

scale in measuring the depth of a narrow slot. Figure 13 shows a set of key seat clamps being used to scribe a line on a cylindrical surface parallel to the axis of the piece. As the name implies, key seat clamps are used mainly in laying out key-ways and splines. Figure 14 shows an application of a hook rule. In this case the hook serves to line up the end of the rule with the edge of the shoulder from which the measurement is taken.

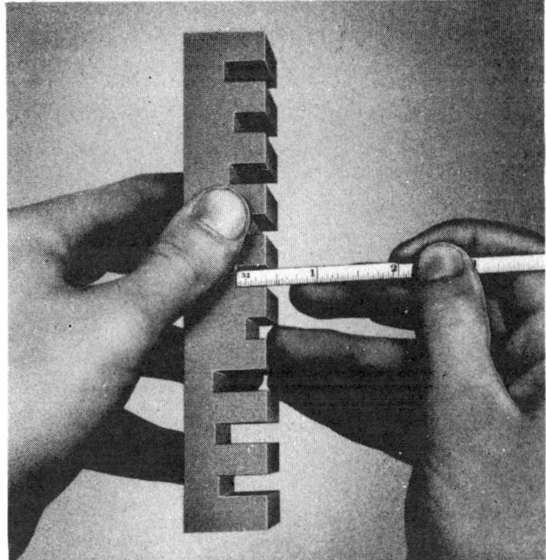

Figure 12. Measuring the Depth of a Narrow Slot

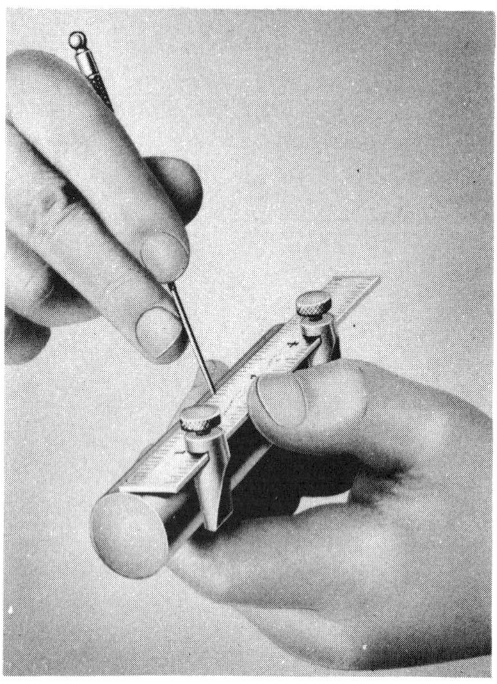

Figure 13. Laying out a Key-way

SHRINK RULE

The shrink rule resembles an ordinary rule, differing only in its scale, which automatically compensates for the shrinkage in castings. In the manufacture of castings, molten metal is poured in a cavity made in sand with a wooden or metal pattern of the part. The impression of the pattern in the sand is of the same size and shape as the pattern. The molten metal fills the cavity, solidifies, and cools. As all metals expand when heated and contract when cooled, the casting shrinks as it cools, and becomes smaller than the cavity. To compensate for this shrinkage, the pattern must be made larger than the part itself. Therefore, in making the pattern from the blueprint of the part, it is necessary for the patternmaker to add the correct amount to every dimension to take care of this shrinkage. Rather than calculate this shrinkage, the patternmaker uses a shrink rule which automatically compensates for it.

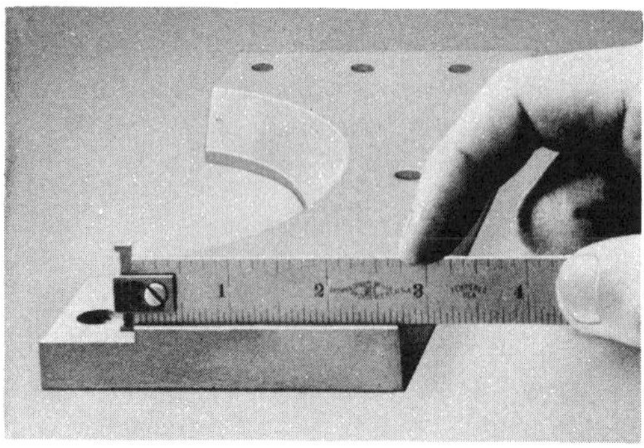

Figure 14. Measuring from a Shoulder

Every metal has its own particular shrinkage value by which expansion or contraction takes place for each degree of change in temperature. The shrinkage values for each of the two common casting metals are:

Iron, cast - 1/8 inch per foot

Brass, cast - 3/16 inch per foot

A special shrink rule is required for each of these casting metals, In order to produce a piece of cast iron twelve inches long, the pattern must be 12-1/8 inches long and to produce the same part in cast brass, the pattern should be 12-3/16 inches long.

In Figure 15 two shrink rules, for brass (A) and for cast iron (C), are shown in comparison with a standard foot rule (B). Note that inch marks on the shrink rules do not match the corresponding marks on the standard rule and that this variation increases with the length of the rule until the total length of the shrink rules exceeds that of the standard by the amount of shrinkage per foot. A foot shrink rule for cast iron is, therefore, 1/8 inch longer than the standard rule. It is divided into 12 parts and each inch is divided into halves, quarters, eighths, and sixteenths, so each fractional part of the rule is compensated for shrinkage.

Figure 16 shows a shrink rule being used to check a dimension on a pattern. This rule eliminates the tedious job of having to calculate the shrinkage on every dimension by automatically allowing for it in the scale. It is an indispensable measuring instrument for the patternmaker.

CALIPER RULE OR SLIDE CALIPER

In the use of rules and instruments which have been described, the position of the edge

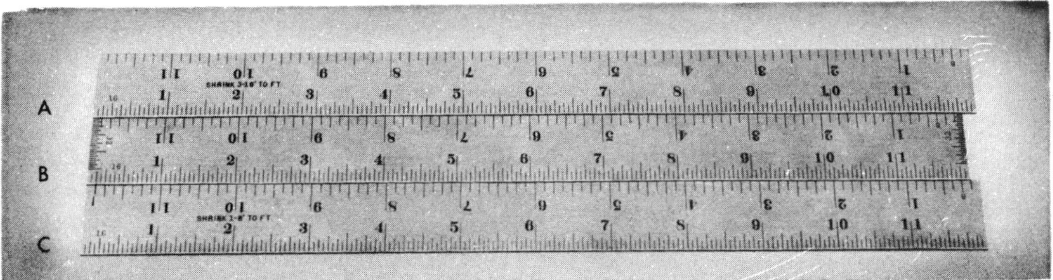

Figure 15. Two Shrink Rules Compared with Standard Foot Rule

or point to be measured in relation to the graduations on the scale is judged by sight. Frequently, two contact points are necessary to measure a dimension more accurately or to reach two surfaces inaccessible to a rule.

The caliper shown in Figure 17 is an ordinary rule with one fixed jaw and one sliding jaw. When the two jaws are brought in contact with surfaces to be measured, the distance between the surfaces may be read from the scale. The ends of the jaws are so shaped that it is possible to measure both inside and outside surfaces. The rule shown in Figure 17 has two lines marked "out" and "in" which enable the user to read either inside or outside measurement directly. If these lines were not on the scale, it would be necessary to add the thickness of the nibs to the reading when using it as an inside caliper.

Caliper rules are usually made in three-inch sizes, although larger instruments up to 48 inches long are commercially available. The larger instruments are known as caliper squares. The standard three-inch pocket caliper rule will measure from zero to about two inches outside and from one-eighth inch to a little more than two inches inside.

All caliper rules are equipped with a locking or clamping device which makes possible the holding of jaws in any desired position. This feature makes it possible to set up a dimension and use the caliper rule as a gage, or to clamp the jaws after making a measurement so the setting is not disturbed in removing the jaws from the piece to read the scale.

Figure 18 shows the use of the caliper rule to check the thickness of a piece of bar stock (top) and the diameter of a hole (bottom).

DEPTH GAGE

A depth gage is a non-precision line-graduated measuring instrument especially adapted to measuring the depth of holes and slots. In its simplest form, a depth gage is a rule with a sliding head designed to bridge the hole or slot, to hold the rule perpendicular to the surface from which the measurement is taken, and to indicate the depth of the hole on the scale.

Depth gages of the type shown in Figure 19

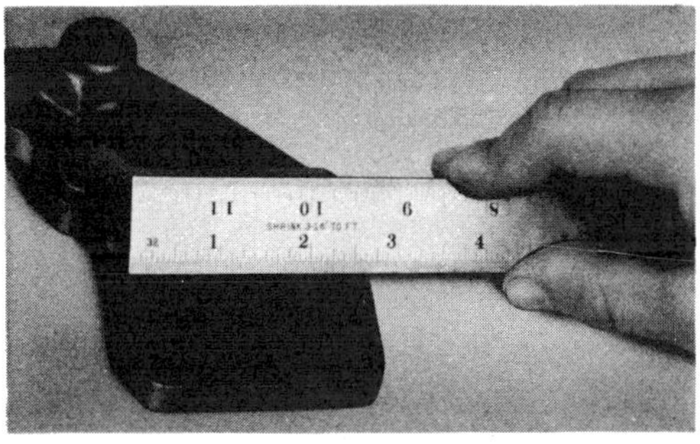

Figure 16.
Checking a Pattern Dimension with a Shrink Rule

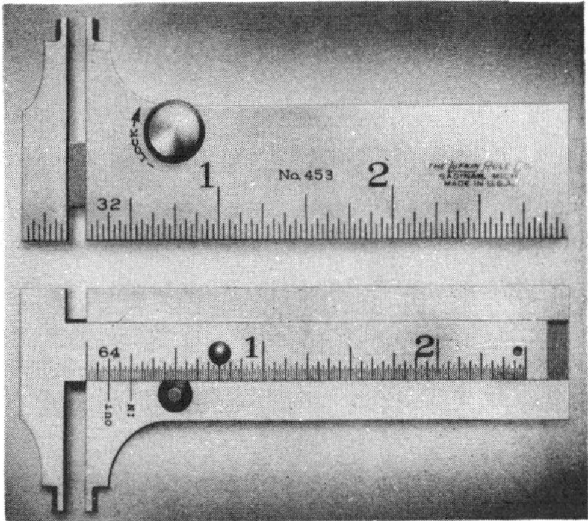

Figure 17. Pocket Slide Caliper Rule

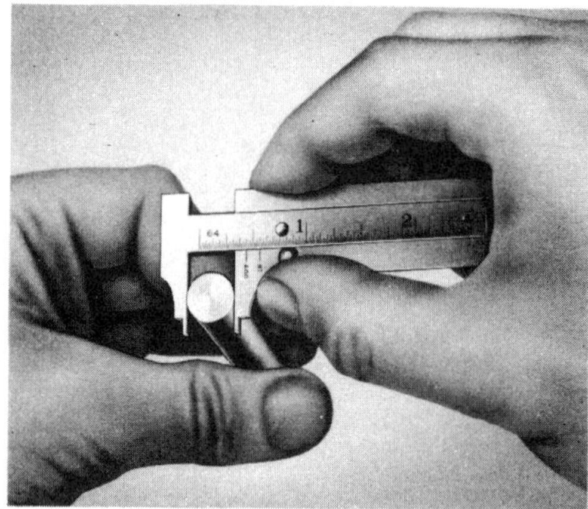

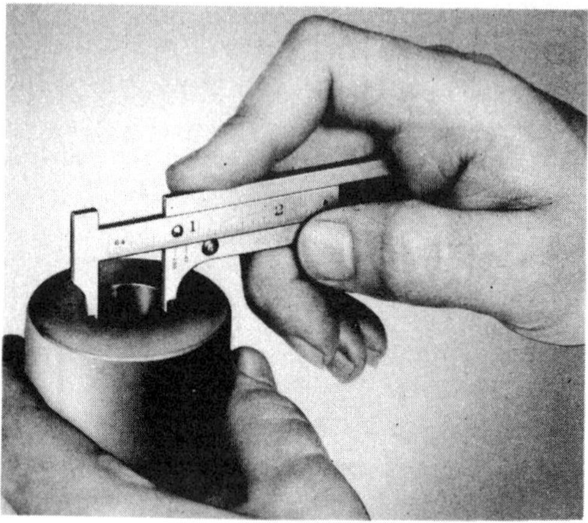

Figure 18. Applications of Caliper Rule

have a measuring range of from zero to five inches. The sliding head has a clamping screw so it may be clamped in any position. Its base is flat and perpendicular to the axis of the rule and will range in size from 1/8 to 1/4 of an inch wide and from 2 to 2-5/8 inches long.

The graduated-rod depth gage shown in Figure 19 is particularly adapted to measuring the depths of small holes. Rule depth gages also are commonly furnished with a slender rod, which may be used to get into small holes. The plain rod replaces the rule, and after the rod has been set to the depth of the hole, it is clamped in place, removed, and measured with a rule. The rod is 5/64 of an inch in diameter and of the same length as the rule. The rule-type depth gage also may be equipped with a hook for measuring the depth of holes going all the way through, and similar applications requiring a depth gage.

The application of the depth gage is increased by the protractor feature on some combination gages, which permits the rotation of the head with respect to the rule. The protractor-rule depth gage shown in Figure 19 is one of this type. The angle scale does not approach the accuracy of a regular protractor but is often convenient for rough-checking the angle of a surface.

Figure 20 shows an application of the depth gage, to the measurement of the depth of a slot from a non-adjacent shoulder. The use of the instrument in this way is limited by the length of the base. This type of depth gage may be adapted to the measurement of the depth of small holes through the use of the auxiliary rod, as shown in Figure 21. In measuring the depth of holes with this rod, the gage is set so the rod "bottoms." The gage is then removed and the length of rod extending from the face of the base is measured with a rule.

Figure 22 shows the use of the depth gage equipped with a protractor to measure an angle on a part.

PLAIN PROTRACTOR

The plain protractor shown in Figure 23 is widely used in the machine shop as well as on the drafting table. The end of the blade moves over the scale which is divided into degrees from 0° to 180° in both directions to facilitate reading complementary degrees. Lines can be drawn along the edge of the blade for laying out angles, or angles can be measured on bevels.

This instrument may be used as a T-square as well as for transferring angles. The four sides are square to one another, an advantage over the semi-circular style because work may be layed out from any of the three sides.

COMBINATION SQUARE OR SET

The combination square shown in Figure 24 varies from the try square in that it has a number of features which increase its application. It consists of a steel rule, a center head, a sliding

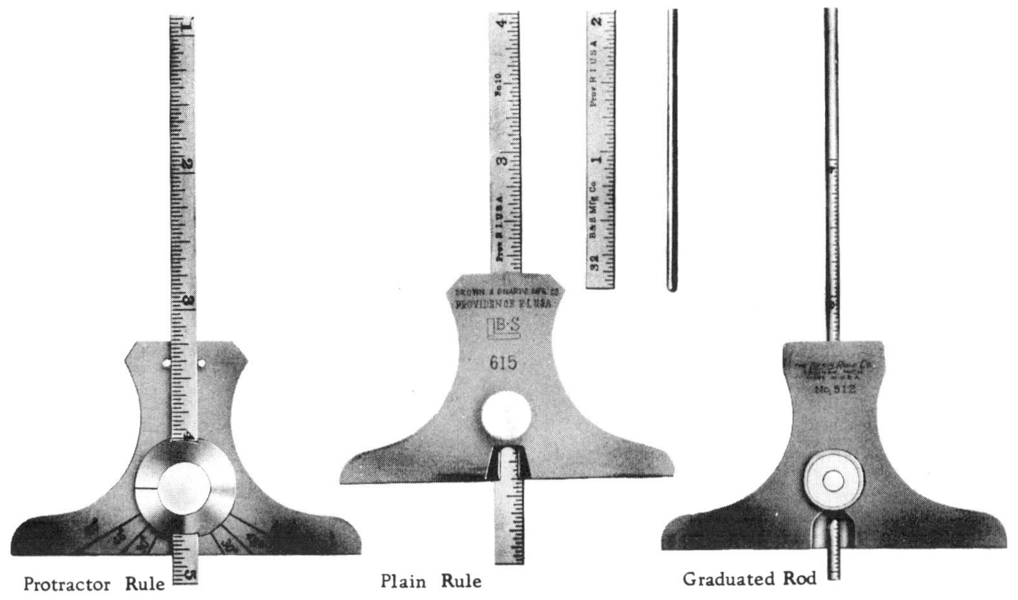

Figure 19. Types of Depth Gages

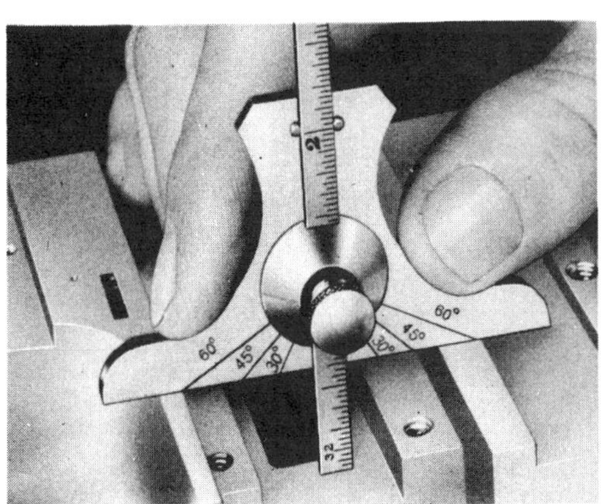

Figure 20. Checking Depth of Slot from Non-Adjacent Shoulder

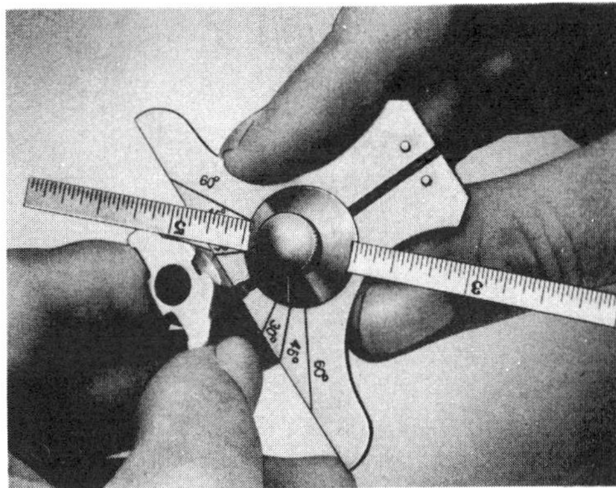

Figure 22. Measuring an Angle with Protractor Rule

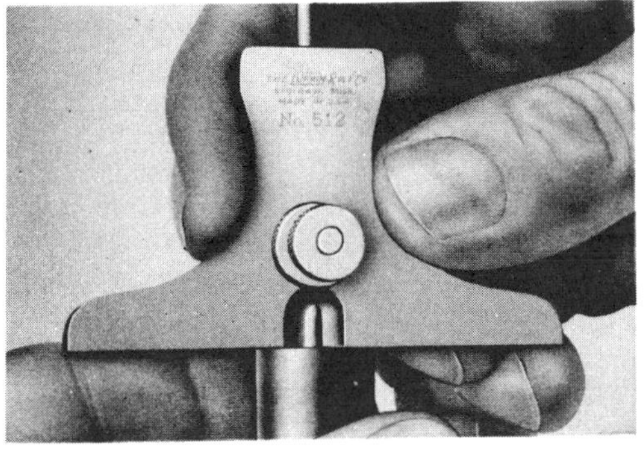

Figure 21. Measuring Depth of a Small Hole with Auxiliary Rod

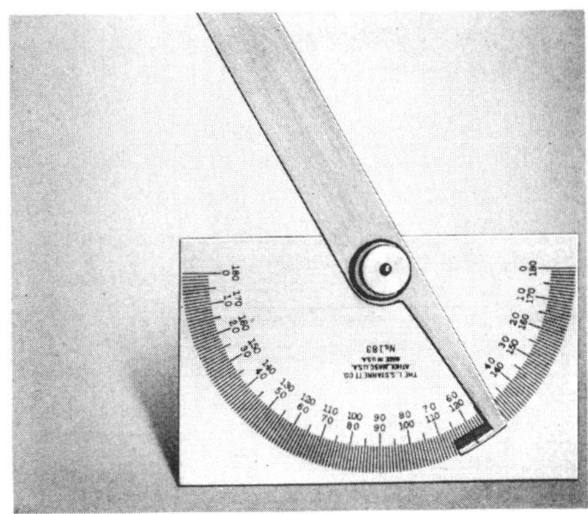

Figure 23. Plain Protractor

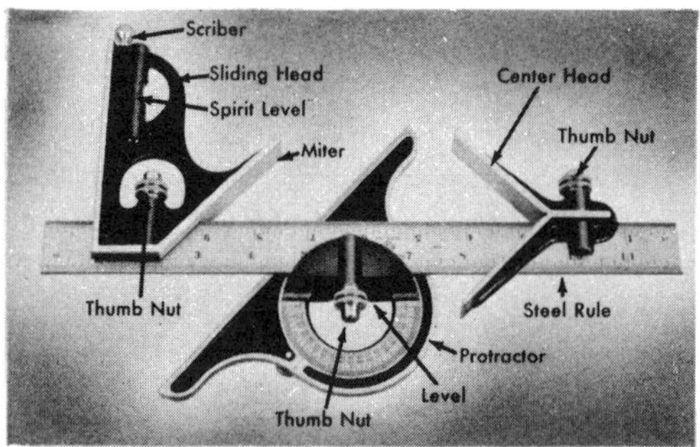

Figure 24. Combination Set

head or beam which also contains a spirit level, a protractor head, and a scriber.

The steel rule and the sliding head may be used to square a piece with a surface and at the same time determine whether one or the other is plumb, as shown in Figure 25A. By using the miter, 45° angles as well as 90° angles with the head may be laid out as shown in Figure 25B. A scriber is inserted conveniently in the head for this purpose.

By setting the steel rule flush with the sliding head and using a right-angle rule clamp and another blade, the square may be used as a height gage as shown in Figure 25C. By loosening the rule the combination may be used as a depth gage as shown in Figure 25D, if microm-eter accuracy is not necessary.

By substituting the center head for the sliding head, a center square is obtained for finding the center line of cylindrical objects. This center head is slotted in the center so that when the rule is inserted, it bisects the 90° angle. In this way, the measuring surfaces become tangent to the circumference of cylindrical work and can be used to locate the center of a bar, as shown in Figure 25E.

The protractor shown in Figure 24 may be inserted in the steel rule in the same manner as the sliding head and center head. The re-volving turret is graduated in degrees from 0° to 180°, or to 90° in either direction. Also, the head contains a spirit level to facilitate the measuring of angles in relation to the horizontal or vertical plane. With the steel rule removed from the head, rule and sliding head may be used separately. The head may be used as an ordinary level.

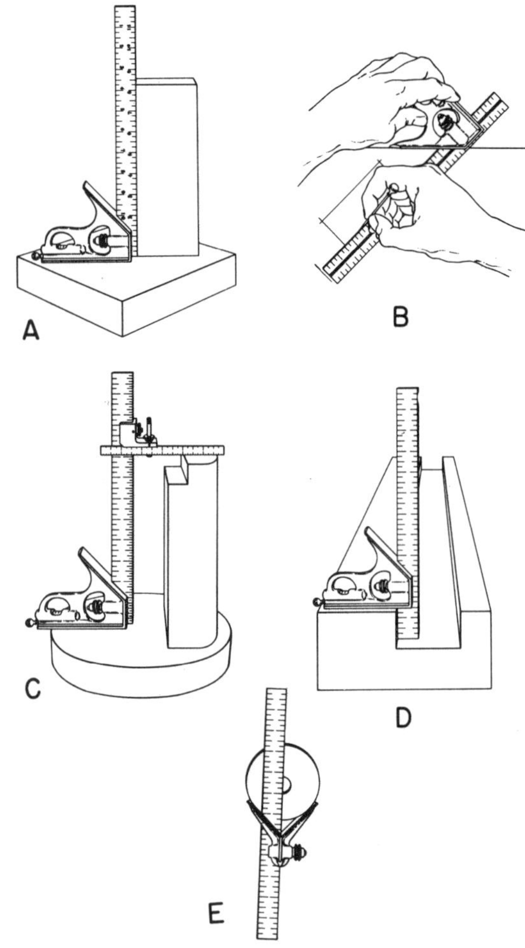

Figure 25. Applications of Combination Square

Shown in Figure 26 are two applications of the combination square. Its use as a depth gage is shown at the left and its use to locate the center line of a cylindrical object, at the right.

The protractor head is being used in Figure 27 to check the angle of a way. With the head in contact with the angle and the blade in contact with the side of the way, the variation, if any, may be read from the protractor.

DIVIDERS

A divider is an instrument used to scribe an arc, radius, or circle, to lay out distances set from a rule, and to transfer distances for measure-ment with a rule. It consists of two sharp points held apart by a spring and adjusted by a screw and nut, as shown in Figure 28.

Checking Location of a Surface

Locating a Center Line

Figure 26. Applications of Combination Square

In setting a divider to a dimension from a scale, the customary procedure is to locate one point in one of the inch graduations of the rule and to adjust the nut so the other point falls easily into the correct graduation (Figure 29). When properly set, the two leg points should locate nicely in the two graduations. Do not set to the end of the rule. Similarly, in transferring a dimension from a part or tool to the scale on a rule, to get an accurate setting it is necessary to use the same care in adjusting the points of the dividers so there will be no pressure tending to spring the points either in or out. The points must be kept sharp to insure accuracy.

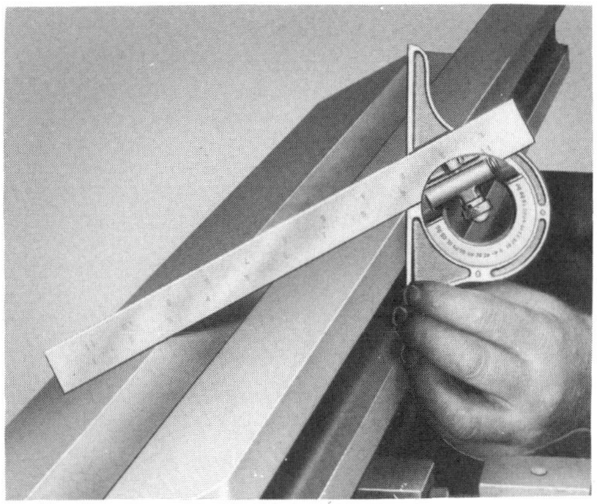
Figure 27. Checking the Angle of a Way

OUTSIDE CALIPERS

Whereas a divider is used to measure distances between points on a surface, to transfer measurements to a scale, or to scribe circles and arcs on a surface, an outside caliper is used to measure distances over and around intervening surfaces and to transfer the measurements to a scale. Several types of outside calipers are shown in

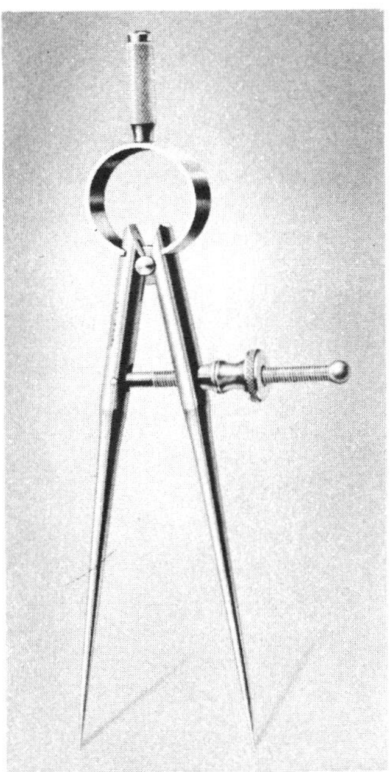
Figure 28. Divider

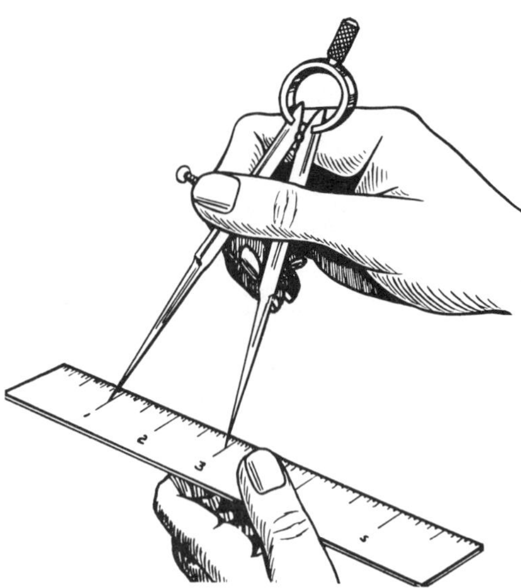

Figure 29. Setting Divider to a Scale

Figure 30. Each type is made in several sizes to accommodate a wide range in measurement. The size of a caliper is expressed in terms of the maximum dimension it can measure. A three-inch caliper, for example, will measure a distance of three inches. Actually, the maximum capacity of the caliper will be greater, often by as much as one third. This means that a three-inch caliper will actually measure up to about four inches.

The toolmaker's spring caliper shown in Figure 30 is made in sizes which range from two to eight inches. This caliper is used where the size of the work falls within the range of the instrument. The action of the spring holds the legs firmly against the nut on the screw; the points are adjusted in or out by turning the nut. To facilitate setting the caliper to the desired dimension, some calipers of this type are equipped with an automatic closing spring nut. Its use saves time in opening and closing the calipers.

The firm-joint caliper shown in Figure 30 is adaptable to larger work. It is made in a number of sizes which range from 3 to 36 inches. It is quicker to adjust than the toolmaker's spring caliper but does not hold its setting as well. The legs hold their position when set because of the friction of the large joint by which they are fastened together.

The transfer caliper shown in Figure 30 is employed in measuring a recessed surface where the legs must be opened to remove the instrument after it has been set. In using the instrument, it is first adjusted to the work in the same manner as an ordinary caliper with the short auxiliary arm engaged with the left leg. When the legs are opened to remove the instrument from the work, the auxiliary arm remains stationary to indicate the original position to which the leg must be returned to obtain the dimension just measured. The thumb screw mounted on the face of the left leg clamps the auxiliary arm to the leg. The thumb screw mounted on the side of the right leg makes possible a fine adjustment.

A caliper is usually used in one of two ways.

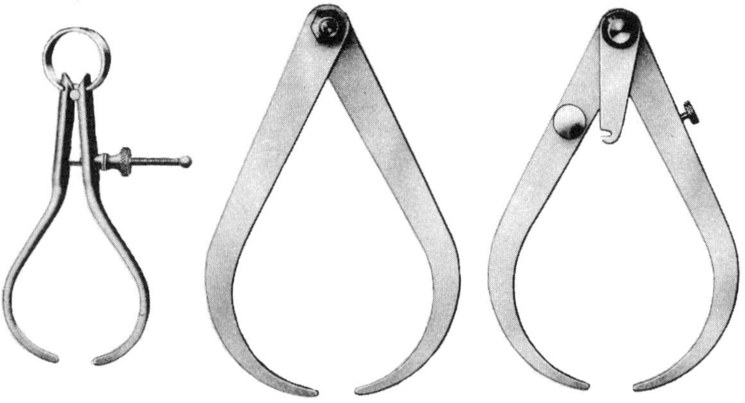

Toolmaker's Spring Caliper   Firm Joint Caliper   Transfer Caliper

Figure 30. Outside Calipers

Either it is set to the dimension of the work and the dimension is transferred to a scale, or the caliper is set on a scale and the work is machined until it checks the dimension set up on the caliper. Figure 31 shows a caliper being used to check work held in the chuck of a lathe.

To adjust an outside caliper to a scale dimension, one leg of the caliper should be held firmly against one end of the scale and the other leg adjusted to the proper dimension, as shown in Figure 32. To adjust the outside caliper to the work, its legs should be opened wider than the work and brought down to the proper dimension, rather than forced open by the work. There is no question but that a sense of "feel" must be acquired to use calipers properly. The sense of "feel" comes through practice and care in using the instrument to eliminate the possibility of error. For example, calipers should never be set from work revolving in a machine. The contact of one leg of a caliper on a revolving surface will tend to draw the other leg over the work because of the friction between the moving surfaces. Only a slight force is necessary to spring the legs of a caliper so that measurements made on moving surfaces are never accurate. Moreover, it is dangerous from a safety standpoint and could result in serious injury. Also, the calipers should be positioned properly on the axis of the work as shown in Figure 33.

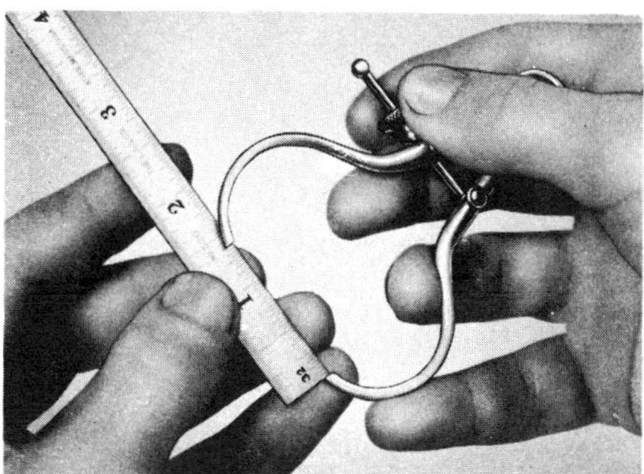

Figure 32. Setting an Outside Caliper to a Scale

INSIDE CALIPERS

Inside calipers have the same general function as outside calipers except that the inside caliper is used in measurement between inside surfaces. Three of the more common types of inside calipers are shown in Figure 34. The spring caliper is commercially available in inch sizes from two to eight inches. The firm-joint type is available in a number of sizes from three to twenty-four inches. The points on both outside and inside calipers are rounded so they are slightly ball-shaped. This definitely establishes the point of contact; in inside calipering where the surfaces are likely to have an inside curva-

Figure 31. Checking Outside Diameter

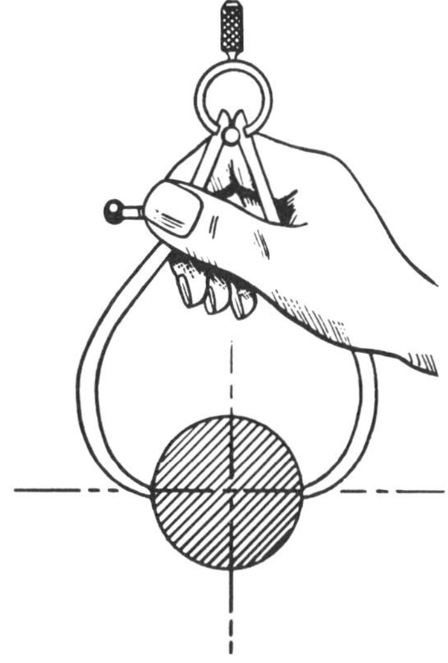

Figure 33. Correct Position of Outside Caliper

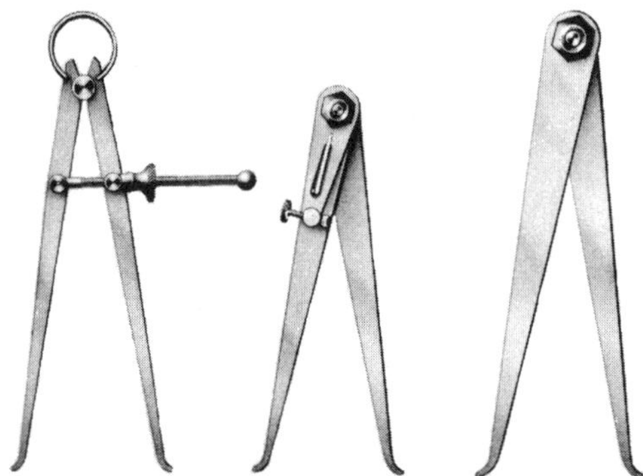

Spring Caliper — Screw Adjusting Firm Joint Caliper — Firm Joint Caliper

Figure 34. Inside Calipers

ture, it eliminates the possibility of error which might occur if the radius of the hole being calipered were less than the radius of the points. The screw-adjusting feature on the caliper in the middle of Figure 34 provides a fine adjustment in setting the legs of the caliper.

The inside caliper may be set to a dimension by putting the end of a scale and one point of the caliper against a solid surface and adjusting the other leg to the proper graduation, as shown in Figure 35A. The use of the inside caliper is illustrated in Figure 35B. It shows a toolmaker's caliper being used to check the inside diameter of a bored hole.

Figure 36 illustrates the correct and incorrect positioning of the calipers with relation to the axis of the work. Figure 37 illustrates the use of the transfer feature on calipers. Note that the diameter being measured is recessed and that the setting cannot be transferred to a scale directly because the legs must be collapsed to get them out of the work. The setting must be reproduced after the calipers are removed, and this is accomplished through the aid of the transfer feature. Figure 38 shows how a micrometer can be used for transferring such a dimension.

HERMAPHRODITE CALIPERS

The hermaphrodite caliper is a cross between a divider and a curved caliper, having one leg of each. Although the use of the instrument is somewhat limited to special work such as scribing a line parallel with an edge, as shown in Figure 39, the hermaphrodite caliper is a very important tool to the mechanic. It is adjusted in the same manner as the outside and inside calipers depending on the position of the caliper leg as shown in Figure 39.

BEAM TRAMMELS

A beam trammel is a tool used for the same purposes as a divider or caliper, but usually for distances beyond the range of either of these two instruments. A steel beam trammel with all of the attachments required in measuring and layout work is shown in Figure 40.

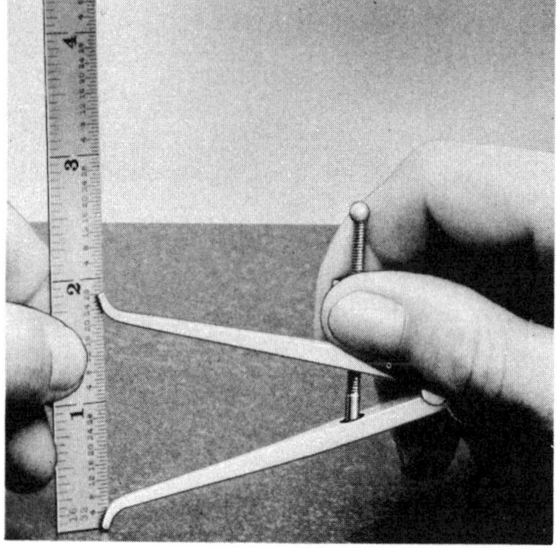

Figure 35A. Setting an Inside Caliper to a Scale

Figure 35B. Measuring an Inside Diameter

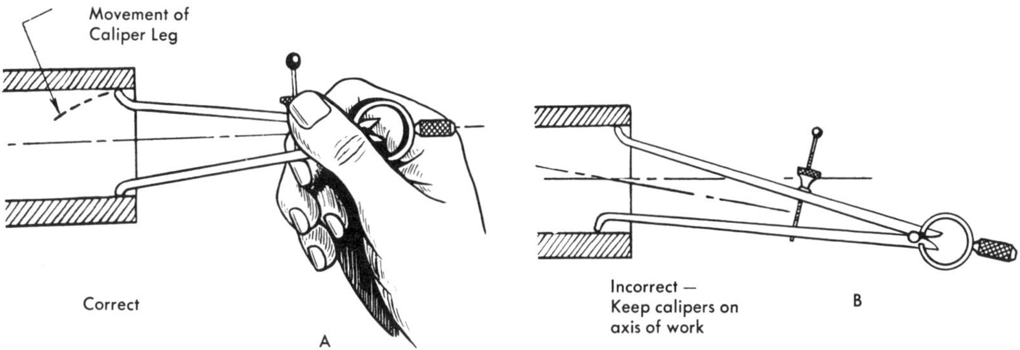

Figure 36

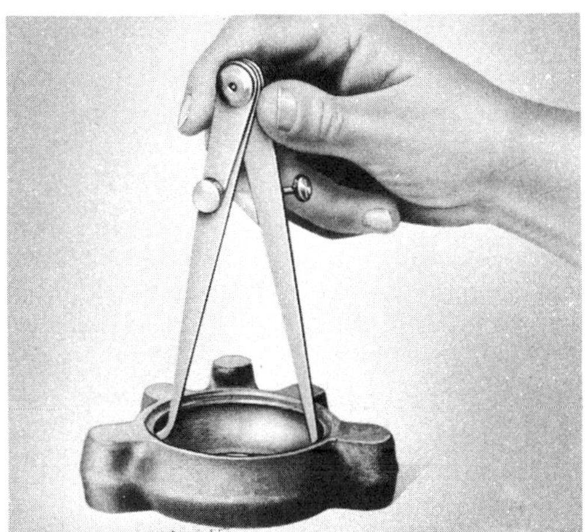

Figure 37. Measuring a Recessea Inside Diameter

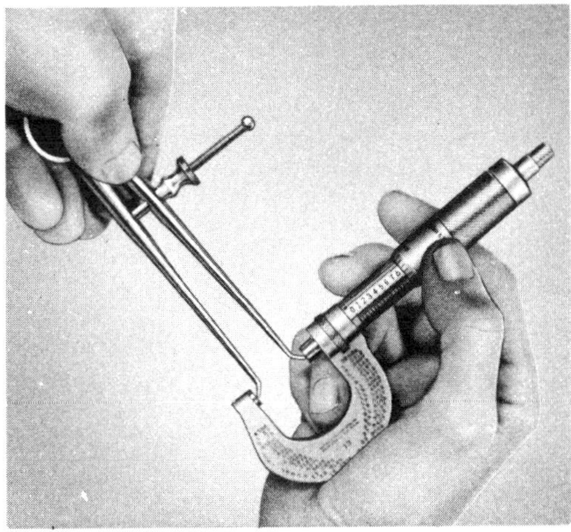

Figure 38. Transferring a Reading to a Micrometer

The instrument consists of a rod or beam to which trams may be clamped. These steel beams may range in length from nine to twenty inches but may be increased further through the use of extensions. Longer beams are often made of wood. The trams carry spring chucks in which divider points, caliper points, and ball points may be inserted so the trammel may be readily converted from a divider to an outside or inside caliper or even to a hermaphrodite caliper. Ball points are used to locate a tram in the center of a hole. By using different size balls or V-points it is possible to locate from a hole of any size up to 1-1/2 inches in diameter. On top of the trams are knurled handles which swivel so the handles may be gripped firmly when describing a circle or an arc. Note that one of the trams is equipped with an adjusting screw which permits a fine adjustment of the points.

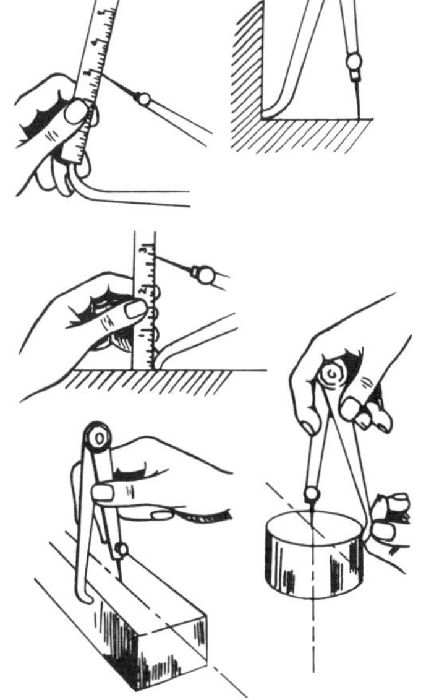

Figure 39. Using a Hermaphrodite Caliper

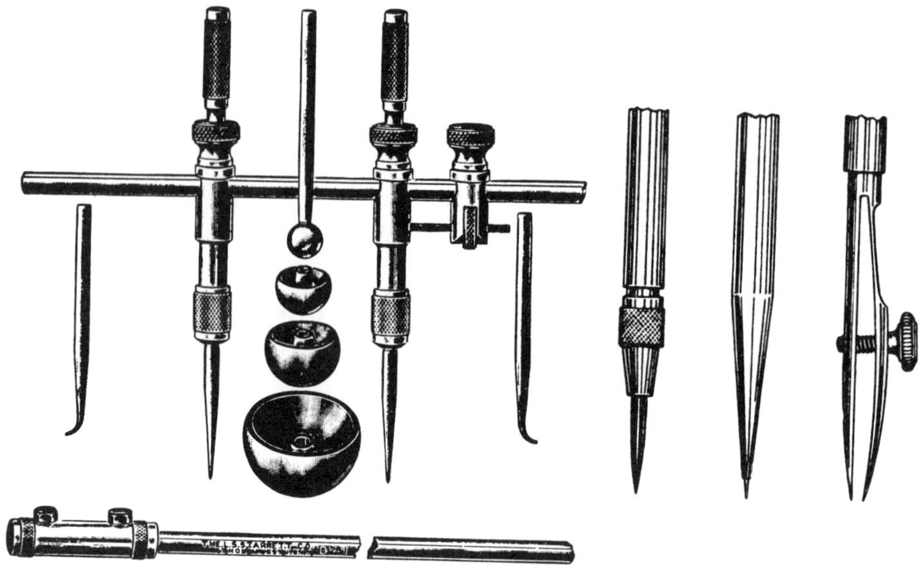

Figure 40. Beam Trammel and Attachments

The use of the beam trammel is confined chiefly to large layout work beyond the range of the instruments previously discussed.

UNIVERSAL SURFACE GAGE

A surface gage is a measuring tool generally used to transfer measurements to a piece by scribing a line, and to indicate the accuracy or parallelism of surfaces. For laying out work and checking the accuracy of surfaces, the surface gage is a very useful instrument, It is used generally in connection with a surface plate.

The surface gage shown in Figure 41 consists of a base with an adjustable upright to which may be clamped a scriber or an indicator. The spindle may be positioned with respect to the base and tightened in place by means of the spindle nut. Further adjustment of the spindle is possible by means of the rocker adjusting screw which operates the rocking bracket and through which the spindle is mounted on the base.

The scriber is fastened to the spindle by means of a clamp. Tightening the scriber nut holds the scriber in any position in which it has been set. The scriber also may be mounted directly in the spindle nut in place of the spindle in order to use it where the working space is limited and the height of the work is within the range of the scriber. The bottom and the front end of the base of the surface gage have deep V-grooves which make the gage adaptable to measuring from a cylindrical surface. Also in the base are two or more gage pins, frictionally held, which may be pushed down to bear against the edge of a surface plate, the side of a slot, or a surface on the piece so the instrument may be traversed in relation to an edge.

Surface gages are made in several sizes and are classified by the length of the spindle -- the smallest spindle being four inches long, the average either nine or twelve inches, and the largest eighteen inches.

A surface gage may be considered either as a non-precision measuring instrument or as a precision instrument, depending on the measuring device used in connection with it. When it is equipped with the scriber, which is considered the standard tool to be used with it, the surface gage is considered a non-precision measuring instrument.

It is often necessary to prepare the surface of the work before using the scriber, so the line will be sufficiently clear cut and visible. In the case of a casting, particularly one having a rough surface, it is customary to chalk the surface and rub in the chalk. With a smooth

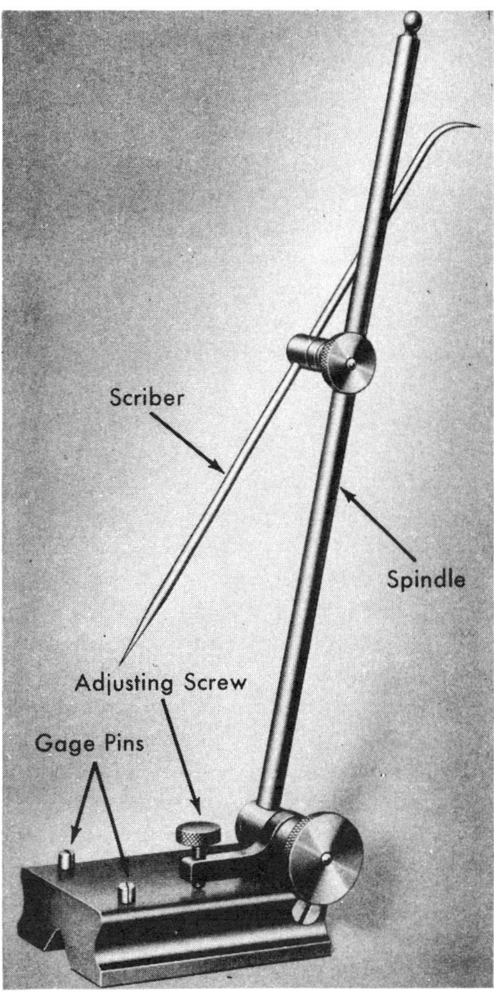

Figure 41. Universal Surface Gage

or finished surface a quick drying layout ink or stain, purple in color, may be used. Scribed lines on this surface will show up clearly against the dark background. Where the temper of a piece need not be considered, the part may be blued by heating it and the lines may be scribed on this surface.

Figure 42 shows an application of the surface gage where it is being used to scribe a line on a casting. The surface has been stained so that the scribed line may be easily distinguished. This is one example of a variety of measuring and layout work for which the surface gage may be used. Figure 43 shows the surface gage being set to the scale of a combination square before indicating a surface. The final adjustment of the scriber is being made with the rocker adjusting screw.

SCRIBERS

Scribers of the type shown in Figure 44, while not classed as measuring instruments, are so frequently used with them that they will be very briefly mentioned. A scriber is to the mechanic who works on steel what the pencil is to the draftsman. He uses it to mark and lay out the pattern to be followed in subsequent machining operations.

The scriber has a knurled handle and a very sharp, hardened steel point. A pocket model may be two or three inches in length; others are twelve inches or more long.

Some mechanics make their own scribers from files or drill rod, but in a shop interested in promoting only safe methods, this practice should be discouraged. A scriber not properly hardened may break off and injure a worker. One not properly knurled may slip out of an oily hand and hurt someone, particularly when used around a machine in operation.

In Figure 45 a scriber is shown being used with a rule to scribe a line on a block of steel.

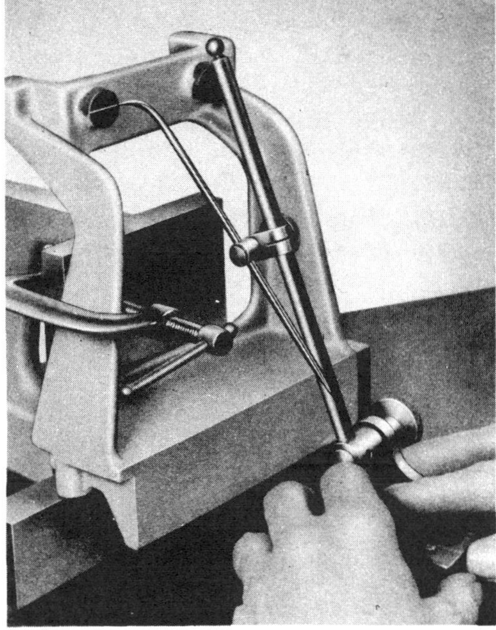

Figure 42. Scribing a Casting

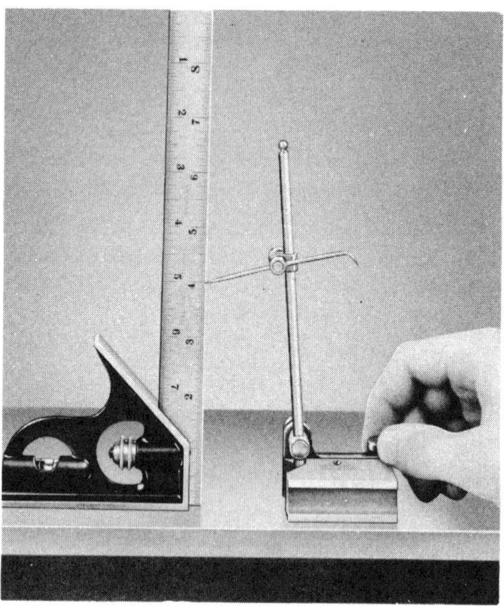

Figure 43. Setting the Scriber of a Surface Gage

CARE OF NON-PRECISION MEASURING INSTRUMENTS

Because non-precision line-graduated measuring instruments need not be as accurate as precision instruments, there is a tendency to abuse them by careless handling. Occasionally, a young man with new tools will try to age them prematurely so as not to call attention to his own newness.

A young craftsman should observe how the experienced tool and die maker uses and cares for the tools in his kit. The manner in which he works is indicated by the way he handles his tools. His instruments get the same careful attention because he knows the quality of his work depends on the condition of his tools and instruments fully as much as on his personal skill.

Someone using a measuring instrument for the first time may unknowingly abuse it. Some of the more common abuses are listed below for those interested in preparing a foundation for quality work with good tools. For the most part the care of measuring instruments is applied common sense, because while the following abuses represent poor shop practice, there are times when a good mechanic does vary from the established method to meet conditions. Some of the more common abuses are:

1. Using an instrument for a purpose other than that for which it was intended; i.e., using a rule as a screwdriver, using one leg of a divider as a scriber, using a square as a hammer, digging grease and chips out of a hole with a rule, scriber, or divider, or scraping off the burr around a hole with a rule.

2. Handling tools and instruments carelessly, such as piling tools in drawers, dropping tools on hard surfaces, using an instrument too close to a moving cutter and nicking it.

3. Forcing an instrument such as springing dividers and calipers beyond their setting or capacity, forcing a bound screw or tightening screws too tight to hold a setting and thus marking the tool.

4. Using an instrument incorrectly such as changing the setting by hammering instead of loosening a clamping screw, bearing down too hard in scribing with a divider or scriber, or wearing measuring surfaces unnecessarily by using

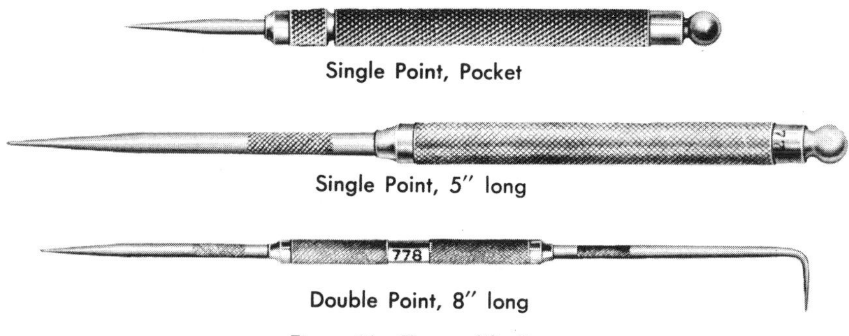

Figure 44. Types of Scribers

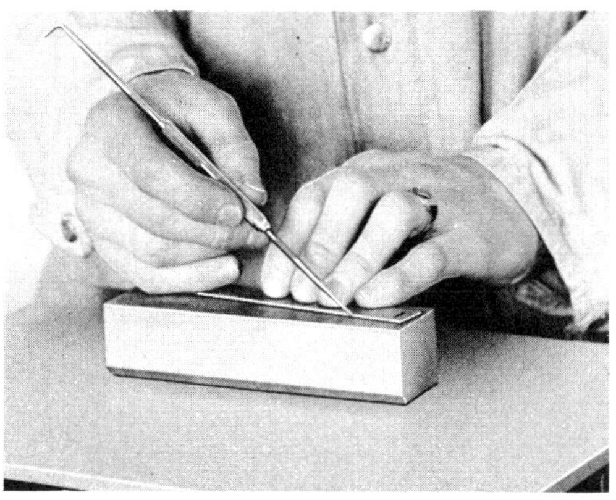

Figure 45. Scribing a Line

a heavy measuring pressure.

5. Neglecting a tool or instrument such as letting tools rust by neglecting to use a protective film of oil, neglecting the cleaning of tools which have become filled with abrasive grit, or neglecting to inspect, oil, and recondition tools and instruments occasionally.

6. Other abuses are changing tools to suit some personal whim, marking tools in a manner that affects their proper functioning, or damaging tools by nervous habits and horseplay such as drumming scales and squares on a hard surface, snapping spring calipers and dividers, and sticking a scriber in the bench.

Non-precision line-graduated measuring instruments are designed to control the accuracy of measurements to the nearest graduation on its scale. They are rugged and will take considerable abuse, and with proper care in storage as well as use will last indefinitely and consistently do a satisfactory job.

The measuring instruments illustrated and discussed in this chapter are those most commonly used. There are many special-purpose tools featuring advantages for some particular application. Wherever the volume of work is sufficient to require steady use of a measuring instrument or where the limitations of the job are such that a standard instrument cannot be satisfactorily used, special instruments which tend to be single-purpose tools may be required.

QUESTIONS AND PROBLEMS

1. What is meant by non-precision and precision instruments?
2. What determines in which class a measuring instrument is placed?
3. What three main factors influence the selection of the proper measuring instrument? What other factors must be considered?
4. A rule graduated in 32nd's will not control which of the following dimensions? 1/16, 1/2, .625, 7/8, 11/32, .956, .125, 15/64.
5. Name three types of steel rules and describe how each is used.
6. Steel rules may be obtained in what lengths; and with what graduations?
7. Why do rules frequently have two or more scales?
8. Name two attachments for rules. For what are they used?
9. For what are shrink rules used?
10. Is the inch on all shrink rules the same length? Why?
11. What is a caliper rule, and how is it used?
12. Why are depth gages more accurate than scales when measuring depth of slots and holes?
13. How closely will a plain protractor measure?
14. Give five uses of the combination set.
15. Name five kinds of calipers and tell where each is used.
16. What precautions must be kept in mind when setting and using calipers?
17. For what are dividers used, and how are they set?
18. What is a beam trammel, and for what is it used?
19. Describe briefly how a surface gage may be used as a non-precision instrument; as a precision instrument.
20. What determines whether a standard measuring instrument or a special instrument developed for the purpose should be used on a job?
21. Name five abuses of measuring instruments.
22. What advantages does the slide caliper have over a plain rule?

Chapter III

FIXED GAGES

STANDARD inspection routines and procedures in a large plant permit the essential inspection function to be performed with a minimum of disturbance to the main objective, production. These inspection routines and procedures must be carried out with gages adapted to the particular job they are required to do. Each gage must be satisfactory from several points of view:

1. It must reliably control the dimension it gages within the limits specified.
2. It must be convenient to use, so that dimensions may be checked quickly and inspection does not interfere with the flow of production.
3. The cost of inspection per piece must be reasonable, considering both the cost of the gage and the time required to inspect the part, as against the time saved in assembly and more reliable performance of the product.

From the many different types of gages available, one may be selected for each application to meet best the requirements described. As conditions change, changes in the method of inspection may be advantageous.

STANDARD AND SPECIAL GAGES

Gages may be further divided into two groups:

1. Standard gages which have sufficiently general use so that they are made commercially available by gage manufacturers.
2. Special gages limited to use with some particular part or group of parts. These gages are usually built specially by the manufacturer.

Some gages considered as standard must be purchased as special gages because the actual size required for a certain job is not among the standard sizes manufactured. This does not necessarily increase the cost of the gage, because it is usually considered a standard gage ordered in a special size.

GAGE STANDARDS

The American Standards Association, in cooperation with leading manufacturers of gages, has set up certain specifications to be followed in gage design. The types of gages discussed in this text will be those conforming to American Gage Design standards, wherever such standards exist.

GAGE ACCURACY

In the average shop, gages are used for a wide range of work, from rough machining to finest tool and diemaking. The accuracy required of even the same type of gage will vary with its application. Recognizing this variety in gage requirements, manufacturers offer several classes of gages and recommend that the gage used have an accuracy of one-tenth of the tolerance on the dimension it is required to control. This means that if the total tolerance on a dimension being checked is .001 inch, the error in the gage should not be greater than .0001 inch.

Plug gages from all manufacturers are grouped in the following classes as to accuracy:

CLASS XX - Precision lapped to laboratory

tolerances. For master or setup standards.

CLASS X - Precision lapped to close tolerances for many types of masters and the highest quality working and inspection gages.

CLASS Y - Good lapped finish to slightly increased tolerances for inspection and working gages.

CLASS Z - Commercial finish (ground and polished, but not fully lapped) for a large percentage of working gages used when tolerances are fairly wide and production quantities are not large.

Table II shows the tolerances for gages of various sizes in each classification.

TABLE II. GAGEMAKERS' TOLERANCES

Nominal Size-Inches		Gagemakers' Tolerances Classes			
Above	To and Including	XX	X	Y	Z
.029	.825	.00002	.00004	.00007	.00010
.825	1.510	.00003	.00006	.00009	.00012
1.510	2.510	.00004	.00008	.00012	.00016
2.510	4.510	.00005	.00010	.00015	.00020
4.510	6.510	.000065	.00013	.00019	.00025
6.510	9.010	.00008	.00016	.00024	.00032
9.010	12.010	.00010	.00020	.00030	.00040

WEAR ALLOWANCE AND GAGE TOLERANCE

Working gages are used for checking parts in production: Inspection gages are used at final inspection. Working gages usually have a larger wear allowance than inspection gages.

Every time a plug gage is pushed into a hole, some of the gage metal is worn off. A wear allowance should be provided to compensate for this loss. This is often done by borrowing a very small percentage of the hole tolerance and transferring it in the form of metal to the gage. For example: a hole is 1.000", with a tolerance of plus .002 minus .000; .0002 is borrowed from the total tolerance of the hole and applied to the "go" plug, whose basic dimension is 1.000", thus making it 1.0002", or .0002" for wear. This limits manufacturing to less than the full part tolerance, but any hole which is passed by the working gage will automatically be passed by the inspection gage.

If holes are produced within specified tolerances there will be a minimum of wear on "not-go" gages. As any wear on "not-go" gages tends to bring them further inside the specified limits, the rejection point of such gages is determined by the amount the manufacturer's tolerance can be reduced. So it is not common practice to apply wear allowance to the "not-go" gage.

Gage maker's tolerance, unless otherwise specified by the user, is applied as follows: plus on "go" work plugs, minus on "not-go" work plugs; minus on "go" inspection plugs, plus on "not-go" inspection plugs.

MANUFACTURE OF GAGES

Manufacture of gages requires use of expensive machines and, more important than machines, expert workmanship and heat-treatment. Most gages are made of hardened alloy steel, the particular type depending mainly on the application.

The material from which gages are made is carefully machined to a dimension slightly larger or smaller than the finished size--larger for plug gages, smaller for ring gages. The gage member is heat treated to harden the measuring surfaces and make them resistant to wear. The gage then is either given another heat-treatment or stored for several months to permit the steel to season or stabilize and relieve internal strains.

The stabilized gage blank then is ground to within a few ten-thousandths of the finished size, after which the grinding marks are removed by polishing with a fine abrasive. The final operation is lapping the gage to size, an operation requiring much care in handling.

Working gages which are subject to a great deal of use or to conditions that shorten their life may have their gaging members chromium-plated to increase their resistance to wear.

Recently cemented carbides have been applied to the wearing parts of gages. These extremely hard and wear-resistant materials originally developed for wire-drawing dies and cutting tools, are practically revolutionizing the manufacture of gages. The long life of carbide gages and their remarkable ability to resist abrasion have more than offset their added cost, and they are being rapidly applied to all gaging work where production or other conditions warrant added expenditure.

The life of a plug gage is dependent on several factors other than the wear-resistance of the gage itself, such as the abrasive action of the material being gaged, the condition of the surfaces, and the care taken in using the gage. Cemented-carbide gages normally can be expected to have a wear life up to 100 times greater than steel or up to 35 times greater than chrome. In addition, a greater portion of the manufacturing tolerance can be utilized, as less wear allowance need be specified.

When ordinary alloy-steel plug gages become worn, they may be returned to the manufacturer to be chromium-plated to size. Chromium-plated gages also may be re-plated to size. Actually, the gage is plated oversize and lapped down to the correct dimension.

CYLINDRICAL-PLUG GAGES

The design of plug gages has been standardized by the American Gage Design Committee to four different designs. Each design is recommended for a specific range of sizes. These are known as:

1. taper lock
2. reversible trilock
3. annular designs
4. wire-type

The taper-lock type of plug gage shown in Figure 46 is available from .059 to 1.510 inches in diameter. This type of gage is simple and inexpensively constructed. The gaging member has a tapered shank that fits snugly into the handle. A hole or slot in the handle is provided for the insertion of a drift pin when the gaging member is to be removed.

When two gaging members are mounted in the same handle, the one at the end opposite the drift hole is removed by running a rod through the hollow handle. The two gaging members are known as the "go" and "not-go" plugs, respectively. The "go" plug measures the lower

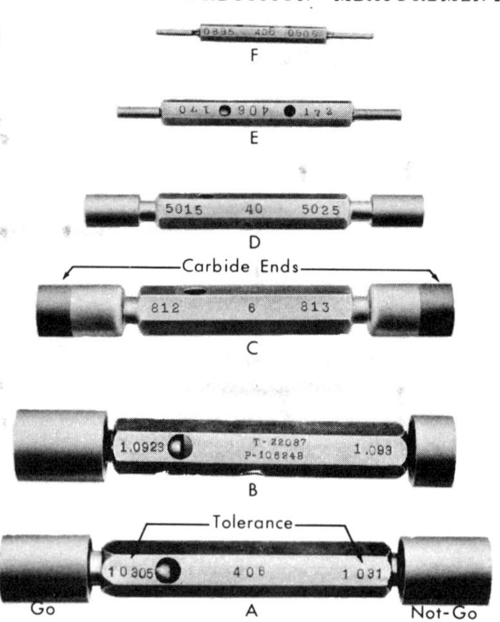

Figure 46. Taper-Lock Type Plug Gages

limit of the hole, whereas the "not-go" plug checks the upper limit. The "not-go" plug is sometimes shorter than the "go" plug, as shown in Figure 46B. Usually the "go" and "not-go" limits are stamped on the handles to designate each plug. However, some plants keep the "not-go" plugs shorter on the larger plugs and cut a groove in the handle near the "not-go" plug on the smaller plugs to distinguish one from the other.

When both gaging members are mounted on the same end, as shown in Figure 47A, the

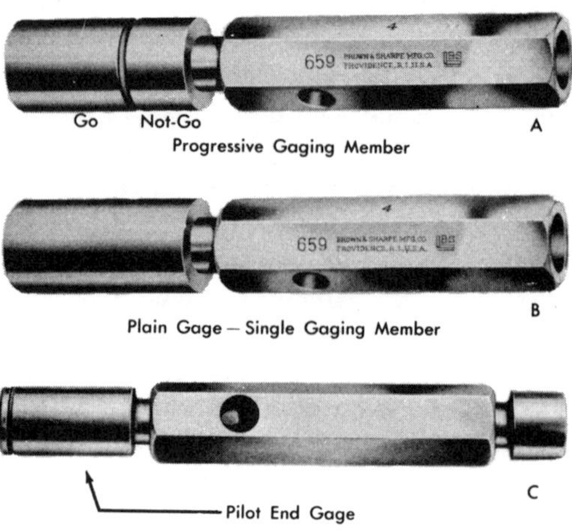

Figure 47

gage is known as a progressive or stepped plug gage. This type of plug gage offers the obvious advantage of being able to gage the "go" and "not-go" dimension in one motion. However, this advantage is, for most applications, offset by the disadvantages of higher cost, fixed relation between the "go" and "not-go" dimension, and the necessity for scrapping both plugs when only the "go" plug is worn out.

Shown in Figure 47B is a plain gage with only one plug or member. This style of gage is not adapted to rapid inspection of both "go" and "not-go" limits.

Another style of plug gage is also illustrated in Figure 47C. This gage has a pilot end on the "go" end to facilitate inserting the plug in soft metals. This prevents the shaving action which often results from the use of standard plugs.

Complete dimensions for the various types of taper-lock type plug gages are given in the standard set by the Committee. The reversible trilock type of plug gage shown in Figure 48A is the recommended design for plug gages from 1.510 to 8.010 inches in diameter. These gages of larger sizes are heavy, and consequently a properly designed rigid connection between the gage and handle is necessary.

The gaging member is a cylinder with a hole through the center, which is counterbored from both ends. Three wedge-shaped locking prongs on the handle are forced into corresponding grooves in the gaging member by a single screw through the center, thus providing a self-centering support with a positive lock and resulting in a degree of rigidity equivalent to that of a solid gage.

Another style of reversible plug gage, called wire or pin-type, is shown in Figure 48B. By loosening the handle, the plugs can be removed and reversed, or the worn end can be cut off, greatly extending their useful life. Wire or pin-type gages are available from .025 to .510 inch in diameter.

For diameters from 8.010 to 12.010 inches, the recommended design is a reversible annular ring, which somewhat resembles a small flywheel. The annular design has a rim and web, which has been bored out to reduce the weight. The web has four equally-spaced threaded holes to receive ball handles or to provide a means of attaching the gage to a face plate.

These standard designs of plug gages are specified so completely in all of the details of their construction that parts made by one gage manufacturer are interchangeable with those made by another.

SPECIAL TYPES OF PLUG GAGES

In addition to the plain cylindrical plug gages described, there are many other special types of plug gages, such as: taper-plug gages, spline-plug gages, flat-plug gages, and special plug gages. Each of the various types of plug gages is better adapted to some application than any other available gage.

Taper-plug gages of the type shown in Figure 49 are used to check the amount of taper and the diameter of a tapered hole. These gages are made with standard or special tapers as required. Their main application is in measuring diameter and amount of taper in the spindle socket of machine tools. The plug gage shown in Figure 49 has a tang similar to that on a

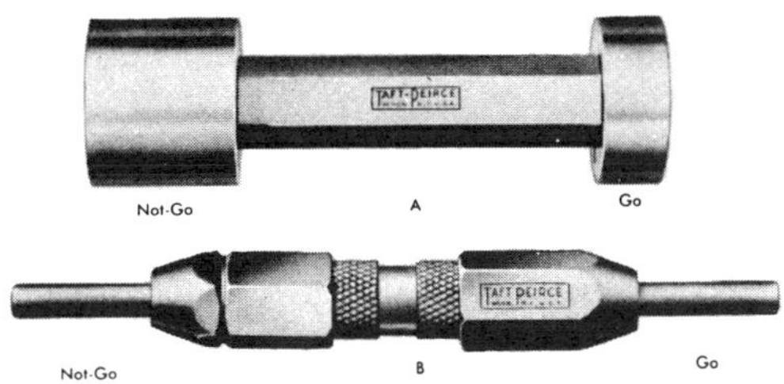

Figure 48. Reversible Plug Gages

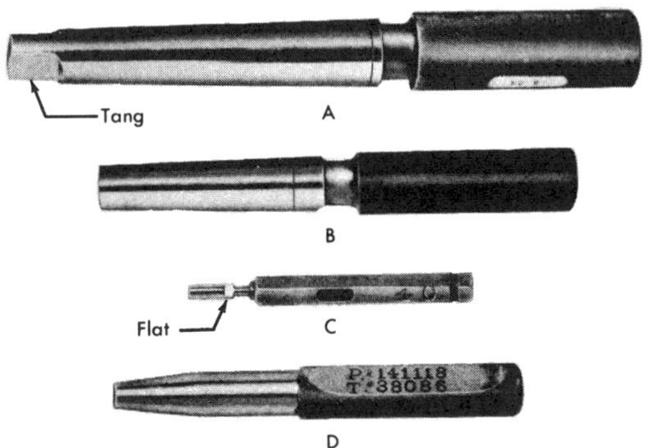

Figure 49. Taper Plug Gages

tapered shank on a drill. This particular type of taper plug is used to check the spindle socket on a drill press. Taper plugs may also be used in checking the taper of an internal pipe thread, or of a tapered hole in a part.

Taper-plug gages with graduated rings along the taper may be used to measure the diameter of cylindrical holes. The difficulty in gaging a cylindrical hole in this manner is that any error at the mouth of the hole, such as a burr or slight bell-mouth, affects the measurement. Only a single mark appears on the gages illustrated in Figure 49A and B. This scribed ring indicates only the basic dimension.

Some taper gages used in the inspection of parts have "go" and "not-go" rings scribed on the gaging member. When the gage is inserted so that it fits the taper snugly, one of the two marks should disappear into the hole, in the manner shown in Figure 50. It should be understood that if the gage fits snugly to either line, the part is considered within limits.

The taper gages shown in Figure 49C and D are special gages for parts inspection. The flat on the gage shown in view C is comparable to the two scribed lines mentioned previously, since a part fitting the flat or the end of the flat is considered within the limits. The taper gage in view D indicates only the basic dimension to the end of the taper.

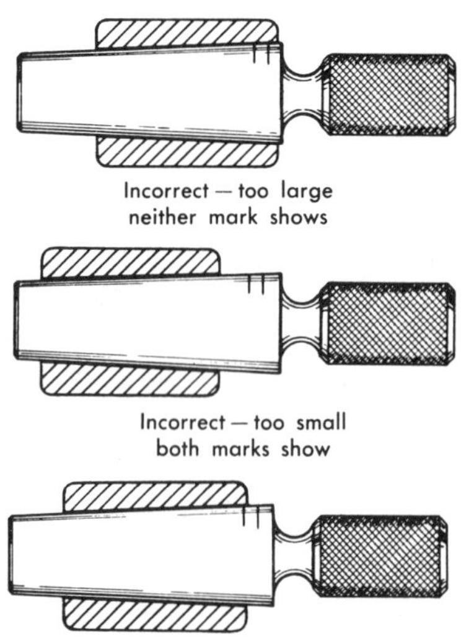

Figure 50

Figure 51. Spline Plug Gages

In Figure 51 are shown some male-spline gages. These gages are really a form of plug gage, and are used to check the splined holes in gears, collars, and other parts which operate on splined shafts. Spline-plug gages are special and usually ordered from the gage maker for the inspection of some particular part.

Flat-plug gages of the type shown in Figure 52 are used to check the width of slots, and to check holes that are not entirely cylindrical, such as the diameter of a splined collar. The "go" and "not-go" flat-gaging members are similar to the plain cylindrical taper shown in Figure 46.

A number of special, single-purpose plug gages are shown in Figure 53. They may be

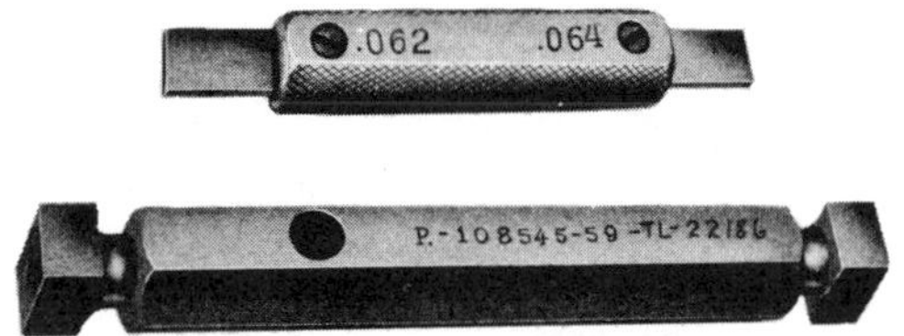

Figure 52. Flat Plug Gages

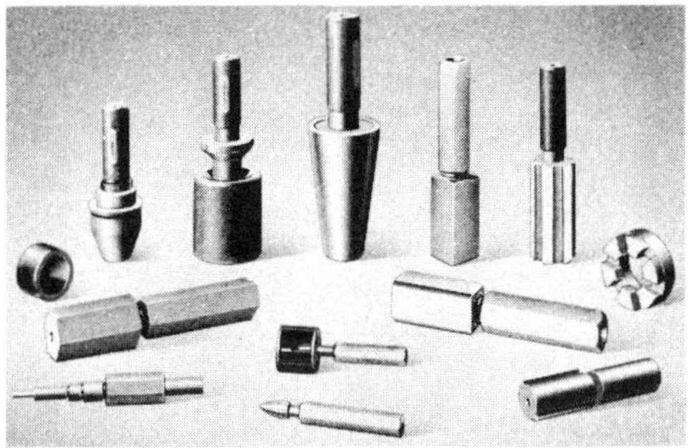

Figure 53. Special Types of Gages

square, hexagonal, splined, tapered, curved, or stepped. They are built by all gage manufacturers to any customer's specifications. Because special gages are more expensive, their use is limited to applications where there is sufficient production or the nature of the job requires that a special gage be used.

The micrometer plug gage shown in Figure 54 is a special type of adjustable plug gage. It is a very accurate internal micrometer adapted to toolroom and special inspection work. This gage may be used to check the roundness, straightness and size of a hole to within one ten-thousandth of an inch. Its gaging members or blades are seated on a hardened cone, and as the barrel is turned the diameter is expanded or contracted over a range of 1/16 inch. This type of gage may be obtained in a number of sizes, ranging from 3/4 of an inch to four inches.

The Woodruff keyway gage shown in Figure 55 is, as its name implies, used to check the width of keyways in shafts. The gaging members are accurately ground and lapped discs and can be set in any of three positions. This affords three times the useful life of the earlier solid types.

APPLICATIONS OF PLUG GAGES

Plug gages are used to check all types of holes, slots, and recesses in parts and tools. In Figure 56 is shown the gaging of the hole in a part with a carbide plug gage. The .312-inch gaging member enters the hole; the .313-inch gaging member does not. By testing the hole on both the "go" and "not-go" members, the inspector has definitely established the diameter of the hole as being between these two limits.

Another application of a special taper gage, shown in Figure 57, is checking the taper of a hole in a cylinder. The taper plug is inserted in the cylinder as far as it will go. The size of the tapered hole must be such that the edge of the cylinder comes within the "go" and "not-go" steps on the gage.

Figure 58 shows a long plug gage being used by an inspector to check the line-reamed holes in a casting for size and alignment. A test indicator is also used with the plug gage to determine the variation from one end to another, as discussed in Chapter V.

Figure 59 shows a plug gage being used by a machine operator to check the diameter of

Figure 54. Micrometer Plug Gage

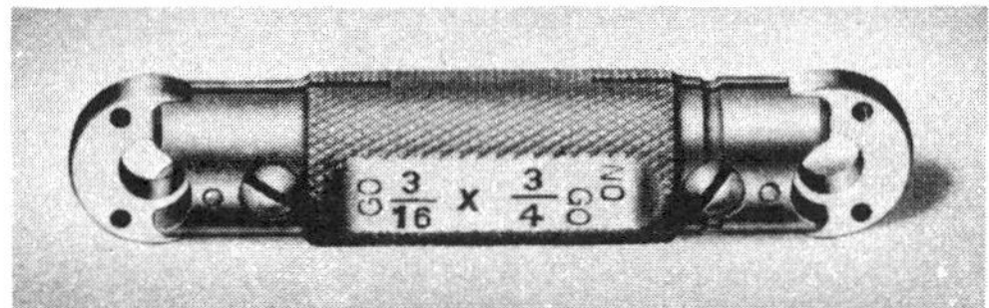

Figure 55. Woodruff Keyway Gage

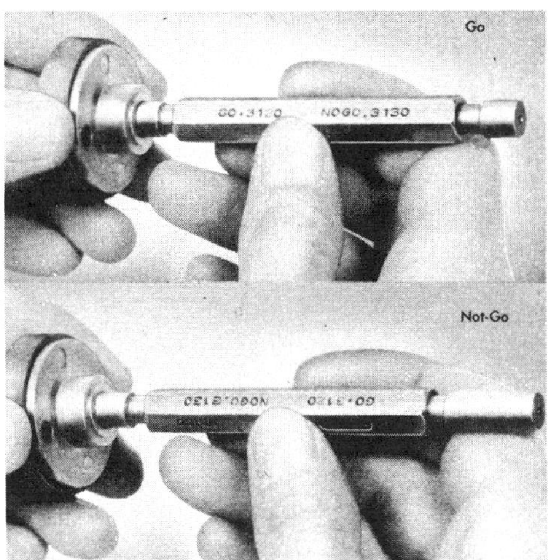

Figure 56. Gaging a Hole

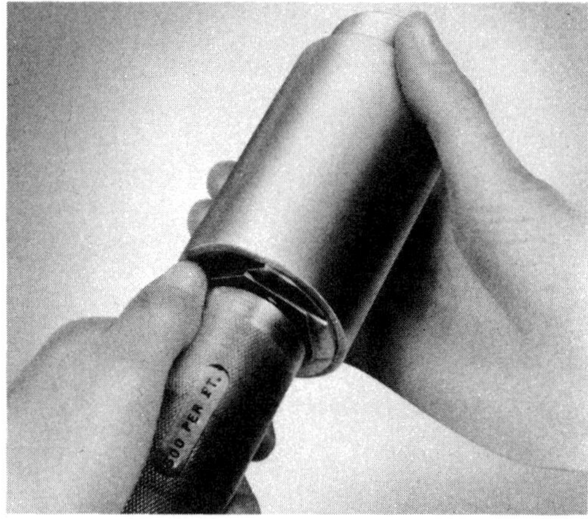

Figure 57. Gaging a Tapered Hole

a hole in a small gear. The "go" gage (.467) enters the hole; so the hole is large enough. The other end of the gage (.468) will be used to determine whether or not the hole is too large. If the gage will not enter, the operator knows that the hole is somewhere between .467 and .468 inch, his limits of accuracy.

CARE AND USE OF PLUG GAGES

The care taken in using a plug gage will determine to a large extent its useful life. The inspector and machine operator must learn to recognize those conditions which affect the life and accuracy of plug gages, and to apply common sense in using them.

One important factor is the abrasive action of the part on the gage. Whenever a plug gage is inserted in a hole, a thin layer of the metal of the gage is worn off, and the operator who is not careful, particularly with some of the more abrasive metals, shortens the life of a plug gage to a fraction of what it might be if the gage were used properly. Cast iron and cast aluminum are both more abrasive than steel, brass, bronze, and non-metals such as plastics. Particular care should be taken not to force the gage into the hole. Special care must be taken when gaging close tolerance holes in aluminum, and other soft metals, to prevent seizing and loading of the gaging member. It is good practice to apply a thin film of light oil to the gage. Special gage lubricants are available for this purpose. It is difficult to describe the proper amount of effort necessary to gage a hole, but in general a plug gage should enter a hole of the smallest correct size with the exertion of no more wrist force than is necessary to wind a spring-driven clock.

Figure 58. Using Long Plug Gage

Abrasives, dirt, and chips from machining

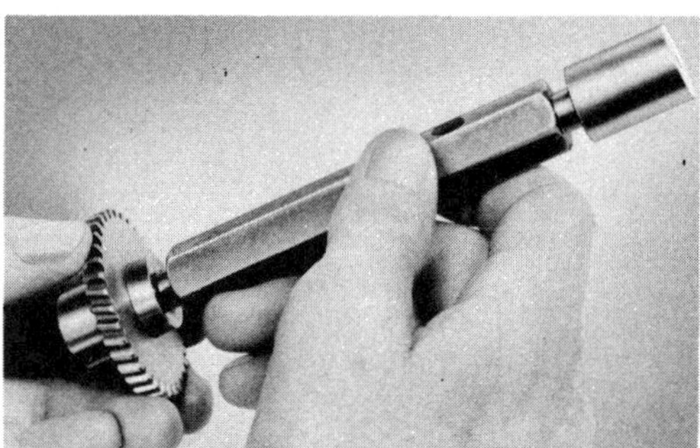

Figure 59. Checking Hole in a Gear

operations tend to collect in holes, and if they are not removed thoroughly they may work under the entering edge of the plug and pack into the space between the plug and the wall. If the plug is forced in under such conditions, it will be worn down rapidly and may even be damaged. Frequently, drilled holes have burrs, which, if not removed before gaging, may scratch the gage surface, or work in between the gage and the wall of the hole and thus contribute to excessive wear or damage.

In general, it is bad practice to clamp a plug gage in a vise or use a wrench on it, because the inspector loses the "feel" of gaging the part, and unknowingly he may exert too much force and shorten the useful life of the gage. The "feel" in gaging is all-important. By sensing side wobble with the fingers, an inspector determines whether or not the hole is tapered or out of round.

The wear on a plug begins at the tip and gradually advances back along the gage. When the limit of allowable wear is reached over about the first quarter of an inch, the plug gage is either reprocessed or scrapped. Plug gages should be inspected frequently either with gage blocks or on a precision measuring machine.

When plug gages are stored, they should be arranged neatly so that possibility of damage due to careless handling is reduced to a minimum. Wherever possible, plug gages should be stored separate from other tools, and if the number stored is large enough, a special drawer should be made to house them, as shown in Figure 60.

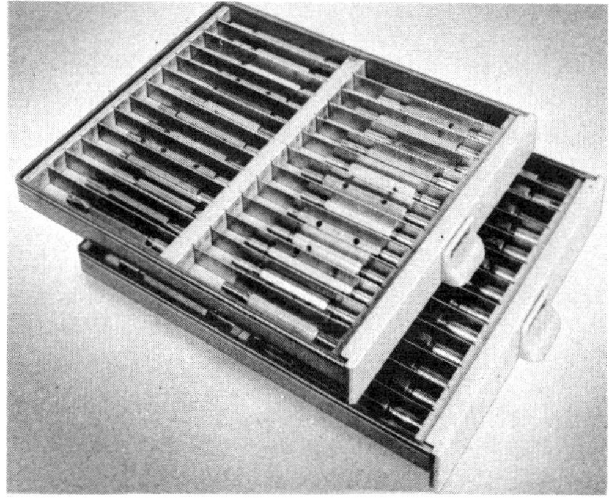

Figure 60. Storing Plug Gages

Figure 61. Plain Ring Gages (small sizes)

PLAIN-RING GAGE

The plain-ring gage is an external gage of circular form used for the size control of external diameters. It may be made from steel, chrome-plate, or cemented carbide. The ring gage designs standardized by the American Gage Design Committee are generally accepted by all gage manufacturers.

In the smaller sizes of plain-ring gages, a hardened bushing may be pressed into a soft gage body in place of the one-piece ring gage. This design is optional in sizes over .059 to and including .510 inch. For ring gages with diameters between .510 and 1.510 inch, the single-piece gage is standard.

Ring gages from 1.510 up to and including 5.510 inches are made with a flange, like the one shown in Figure 62. The purpose of the change in design from the one illustrated in Figure 61 is to reduce the weight of the gage and make the larger sizes easier to handle.

The "go" and "not-go" ring gages are separate units. The two are readily distinguished from each other by an annular groove cut in the knurled outer surface of the "not-go" gage as shown in Figures 61 and 62. Gages larger than 5.510 and up to 12.260 inches in diameter are constructed similarly to the one shown in Figure 62, except that they are equipped with two handles, which project radially from the flange, to aid the user in handling them.

These gages are made in four classes of finish and tolerance: XX, X, Y and Z. The tolerances for these classes of accuracy are the same as for plug gages as given in Table II. On working and inspection ring gages, the gage tolerance is applied in the direction opposite to that on the plug gage, namely, minus on the "go" gage and plus on the "not-go" gage, and split plus and minus on setting rings.

The practice of applying the tolerance on a "not-go" gage in a direction which may place the size of the gage outside of the part limits has not been approved by any standardizing body in this country. Certain large users of gages have adopted this practice, but when the part is subject to inspection by a purchaser, such practice is practically certain to give rise to disputes between the manufacturer and purchaser.

CLASS XX ring gages should be used only as masters to set comparators and calibrate other accurate internal measuring devices, or as reference gages.

CLASS X ring gages are generally used for close tolerance inspection.

CLASS Y ring gages are used in the inspection of parts when the limits of accuracy are fairly close and the production large.

CLASS Z ring gages are used for general inspection when a ground gage is adequate, production is not too large, and tolerances are liberal.

Figure 62. Plain Ring Gage (large size)

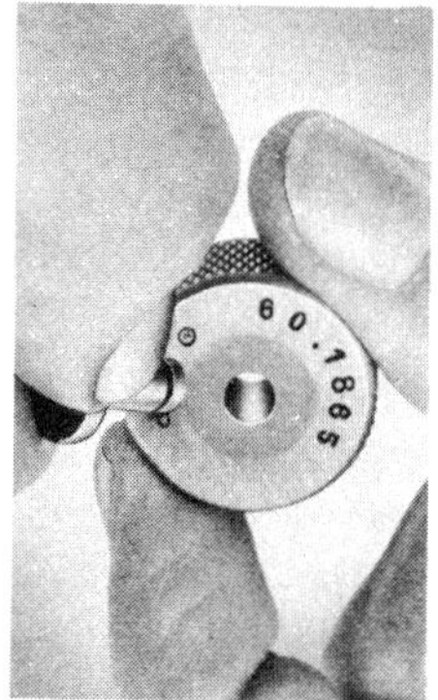
Figure 63. Gaging the Diameter of a Pivot Stud

APPLICATIONS OF PLAIN-RING GAGES

Ring gages are used largely to check the external diameter of shafts, arbors, pins, studs, and balls at final parts inspection before the part goes into an assembly, and as setting rings for precision gages. Ring gages are used more often in the inspection of finished parts than of parts in process. The reason is that finished parts are usually in a container and can be picked up and gaged, whereas parts in a machine would probably have to be removed.

The advantage in using a ring gage is that it checks the part for size, taper, and out-of-roundness at one trial. Other types of gages do not give a complete inspection, unless they are tried in several places along the shaft.

In Figure 63 a "go" ring gage is shown being used to check the shank diameter of the pivot stud. The stud is first lined up with the hole and pressed in gently. If the stud refuses to enter the hole with hand pressure, the shank is too large. If the stud does enter the hole, it is not oversize. With the stud in the hole, the inspector can also check the piece for taper and out-of-roundness by discovering any wobble that may be present.

The part now must be tried in a "not-go" ring gage, and the stud must not enter the hole in this gage. If it does not enter, the diameter of the shank is established as being between the limit specified.

The "go" ring gage controls the maximum dimension of a cylindrical object and the "go" plug gage controls the minimum dimension of the hole. "Go" gages control the tightness of fit of mating parts, whereas "not-go" gages control the looseness of fit of mating parts.

CARE OF RING GAGES

As with plug gages, the user of a ring gage should take care to see that the surface of the parts gaged and the gage itself are kept free from abrasives, dirt, grit, and chips. It also is important to consider the abrasive action of some metals and to use particular care when gaging parts made from these materials. Ring gages should be held in the hand when being used and not clamped in a vise. Clamping a ring gage in a vise may distort the hole and destroy the feel of gaging.

In general, ring gages can take more abuse and careless handling than plug gages, because the gaging surfaces are protected by surrounding metal. Being precision gages, however, they should be given the same care as plug gages.

Ring gages must be checked at frequent intervals to insure accuracy. In general, internal gaging surfaces are a little more difficult to measure accurately than external surfaces, but ring gages may be checked with gage blocks, accurate internal indicator gages, or master plug gages.

SPECIAL TYPES OF RING GAGES

Special types of ring gages are required to gage non-cylindrical, splined, and tapered shafts, and studs, or to make the gage more applicable to the work it is required to do.

TWIN-RING GAGES

The gage shown in Figure 64 is known as a twin-ring gage. It combines into one piece the "go" and "not-go" ring gage bushings. This gage was considered sufficiently convenient for the rapid inspection of a certain type of small precision parts by the American Gage Design Committee and was adopted as a standard design in the range from .059 to 1.135 inches, inclusive.

The gage consists of an unhardened steel blank bored to take two hardened steel bushings. The "not-go" gage has been designated by

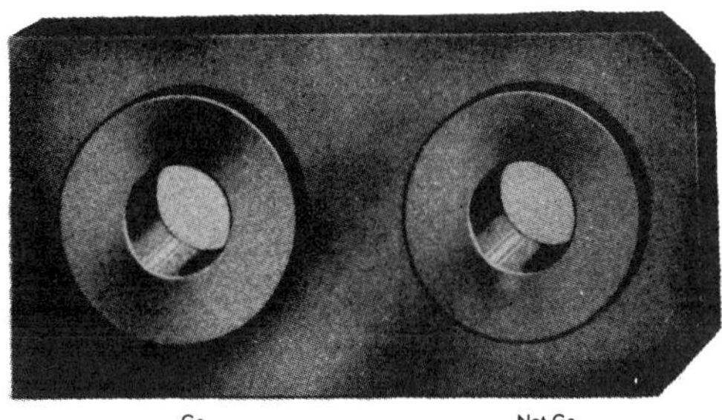

Figure 64. Twin Ring Gage

chamfering the corners of the flat blank in the manner shown in Figure 64.

TAPER-RING GAGES

Taper-ring gages of the type shown in Figure 65 are used to test both the accuracy of taper and a diameter at some point on the tapered section. The male taper to be tested is pressed into the gage by hand. After the piece is firmly seated in the gage, the large and small diameters are inspected for clearance. If there is any clearance, the part can be wobbled, indicating the taper is not correct.

A more accurate method of checking the taper is to use Prussian blue. By coating the male taper with a very thin layer of blue and inserting it in the gage, it is possible to find exactly where the taper is touching the gage. If the gage and taper are rotated back and forth slightly, when the two are in light contact,

Figure 66. Taper Test Gage

the blue will be rubbed off where the taper touches the gage. A perfect taper would be marked uniformly over the entire length gaged.

The diameter of the tapered section is checked by measuring the distance the male taper enters the gage. In Figure 65 a section of the gage has been cut away. On the flat surface parallel to the axis of the gage, two lines are scribed in the manner shown. If the diameter of the male taper is too small, it will enter the gage too far, and if it is too large, it will not enter far enough. To be correct, the edge or some designated line on the male taper must come between the two scribed lines when the gage and the taper are fitted snugly together.

In checking milling arbors, one application of this type of gage, the inspector is interested only in the amount of taper, and gages used for this and similar applications will not have these scribed "go" and "not-go" lines. Gaging tapers accurately is a rather lengthy procedure, particularly if the tolerances are so close or the taper is so slight that Prussian blue must be used.

Another form of taper gage for checking male tapers is shown in Figure 66. This taper test gage is adjustable and may be used to check the taper of drill and reamer shanks, taper reamers, and taper pins. It has a handle on the side so that it can be clamped in a vertical position. The taper test gage is set up to the correct taper with a master taper plug gage, precision gage blocks, or a master part.

An adaptation of the principle of the taper test gage in Figure 66 is made in the small gage shown in Figure 67. The diameter and slight taper on a contact plug is shown being checked by the gage. The diameter is checked by noting the position of the end of the plug in relation to the two scribed lines, and any error in taper is detected by "feeling" for any side movement.

RECEIVER GAGES

In general, a receiver gage is a special form of ring gage with inside measuring surfaces arranged to check the size and contour of some manufactured part. This type of gage is, in effect, a mate to the part being gaged, and its principal function is to determine whether the part will fit mating parts in the assembly.

Some receiver gages, such as those employed in the inspection of gun parts, may be constructed so that certain important dimensions can be checked as being within the minimum as well as the maximum limits. This is done with flush pins or with plug or feeler gages inserted at particular locations. Most receiver gages, however, are not designed to check dimensions, but to check the suitability for assembly of the part as a whole, Any important dimensions are checked with some other gage.

The most common type of receiver gage is the female spline gage. This gage, shown in Figure 68, is used to check all principal "go" dimensions of a spline shaft such as the width of the lands, the spacing and the base diameter, with the exception of the outside diameter. If the spline gage will go on the shaft, the part will fit in assembly, even though there is con-

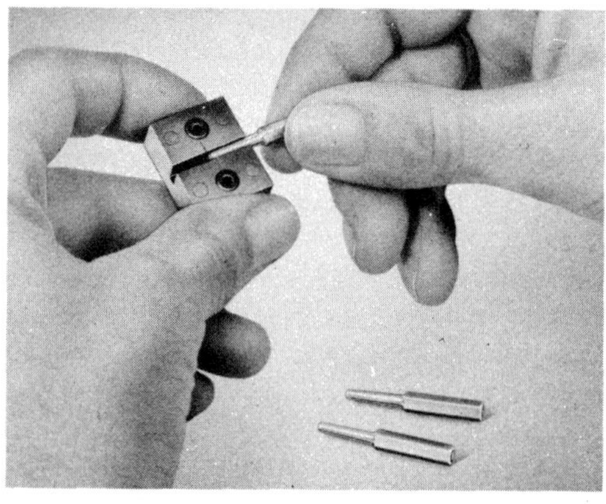

Figure 67. Checking Diameter and Taper

Figure 68. Female Spline Gages

siderable error in some of the dimensions.

Receiver gages are made to fit the contour of some part, and because of internal machining operations involved in making one of these gages, they are quite expensive. Because these gages are special and expensive, persons using them should be very careful not to abuse them. These gages may be checked at intervals with gage blocks as shown in Chapter VII with master plug gages, or with master parts.

PLAIN SNAP GAGES

There are two general types of plain snap gages, the non-adjustable and the adjustable. The American Gage Design Committee defines the snap gage as follows: "A snap gage is a complete external caliper gage employed for the size control of plain external dimensions, comprising an open frame, in both jaws of which gaging members are provided, one or more pairs of which can be set and locked to any predetermined size within the range of adjustment."

PLAIN SOLID SNAP GAGES

Solid snap gages are one of the earliest types, and, while still in use today, they have in most

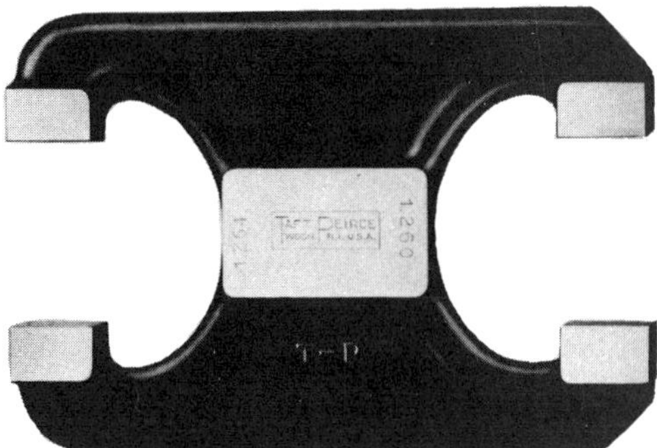

Figure 69. Plain Solid Snap Gage

shops been replaced by the more practical and longer-lived, adjustable snap gage. Snap gages were originally made to the basic dimensions they were to gage, and it was necessary for both the machine operator and the inspector to estimate the difference between the gage dimension and that of the part. It took much longer than necessary to machine parts and to inspect them.

The limit snap gage was developed to give both machine operator and inspector a range in which to work. To control this range, two gaging dimensions were established as limits, and gages of these dimensions were used instead of the single gage. A limit snap gage of this type is shown in Figure 69.

In using this type of gage, the part to be inspected is first tried on the "go" side; then the gage is reversed and the part is tried on the "not-go" side. In Figure 69 the "not-go" side

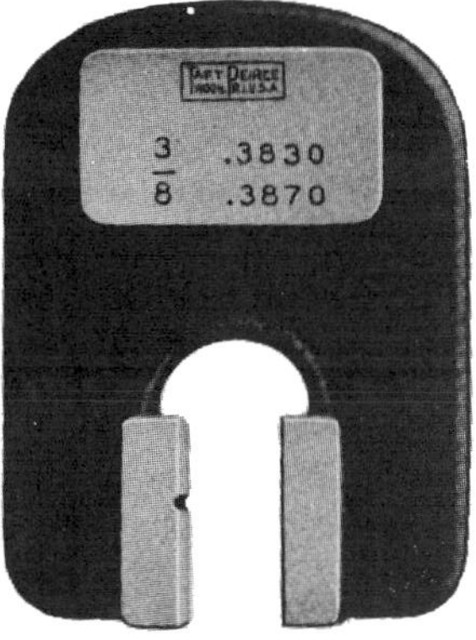

Figure 70. Solid Snap Gage (progressive type)

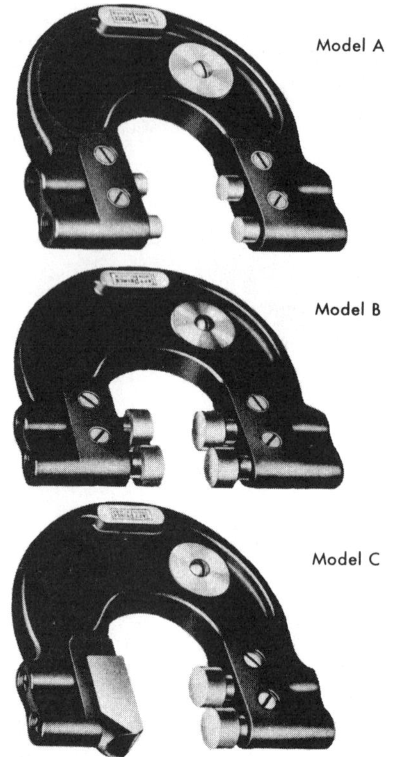

Figure 71. Standard Types of Snap Gages

of the snap gage has been designated by chamfering the corners of that side.

When the two gaging members are combined in the same set of jaws, as in Figure 70, the snap gage is known as a progressive type. The outer set of jaws gage the "go" dimension, and the inner set the "not-go" dimension. For this type of gage the distance between the outer pair of jaws is always greater than that between the inner pair.

In the manufacture of these gages, the tolerance is distributed the same as for ring gages.

PLAIN ADJUSTABLE SNAP GAGES

The increased use of adjustable snap gages in the inspection of production parts has led to the standardization of their design. Three recommended designs for adjustable snap gages are shown in Figure 71.

The adjustable snap gage consists of:

A frame designed for structural rigidity, lightness, and balance.

Adjustable gaging pins, buttons, or anvils, which may be securely locked in place after adjustment.

Locking screws, which may be tightened to hold the gaging dimensions.

A marking disc, which is marked to identify the gage as to dimension, tool, or part number.

The Model A adjustable snap gage is made in sizes ranging from 0 to 12 inches. It is equipped with four gaging pins and is suitable for checking the dimension between surfaces where the measuring surface is restricted by a shoulder or boss.

The Model B adjustable snap gage is made in sizes ranging from 1/2 to 11-1/4 inches. It is equipped with four gaging buttons and is suitable for checking flat or cylindrical pieces. The larger area of the gaging button distributes the wear so that more parts may be checked before the button needs to be readjusted.

The Model C adjustable snap gage is made in sizes ranging from 0 to 11-5/8 inches. It is equipped with two gaging buttons and a single block anvil, and is especially suitable for checking the diameter of shafts, pins, studs, and hubs.

Model MC is a miniature snap gage similar to Model C. Its range is from 0 to .760 of an inch.

ADJUSTABLE LENGTH GAGE

Figure 72 shows an adjustable length gage which has all the desirable features of the snap gage and also can control dimensions beyond snap-gage range. By using gage-spacing bars of different lengths, or standard size blocks, the range of the gage can be made any length required. The gage heads are made in two styles, the progressive model with both buttons

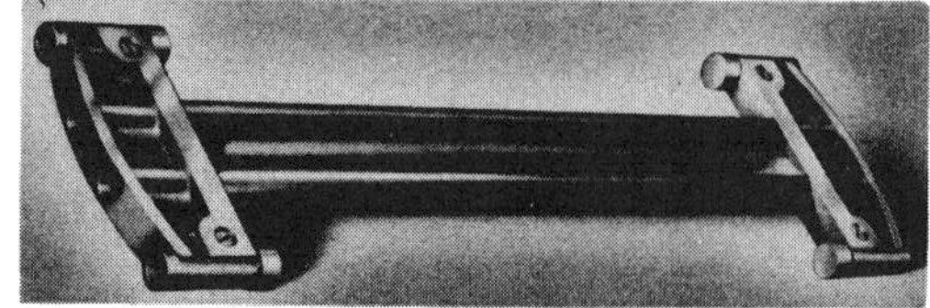

Figure 72. Adjustable Length Gage

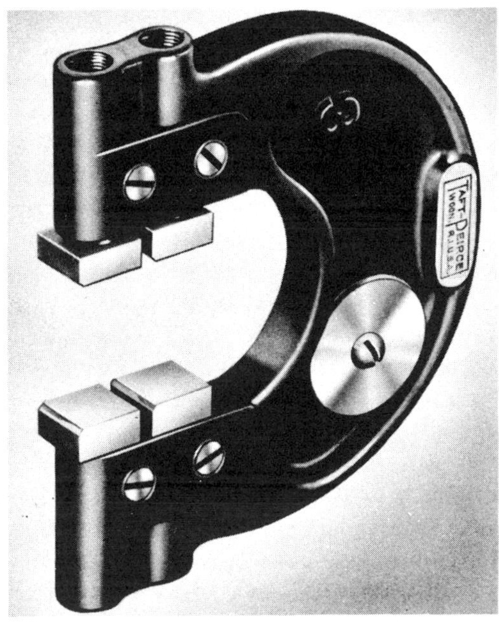

Figure 73. Snap Gage with Square Buttons

on the same side, and the double-sided model with "go" and "not-go" on opposite sides of the spacing bar, as shown in Figure 72.

SPECIAL BUTTONS OR ANVILS

For some types of work it is often desirable to have a larger gaging surface than is provided by the standard type of gaging pin or button. Square buttons of the type shown in Figure 73 are used for applications of this nature. To minimize the wear on the anvils when gages get considerable use, or are subject to rapid wear, the buttons may be chromium-plated or tipped with carbide.

SETTING THE SNAP GAGE

Before an adjustable snap gage can be used to check parts, the "go" and "not-go" buttons must be set to the proper dimensions. The snap gage is first clamped in a holder as shown in Figure 74. The locking screws are then loosened, and the adjusting screw is turned in the manner shown in Figure 74 to set the button to the proper dimension. This dimension may be taken from a master disc, a precision gage block, or a master plug. With the gage block between the anvil and button, the locking screw is tightened, and after checking to see that the dimensions have not changed, the snap gage is ready to be used. Usually wax is melted into the head of the screws to prevent or indicate tampering.

APPLICATIONS OF ADJUSTABLE SNAP GAGES

The plain, adjustable-limit snap gage can be used wherever ordinary calipers, micrometers, ring gages, and similar instruments are applicable. These gages are commonly used by machine operators to check the part at intervals, by process inspectors to check accuracy of parts as they come off the machine, and by final inspectors to check parts before they are assembled.

The adjustable-limit snap gage is widely

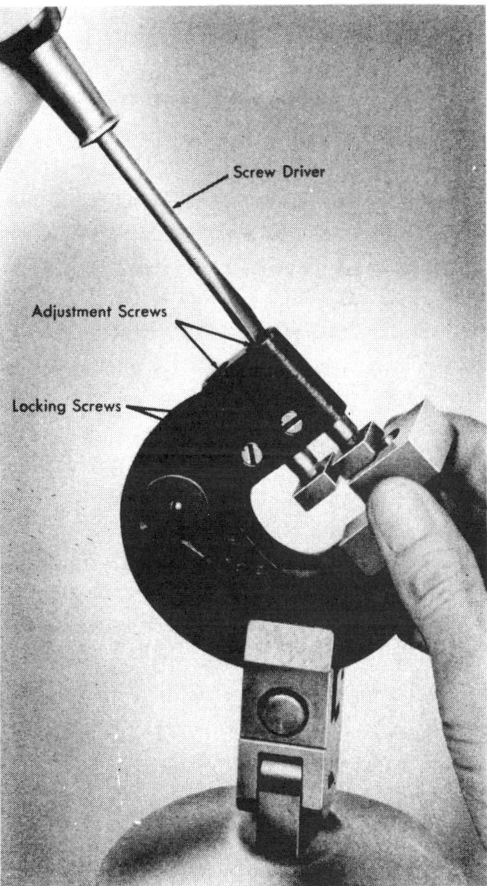

Figure 74. Setting Snap Gage

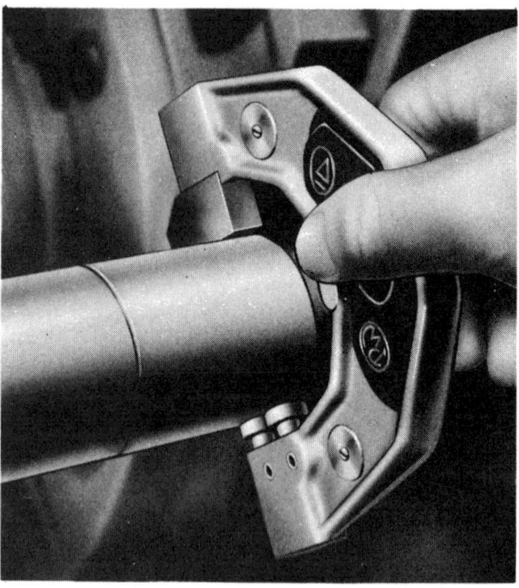

Figure 75. Checking a Shaft Diameter

used because it is fast, accurate, and convenient to use in inspecting parts. However, every type of gage has limitations which confine its field of application. The adjustable snap gage is not a practical gage to set up and use for less than approximately fifty parts. Since the "go" and "not-go" dimensions must be set up, there must be a sufficient number of parts to warrant the time required to do this. For very small lots of parts it may be more practical to use a micrometer or vernier caliper.

Another limitation is that a snap gage checks a part at only one point at a time. If it is necessary to check the part in several places to control its accuracy, it may be advisable to use another type of gage. For example, if it is necessary to check the diameter of a shaft for out-of-roundness and taper as well as diameter, it may be more practical to use a ring gage instead of a snap gage, because a ring gage will check all three in a single gaging operation.

Other limitations are that the depth of the throat limits use of the gage to around the edge of large, flat parts. When checking a part in a machine, the operator must know how many thousandths remain to be removed; a snap gage shows that the part is oversize but not by how much. Figure 75 shows the use of a snap gage to check the diameter of a shaft which is being ground on a cylindrical grinder. Figure 76 shows a pin being checked in an adjustable snap gage. The pin checks "OK," since it passes the "go" button and is stopped by the "not-go" button.

GAGING WITH A SNAP GAGE

The correct method of gaging a flat part is shown by the series of drawings in Figure 77. The snap gage approaches the part directly so that the buttons are square with the surfaces on

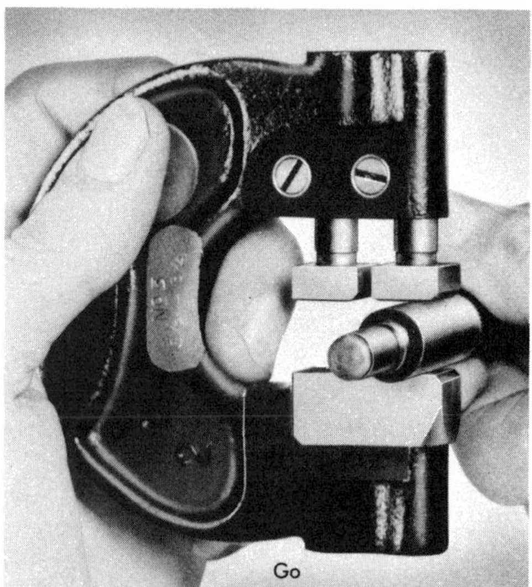

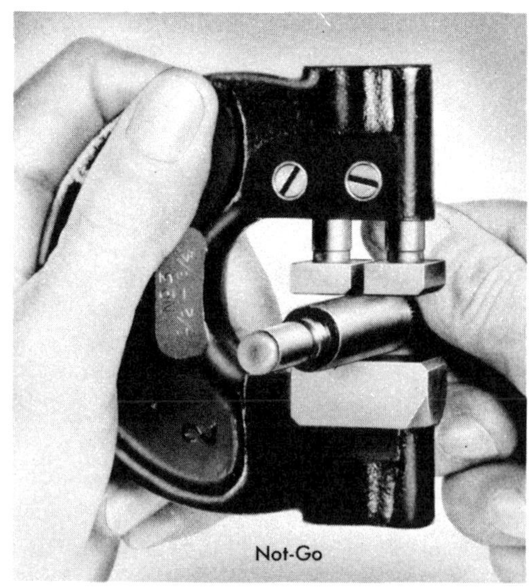

Figure 76. Gaging a Pin

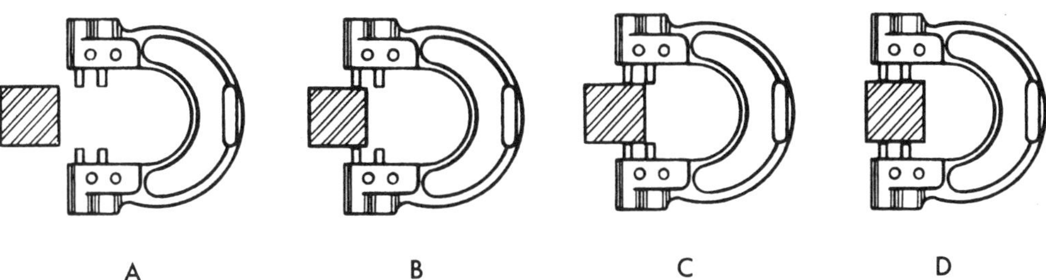

Figure 77. Snap Gaging an Object with Parallel Surfaces

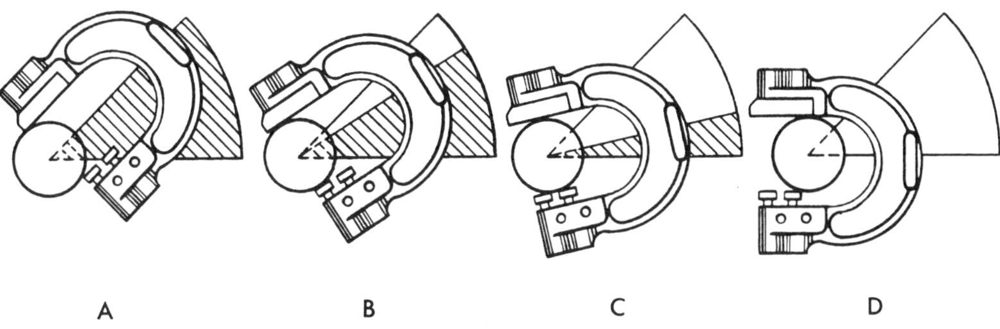

Figure 78. Snap Gaging a Cylindrical Object

the part to be gaged, as in Figure 77A. The "go" pins are pushed over the part as illustrated in Figure 77B. The gage pins must be kept square with the surface of the part, and only a light hand pressure should ever be necessary to pass the "go" pins, If the part is within the limits set up in the gage, it will be stopped by the "not-go" pins, as in Figure 77C. If the part is undersize, it will be possible to push it past these pins, as in Figure 77D.

The correct procedure for gaging a cylindrical piece is illustrated by the series of diagrams in Figure 78. This applies to larger pieces and not to small cylindrical parts, like the one shown in Figure 76, which are rolled between the buttons.

The gage is located on the part with the solid anvil on top, as shown in Figure 78A. The gage is then rocked, as shown by the shaded segment in Figure 78B, where the "go" dimension is checked. If the shaft is not oversize, the first button will pass over it easily. The gage should then be advanced to the position shown in Figure 78C. If the "not-go" button stops the gage, the shaft is within the limits. If the gage can be rocked further to the position shown in Figure 78D, the diameter is too small because it has passed the "not-go" gage.

In using a snap gage on a part in a machine, the operator should be certain that the temperature of the part is not very different from room temperature. Parts are heated by machining and often cooled by the coolant. Also, a snap gage should never be used on a moving part. To do so not only is very dangerous to the operator, but also hard on the instrument and will nearly always spring the gage so that the measurement is of no value.

THICKNESS OR FEELER GAGES

These gages, shown in Figure 79, are fixed gages in leaf form, which permit the checking and measuring of small openings such as throats, contact points, and narrow slots. They are widely used to check the flatness of parts in straightening and grinding operations and in squaring objects with a feeler gage as described in Chapter VIII.

The set shown in Figure 79 ranges from .0015 to .025 inch. By combining leaves, it

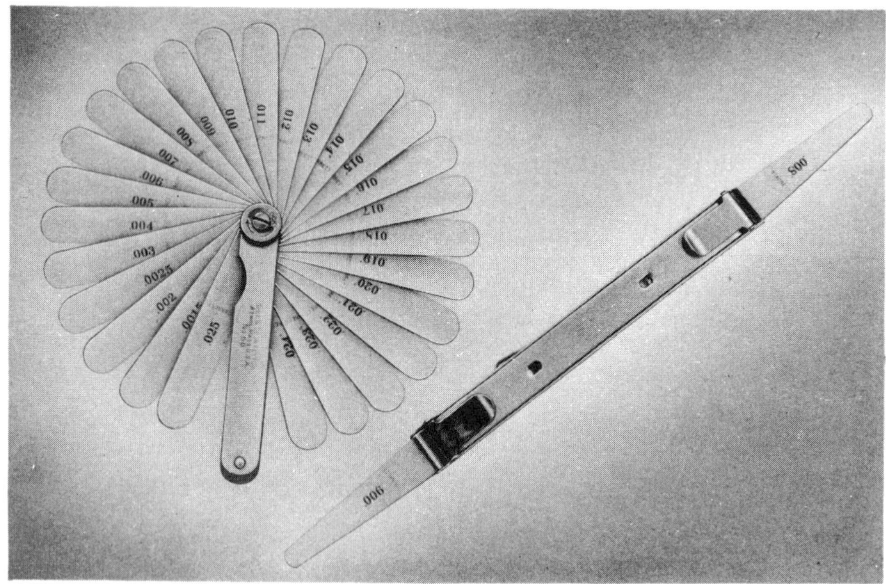

Figure 79. Feeler Gages

is possible to obtain many thicknesses. Available sets range in thickness from .0015 to .200 inch and in length from approximately 2 to 12 inches.

These gages are used in one of two ways: as a means for determining a measure, or as a means for adjusting to a definite limit, as shown in Figure 80. Here a knife throat is being adjusted by using two blades. The knife is adjusted vertically to permit the free passage of an .008 inch leaf but the snug fitting of a .010 inch leaf.

Care should be exercised in using these gages, particularly for the purpose just described. The knife should not be lowered on a blade which is then removed, as this may shave the leaf if it is too tight. Also, the blades should not be used for cleaning slots or holes.

A leaf which has become damaged or is no longer of proper size can be replaced by a new one. The style in Figure 79B holds individual leaves of any length, which is particularly desirable when one or two sizes are ordinarily used.

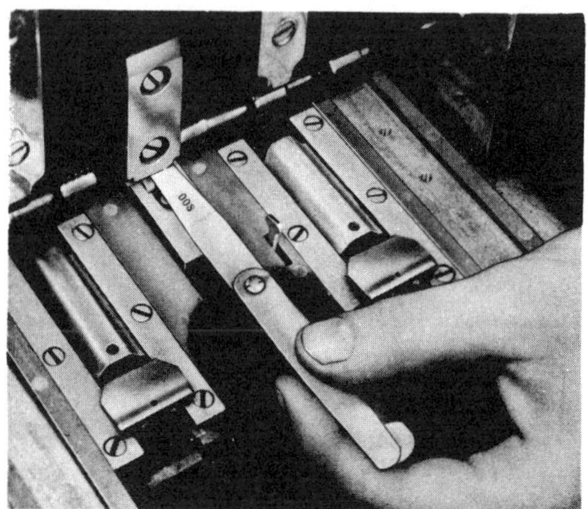

Figure 80. Checking Clearance with Feeler Gage

WIRE AND DRILL GAGES

These gages are widely used in the fields implied by their names. The wire gage, Figure 81 is commonly used for gaging iron wire but is also used for hot and cold-rolled steel, music wire, and sheet iron.

The use of the drill gage to determine the size of a drill is shown in Figure 82. A chart

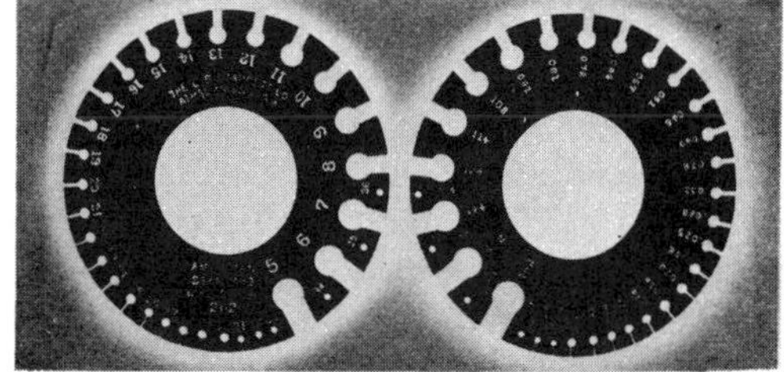

Figure 81. Wire Gage

Figure 82. Drill and Wire Gage

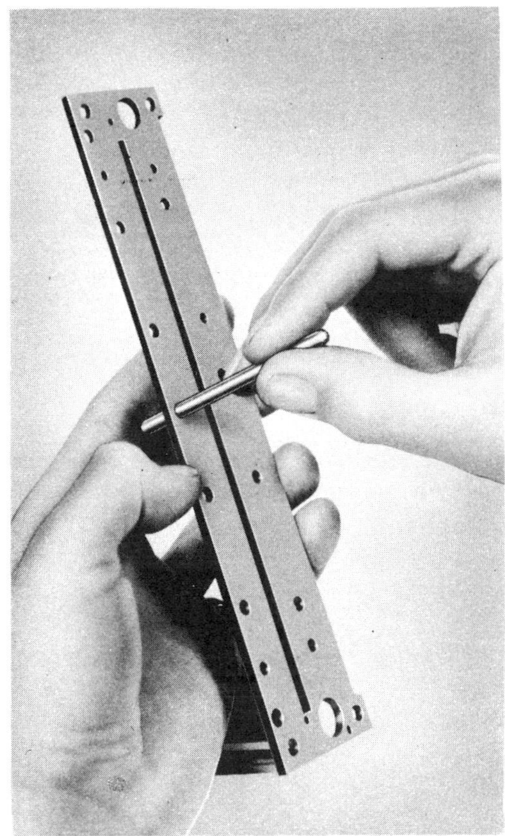

Figure 83. Checking Hole with Drill Blank

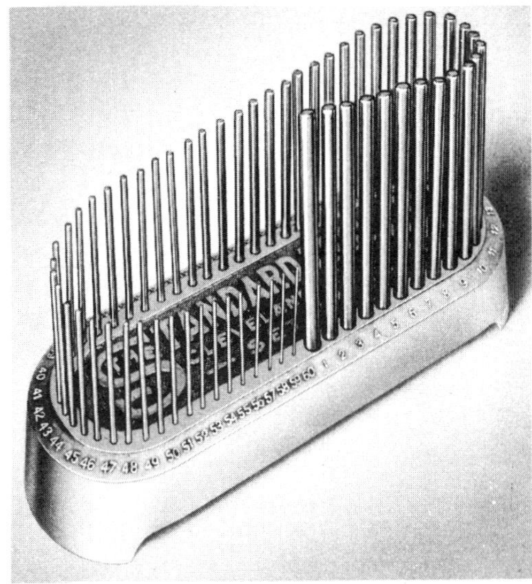

Figure 84. Drill Blanks

is included on the gage which indicates the correct size of drill to use for a given tap size. The drill number and decimal size are also given.

DRILL BLANKS

Drill blanks are used on line inspection to check the size of drill holes, as shown in Figure 83. They are also used for setup inspection to check the location of holes in the same manner as plug gages.

These blanks come in sizes from 1 to 60. They are made of hardened steel and ground with a taper of approximately .002 inch, the basic dimension being in the center of the blank. One end is .001 inch oversize, and the other is .001 inch undersize. In this way, the blanks will tighten in the hole and be parallel to facilitate checking of holes in setup inspection. Figure 84 shows a set of drill blanks in a holder.

TAPER-LEAF GAGES

These gages are used to measure the width of slots and diameter of holes. They are particularly adaptable for odd-size holes and slots for which plug gages are not available. The set shown in Figure 85 is graduated in thousandths from .100 to .500 inch. The gages are being used to measure the width of a slot. These gages also are available in a set ranging from

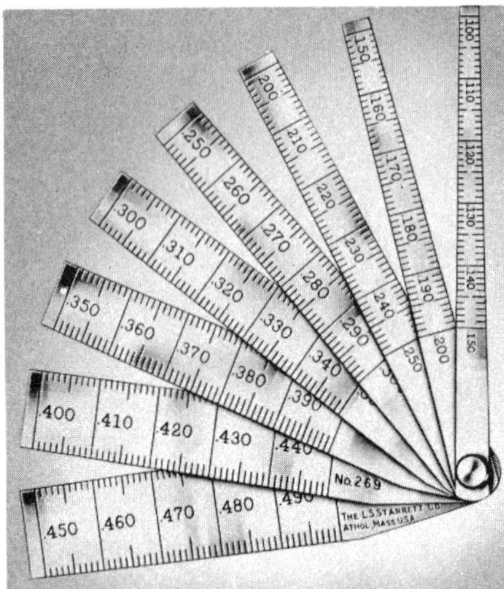
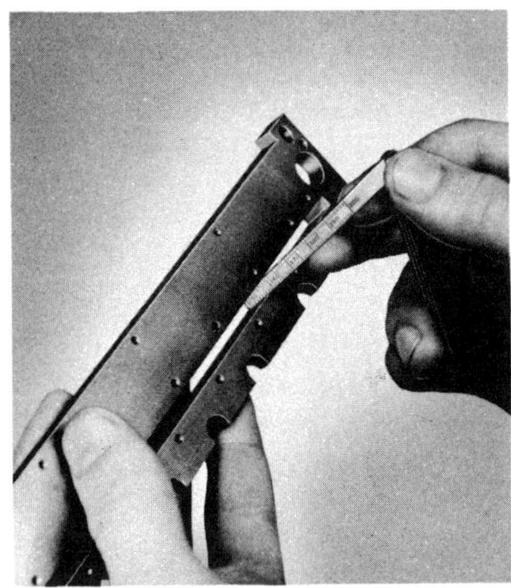

Figure 85. Taper Leaf Gages

.500 to 1.000 inch. It is important that these gages be properly used. They should not be inserted at an angle or forced. They should not be used to check holes having heavy burrs or deep counter-sinks.

RADIUS GAGES

The radius and fillet gages shown in Figure 86 consist of a set of blades marked in fractions of an inch, 1/32 to 1/2 inch, with the corresponding radius formed in them. Radius gages are also available marked from .020 to .400 inch, inclusive. Figure 87 illustrates how they are used to facilitate the checking of male and female radii (a fillet is a female radius). These gages also come in leaf sets.

View A shows use of the gage to determine the radius of inside corners or fillets for one-fourth or less of a circle. Straight sides of the gage are at 90 degrees, and can be used for checking the location of the radius. View B shows use of the gage to determine the radius of outside corners, and also whether the sides are at 90 degrees and tangent to the circle. In view C, work is being checked on a piece of glass. This application also checks any other convex parts, where the radius is one-fourth or more of the circle, that have projections which will not permit the use of the gage as in B and E. View D shows use of the gage on a concave cutter of one-half or less of a circle. This gage can be used to check the radius shown in view A, but it will not show the relation of the radius to the sides. View E shows the gage being used to check one-half of a circumference.

GAGING

In general, the amount of process gaging required on a job depends on the tolerance on the dimension, the type of machine tool, and the operator's skill and judgment. A dimension with a tolerance of .005 inch ordinarily does not need to be gaged so often as one carrying a tolerance of .002 inch.

It should not be necessary to gage a part on a milling machine so frequently as a part

Figure 86. Radius Gages

on a grinder, since a milling cutter can be set in one position for machining many parts, whereas a grinding wheel must be reset after grinding each part. Constantly resetting the wheel requires that each part be gaged.

As an operator becomes more experienced in running some particular part, he is able to machine the piece down to within a few thousandths before having to check it even once, whereas a new operator or an experienced operator working on a new job should gage the part frequently.

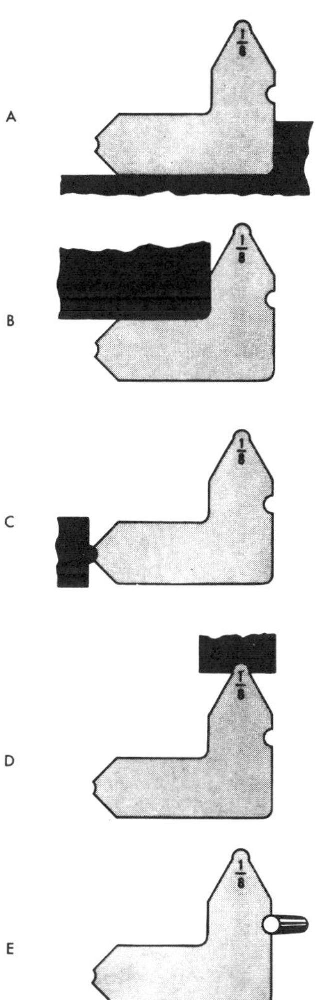

Figure 87. Applications of Radius Gages

QUESTIONS AND PROBLEMS

1. What is the difference between a standard and a special gage?
2. What relation should there be between the tolerance on a part and the tolerance on a gage?
3. How is gage wear compensated for in plug gages?
4. If a dimension on a part is 1.500 + .0005, - .001, how accurate should the gage be that is used to check this part?
5. A two-inch plug gage of "Y" accuracy will have what tolerance? What is the minimum tolerance on a part that can be checked with this gage?
6. For what accuracy of work are plug gages of "X" accuracy used?
7. Approximately what is the wear life of a cemented-carbide plug gage compared to hardened steel and chromium-plated plug gages?
8. What can be done with a worn steel plug gage? a cemented carbide plug gage?
9. Name three standard designs of plug gages. For what range of work is each used?
10. What is a progressive plug gage? What are the advantages and disadvantages of this type of gage?
11. How can the following be determined when using a taper plug gage: whether or not the taper of a tapered hole is correct; whether or not the diameter of the tapered hole is correct?
12. What are flat plug gages, and for what are they used?
13. What is a spline gage?
14. What important factors should be remembered when using plug gages?
15. What does the ring gage check besides the diameter of a piece?
16. What is a twin ring gage?
17. Explain how the correctness of the taper and the diameter of a taper part are determined with a taper ring gage.
18. What important features does the taper test gage have that the taper ring gage does not have?

19. What does a receiver gage measure?
20. What is meant by the term limit gage?
21. Name two types of snap gages.
22. What are adjustable length gages, and when are they used?
23. Name the advantages and disadvantages of adjustable snap gages.
24. When using a snap gage, how is the gaging of cylindrical parts different from the gaging of flat parts?
25. How is an adjustable snap gage set?
26. Upon what factors does the amount of gaging during machining depend?
27. What are feeler gages and for what are are they used?
28. What are radius gages?
29. What is the taper of drill blanks, and why is the basic dimension in the center?
30. How are taper leaf gages graduated and for what are they used?

Chapter IV

SCREW-THREADS AND THREAD GAGES

THE screw thread is one of man's earliest and most fundamental mechanical inventions. The ancient Egyptians invented a water screw to raise water from wells for the irrigation of their fields. The principle of the screw was later applied in the invention of the screw press for squeezing grapes and olives and for pressing cloth. Leonardo da Vinci, celebrated engineer of the 16th century, made many sketches of machines using screw threads, and was himself the inventor of the square or buttress thread.

For many centuries, however, every craftsman who made a screw used the particular tooth form and lead that happened to appeal to him at the moment. The screws he made were not particularly accurate; all that was required was that they do the work they were built to do.

It was not until 1841 that any attempt was made to standardize screw thread designs. The Whitworth system, devised by Sir Joseph Whitworth, was rapidly accepted by manufacturers of the day, and it looked for a time as though it would become the single universal standard. However, the International Standard, the United States Standard, and other systems soon came into existence and were accepted by certain groups or nationalities.

Although England has used the Whitworth System since 1841, British industry is now changing from that system to the Unified Thread System, which will be described later in this chapter. Basically, this change from the 55-degree angle (Whitworth Standard) currently involves coarse threads in sizes 1/4 to 4 inches, inclusive, and fine thread in sizes 1/4 to 1-1/2 inches, inclusive. For screws in sizes below 1/4 inch, British industry continues to use British Association (B. A.) threads, sizes 6 to 1.7 mm, inclusive.

France uses the French and International thread, which in thread form is identical to the U.S. Standard thread. Because all dimensions are given in terms of millimeters instead of inches, the threads, although similar, are not interchangeable. Currently through the International Standards Organization (ISO) the Unified Thread Profile has been recommended for world-wide use, and problems of diameter-pitch combinations and tolerances are being studied.

The Loewenherz thread is still another thread form, developed in Germany for use in fine instruments and chronometers. This series is based on the metric system and is used extensively in Germany. Its principal point of deviation from the International and the American National forms is in the included angle, which is 53 degrees and 8 minutes, instead of 60 degrees.

In the United States a special committee appointed by the Franklin Institute investigated screw thread standards, and on the basis of their studies the William Sellers thread system was accepted in 1868 by the U. S. Navy. It was known as the Sellers, Franklin Institute, and later as the United States Standard thread.

While this thread system was satisfactory as far as it went, it did not apply to diameters of less than one-fourth of an inch. Between the years 1868 and 1928, three distinct thread systems were developed to supplement the United States Standard system. These were known as the S. A. E. (Society of Automotive Engineers), the A. S. M. E. (American Society of Mechanical Engineers), and the U. S. S. (United States Standard) thread series. All of these systems used the 60-degree included angle of the U. S. S. thread, but differed in the range of pitch diameters, pitch, and thread depth. See Figure 88.

In 1928, the National Screw Thread Commission and Sectional Committee B1, sponsored jointly by A. S. M. E. and S. A. E., changed the name "U. S. Standard" to "American National Form" of thread, and incorporated in the Standard thread system for this country the A. S. M. E., the S. A. E., and the U. S. S. systems. The present general-purpose standard threads for the United States established by the Commission are:

1. American National coarse-thread series
2. American National fine-thread series

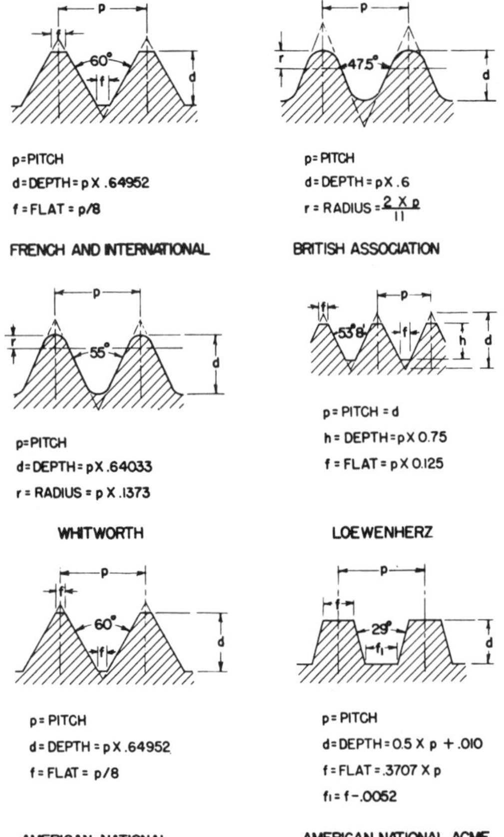

Figure 88. Standard Thread Forms

3. American National extra-fine-thread series
4. American National 8-pitch thread series
5. American National 12-pitch thread series
6. American National 16-pitch thread series
7. American National pipe-thread series
8. American National acme-thread series

In addition to these general-purpose threads, recognized standards have been established in the following fields:

American National hose coupling thread series.

American National fire hose coupling thread series.

American Petroleum Institute of oil-well drilling equipment thread series.

American National rolled-thread series for electric sockets and lamp bases.

Each of these thread series has a field of application of its own, and for the most part they were first developed by an association or society representing the industry. When the National Screw Thread Commission was appointed, these standards were well established in their fields and so became a part of the national standardization of screw threads.

APPLICATION OF DIFFERENT SERIES

The fields of application for the various thread series are quoted from the Report of the National Screw Thread Commission:

"The American National coarse-thread series is recommended for general use in engineering work, in machine construction where conditions are favorable to the use of bolts, screws, and other threaded components where quick and easy assembly of the parts is desired, and for all work where conditions do not require the use of fine-pitch threads.

"The American National fine-thread series is recommended for general use in automotive and aircraft work, for use where the design requires both strength and reduction in weight, and where special conditions require a fine thread.

"The American National extra-fine thread series is intended for special uses where (1) thin-walled material is to be threaded, (2) thread depth of nuts clearing ferrules, couplings, flanges, etc., must be held to a minimum, and (3) a maximum practicable number of threads required within a given thread length.

"American National 8-Pitch Thread Series--bolts for high-pressure pipe flanges, cylinder-head studs, and similar fastenings against pressure require that an initial tension be set up in the fastening by elastic deformation of the fastening and the components held together, such that the joint will not open up when steam or other pressure is applied. To secure a proper initial tension, it is not practicable that the pitch should increase with the diameter of the thread, as the torque required to assemble the fastening would be excessive. Accordingly, for such purposes, the 8-pitch thread has come into general use.

"American National 12-Pitch Thread Series--sizes of 12-pitch threads from one-half inch to and including one and three-fourths inches are used in boiler practice, which requires that a worn stud hole be retapped with a tap of the next larger size.

"The 12-pitch threads are also widely used in machine construction, as for thin nuts on shafts and sleeves. From the standpoints of good design and simplification of practice, it is desirable to maintain shoulder diameters to one-eighth inch steps. The 12-pitch is the coarsest in general use, which will permit a threaded collar which screws onto a threaded shoulder to slip over a shaft, the difference in diameter between shoulder and shaft being one-eighth inch.

"American National 16-Pitch Thread Series--the 16-pitch thread series is a uniform-pitch series for such applications as require a relatively fine thread. It is intended primarily for use on thread-adjusting collars and bearing retaining nuts.

"American National Pipe Threads--taper external and internal pipe threads are recommended for threaded pipe joints and pipe fittings for any service. Straight external pipe threads are recognized only for special applications, such as long screws and tank nipples. Straight internal pipe threads may be used, with a taper threaded pipe for ordinary pressures, as they are sufficiently ductile to adjust themselves to the taper external thread when properly screwed together.

"American National Acme Screw Threads--when formulated, prior to 1895, acme screw threads were intended to replace square threads and a variety of threads of other forms used chiefly for the purpose of providing transversing motions on machines and tools. Acme screw threads are now extensively used for a variety of purposes."

UNIFIED SCREW THREAD STANDARDS

Effort toward reaching agreement on a common system of screw threads for use in English-speaking countries was started as a result of experience in World War I. Little was accomplished, however, until World War II had again demonstrated the overwhelming need for a unified thread system. The Combined Production and Resources Board (British-Canadian-American) paved the way with conferences in New York (1943), London (1944), and Ottawa (1945). Final work was accomplished through the Sponsors Council and representatives of A.S.M.E., S.A.E., and the American Standards Association, which worked closely with representatives of the standards associations, industries, and armed forces of all three nations. The U.S. National Bureau of Standards played a most important role in the planning.

On November 18, 1948, the United States, Great Britain, and Canada closed a 100-year gap in industrial co-operation with a formal signing of an agreement for the standardization of screw threads. Heretofore, U.S. nuts would not screw on British bolts, and vice versa, while Canada necessarily was on a double standard, having to produce both British and American threads. The new Unified Screw Thread system will be common in all English-speaking, inch-using countries.

The new system does not supersede the American Standard; it becomes a part of it. Present classes 2 and 3 are retained in the American Standard because of their long-established and widespread use, but are not among the

Unified classes.

Agreement between the three nations achieves interchangeability of threaded parts by attaining unity in the four elements of screw thread geometry and dimensional limits required for mating an external with an internal thread--(1) thread angle, (2) thread form, (3) pitch (number of threads per inch), and (4) tolerances and allowances.

The new Unified thread standards comprise a basic screw thread form, the several series of diameter-pitch combinations used most commonly, and completely detailed limits of size for three classes of tolerances, two of which are provided with positive allowance. The 60-degree angle is now established for common, straight threads used in the United States, Canada, and Great Britain.

THREAD FORM

An early idea of the American V thread was a shaft turned to basic diameter (that is, within about 1/16 inches of the intended size), and then grooved with a 60-degree sharp tool until the land formed a sharp peak. Similarly, the hole was bored and grooved until it would assemble on the external thread (Figure 89A). No consideration was given to interchangeability. The sharp crests were too difficult to hold, and soon the form was modified to the form shown in Figure 89B. Sellers experiments proved that a much greater flat (reduced flank contact) was just as serviceable and much less costly to produce. The resulting U.S.S. form (Figure 89C) is still the basis for our modern screw thread standards. Great economies in tapping now are obtained by use of maximum metal form for external threads (Figure 89D).

This brief review indicates the natural progression to the present maximum metal form. The rounded root is not mandatory nor is it the subject of rigid inspection; it approximates the form resulting from a worn tool.

Almost any normal contour which clears the 1/4 p flat in a ring or thread-roll snap gage is acceptable unless something more specific is required for a particular part. The added material which is permitted adds strength to the weaker member. The enlarged minor diameter of the internal thread aids in tapping, which is highly desirable because the internal thread is more expensive to produce.

The form agreed upon satisfies all groups. Either a flat or rounded root is specified for the internal thread. A rounded root is specified for the maximum metal form external thread; however, a flat root as produced with new or unworn tools is also specified for minimum metal form, as shown in Figure 90.

In practice, the British plan to continue using rounded crests and roots on the external threads and rounded roots on internal threads, contrasted with the American practice and preference of using flat and truncated crests and roots.

In actual practice, there is far less difference between flat and rounded roots and crests than appears in the specifications. New tooling may cut or grind a flat on the first few pieces, but sharp corners soon wear or break off; so most of the threads produced by tools are rounded. Because of the process, rolled threads naturally tend to be rounded. Therefore, interchangeability of threads will not be hindered by actual practice in any of the countries adopting the Unified screw thread system.

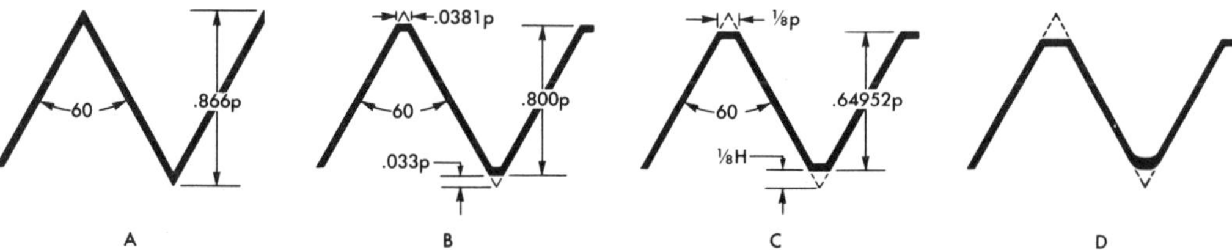

Figure 89

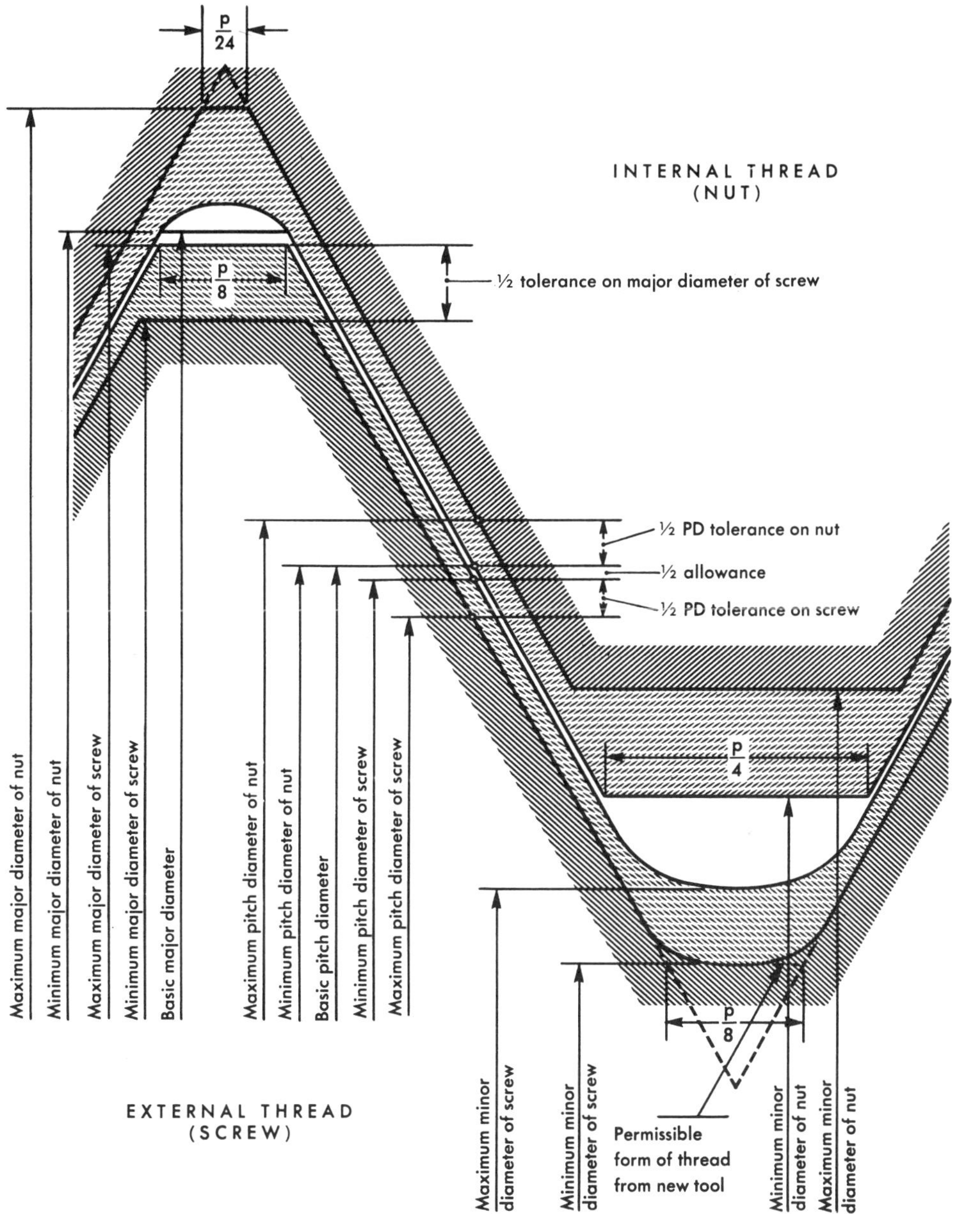

Figure 90. Unified Thread Form

PITCH (THREADS PER INCH)

Unification has been agreed upon for diameter-pitch combinations in sizes 1/4 to 4 inches in diameter in the coarse series, and for sizes 1/4 to 1-1/2 inches in the fine series, and in many sizes of the extra fine, 12-thread and 16-thread series. While classes 1A, 2A, 3A, 1B, 2B, and 3B are not yet unified in sizes smaller than 1/4 inch, classes 2A and 2B are used in the United States. "A" designates the external (screw) thread; "B" designates the internal (nut) thread.

For numbered sizes and for diameter-pitch combinations not unified, the designations NC, NF, NEF, N, and NS will be continued. Effort will be continued to complete the unification.

TOLERANCES AND ALLOWANCES--UNIFIED THREADS

The Unified tolerances represent a refinement of, and an improvement over, former American practice. The Unified thread standards provide for three classes of tolerance for the external thread (screw) 1A, 2A, and 3A, and three classes for the internal thread (nut) 1B, 2B, and 3B.

Classes 1A and 1B are adaptable for ordnance and special parts where quick and easy assembly is necessary and where liberal allowance is required to permit ready assembly even with slightly bruised or dirty threads.

Classes 2A and 2B are suitable for a wide variety of applications and are the recognized standards for bolts, nuts, screws, and similar thread fasteners. The moderate allowance on class 2A lessens the danger of galling and seizing in assembly and use. It also provides some clearance for plating or coatings of similar thickness. Tolerances on class 2A are realistic for mass production of commercial nuts. Both classes are ideal for production line assembly where speed wrenches are used, and include the great bulk of screw thread work.

Classes 3A and 3B are useful for such applications where closer tolerances than 2A and 2B are required.

Class 4 has been discontinued in the Unified and American Standards.

Classes 1A and 2A external threads have an allowance; that is, the largest external thread (bolt or screw) is always smaller than basic, and should assemble freely in its mating internal thread.

Under the Unified system, tolerance is greater for a larger diameter than for a small one having the same pitch. Heretofore, in the coarse and fine series, a fixed tolerance was arbitrarily set by formula which disregarded diameter.

The general assumption under the old standard has been that a class 2 screw must be used with a class 2 nut. It is now recognized that the type of assembly and the requirements of the fastener determine the class requirements for both the screw and the nut. Therefore, it will be practical to specify a 3A screw in a 1B tapped hole as well as a 2A screw in a 2B tapped hole, or any combination of classes best meeting all conditions.

CLASSIFICATION OF THREAD TOLERANCES FOR AMERICAN NATIONAL FORM

While it is practical and economically advantageous to standardize the thread form, it would not be practical to have only one class of accuracy. Obviously, one quality of thread would not fulfill the requirements of all thread applications. For example, the same quality of thread is not required in fastening together the parts of a stove as would be required in the assembly of an aircraft engine.

The various degrees of accuracy have been classified into three classes of tolerance. Each class and its field of application is defined by the Interdepartmental Screw Thread Committee as follows:

1. Includes screw-thread work of rough commercial quality, where the threads must assembly readily, and a certain amount of shake or play is not objectionable.

2. Includes the great bulk of screw thread work of ordinary quality, of finished and semi-finished bolts and nuts, machine screws, etc.

3. Includes the better grade of interchangeable screw-thread work.

For similar data on Unified classes, refer to immediately foregoing section on Unified threads.

STRENGTH AND INTERCHANGEABILITY

The machine screw is undoubtedly the most used device for fastening parts together. Nearly always the screws used for this purpose are standard parts commercially available from suppliers, who manufacture them in tremendous quantities.

Specialization in the manufacture of standard parts such as screws has been possible only because standardization has been successful in this field. It has benefited manufacturers in a number of ways:

1. Standard parts are usually cheaper to buy than they are to make.

2. They are usually of more uniform quality.

3. Inventories can be reduced by stocking smaller amounts of fewer sizes.

4. Standard parts lost, worn, or broken are easily replaced.

The old assumption that the classes of threads determine the holding power of the screw and nut has been replaced by the theory that classes of threads determine the ease of assembly and that a freer fit has the advantages of better mating parts throughout the entire length of engagement. Changing from class 2 to classes 2A and 2B does not reduce the strength nor does it affect interchangeability. Components are mechanically interchangeable in any combinations of the old and new classes.

The new standard has not changed the basic pitch diameter of the screw or nut. Therefore, a changeover to the new standard can be made by using a class 2A screw and a class 2B nut to replace the present class 2 screw and nut. For example, the minimum pitch diameter of a 1/4-20 UNC 2A screw has been decreased from minus .0036 to minus .0048 from the basic, which provides for the allowance of .0011 with no reduction in tolerance. The maximum pitch diameter tolerance on a 1/4-20 UNC 2B nut also has been increased from plus .0036 to plus .0048 from the basic, or an increase in tolerance of .0012. (The basic pitch diameter of a 1/4-20 UNC 2 screw is 0.2175.)

THREAD TERMINOLOGY

It is difficult for shop men to keep up with the increasingly technical subject of thread cutting and thread measurement. In standardizing the various thread systems, however, the National Screw Thread Commission developed a number of non-technical definitions for thread terms. These are quoted exactly to familiarize shop men with their meanings:

SCREW THREAD. A ridge of uniform section in the form of a helix on the external or internal surface of a cylinder, or in the form of a conical spiral on the external or internal surface of a cone.

EXTERNAL AND INTERNAL THREADS. An external thread is a thread on the outside of a member. Example: a threaded plug. An internal thread is a thread on the inside of a member. Example: a threaded hole.

MAJOR DIAMETER (formerly known as Outside Diameter). The largest diameter of the thread of the screw or nut. The term "major diameter" replaces the term "outside diameter" as applied to the thread of a screw and also the term "full diameter" as applied to the thread of a nut.

MINOR DIAMETER (formerly known as Core Diameter). The smallest diameter of the thread of the screw or nut. The term "minor diameter" replaces the term "core diameter" as applied to the thread of a screw and also the term "inside diameter" as applied to the thread of a nut.

PITCH DIAMETER. On a straight screw thread, the diameter of an imaginary cylinder, the surface of which would pass through the threads at such points as to make equal the width of the threads and the width of the spaces cut by the surface of the cylinder. On a taper screw thread, the diameter at a given distance from a reference plane perpendicular to the axis of an imaginary cone, the surface of which would pass through the threads at such points as to make equal the width of the threads and the width of the spaces cut by the surface of the cone.

PITCH. The distance from a point on a screw thread to a corresponding point on the next thread, measured parallel to the axis. The pitch in inches equals:

$$\frac{\text{One}}{\text{Number of Threads per inch}}$$

LEAD. The distance a screw thread advances axially in one turn. On a single-thread screw the lead and pitch are identical; on a double-thread screw the lead is twice the pitch; on a triple-thread screw the lead is three times the pitch, etc.

ANGLE OF THREAD. The angle included between the sides of the thread measured in an axial plane.

HALF ANGLE OF THREAD. The angle included between a side of the thread and the normal to the axis, measured in an axial plane.

HELIX ANGLE. The angle made by the helix of the thread at the pitch diameter, or conical spiral, with a plane perpendicular to the axis.

CREST. The surface of the thread corresponding to the major diameter of the screw and the minor diameter of the nut.

ROOT. The surface of the thread corresponding to the minor diameter of the screw and the major diameter of the nut.

SIDE OR FLANK. The surface of the thread which connects the crest with the root.

AXIS OF A SCREW. The longitudinal central line through the screw.

BASE OF THREAD. The bottom section of the thread; the greatest section between the two adjacent roots.

DEPTH OF THREAD. The distance between the crest and the base of thread measured normal to the axis.

NUMBER OF THREADS. Number of threads in one inch of length.

LENGTH OF ENGAGEMENT. The length of contact between two mating parts, measured axially.

DEPTH OF ENGAGEMENT. The depth of thread contact of two mated parts, measured radially.

PITCH LINE. An element of the imaginary cylinder or cone specified in definition of pitch diameter.

THICKNESS OF THREAD. The distance between the adjacent sides of the thread, measured along or parallel to the pitch line.

MEAN AREA. The term "mean area of a screw," when used in specifications and for other purposes designates the cross-sectional area computed from the mean of the basic pitch and minor diameter.

Figure 91 illustrates the use of thread terminology, and relating dimensional symbols are given in Table III. Tables IV and V from A.S.A. B1-1-1949 are included for reference; for more complete information on screw threads, see Handbook H-28, Screw-Thread Standards for Federal Services, and American Standards Association Bulletins A.S.A. B1-1 and A.S.A. B1-2.

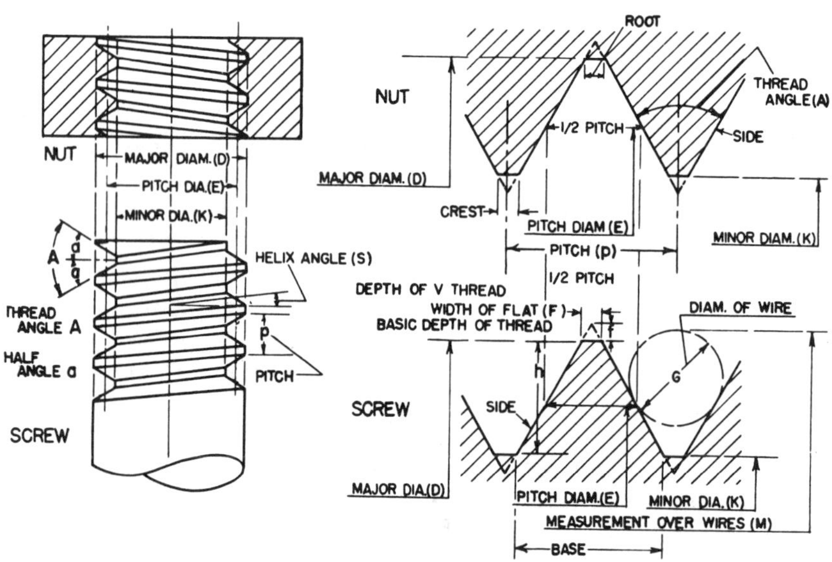

Figure 91. Screw Thread Notation

TABLE III
DIMENSIONAL SYMBOLS

Major diameter	D
Corresponding radius	d
Pitch diameter	E
Corresponding radius	e
Minor diameter	K
Corresponding radius	k
Angle of thread	A
One-half angle of thread	a
Number of turns per inch	N
Number of threads per inch	n
Lead	$L = \frac{1}{N}$
Pitch or thread interval	$p = \frac{1}{n}$
Helix angle	s
Tangent of helix angle	$S = \frac{L}{3.14159 \times E}$
Width of basic flat at top, crest, or root	F
Depth of basic truncation	f
Depth of sharp V thread	H
Depth of American National form of thread	h
Length of engagement	Q
Included angle of taper	Y
One-half included angle of taper	y

TABLE IV. COARSE-THREAD SERIES -- UNC AND NC
(Basic Dimensions)

Sizes	Basic Major Diameter,	Thds. per Inch,	Basic Pitch Diameter,§	Minor Diameter Ext. Thds.	Minor Diameter Int. Thds.	Lead Angle at Basic Pitch Diameter,		Section at Minor Diameter	Stress Area
	D	n	E	K_s	K_n	λ		at $D-2h_b$	
	Inches		Inches	Inches	Inches	Deg	Min	Sq In.	Sq In.
1 *(.073)*	0.0730	64	0.0629	0.0538	0.0561	4	31	0.0022	0.0026
2 *(.086)*	0.0860	56	0.0744	0.0641	0.0667	4	22	0.0031	0.0036
3 *(.099)*	0.0990	48	0.0855	0.0734	0.0764	4	26	0.0041	0.0048
4 *(.112)*	0.1120	40	0.0958	0.0813	0.0849	4	45	0.0050	0.0060
5 *(.125)*	0.1250	40	0.1088	0.0943	0.0979	4	11	0.0067	0.0079
6 *(.138)*	0.1380	32	0.1177	0.0997	0.1042	4	50	0.0075	0.0090
8 *(.164)*	0.1640	32	0.1437	0.1257	0.1302	3	58	0.0120	0.0139
10 *(.190)*	0.1900	24	0.1629	0.1389	0.1449	4	39	0.0145	0.0174
12 *(.216)*	0.2160	24	0.1889	0.1649	0.1709	4	1	0.0206	0.0240
1/4	**0.2500**	**20**	**0.2175**	**0.1887**	**0.1959**	**4**	**11**	**0.0269**	**0.0317**
5/16	**0.3125**	**18**	**0.2764**	**0.2443**	**0.2524**	**3**	**40**	**0.0454**	**0.0522**
3/8	**0.3750**	**16**	**0.3344**	**0.2983**	**0.3073**	**3**	**24**	**0.0678**	**0.0773**
7/16	**0.4375**	**14**	**0.3911**	**0.3499**	**0.3602**	**3**	**20**	**0.0933**	**0.1060**
1/2	0.5000	13	0.4500	0.4056	0.4167	3	7	0.1257	0.1416
1/2	**0.5000**	**12**	**0.4459**	**0.3978**	**0.4098**	**3**	**24**	**0.1205**	**0.1374**
9/16	**0.5625**	**12**	**0.5084**	**0.4603**	**0.4723**	**2**	**59**	**0.1620**	**0.1816**
5/8	**0.6250**	**11**	**0.5660**	**0.5135**	**0.5266**	**2**	**56**	**0.2018**	**0.2256**
3/4	**0.7500**	**10**	**0.6850**	**0.6273**	**0.6417**	**2**	**40**	**0.3020**	**0.3340**
7/8	**0.8750**	**9**	**0.8028**	**0.7387**	**0.7547**	**2**	**31**	**0.4193**	**0.4612**
1	**1.0000**	**8**	**0.9188**	**0.8466**	**0.8647**	**2**	**29**	**0.5510**	**0.6051**
1 1/8	**1.1250**	**7**	**1.0322**	**0.9497**	**0.9704**	**2**	**31**	**0.6931**	**0.7627**
1 1/4	**1.2500**	**7**	**1.1572**	**1.0747**	**1.0954**	**2**	**15**	**0.8898**	**0.9684**
1 3/8	**1.3750**	**6**	**1.2667**	**1.1705**	**1.1946**	**2**	**24**	**1.0541**	**1.1538**
1 1/2	**1.5000**	**6**	**1.3917**	**1.2955**	**1.3196**	**2**	**11**	**1.2938**	**1.4041**
1 3/4	**1.7500**	**5**	**1.6201**	**1.5046**	**1.5335**	**2**	**15**	**1.7441**	**1.8983**
2	**2.0000**	**4½**	**1.8557**	**1.7274**	**1.7594**	**2**	**11**	**2.3001**	**2.4971**
2 1/4	**2.2500**	**4½**	**2.1057**	**1.9774**	**2.0094**	**1**	**55**	**3.0212**	**3.2464**
2 1/2	**2.5000**	**4**	**2.3376**	**2.1933**	**2.2294**	**1**	**57**	**3.7161**	**3.9976**
2 3/4	**2.7500**	**4**	**2.5876**	**2.4433**	**2.4794**	**1**	**46**	**4.6194**	**4.9326**
3	**3.0000**	**4**	**2.8376**	**2.6933**	**2.7294**	**1**	**36**	**5.6209**	**5.9659**
3 1/4	**3.2500**	**4**	**3.0876**	**2.9433**	**2.9794**	**1**	**29**	**6.7205**	**7.0992**
3 1/2	**3.5000**	**4**	**3.3376**	**3.1933**	**3.2294**	**1**	**22**	**7.9183**	**8.3268**
3 3/4	**3.7500**	**4**	**3.5876**	**3.4433**	**3.4794**	**1**	**16**	**9.2143**	**9.6546**
4	**4.0000**	**4**	**3.8376**	**3.6933**	**3.7294**	**1**	**11**	**10.6084**	**11.0805**

§British: Effective Diameter.

Bold type indicates Unified threads—UNC.

TABLE V. FINE-THREAD SERIES -- UNF AND NF
(Basic Dimensions)

Sizes	Basic Major Diameter, D	Thds. per Inch, n	Basic Pitch Diameter,$ E	Minor Diameter Ext. Thds. K_s	Minor Diameter Int. Thds. K_n	Lead Angle at Basic Pitch Diameter, λ		Section at Minor Diameter at $D-2h_b$	Stress Area
	Inches		Inches	Inches	Inches	Deg	Min	Sq In.	Sq In.
0 (*.060*)	0.0600	80	0.0519	0.0447	0.0465	4	23	0.0015	0.0018
1 (*.073*)	0.0730	72	0.0640	0.0560	0.0580	3	57	0.0024	0.0027
2 (*.086*)	0.0860	64	0.0759	0.0668	0.0691	3	45	0.0034	0.0039
3 (*.099*)	0.0990	56	0.0874	0.0771	0.0797	3	43	0.0045	0.0052
4 (*.112*)	0.1120	48	0.0985	0.0864	0.0894	3	51	0.0057	0.0065
5 (*.125*)	0.1250	44	0.1102	0.0971	0.1004	3	45	0.0072	0.0082
6 (*.138*)	0.1380	40	0.1218	0.1073	0.1109	3	44	0.0087	0.0101
8 (*.164*)	0.1640	36	0.1460	0.1299	0.1339	3	28	0.0128	0.0146
10 (*.190*)	0.1900	32	0.1697	0.1517	0.1562	3	21	0.0175	0.0199
12 (*.216*)	0.2160	28	0.1928	0.1722	0.1773	3	22	0.0226	0.0257
1/4	**0.2500**	**28**	**0.2268**	**0.2062**	**0.2113**	**2**	**52**	**0.0326**	**0.0362**
5/16	**0.3125**	**24**	**0.2854**	**0.2614**	**0.2674**	**2**	**40**	**0.0524**	**0.0579**
3/8	**0.3750**	**24**	**0.3479**	**0.3239**	**0.3299**	**2**	**11**	**0.0809**	**0.0876**
7/16	**0.4375**	**20**	**0.4050**	**0.3762**	**0.3834**	**2**	**15**	**0.1090**	**0.1185**
1/2	**0.5000**	**20**	**0.4675**	**0.4387**	**0.4459**	**1**	**57**	**0.1486**	**0.1597**
9/16	**0.5625**	**18**	**0.5264**	**0.4943**	**0.5024**	**1**	**55**	**0.1888**	**0.2026**
5/8	**0.6250**	**18**	**0.5889**	**0.5568**	**0.5649**	**1**	**43**	**0.2400**	**0.2555**
3/4	**0.7500**	**16**	**0.7094**	**0.6733**	**0.6823**	**1**	**36**	**0.3513**	**0.3724**
7/8	**0.8750**	**14**	**0.8286**	**0.7874**	**0.7977**	**1**	**34**	**0.4805**	**0.5088**
1	1.0000	14	0.9536	0.9124	0.9227	1	22	0.6464	0.6791
1	**1.0000**	**12**	**0.9459**	**0.8978**	**0.9098**	**1**	**36**	**0.6245**	**0.6624**
1 1/8	**1.1250**	**12**	**1.0709**	**1.0228**	**1.0348**	**1**	**25**	**0.8118**	**0.8549**
1 1/4	**1.2500**	**12**	**1.1959**	**1.1478**	**1.1598**	**1**	**16**	**1.0237**	**1.0721**
1 3/8	**1.3750**	**12**	**1.3209**	**1.2728**	**1.2848**	**1**	**9**	**1.2602**	**1.3137**
1 1/2	**1.5000**	**12**	**1.4459**	**1.3978**	**1.4098**	**1**	**3**	**1.5212**	**1.5799**

$British: Effective Diameter.

Bold type indicates Unified threads—UNF.

SPECIFICATION OF THREAD FORM

Three thread forms are generally recognized as standard in the United States:

1. American National Form
2. American National Acme Form
3. American National Pipe Thread Form

The pitch and the range, and steps in sizes, may be different with each series, but the form will remain the same.

SPECIFICATIONS FOR AMERICAN NATIONAL FORM

ANGLE OF THREAD. The basic angle of thread (A) between the sides of the thread measured in an axial plane is 60 degrees. The line bisecting this 60-degree angle is perpendicular to the axis of the screw thread.

FLAT AT CREST AND ROOT. The flat at the root and crest of the basic thread form is 1/8 x p or 0.125 p.

DEPTH OF THREAD. The depth of the basic thread form, h, is

$$0.649519 \times p \text{ or } \frac{0.649519}{n}$$

where p equals pitch in inches, n equals number of threads per inch, and h equals basic depth of thread.

CLEARANCE AT MINOR DIAMETER. A clearance shall be provided at the minor diameter of the nut by removing from the crest of the basic thread form an amount such as to provide a depth of thread not less than 62 to 75 percent (depending on the size), and not more than 83-1/3 percent of the basic thread depth.

CLEARANCE AT MAJOR DIAMETER. A clearance should be provided at the major diameter of the nut by removing from above the basic thread form an amount such that the width of the flat shall be less than 1/8 x p, but not less than 1/25 x p.

MACHINING OF SCREW THREADS

Screw threads are produced by several machining processes, and the selection of the process usually involves such factors as the accuracy and quantity required. A screw thread may be machined by any one of the following methods:

1. In a lathe using a single-point threading tool (the shape of the thread) advanced by the lead screw.

2. In a lathe using a threading roll--a roller with threads cut in the surface, which is forced into the revolving stock.

3. In a lathe using a multiple-point tool known as a chaser.

4. In a screw machine using a threading die of the fixed, or self-opening type for external threads and a solid, an adjustable, or a collapsible tap for an internal thread.

5. In a thread milling machine using a single revolving form cutter to machine a more slowly revolving part.

6. In a hobbing machine using a thread hob.

7. In a thread rolling machine, for forming screw threads. The blank is rolled between reciprocating dies.

8. In a grinding machine using a formed wheel of the same shape as the thread form.

CLASSIFICATION OF THREAD, PLUG, AND RING GAGES

"Object of Gaging--The final results sought by gaging are to secure interchangeability, that is, the assembly of mating parts without selection or fitting of one part to another, and to insure that the product conforms to the specified dimensions within the limits of variation establishing the closest and loosest conditions of fit permissible in any given case. This requires the use of gages representing the limit of maximum metal, known as "go" gages, which control the minimum looseness or maximum tightness in the fit of mating parts, and which accordingly control interchangeability, and the use of gages representing the limit of minimum metal, known as "not-go" gages, which limit the amount of looseness between mating parts, and thus control in large measure the proper functioning of parts." (From Screw-Thread Standards for Federal Services, 1944, Hand Book H-28, U. S. Department of Commerce, National Bureau of Standards.)

"Not-go" as applied to minimum metal limit thread gages is being supplanted in many instances by "HI-LO" - "HI" for the maximum plug and "LO" for the minimum ring. The fact that minimum metal limit gages may enter or be entered, provided a definite drag results on or before the third turn of entry, requires the use of words which more accurately reflect the condition.

Working gages should be used on all thread-producing operations. After the threading operation is completed, inspection gages are often used to cull out unacceptable parts that may have escaped detection while being produced.

A threaded part in the process of manufacture must be checked in several stages. As the machine operator is generating the thread on a part, he needs to know when he has the thread to size; or if the machine controls the size, how close the thread is to the dimensional limits. For this purpose, he uses "working" gages--thread gages that are convenient to use and sufficiently accurate to hold the dimensions within the specified tolerances.

Finished parts are inspected before assembly, and the inspector uses a class of gage known as an "inspection" gage. With these gages he can check parts quickly and accurately.

Master setting thread-plug gages are used to set and check adjustable thread-ring gages, thread snap gages, and other thread comparators. They are of two standard designs, which are designated as full-form setting plugs, and truncated setting plugs. The truncated setting plug is the same as the full-form setting plug except that the crest of the thread is truncated for about one-half its length, giving a full portion and a truncated portion.

In setting thread-ring gages to size, the truncated portion controls the pitch diameter, and the full portion assures that the proper clearance is provided at the major diameter of the ring gage. Ring thread gages should be set to fit the full-form portion of the setting plug and then tried on the truncated portion. There should be only a slight difference in the fit. The presence of shake or play on the truncated portion indicates that the sides of the threads are no longer straight and the gage should be repaired or discarded.

It is always desirable to have the limits on a working gage set within the limits of the inspection gage so that all parts which pass the working gage will surely pass the inspection gage.

Thread gages are further classified according to their accuracy, W or X, the class W being the most accurate. Class W gages are used for certain close limit inspection gages, master gages, and are recommended in many instances for truncated setting plugs, particularly because of the closer lead and angle tolerances. For example, a Class W thread gage having 28 threads per inch will have tolerance of:

.0001 on the pitch diameter
.00015 on the lead
0° 8' on the half angle of the thread
.0005 on the major and minor diameters

X gages are used for both inspection and working gages for all thread classes not closer than class 3, a wear allowance being provided for working gages when so specified. In all thread gages the tolerances on a gage should never permit the gage to fall outside of a tolerance on a part it is to check. This means that if the tolerance on a part is .0000 in one direction, the "go" gages used to check this dimension must not exceed the basic dimension in this direction.

A class X thread gage having 28 threads per inch will have a tolerance of:

.0003 on the pitch diameter
.0003 on the lead
0° 15' on the half angle of the thread
.0005 on the major and minor diameters

TOLERANCE ON LEAD OF THREAD GAGES

The tolerance on lead is defined as the permissible variation between any two threads not farther apart than the length of the standard gage, as shown in Table VI, omitting one full thread at each end of the gage.

TABLE VI. LENGTHS OF STANDARD THREAD-PLUG GAGE BLANKS

Thread Sizes				Thread Plug Gages		Large Instrument Thread Plug Gages	
Nominal Range, Inclusive		Decimal Range					
From	To	Above	To and Including	GO (See Notes)	HI (Not Go)	GO	HI (Not Go)
1	2	3	4	5	6	7	8
# 0	# 3	0.059	0.105	1/4	3/16	3/16	1/8
# 4	# 6	0.105	0.150	5/16	7/32	7/32	5/32
# 8	#12	0.150	0.240	13/32	9/32	9/32	7/32
1/4	5/16	0.240	0.365	1/2	5/16	5/16	1/4
3/8	1/2	0.365	0.510	3/4	3/8	3/8	5/16
9/16	3/4	0.510	0.825	7/8	1/2	1/2	3/8
7/8	1 1/8	0.825	1.135	1	5/8	5/8	7/16
1 1/4	1 1/2	1.135	1.510	1[1] 1 1/4[2]	3/4	3/4	1/2
1 1/2	2	1.510	2.010	1 7/8[3] 1 1/4[4] 7/8[5]	7/8	3/4	5/8
2	2 1/2	2.010	2.510	2[3] 1 3/8[4] 7/8[5]	7/8	3/4	5/8
2 1/2	3	2.510	3.010	2 1/8[3] 1 1/2[4] 1[5]	1	......	......
3	12	3.010	12.010	2 1/4[3] 1 1/2[4] 1[5]	1	......	......

[1]For 12 threads per inch and finer.
[2]For threads coarser than 12 per inch.
[3]For 7 threads per inch and coarser.
[4]For threads finer than 7 and coarser than 16 per inch.
[5]For 16 threads per inch and finer.

(From American Standards Association Bulletin, A.S.A. B1-2-1951, Table 7.)

TOLERANCE ON ANGLE OF THREAD OF THREAD GAGES

This tolerance limits the irregularities in thread form, such as convex or concave sides, rounded crests, or slight projections on the sides. The equivalent deviation from the true thread form caused by these irregularities should not exceed the tolerances permitted on angle of thread. The tolerance on X gages is plus or minus 15 minutes for the representative thread chosen.

TOLERANCE ON MAJOR, MINOR, AND PITCH DIAMETERS OF THREAD GAGES

The tolerance on the major and minor diameters of thread gages of W and X quality is the same for the same thread. For example, it is .0005 for all 28 threads-per-inch gages to and including 4 inches diameter.

The tolerance on the pitch diameter varies with each class of gage, and usually it is the most important measurement made on a thread. In the example chosen above, while the tolerances on other dimensions on the thread gage are fairly uniform, the tolerance on the pitch diameter practically determines the class of the gage.

DISTRIBUTION OF GAGE TOLERANCE

The gage tolerances for lead and angle are distributed either plus or minus the theoretical figure.

The gage tolerances on the pitch diameter are distributed as much as possible on the plus side of the "go" working plug gage to give the gage additional wear. For the same reason it is distributed as much as possible on the negative side of the "go" working ring gages. Maximum metal limit gages should always be within product limits. Holding minimum metal limit gages within the product limit may result in rejection of acceptable parts. If gage tolerance is outside product limits, they may accept parts slightly outside the specified tolerances. A possible solution is to specify gages within product limits as working gages; inspection or final acceptance gages might have gage tolerances outside the product limit.

The gage tolerance on "go" setting plugs is distributed on the minus side because setting plugs are used to control the accuracy of thread-ring gages, and thread-ring gages are kept on the minus side for the reasons given above.

"Wear on Gages--'go' gages may be permitted to wear to the extreme product limits. It is desirable, however, that working and inspection gages be so selected that the dimensions of the working gages are inside of the limiting dimensions represented by the inspection gages, in order that all parts passed by the working gage will be accepted by the inspection gage." (From Screw-Thread Standards for Federal Services, 1939 Hand Book H-28, U.S. Department of Commerce, National Bureau of Standards.) The wear allowances provided in A.S.A. Standard B-1-2, Screw Thread Gages and Gaging, if applied to working gages, are helpful; these allowances are equal to one-half of the X pitch diameter tolerances.

"GO" AND "NOT-GO" THREAD GAGE

The "go" thread-plug and ring gage shown in Figure 92 gages the whole thread, that is, the major, pitch, and minor diameters, and the lead. The "go" thread-plug gage checks the basic major diameter of the minimum nut, and the "go" thread-ring gage checks the minor diameter of the minimum screw for sufficient clearance. The radius at the bottom of the tooth shows how the gage is relieved so that only the proper dimension is checked.

The "not-go" thread-plug gages are truncated to two-thirds of the basic depth of thread in order to check only the pitch diameter. By truncating or flattening the "not-go" threads, contact with the sides of the threads, rather than either the crest or the root, is assured.

THREAD-PLUG GAGES

"The thread-plug gage," according to the American Gage Design Committee, "is a complete internal thread gage of either single- or double-ended type, comprising a handle and threaded gaging member or members, with suitable locking means."

Thread-plug gages are standardized into

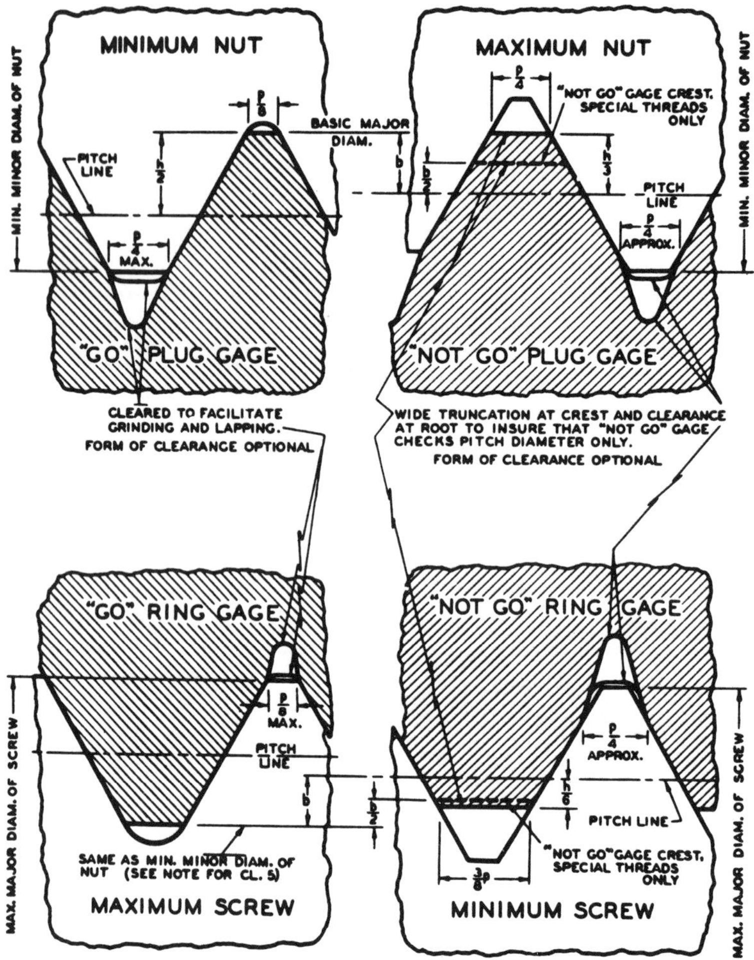

Figure 92. Thread-form "Go", "Not-go" Gage

three designs:

1. The taper lock design shown in Figure 93 is recommended for thread-plug gages between .059 of an inch and 1.510 inches in diameter.

2. The reversible tri-lock design shown in Figure 93 is recommended for thread-plug gages between 1.510 and 8.010 inches in diameter.

3. The annular design shown in Figure 94 is recommended for thread-plug gages between 8.010 and 12.010 inches in diameter.

The "go" member of the thread-plug gage is identified by its length; it is always longer than the "not-go" member. Either end may be replaced when worn.

The wear that may be obtained from a gaging member is doubled when it is constructed so that the ends are reversible. Although more expensively made, it is recommended for thread-plug gages over 1.510 inch.

In some applications, it is desirable to use a progressive thread-plug gage with the "go" and "not-go" members combined in the manner shown in Figure 95. This type of thread-plug gage offers the advantage of being used to gage a hole with a single insertion of the plug; but it costs more and is useless as soon as the "go" section is worn out.

Thread-plug gages are usually equipped with some device for removing loose chips from threaded holes. On some makes the first thread

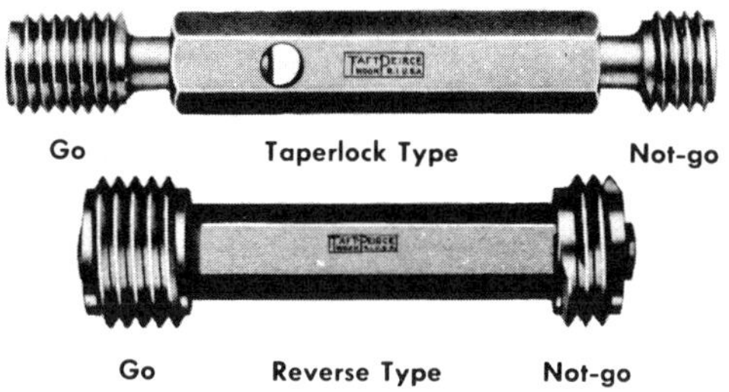

Figure 93. Thread-plug Gages

has a square end, and any loose chips are pushed out ahead of the gage. Other makes accomplish the same purpose with a chip groove similar to the one shown on the gaging member in Figure 95. On all thread-plug gages the end thread is cut off to eliminate a sharp edge which may injure the hands.

SPECIAL TYPES OF THREAD-PLUG GAGES

The thread-plug gages described previously, being the recommended designs of the American Gage Design Committee, are generally accepted as standard. However, for particular applications, or where there are special specifications on the job, other than standard design gages may be used.

For example, the threaded setting blocks shown in Figure 96 are for setting thread snap gages. Rather than maintain threaded setting plug gages of every diameter, with these accessories to a standard gage block set, the inspector may build any desired pitch diameter by using

Figure 94. Annular Thread Gage

Figure 95. Progressive-type Thread Gage

gage blocks as spacers in the manner shown. These particular blocks are limited to pitch diameters over one inch, and one set of thread blocks is required for each pitch of thread.

MARKING OF THREAD-PLUG GAGES FOR IDENTIFICATION

A thread-plug gage is marked on the handle with all the pertinent identifying information. The identification system has been standardized by the National standardization bodies. For example, a thread gage with a major diameter of one inch, eight threads per inch, and designed to gage threads with class 2 tolerances, is marked as follows:

1" - 8 NC - 2.

The first number indicates the major diameter of the thread. For threads under one-fourth of an inch, a numerical symbol is used to represent the major diameter. For example, a number 3 thread has a major diameter of .099 of an inch. For threads over one-fourth of an inch, the number and fraction, if any, indicate the major diameter of the gage.

The second number indicates the number of

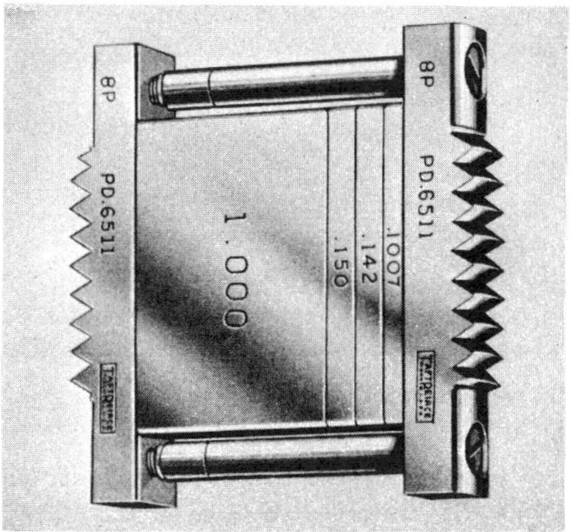

Figure 96. Thread Snap-gage Setting Block

threads per inch. This number may range from four to eighty threads per inch, and it varies with the major diameter of the thread and the thread series. As the major diameter increases, the number of threads per inch decreases; the same screw will have more threads per inch in the fine-thread series than in the coarse-thread series.

Thread-plug gages for unified threads are marked UNC or UNF, depending on the thread series they are to check. For example, a thread plug gage which is to be used to check a one-inch, eight-threads-to-the-inch, class 2B threaded hole in the coarse-thread series will be marked 1" - 8 UNC - 2B. The letters in the marking indicate the series.

The letters of the various series follow:

UNC Unified coarse thread
UNF Unified fine thread
UNEF Unified extra-fine thread
NC American National coarse thread
NF American National fine thread
NEF American National extra-fine thread
N American National 8-, 12-, and 16-pitch thread
NA American National acme thread
NPT American National taper-pipe thread
NPS American National straight-pipe thread
NH American National hose-coupling thread
NS American National form-thread--special pitch

The last number indicates the class of tolerance the gage is designed to control. The several classes of tolerance have been defined; thus, the number (1, 2, 3, for American National or 1A, 1B, 2A, 2B, 3A, 3B for Unified and American National) indicates the type of work on which the gage should be used.

Any special information may be added to the above number. For example, letters LH are added to indicate a left-hand thread. The thread is assumed to be right-hand unless identified otherwise.

The following examples show how marks are read:

12-24 NC - 2 LH. (1) This thread is a number 12, which is one size smaller than one quarter of an inch. (2) It has 24 threads to the inch. (3) It is a standard thread from the American National coarse-thread series. (4) The limiting "go" and "not-go" dimensions are such that it should be used to check a class 2 thread tolerance. (5) The thread is left-hand.

1-1/4 - 20 NS - 3. (1) This thread has a major diameter of 1-1/4 inches. (2) It has twenty threads per inch. (3) While it has the American National thread form, it is a special pitch. (4) The gage is suitable for use in checking a thread with class 3 tolerance. (5) The thread is right-hand.

APPLICATION OF THREAD-PLUG GAGES AND GAGING

Thread-plug gages are in general use for checking the accuracy of tapped holes. The usual checking procedure is first to try a "go" plug, and if it screws in easily with a light wrist motion, then to try the "not-go" plug. Threads are acceptable as within minimum metal limits if the "not-go" plug does not enter. Threads are also acceptable if all complete threads can be entered, provided that a definite drag results from metal-to-metal contact on or before the third turn of entry. Requirements of extreme applications such as exceptionally thin or ductile material or a small number of threads, etc., may require modification of this practice. (From A.S.A. - B1-2-1951.)

In some classes of work where there is considerable latitude, the "not-go" gage is omitted entirely, and only the "go" gage is used. Where this is done, the machine operator or inspector "feels" the play of the gage in the hole and decides whether or not the pitch diameter is too large.

The "go" thread-plug gage checks the effective size, considering the errors in lead, thread angle, and major, minor, and pitch diameters against the maximum metal limit. Distribution of the error among these various factors can vary, but as long as they do not add up in one direction but pass the "go" plug, the thread is acceptable to the maximum metal limit.

The "not-go" gage has truncated threads, and the bottoms of the threads are relieved

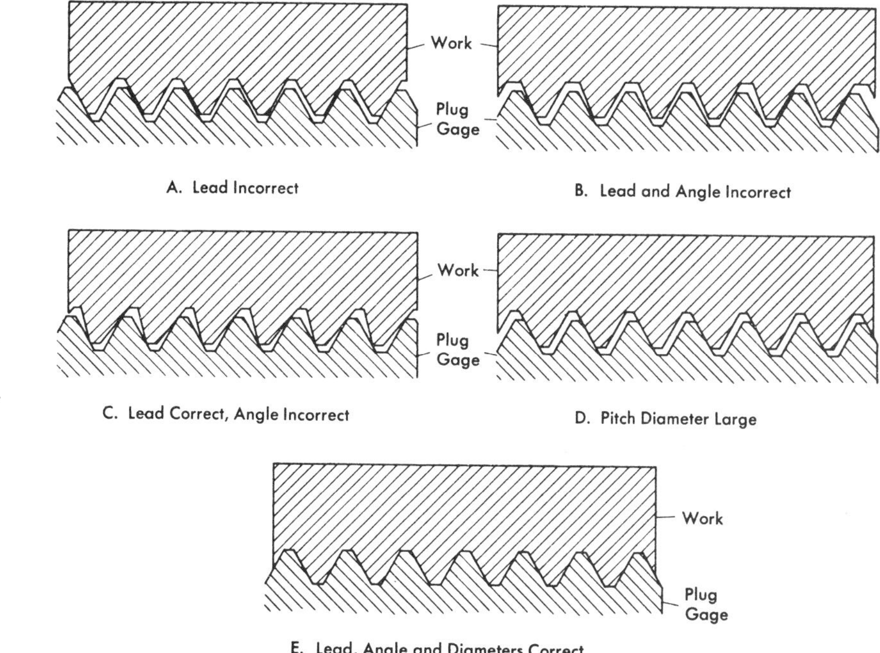

Figure 97. Common Thread Errors

so that it checks only the pitch diameter as far as that is practical.

The relative importance of the more common thread errors is illustrated in Figure 97. All threads shown there should be rejected except E.

Figure 97A shows the condition resulting from a lead error in a thread. The end threads are the only ones in contact; the remaining threads float. This means all the load is taken by the end threads, and only as these wear or break down are the others brought into contact. When the binding effect of either of these two threads is destroyed through wear or fatigue, the assembly is loosened or the thread fails altogether.

Should the angle of the thread also be in error, as shown in Figure 97B, the above condition is further aggravated. The contact between the work and the plug occurs on just two points, one on each thread. The thread has practically no resistance to wear or vibration and very little strength.

If only the angle of the thread is in error, as shown in Figure 97C, the condition is somewhat improved. Point contact still exists, which reduces the resistance of the thread to wear and may cause it to loosen in service. The thread is stronger than before, since the load is distributed evenly between the threads. This shows why an error in lead is so serious when the strength of the thread is important to its function in a machine.

Figure 97D shows the effect of a large pitch diameter. All the threads are in good contact, and the resistance to wear and loosening is practically the same as for a perfect thread. The load on the thread is distributed evenly not only between the threads, but also on each thread; also strength is practically the same as for a perfect thread.

Figure 97E shows a theoretically perfect thread. This series of diagrams emphasizes the fact that lead and angle of thread are more important than the minor, major, or pitch diameter.

Application of the thread-plug gage is illustrated in Figure 98. The part is a bearing bracket for a shaft on an accounting machine. The threaded hole in the base of the bracket is being checked with a "go" and "not-go"

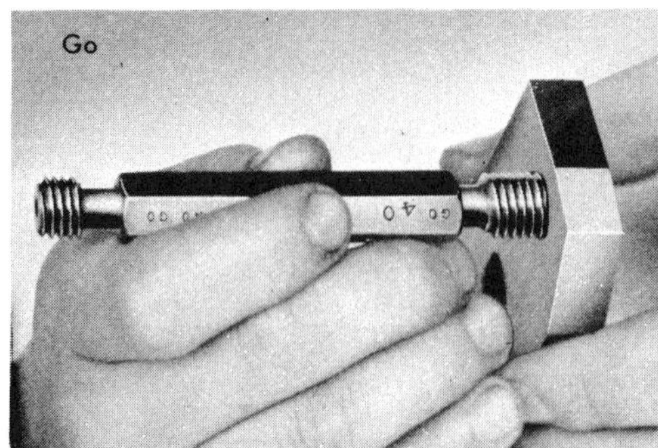

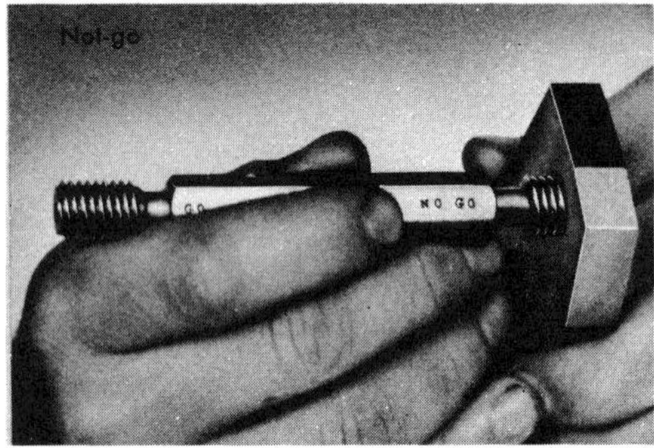

Figure 98. Checking a Threaded Hole

gage. To check this hole properly, the "go" plug must be screwed in until it is at least flush with the other side. The "not-go" plug should just start in the hole or should reach a definite drag on or before the third turn of entry.

Every thread-plug gage has a chip groove or some means for clearing the thread of loose chips. However, in the gaging of a hole, the thread gage should never be forced, for it is possible to get a chip wedged in the thread. To remove any burr that remains after machining, an old tap or an old drill equal to the minor diameter can be run through by hand before the gage is inserted.

There is no doubt that the life of a thread gage is determined more by the handling it receives than by any other single factor. Careful use of the gage will prolong its life. The user should remember that a thread-plug gage is not a tap, and besides being a destructible article, its original cost is three to four times that of a plain plug. Use of a thin coating of light oil serves both to lubricate and to minimize wear.

THREAD-RING GAGES

A thread-ring gage is an external thread gage designed to check whether a thread can later be assembled to its mate. The American Gage Design Committee, determining that manufacturers generally preferred the adjustable thread-ring gage shown in Figure 99, describe it as standard for this type of gage.

This gage consists of a knurled gage blank with a threaded hole in the center so designed that it may be adjusted, in a limited amount, to fit the small differences that exist in two different set plugs of the same theoretical size. The gage shown in Figure 99 has three slots cut radially to permit this adjustment; two of the slots terminate in holes, while the third is cut all the way through. This adjustable ring gage is one of a series of four recommended designs for gages of this type. The four designs are all similar, differing only in these respects:

1. Sizes 0 to and including .150 of an inch may have either two or three radial slots.

2. Sizes .150 to and including .510 of an inch have three radial slots.

3. Sizes .510 of an inch to and including 5.510 inches have three radial slots; blanks for coarse pitches are flanged.

4. Sizes 5.510 to and including 12.260 inches have six radial slots; blanks for coarse pitches are flanged, and all have four radial ball handles.

All "go" thread ring gages have a continu-

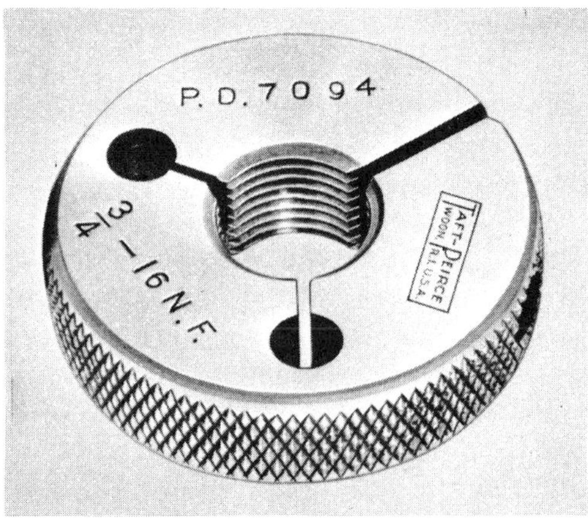

Figure 99. Adjustable Thread-ring Gage

ous knurled diameter, whereas all "not-go" gages are distinguished by an annular ring cut in the knurled surface.

The construction of the thread-ring gage, including the mechanism for adjusting the gage and for locking the adjustment in place is shown in Figure 100. By opening and closing the one slot, the size of the threaded hole may be adjusted to its setting plug. The normal position of the slot, without any screw pressure, is slightly closed; therefore, the adjustment is to open the jaws instead of to pull them together.

The adjustment is made with the screw marked 1 in Figure 100. This screw operates against the sleeve marked 2, which extends into the opposite side of the slot and operates against a shoulder. As screw 3 is advanced into the threaded hole, it pushes the sleeve against the shoulder and opens the slot. The screw

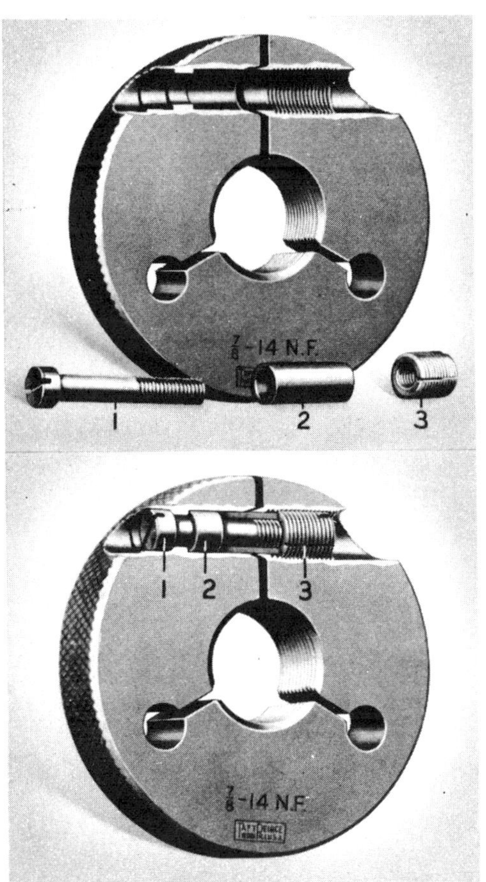

Figure 100. Construction Details of Thread-ring Gage

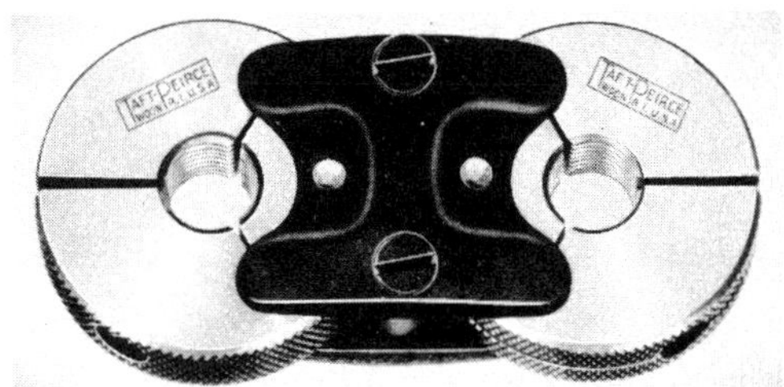

Figure 101. Thread-ring Gages in Holder

marked 1, which extends through the sleeve 2, expands hollow screw 3 so that the adjustment is locked. Tightening of screw 1 exerts a pull between the shoulders immediately under its head and the internal threads of screw 3; this causes it to expand into the threads in the wall of the gage without exerting any extra pressure on the sleeve 2. Sleeve 2, being accurately fitted, serves as a large dowel to maintain alignment of the gage.

As the gage is used, wear may be taken up by adjustment of the hollow screw. The amount of wear which may be compensated for usually is limited by the major diameter of the gage. The minor diameter and the pitch diameter wear as the gage is used, but the flanks near and above the major diameter do not; this limits the compensation that may be made.

A "go" and "not-go" gage must be used to check a thread, and the two gages are often assembled into a single tool or holder as shown in Figure 101. The holder is more convenient and easier to handle in checking a thread than two separate gages.

APPLICATION OF THREAD-RING GAGES

Thread-ring gages are used to check the lead, thread form, and major, minor, and pitch diameters of male threads. Like the thread plug, the ring gage checks a combination of all these factors, or the ability of the thread to fit in an assembly, Because with a ring gage, the user cannot distinguish between the various kinds of errors, it is sometimes supplemented or replaced by the thread snap gage or other checking means on jobs where distinction is required.

Figure 102 shows the use of the thread-ring gage to check the thread on a bolt. The "go"

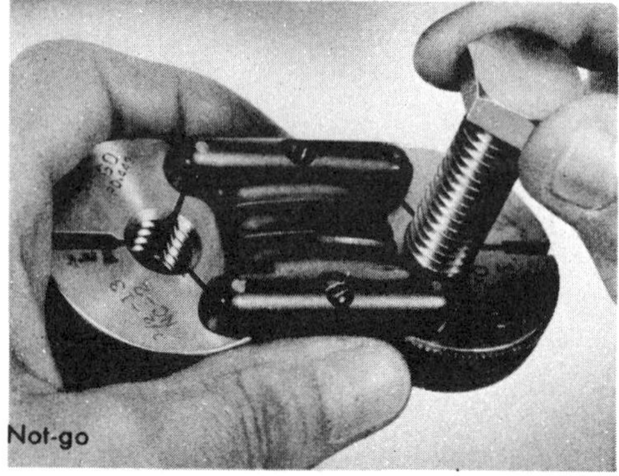

Figure 102. Checking the Thread on a Bolt

and "not-go" gages are held in a holder. The part illustrated passes inspection because it fully enters the "go" gage and enters the "not-go" gage less than one and one-half turns.

The ring gage is less likely to be damaged than the plug, because the threads are protected in the body of the gage. It is subject to wear, however, and anyone using this gage should carefully clean and remove any burr from the threads on the work and any chips, grit, or dirt lodged in the threads. As with the plug, the life of this gage is largely determined by the care it receives in use.

ADJUSTABLE THREAD SNAP GAGES

The adjustable thread snap gage is an adaptation of the plain adjustable snap gage. The flat buttons of the latter are replaced with cone-shaped anvils, wedge-shaped prisms, grooved anvils, or threaded rollers. This type of gage is used quite widely; it makes possible the checking of the thread more analytically, it is easier to use, it checks the thread faster,

Figure 103. Thread Snap Gage (Fixed Anvil Type)

and it is less expensive because it lasts longer.

Thread snap gages are not new, but for some reason, possibly faulty construction or improper application of earlier types, their acceptance by industry has been somewhat delayed. Construction details of adjustable thread snap gages have not yet been standardized to the same extent as for thread plug and ring gages. A few representative makes will be described.

Figure 103 shows an adjustable thread snap gage with grooved anvils. These grooves must have the same form as the thread, and the spacing (number of grooves per inch) must be the same as the pitch of the thread. In this

Figure 104. Ring-snap Thread Gage

Figure 105. Thread Snap Gage (Movable Anvil Type)

type of gage the anvils are clamped solidly in place and held in alignment by the clamping bar shown in Figure 103. The upper and lower jaws must be in correct relation to each other to fit the thread properly, and the "go" and "not-go" jaws must also be in correct alignment so that the piece being checked may be moved directly from the "go" to the "not-go" anvils. The front and wider anvils are the "go", and the rear and narrower anvils are the "not-go" gage. In use the thread is inserted sideways, and all threads covered by the length of the anvil are checked simultaneously.

Figure 104 shows a ring-snap thread gage. The upper "go" members are movable, threaded segments which are closed by the screw being gaged as it is passed through the segments. The lower "not-go" members are threaded rolls which check pitch diameter only. The rolls turn to minimize wear. This type of gage gives a composite check of several thread elements collectively and reveals whether a screw can be assembled with its component. A part which passes through the "go" segments and is stopped by the "not-go" rolls is acceptable.

Another type of thread snap gage is shown in Figure 105. On this gage the lower anvil is common to both the "go" and the "not-go" gages. The upper anvil is split: half of it checks the "go" dimension, while the other half is set at the "not-go" limit. The lower anvil is movable; that is, it has a small amount of axial movement in the dovetail slide, which permits the anvil to adjust itself to the lead of the thread. It returns to a normal position after the gaging operation.

A fourth type of adjustable snap gage is shown in Figure 106. The gaging members are threaded rolls mounted on lapped shafts so that they rotate with the thread being checked. The roll shaft is a hollow eccentric stud, which may be turned to adjust distance between the rollers to a desired dimension.

The rollers have sufficient axial movement to adjust themselves to the lead of the thread. As the thread is checked with the gage, the rollers turn so that any wear on the roller is distributed over the entire circumference.

Roll thread snap gages are made in different styles for different types of work and lengths of thread engagement, as shown in Figure 107. For ordinary work where there is clearance on each side of the thread, the closed-face types

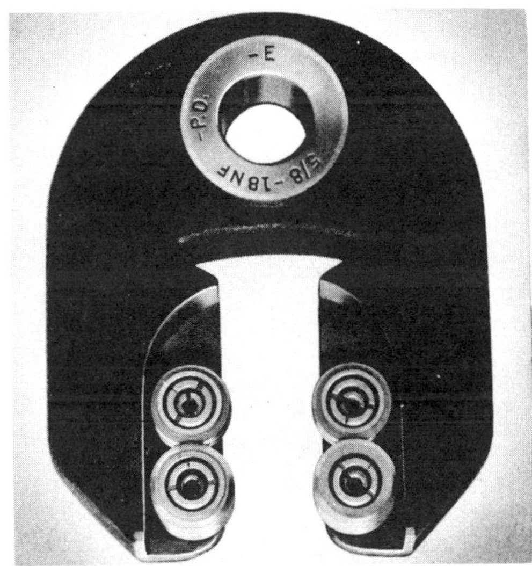

Figure 106. Roll-thread Snap Gage

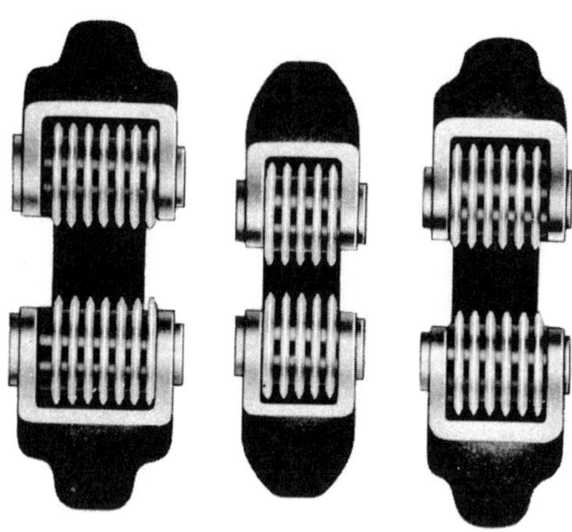

Closed Face Type

Open Face Type

Figure 107. Styles of Roll-thread Snap Gage

are used, but where a thread terminates against a shoulder, the open-face types of gage must be used. The width of the gage or the number of threads to be checked by the gage is determined by the length of engagement of the thread and the mating part. To check too many threads may cause the rejection of acceptable parts, and to check too few may allow parts to pass which will later give trouble in assembly.

SPECIAL TYPES OF THREAD SNAP GAGES

The principle of the thread snap gage has been applied to special gages for special machines or types of work. A type of thread gage using thread rolls is shown in Figure 108. The guide plates used in this type of gage are replaceable; by inserting a special guide plate for the part, the threads to be checked are lined up with the thread rolls of the gage.

Special gages and fixtures are often designed to check work in a machine. A special roll thread gage designed for use on a thread grinder is shown in Figure 109. This gage measures the pitch diameter of the thread in the machine and indicates the amount of stock that must be removed to bring the thread to size. This gage, using a dial indicator, is a type that will be discussed in another chapter.

APPLICATION OF THREAD SNAP GAGES

Thread snap gaging has increased in recent years because of the greater attention to accuracy

Figure 108. Special Type of Roll-thread Snap Gage

Figure 109. Roll-thread Gage with Indicator

in screw threads. As physical requirements of machines have increased, stronger and more dependable fastenings have become necessary. Experimentation has shown that a thread with an error in lead will not sustain even a fraction of the load that can be put on a perfect thread. Threads that are out-of-round, or threads that are tapered and do not have the proper tooth form, come loose in machines in service, wear quickly, and frequently fail in their function. Thread snap gages simplify the control of thread accuracy during machining and before assembly by checking the accuracy of a thread in a simple operation.

Figure 110 shows a machine operator checking an acme thread just produced on a turret lathe. The thread enters the "go" but not the "not-go" rolls, which indicates that the threading tool is producing a correct thread. By such periodic inspection with a thread gage, the machine operator can control the quality of the thread he produces and correct any condition which may later result in a defective thread.

Figure 111 shows an inspector checking the threads on a special bolt. The roll thread snap gage is held in a mount so that both hands are free to handle parts. Also, the gage is held in such a position that the inspector can sight-check the fit of the thread in the rollers.

The inspection setup in Figure 112 shows how large threaded parts may be checked. The roll thread snap gage in this case is mounted so that it can adjust itself to the part. The gage is attached to the base only at the stem of the back, and an arrangement of springs in the stem allows the gage to float in a plane perpendicular to the gage table. The work rests on the table and is slid into the gage so that the threads approach the rollers correctly. This eliminates the danger of cross-threading or damage to a fine thread.

For the average run of work, if a thread passes the "go" and is stopped at the "not-go" gage, it is satisfactory and will assemble. In some manufactured products where the failure of a thread is serious, a further check on the accuracy of the thread is desirable.

To check the lead with a thread snap gage, the fit between the thread and the gage is sighted. Any excessive error in lead is evident if the crests of the threads do not line up with the roots of the gage. It will appear that the thread makes contact on the right side of some threads and on the left side of others. At the same time, the inspector may check the form of the thread by noting how well it matches the thread form of the gage.

If the pitch diameter is at the low limit for the thread, the lead can be considerably in error and the thread will still pass the "go" gage. Lead and the thread form must be checked where a more accurate thread is required. (For a more accurate check of thread form and lead it is necessary to magnify the thread, as discussed in Chapter XI.)

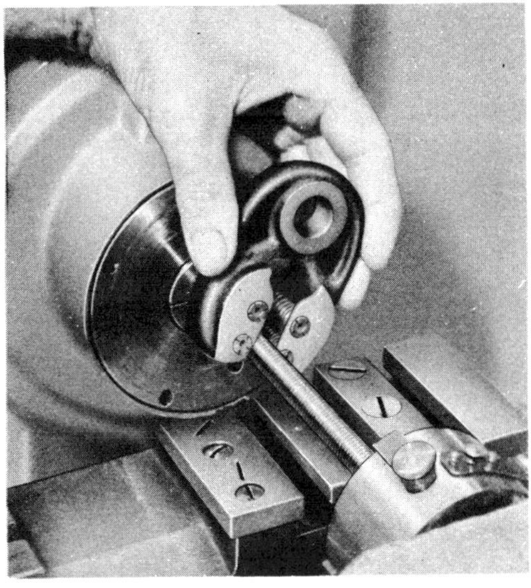

Figure 110. Gaging a Thread in a Lathe

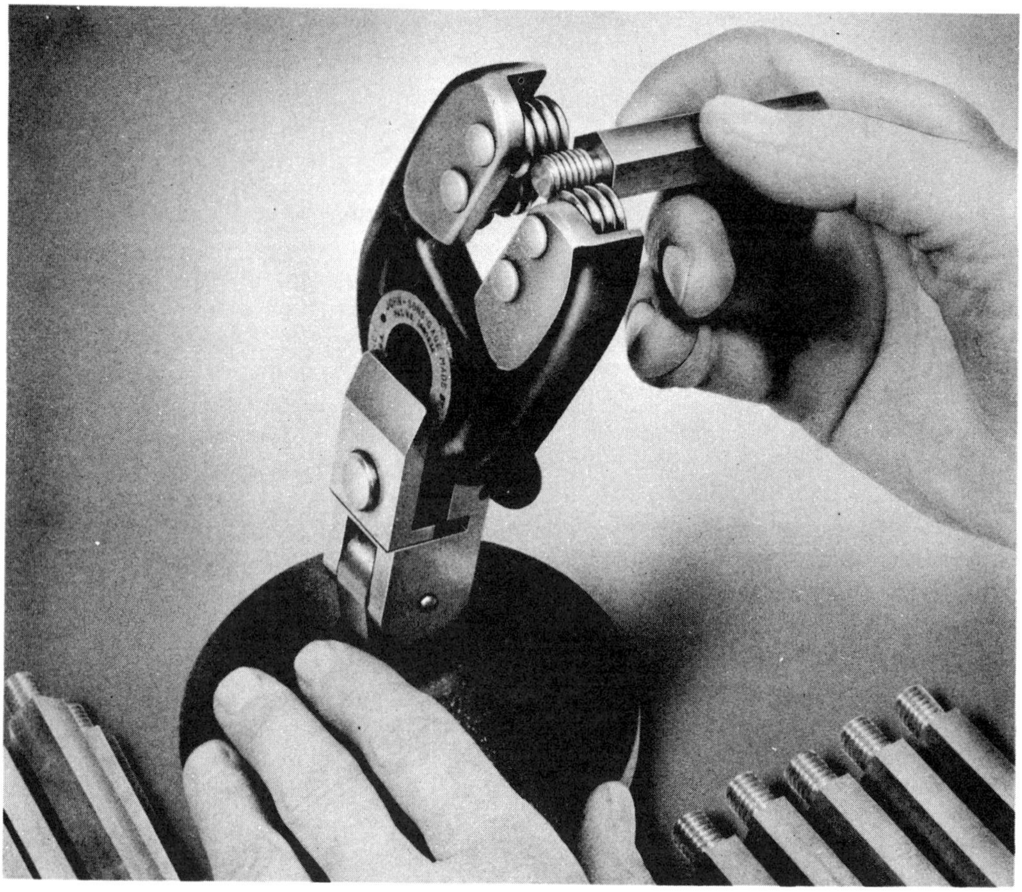

Figure 111. Thread Gage in Holder

Occasionally a thread will be out-of-round, a condition contributing to thread failure. Any out-of-roundness may be checked by turning the thread a quarter of a turn with the work held against the "not-go" gage. If the thread is undersize at any diameter, it will drop through the gage.

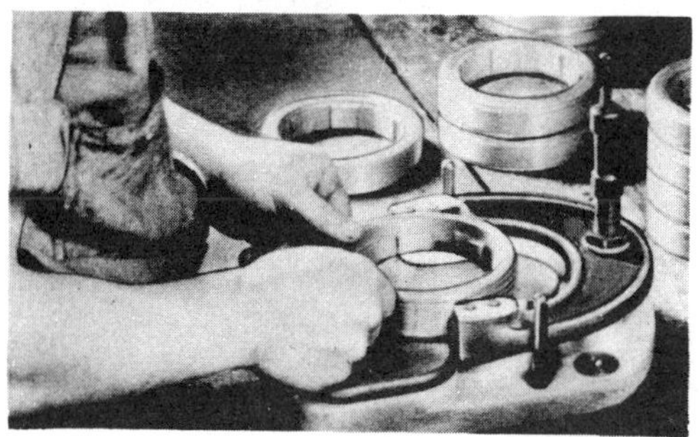

Figure 112. Checking Threads on a Large Part

Another thread fault may be taper, or a variation in diameter along the thread. This condition may be checked by inserting only a few threads into the gage, as shown in Figure 105A, and then inserting all the threads, as shown in Figure 105B. Any appreciable difference in diameter will be evident from the feel of the thread in the gage. Where the "not-go" gage extends over just a few threads, any differences in pitch diameter can be determined by testing the thread at several points.

TAPER THREAD GAGES

Taper pipe threads are used as a means of joining sections of pipe and pipe fittings, on plugs to seal drain holes in oil-filled gear housings and crankcases or, in general, wherever it is necessary to form a tight, leak-proof threaded joint without the aid of gaskets and

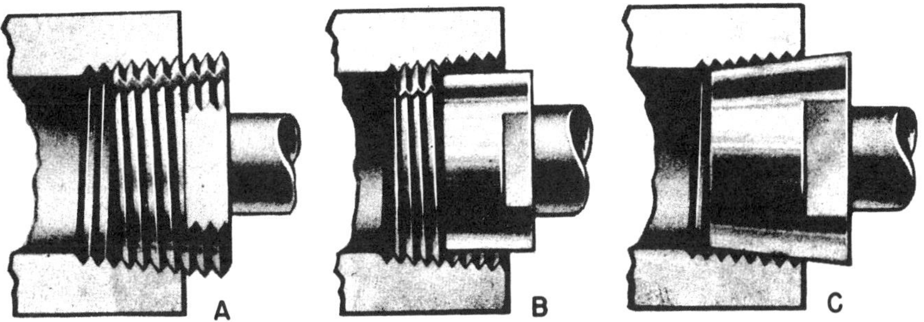

Figure 113. Inspection of an Internal Taper Thread

packing. The form of the pipe thread is similar to the American National form except that it has a sharper crest and root, which makes the thread a better seal. However, the taper also is important and must be controlled to get a joint tight enough to retain fluids and gases under pressure. The standard taper on pipe threads is one in sixteen, or .750 of an inch per foot measured on the major or pitch diameter and along the axis.

The standard taper thread plug gage is shown in Figure 113A. This gage consists of only the single member, on which a flat has been ground, as shown in the figure. In gaging a threaded hole, the plug is screwed in until it is hand-tight. With the gage snugly in place, the edge of the flat should line up with the face of the hole. The thread is still acceptable if it falls within the standard tolerance. The standard working tolerance is plus or minus one thread. This means that a machine operator should hold the size of the thread to such limits that when the gage is inserted, the front edge of the flat will not be more than one thread in the hole nor more than one thread away from the face of the hole. The standard inspection tolerance is plus or minus a thread and a half, which means that all threads which pass the working standard must pass inspection.

The standard gage is supplemented by two other gages shown in Figure 113B and 113C. The first of these is designed to check the length of tapeı, as well as the correctness of taper. The main purpose of this gage is to make sure that when it is made wrench-tight the joint will be correct.

The plain taper plug illustrated in Figure 113C checks the minor thread diameter and the amount of thread truncation and taper. The minor diameter is not checked by the two threaded gages because it is customary to undercut the root of the threads on the gages to permit grinding the tooth form. A plain taper gage must be used to check this dimension.

The accuracy of taper thread gages is easily destroyed because they are screwed hand-tight in checking a thread. It is customary to check working and inspection taper thread ring gages at frequent intervals. Wear on the gage shows when it is checked in the thread ring gage, and an allowance can be made for this wear in checking parts. Similarly, all taper thread ring gages frequently are checked with plugs.

External taper pipe threads may be checked with a single ring of the type shown in Figure 114. The gage is run on the end of the thread hand-tight. The end of the pipe thread should be flush with the surface of the gage. A working tolerance of one thread and an inspection tolerance of one and one-half threads, either plus

Figure 114. Taper-thread Ring Gage

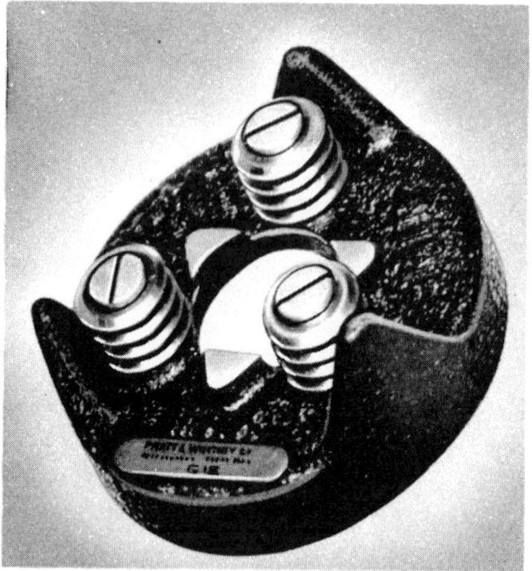

Figure 115. Three-roll Taper-thread Gage

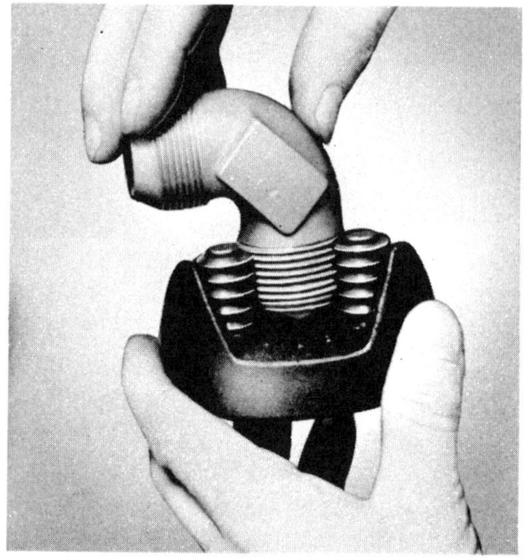

Figure 116. Checking a Tapered Pipe Thread

or minus, are acceptable in general practice.

The root of the thread in a taper thread gage also is cut away to facilitate finishing. This means that the major diameter of the work is not held by this gage; where a complete inspection is necessary, the regular gage is supplemented by a plain taper ring which checks the major diameter and taper of the thread.

A recent development in gages of this type is the three-roll thread gage shown in Figure 115. This gage is similar to a roll-thread snap gage in that adjustment is obtained by turning an eccentric pin on which one of the rollers is mounted.

This type of gage permits a more complete check of the thread because of the increased visibility of the work in the gage. Excessive errors in lead, form of thread, and

Figure 117. Set of Taper-thread Gages

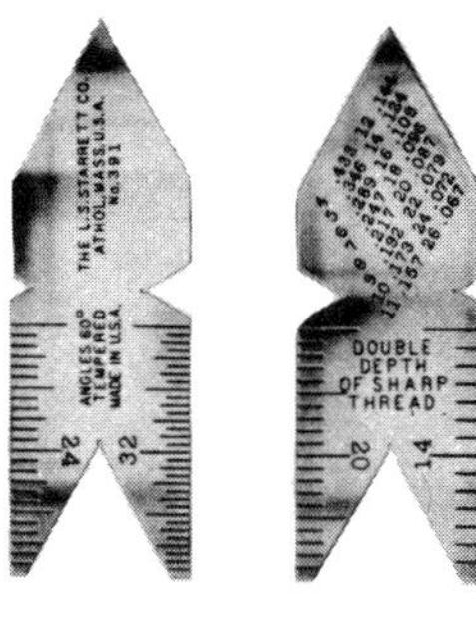

FRONT BACK

Figure 118. Center Gage

taper can be individually detected with this type of gage, while they cannot with the ordinary taper thread ring gage.

The tolerance on the diameters is definitely set on the three-roll gage by three buttons between the rolls. These buttons are marked "min.," "basic," and "max." The maximum and minimum buttons set the lower and upper limits of the thread diameters, and the basic button is used when setting the adjustable roller with the master setting plug. The height of the minimum and maximum buttons is compared to the end of the thread when the work is inserted in the gage hand-tight, and if the end of the pipe is between the two, then the pitch diameter is within the specified tolerance.

The lead, taper, and thread form are checked by sighting the line of contact between the roller and the thread. Figure 116 shows a threaded elbow inserted in a three-roll gage. From the illustration, it is clearly seen how easily thread errors can be detected and analyzed. Such a complete inspection is not possible when using a taper-thread ring gage.

A complete set of gages for the inspection of three-quarter inch pipe threads is shown in Figure 117. The three plug gages (1, 2, and 3) are employed to check internal threads, and the solid thread ring gage (4) is a master for verifying the accuracy of the thread plug (1). The three-roll gage (5) is employed to check external threads, and the plain taper ring gage (6) supplements it and checks the major diameter of the male pipe thread and assures sufficient thread depth. The three reference flats on the ring gage (6) represent the basic, minimum,

Figure 119. Setting a Lathe Tool

and maximum dimensions on the major diameter.

MISCELLANEOUS THREAD GAGES

Other gages are occasionally used in connection with machining and inspection of threads, such as the center gage and the screw pitch gages. The center gage shown in Figure 118 is used to set the cutter on a lathe. The manner in which the gage is applied is shown in Figure 119. On the center gage there are four scales, graduated in 14ths, 20ths, 24ths, and 32nds of an inch, which are useful in measuring threads per inch. On the back of the center gage is a table giving the double depth of the thread in thousandths of an inch for each pitch. This information is useful in determining the size of tap drills.

Screw pitch gages, shown in Figure 120, are used to determine the pitch of an unknown thread. The pitch is established by selecting

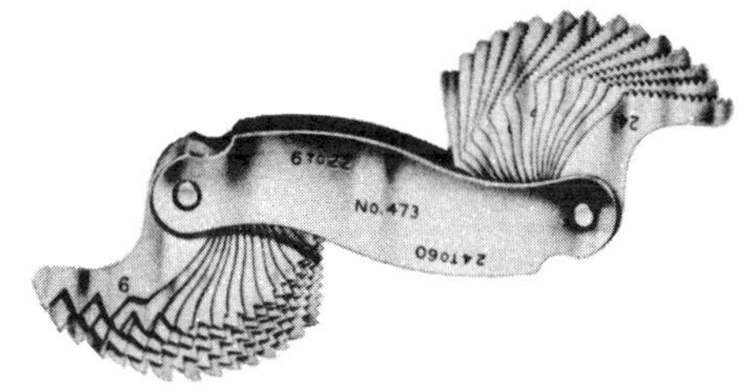

Figure 120. Screw-pitch Gage

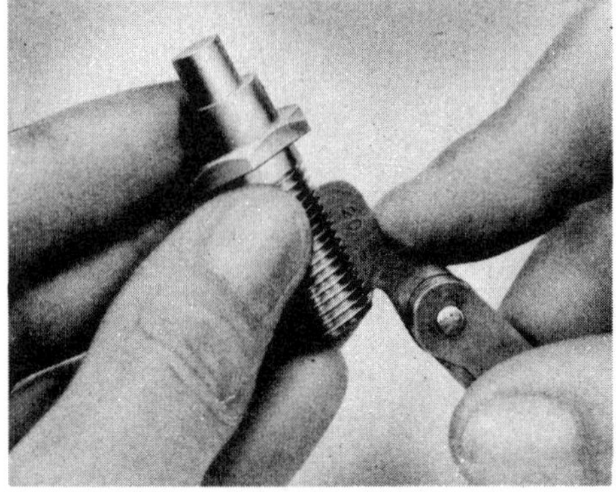

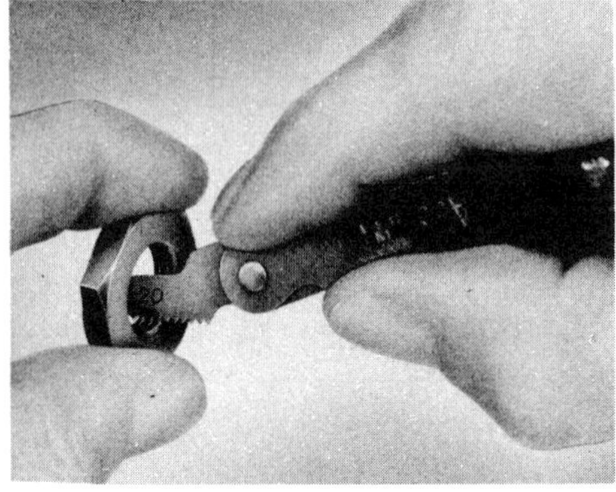

Figure 121. Checking the Pitch of a Thread

various blades from the gage and trying them on the thread as shown in Figure 121. When a match is obtained, the pitch of the thread is the number stamped on the side of the blade.

SPECIAL GAGES

Only the more commonly used gages and gaging methods have been covered in this chapter. Other chapters will consider a number of instruments and gages which have an application in thread gaging; among these will be a discussion of the operation and use of:

The thread micrometer, a micrometer caliper with special anvils for measuring pitch diameters.

The gear-tooth vernier caliper, a special form of vernier caliper with two scales at right angles. The instrument may be used to measure the width of the thread at the pitch diameter and the distance from the crest to the pitch diameter of large V or acme threads.

The projection comparator, an optical projection machine which can project an enlarged profile of a thread. The greatly magnified silhouette of the thread form may be compared to a chart layout on the screen.

The toolmaker's microscope, an optical instrument similar to an ordinary microscope, but equipped with special eyepiece charts for threads, micrometer screws on the cross slides, and fixtures for positioning and clamping objects on the stage.

The micrometer-caliper, precision bench micrometer, and measuring machine, using the three-wire method of thread measurement of the pitch diameter.

THREE-WIRE METHOD OF THREAD MEASUREMENT

The three-wire method of thread measurement was developed to adapt a caliper-type instrument with flat anvils to the measurement of threads. The method involves the use of three hardened cylindrical wires to fit in the thread to be measured. The wires are placed in the thread, two on one side and one on the other, in the manner shown in Figure 122. The measurement over the wires may then be made with a micrometer or measuring machine, and from this measurement the pitch diameter

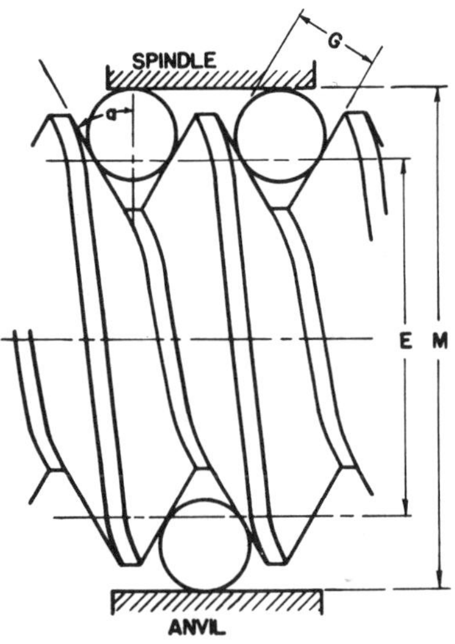

Figure 122. Three-wire Method of Thread Measurement

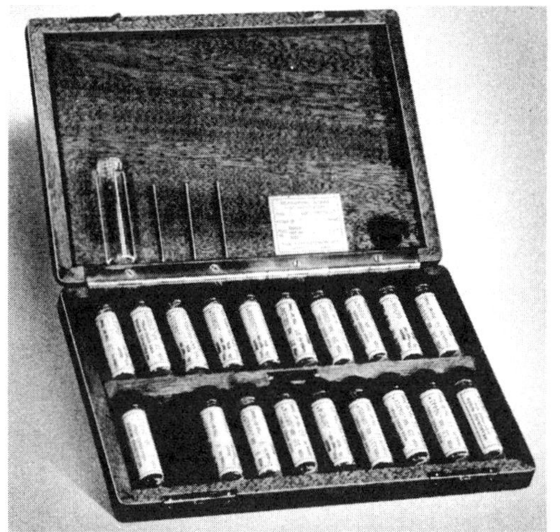

Figure 123. Set of Thread-measuring Wires

is computed by formula. Wires of different size are used in the measurement of different threads; the larger the thread, the larger the wire that should be used in measuring it.

A set of measuring wires is shown in Figure 123. Each bottle contains three wires of a size, made of hardened tool steel. They should be round to within .00002 of an inch, straight to within .00002 of an inch over any quarter-inch interval, and accurate to size within .000025 of an inch.

Each group of wires is the best size for some pitch of thread because it is this size of wire that will touch the thread at the pitch diameter.

Table VII indicates the best wire size for National Form threads.

This does not mean that a size other than the one specified cannot be used, or that using other than the best wire size leads to an inaccurate answer. With other than the best size of wire, the measurement of the pitch diameter is made with the wires in contact at some other point, and any errors in the thread angle will be included in the measurement. If the thread has perfect form, the same result will be obtained as long as the wires are small enough to be tangent to the sides of the thread and large enough that a measurement can be made without striking the crest of the thread.

The formula for computing the pitch diameter from the measurement made over the wires, for the American National form (60 degrees included angle) is:

$$E = M + \frac{.86603}{N} - 3G$$

Where E is the pitch diameter, N is the number of threads per inch, G is the diameter of the

TABLE VII. BEST-WIRE SIZES FOR THREADS

N = No. Thds. per Inch	$G = \frac{.57735}{N}$ Dia. Best Wire	N = No. Thds. per Inch	$G = \frac{.57735}{N}$ Dia. Best Wire	N = No. Thds. per Inch	$G = \frac{.57735}{N}$ Dia. Best Wire
80	.00722	28	.02062	11-1/2	.05020
72	.00802	26	.02221	11	.05249
64	.00902	24	.02406	10	.05774
56	.01031	22	.02624	9	.06415
50	.01155	20	.02887	8	.07217
48	.01203	19	.03039	7	.08248
44	.01312	18	.03207	6	.09622
40	.01443	16	.03608	5-1/2	.10497
36	.01604	14	.04124	5	.11547
32	.01804	13	.04441	4-1/2	.12830
30	.01924	12	.04811	4	.14434

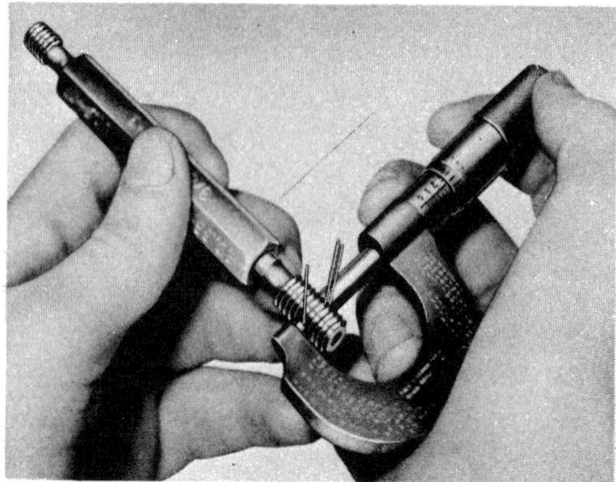

Figure 124. Checking a Thread-plug Gage (Three-wire Method)

wires, and M is the measurement over the wires.

The application of this formula is illustrated for the setup for checking a thread plug gage with a micrometer, using the three-wire method (Figure 124). With the wires in position, as shown, the inspector reads .384 on the micrometer. The best wires (.02406) for a 24-thread-per-inch gage were chosen. Substituting these figures in the formula:

$$E = .384 + \frac{.86603}{24} - 3\,(.02406)$$

$$= .384 + .03608 - .07218$$

$$= .3479 \text{ of an inch}$$

The pitch diameter of this gage is .3479 of an inch.

The three-wire method of thread measurement is also applied to the measurement of American National taper and American National acme threads. Special formulas have been developed so that the pitch diameter may be computed from a measurement over the wires. A complete discussion of this subject is given in the Report on Screw Thread Standards for Federal Services, 1944, Handbook H-28, published by the National Bureau of Standards.

The control of thread accuracy has been particularly important in those fields where the performance of the product depends on the ability of threads to withstand severe service and to maintain their adjustment over long periods of time.

Improved machine performance has, in no small measure, been aided by refinement and control of manufacturing processes through the thread measurement methods described in this chapter. And the progress made in more efficient manufacture has been due to efforts at standardization of parts, such as screws, nuts, and bolts, commonly used in industry. The National Bureau of Standards, the Interdepartmental Screw Thread Committee and the sectional committees of A.S.A. sponsored by A.S.M.E. and S.A.E. have rendered invaluable service to industry in this field.

QUESTIONS AND PROBLEMS

1. What were some of the early uses of screw threads?
2. In what country was the first thread system developed and by whom?
3. In what country was the Loewenherz thread developed and for what is it used?
4. Name five standard thread terms and define each.
5. Define the following markings on thread gages:
    1. #1-64 NC - 1 LH
    2. 5/16 - 18 UNC 3B
    3. #10 - 32 NF 2
    4. 1/2 - 10 NA 3
    5. 1/4 - 20 - NC 2A
    6. 1/4 - 28 UNF 3B
6. The lead of a single thread bears what relation to the pitch? The lead of a double thread?
7. What is the thread angle of the Unified Thread Form?
8. What is the principal difference between the Unified Thread Series and the
    American National Form?
    Whitworth Standard Form?
9. The thickness of a thread is measured at what point?
10. What is the relation between the thickness of a thread and the space between threads?
11. How many classifications of thread tolerances are there? For what type of work are they used?

12. What are some of the advantages to manufacturers as a result of the standardization of screw threads?
13. Name three ways in which screw threads can be produced.
14. What is the difference between a "working" gage and an "inspection" gage? Which of these types of gages should be more accurate?
15. How are thread gages classified? For what is each class of gage used?
16. What does the "go" thead gage check?
17. What does the "not-go" thread plug gage check? Why is it truncated?
18. Why is a lead error more serious than an error in pitch diameter?
19. Why is an error in the thread angle particularly serious?
20. What precaution must be observed when using and storing thread plug gages?
21. What are some of the advantages of the roll thread snap gage?
22. How does the American taper pipe-thread form differ from the American National?
23. What is the standard taper on the taper pipe thread?
24. What are the standard limits for the taper pipe threads on the working gage? On the inspection gage?
25. Why is it necessary to have a plain taper ring gage to check the major diameter of a male taper thread?
26. What are some of the advantages of the three-roll taper thread gage?
27. For what purpose are screw pitch gages used?
28. The three-wire method of thread measurement is employed to check what dimension on a thread?
29. What limits the size wires that may be used?
30. Compute the pitch diameter of each of the following threads. Assume that in each case the best size wires are used.
    (a) A 5/16 - 18 thread plug measuring .3240 over wires.
    (b) A No. 12 - 24 thread plug measuring .2250 over wires.

Chapter V

DIAL GAGES AND TEST INDICATORS

THIS chapter is devoted to the description and use of dial indicators, simple comparators and test indicators. Various types of special comparators will be discussed in Chapters X and XI.

A dial or test indicator amplifies and measures small displacements. It consists of a contact point attached to a multiplying lever system, gear train, or other amplifying means of operating a pointer in relation to a graduated dial, dial sector, or scale; it has suitable means for rigid attachment to a support, a gage, or a measuring instrument. Dial and test indicators are used for comparing the dimensions of a piece of work with length standards, or measuring variations from a standard measurement. Sharp distinction between dial indicators and test indicators cannot be made, but in general a dial indicator operates through a longer range than a test indicator.

DIAL INDICATORS

A dial indicator is commonly a measuring instrument in which linear movement of a spindle is amplified by a gear train; the movement rotates a pointer over a graduated dial; spindle movement of a thousandth of an inch or less is readily seen on the dial.

The dial indicator in itself is not a complete gage: it cannot measure, but must be supported on an arm from the frame of a machine, or mounted on some type of pedestal or gaging fixture. The dial can then measure or indicate the difference between the work being checked and the master or standard from which it was set.

Dial indicators are made in various sizes (Figure 125). They range from about 1-1/4 inch to 3-5/8 inches in diameter, and should be selected for size according to the visibility desired or the space available. The larger the dial, the larger the subdivisions will be for the same range in one revolution--and, in general, the more rugged the mechanism. Dials are classified according to the value of their graduations and should be selected according to the tolerances to be inspected. Dials are graduated in .00005, .0001, .00025, .001, or .010 inch. Metric dials are available in various graduations and dial sizes. Special dials made to the user's specifications are also available.

The indicator dial scale is usually the balanced type, reading both to the right and to the left of the zero; when properly set up this indicates the plus-or-minus variation from the basic dimension. The continuous-reading dial has a single scale, which runs in a clockwise direction; this scale is used where the dial indicator serves as an instrument to measure a dimension (Figure 126). Length of the scale or range of measurement on the standard models may vary from .002 inch to 1.000 inch. The value of the individual graduations may, at the same time, range from .00005 to .010 inch.

The American Gage Design Committee has standardized some of the principal mounting dimensions so that all dial indicators, regardless of make, will be interchangeable in manufacturers' gaging fixtures. The dial indicator is usually attached to a machine, pedestal, surface gage, or fixture by means of a lug or post on the back. The standard lug is shown in Figure 127; other styles of backs are available. A

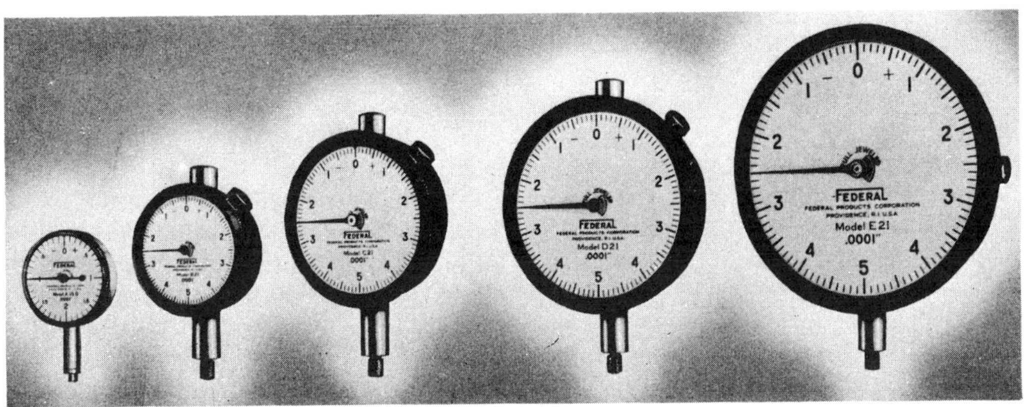

Figure 125. Sizes of Dial Indicators

screw, post, offset, or adjustable bracket may be obtained, and a selection of a particular style depends largely on the application of the indicator.

The committee has further standardized operation of the dial indicator by relating the range or point-travel to the magnification. A standard-type indicator should have a point travel of 2-1/2 revolutions of the hand, unless a special application requires a longer travel. This means that an indicator with a scale 0-10-0 will have a 2-1/2 by .020 or .050 inch travel, while a 0-50-0 indicator will have 2-1/2 by .100 or .250 inch travel.

Dial indicators used for measuring distances usually require a longer travel. If the hand makes more than one revolution, however, the operator may fail to count correctly the number of turns the hand makes; so indicators with long spindle travel are equipped with a second dial. This small inner dial (Figure 126) counts the number of revolutions of the large hand.

A third standard which has been universally

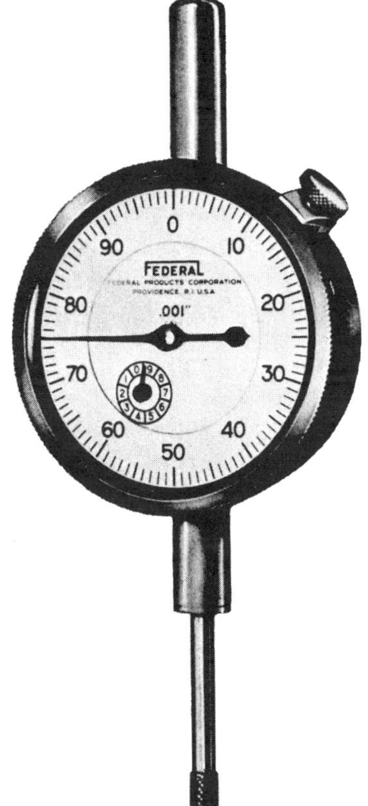

Figure 126. Dial Indicator with Revolution Counter

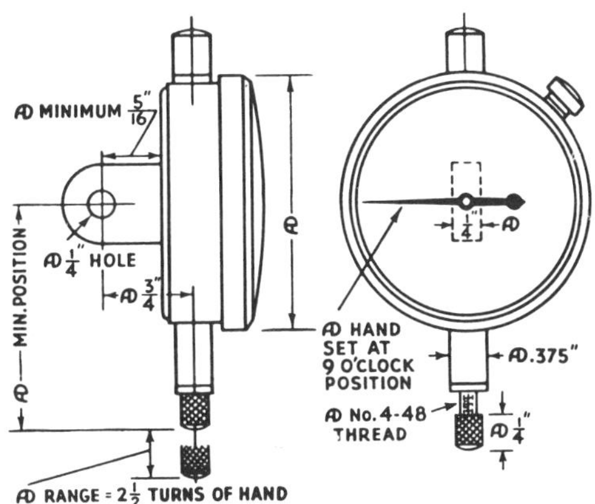

Figure 127.
AGD Standard Dimensions for Dial Indicators

adopted is to set the return stop so that the hand returns to the nine-o'clock position on the dial or 1/4 of a turn to the left of 0. By starting with the hand in this position, the operator can put tension on the mechanism and still allow the hand to operate at either side of 0 when the latter is in its vertical position. To do this with old-style indicators it was necessary to make one complete revolution of the hand.

The dial on a dial indicator is mounted to the bezel and may be adjusted to any position by turning the knurled rim. On the larger indicators, a locking screw is provided to hold the dial in place after the adjustment has been made. The procedure for setting up a dial indicator will be discussed later under "Applications."

The commonest inspection application of the dial indicator is to indicate the variation in a dimension. If the part falls within some predetermined allowable variation it is acceptable; if it does not, it must be rejected. To aid in checking parts, tolerance pointers of the type shown in Figure 128 may be used. The tolerance also may be indicated on the dial by shutters which mask out portions of the dial beyond the limits on both sides. Shutters are more often used where two or more dimensions on a part are being checked with the same dial indicator. The shutters clearly indicate the tolerance range for each of the dimensions to be checked.

Figure 128. Dial Indicator with Tolerance Pointers

Where the same part is checked day after day and the tolerance never changes (or where a stock of indicators is kept and it is possible to classify them by the tolerance marked on their dials), the limits are sometimes marked on the dial in ink. In the latter case, the gage crib will issue with the inspection fixture or apparatus the dial indicators to be used on the job, and the dial will be marked in red ink with the tolerance plus and minus, both sides of zero.

Zero on the dial represents the basic dimension.

The .001-inch indicator is the one most commonly used on inspection fixtures, and can easily control the accuracy of a dimension to within .001 inch.

The .0001-inch indicator is useful in checking dimensions to an accuracy beyond the range of the .001-inch model. Because these indicators do not have the same length of spindle travel, they are limited in application.

Application of the indicator is extended by a number of special features and accessories. At one time the dial indicator was a delicate mechanism and was not adaptable where there was a possibility of the spindle receiving a sharp

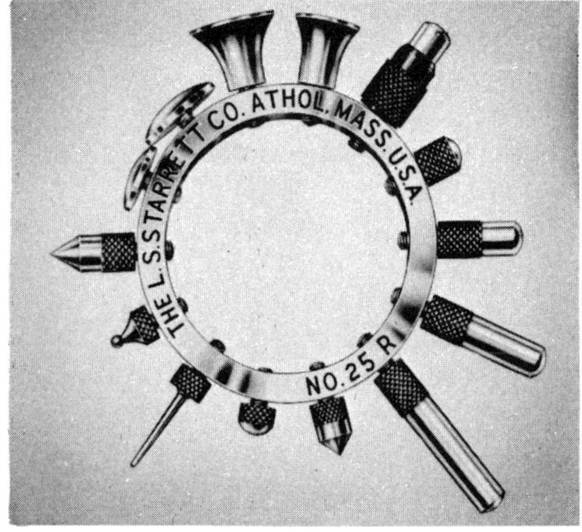

Figure 129. Indicator Contact Points

Figure 130. Maximum Hand; Bell-Crank Attachment

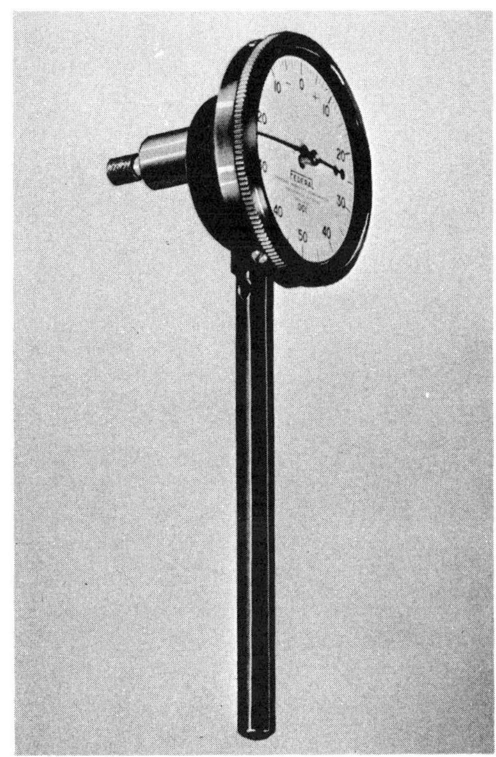

Figure 131. Perpendicular-Spindle Indicator

blow. Development of the shock-proof and cushioned-movement dial indicator satisfactorily meets this condition and makes possible application on such work. The shock-proof feature is obtained by a heavy spring between the spindle and the dial mechanism.

To make the indicator adaptable to various materials, and to different sizes and shapes of parts, a variety of contact points is employed. Figure 129 shows a selection of special points. They may be chromium-plated, cemented carbide, or diamond-tipped.

In some applications of the indicator it is desirable to mark the farthest point or maximum deflection of the hand--for example, a measure of wobble or eccentricity in a piece revolving in a machine. The maximum hand shown in Figure 130 moves forward with the gage hand and stops at the farthest point forward. It is returned to zero by the knurled knob in the center of the dial.

Figure 130 shows the bell-crank attachment, which makes possible the use of the indicator to check surfaces on a part normally inaccessible to the spindle, such as small holes, shoulders, and back recesses.

The perpendicular spindle indicator shown in Figure 131 in effect shifts the position of the dial in relation to the spindle so that it may be more easily read on certain applications where the regular type of indicator would be in an inconvenient reading position.

CONSTRUCTION OF THE DIAL INDICATOR

The dial-indicator gage gets its amplification of small linear increments through an accurately made rack and pinion, and a train of four gears. Figure 132 is a phantom view of the mechanism.

Note that the spindle extends through the indicator and that its top bearing is a hole in the instrument case. The right side of the spindle has a fine-tooth rack cut in it, which meshes with the small pinion. This pinion runs in jeweled bearings and is mounted on

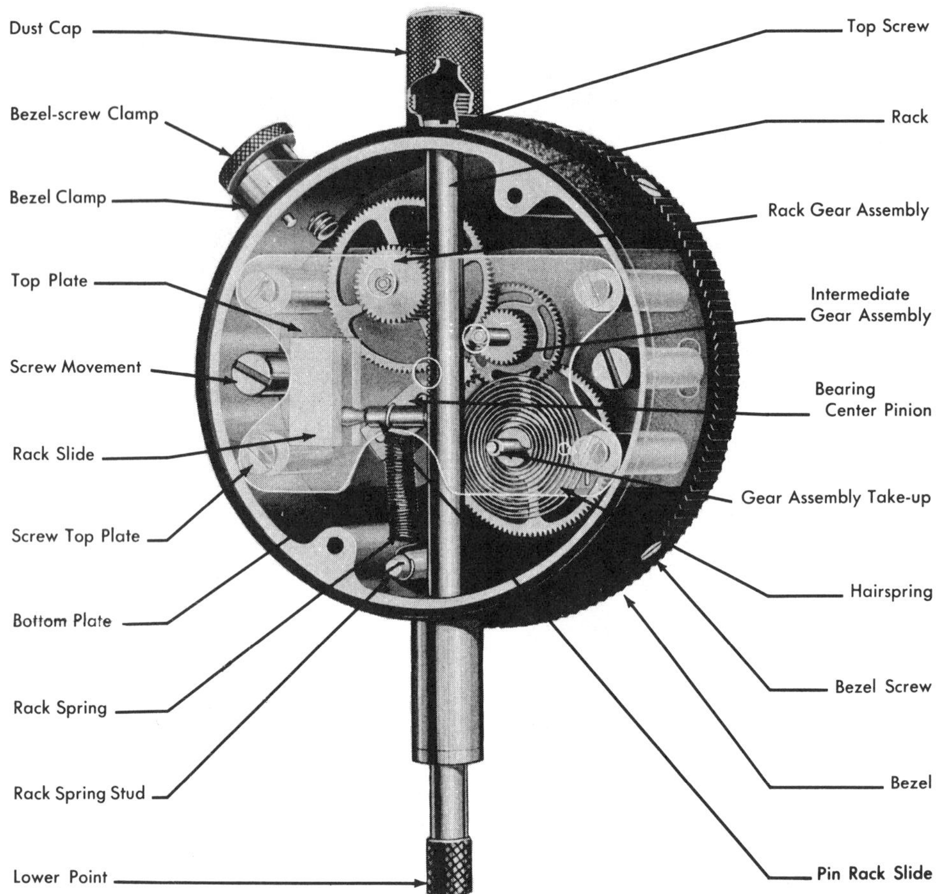

Figure 132. Mechanism of Dial Indicator

the same shaft as the larger gear driving the small pinion on the left. This second pinion in turn drives a larger gear, which meshes with the pinion mounted on the same shaft as the hand. So a very small movement of the spindle is magnified to a sizeable rotation of the hand.

The mechanism is returned to normal by a helical spring, called the rack spring, attached to the spindle. A second flat spiral spring, the hair spring, is attached to a shaft on which is mounted a gear meshing with the center pinion. This spring is to the left of the spindle. Its purpose is to take up the backlash so that any lost motion will be removed from the train of gears and thus will not affect the accuracy of the gage. But when the motion of the spindle is reversed, a portion of the motion of the spindle is not transmitted to the pointer. The amount of this lost motion is known as the hysteresis of the indicator. It is easily measured by a micrometer screw, with the indicator and screw arranged so that the screw and spindle axes are vertical. Hysteresis arises from friction.

Wear occurs in the rack and pinion and ultimately in the spindle bearings, and in most shops a stock of parts is kept to replace them as they break or become worn. All of the dial indicators are tested periodically with precision gage blocks to see that they are within the limits of accuracy required on the work they are used to check. Indicators sometimes get sluggish and bind because dirt and grit gets between the spindle and the bearing. When this occurs, the gage should be taken apart and cleaned. Dial indicators should never be oiled. If the spindle is oiled, the oil may act as a carrier for dirt and grit that would work into the mechanism.

If an indicator becomes sticky, it may have dirt or grit in the mechanism or around the plunger; or the pointer may be rubbing on the crystal. The celluloid crystal may shrink with age, and the crown may flatten so that the pointer touches it. This should be checked before the indicator is taken apart and cleaned because of suspected stickiness.

In time, the ball point on the plunger may become worn and develop a flat. Change the point when this occurs, because it may become a source of error in measurement.

The indicator is a substantial and durable instrument, but it needs reasonably careful handling to preserve its accuracy. Its mechanism is delicate and in many respects similar to that in a pocket watch.

APPLICATIONS OF DIAL INDICATORS

Figure 133 shows a small, general-purpose type of bench comparator which can be used to check a great variety of small parts. It may be used either as a comparator or as a measuring gage. When used as a comparator, a master the exact size of the part's basic dimension is used to set the dial to zero. The master is then replaced by the part to be checked. The difference between the dimension of the part and that of the master is indicated, plus or minus, on the scale.

Figure 133. Dial Comparator

This comparator consists of a base (C) and a column (D) on which are mounted a dial indicator (E) and an adjustable table (F). The dial indicator is rigidly mounted on the column, but the table is adjustable vertically to accommodate work of different sizes. The comparator is set to the basic dimension by means of a master or gage blocks.

For coarse adjustment of the table, release lockscrew (G) on the column and move the table by hand until the gage blocks touch the contact point (H). Then clamp the bracket, supporting the table, to the column; for fine adjustment of the table, release the lockscrew (J) and adjust the table by means of the table-adjusting screw (K) until the pointer of the dial gage is on zero. Lock the table in this position, remove the gage blocks, and the comparator is ready to make comparison measurements. For final adjustment to zero, release the thumbscrew (L) and turn the face or bezel of the dial gage until the pointer lines up with zero of the scale.

The lifting lever (M) raises the contact point for convenience when master or parts are inserted or removed. After the table has been locked in place, the lifting lever should be raised several times while the gage blocks are shifted about on the table, to make sure the dial gage is properly zeroed.

Parts can now be inspected by lifting the contact point and inserting the parts between the contact point and table.

Figure 134A illustrates how the indicator shows the differences between the part being checked and the master to which the gage was set. The part is .0011 oversize but within the tolerance.

Figure 134B shows how a comparator may be used for taking direct measurements. Dial indicators used for this purpose should be of the continuous-reading type having a revolution counter (Figure 126). To measure the thickness of a piece, set the indicator to zero on the gage table or on some surface on the part from which the measurement is to be taken. The example (Figure 134B) shows use of the revolution counter; it indicates nine revolutions or .0900 inch, which when added to the indicator reading of .0027 inch, shows the part to be .0927 inch thick. A dial indicator used for direct measurement should be carefully checked to verify its accuracy over the range of its spindle travel.

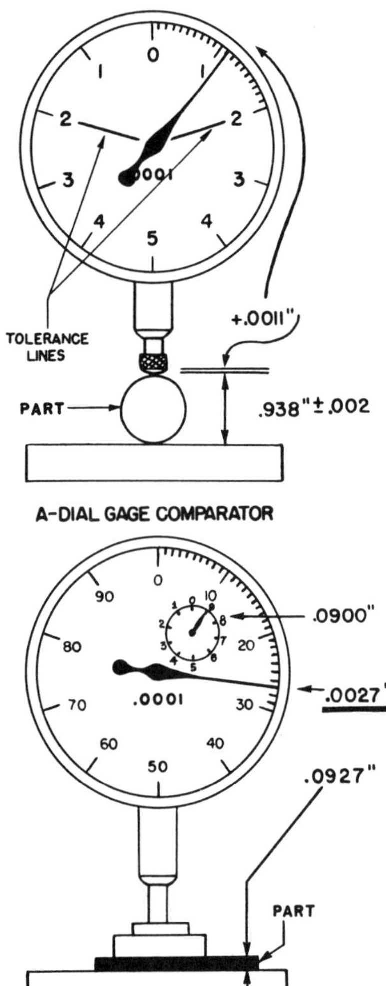

Figure 134. Application of Dial Gage

ADVANTAGES OF DIAL INDICATORS

Dial-indicator gages have many advantages over the conventional fixed-type gages. The dial-indicator gage makes it unnecessary to depend on the uncertain ability to feel or judge the variations in pressure of a conventional gage when applied to the part to be inspected. Dial indicators amplify small dimensional variations so that they can easily be seen. Classified inspection is made possible by use of dial indicators: parts can be segregated into groups for selective assembly purposes; rejected parts can be sorted for re-working. Various conditions affecting the accuracy of a specified dimension and which could not be detected by means of a conventional gage can readily be discovered by dial-indicator gages: for example, out-of-roundness, taper, lack of uniformity in the diameter of a hole, and many others.

Dial-indicating gages are an indispensible aid in dimensional quality control by statistical methods. By using indicating gages of various types, it is possible to obtain the necessary data to build quality into a product by controlling the variables which exist during its production.

Thickness gages of the dial-indicator type, known as dial micrometers, are well suited to measurement of paper, rubber, mica, fabric, cardboard, textiles, sheet metals, plastics, and other compressible materials. To take into account this compressibility, standard industrial measuring pressures have been specified; by using the correct weight on a gage of the type shown in Figure 135, comparable tests may be made by the supplier and by the user. By varying either the weight or the contact-point area, any desired measuring unit pressure can be obtained. For detailed information on the measurement of paper, see the National Bureau of Standards Technologic Paper No. 226.

The caliper gage shown in Figure 136 is particularly useful in checking the thickness of the web of castings, forgings, plastic moldings, etc. The field of application of the gage is similar to that of the transfer caliper. It is a handy gage for checking dimensions inaccessible to ordinary gages.

The particular gage shown in the figure has a range up to three inches. The dial is calibrated

Fig. 135. Dial Micrometer for Compressible Materials

to read in .010 inch, and one revolution of the hand represents 1.000 inch at the caliper points. In measuring a piece, the jaws are opened wide by grasping the lever below the handle and are inserted over the work. When released, they come together, and the reading on the dial indicates the thickness of the web at that point.

Dial-indicating snap gages of the type shown in Figure 137 are used for measuring cylindrical parts (such as shafts, studs and rolls) and other shapes and sizes of parts where an ordinary snap gage can be applied. The gage is equipped with an adjustable backstop against which cylindrical parts may be rested and revolved to check out-

Figure 136. Caliper Gage

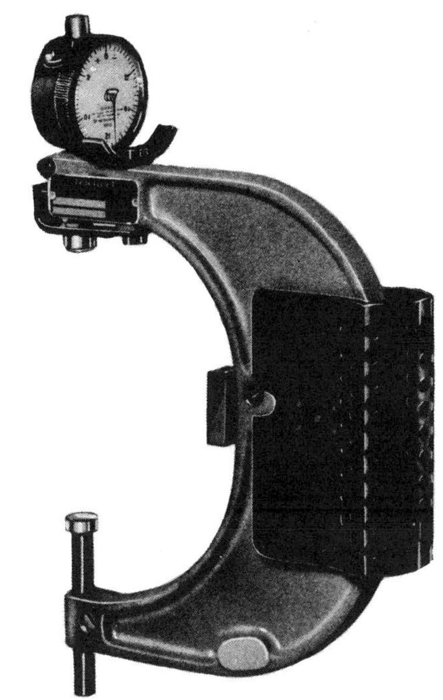

Figure 137. Dial Snap Gage

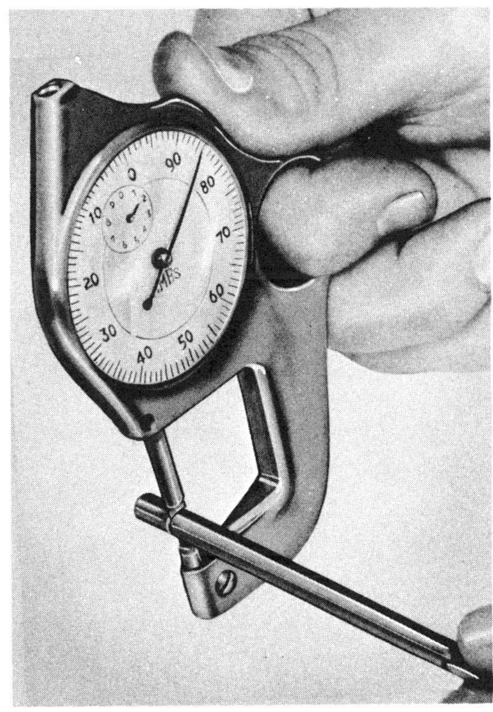

Figure 138. Dial Caliper Gage

Hydraulic Mounting

Caliper Jaw

Figure 139. Automatic Grinding Gage

of-roundness as well as diameter. An insulated finger grip protects it against temperature changes from handling. Approximate setting is made with the adjustable lower anvil and final adjustment is made by setting the indicator dial.

Indicating snap gages of this type are available in a variety of sizes for checking dimensions from 0 to 14 inches. The gage shown has a measuring range of 2 inches and will check dimension from 4 to 6 inches.

A small dial indicator of the caliper type is shown in Figure 138; this instrument is adaptable to small recesses. The indicator is graduated in thousandths and is equipped with a revolution counter. The recess being measured is .185 inch, because the counter hand is not quite on 2.

Figure 139A and B shows an application of dial indicators on an automatic grinding gage. These gages are mounted on the machine and permit the gaging of work while the machine is in operation. Several different diameters can be gaged on the same work-piece by interchanging gages of required diameter. Grinding a diameter and a width of shoulder at the same setup can also be controlled with this gage (Figure 139B).

Another feature of this gage is a hydraulic device, which lifts it up out of the way when not in use. Gaging calipers are 1/8 inch thick, permitting use on narrow shoulders and grooves. Special calipers can be made to a minimum of .050 inch thickness. Attachments are available for gaging parts having keyways, splines, tapers and narrow flanges. Tolerances to .0001 can be indicated.

Figure 140A shows a dial indicator adapted to screw-thread measurement. This gage is used for checking the pitch diameter of the threaded parts by means of roller contacts or anvils which are set to master plugs. Only one thread is contacted, thus eliminating the effect of possible lead errors in pitch diameter reading (Figure 140B).

The upper-roller anvil, the sensitive contact, is mounted on a frictionless unit which transfers any size variation to the indicator. The roller has a single annular rib, ground to fit between two adjacent threads, and contacting the sides of each. The lower or reference roller anvil has a double annular rib, ground to contact on either side of the thread. It is adjustable and can be set to receive work of any diameter within the capacity of the gage. The anvils are free to rotate and slide laterally, permitting accurate positioning on the thread, and can be changed for checking different pitches. Each set of roller anvils will cover several pitches. Pitches from 4-1/2 to 80 can be checked on this gage, and it is available in capacities from 1/8 to 4 inches in diameter. The backstop is adjustable vertically, laterally, and angularly. The dial is graduated in .0005 inch.

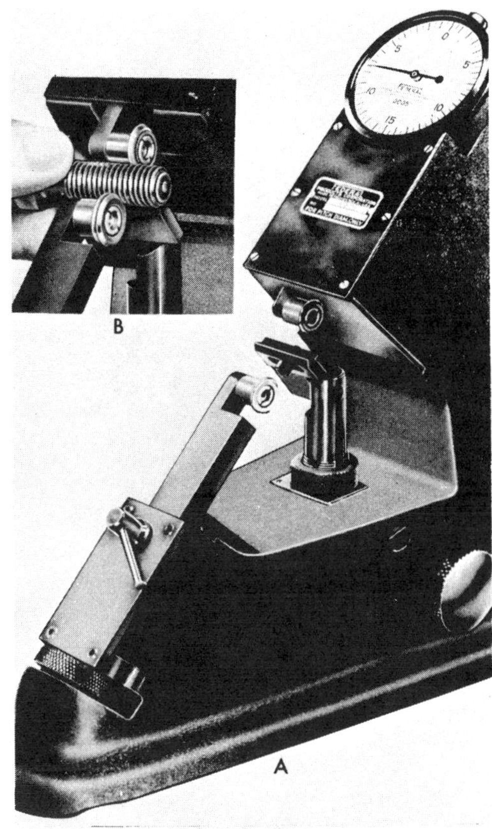

Figure 140. Dial Indicator Checking Screw Threads

Figure 141 shows another type of dial-indicating gage adapted to the measurement of screw threads. It is called a ring-snap thread comparator. The upper "go" gaging member gives a composite reading of all thread elements

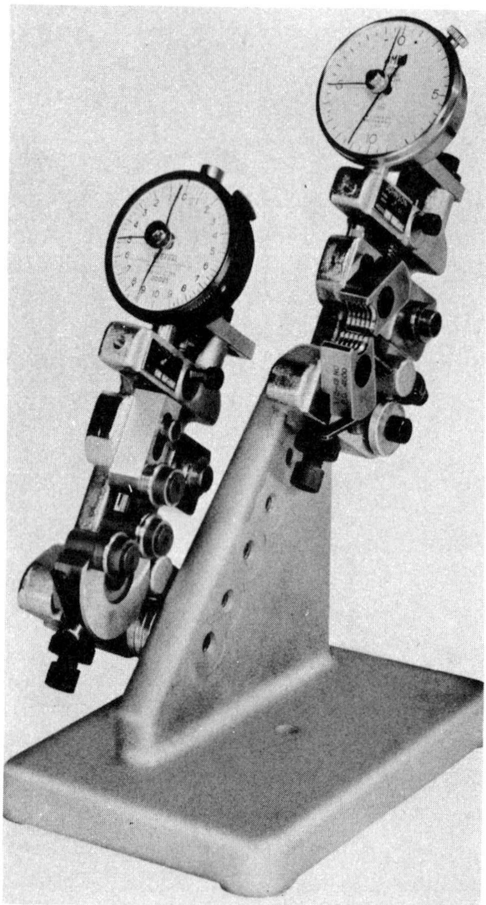

Figure 141. Ring-Snap Thread Comparator

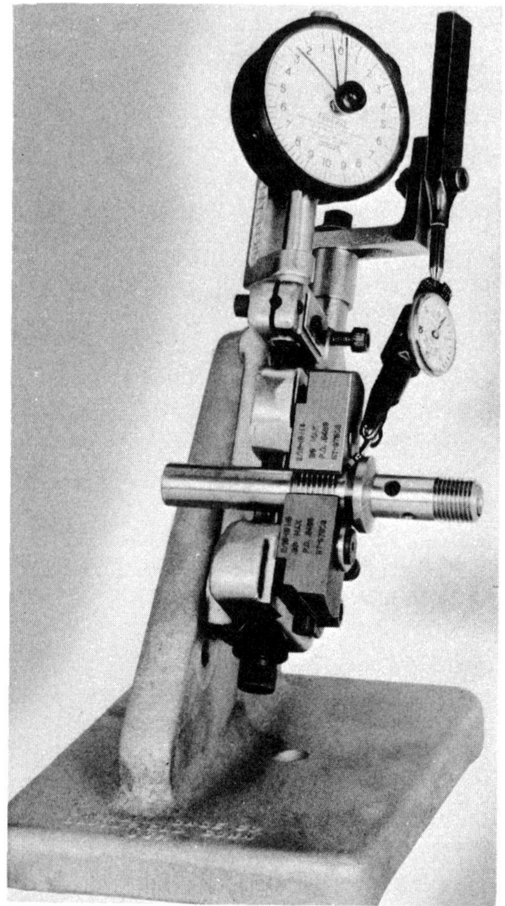

Figure 142. Thread Comparator Checking Concentricity of threads.

collectively, while the lower "not-go" member gives a single pitch diameter check only. This gage is set to setting plugs, and the indicators show the deviations from the set dimensions. This type of gage is suitable for general inspection of external threads, and for grading for selective assembly. Segments for each thread size and pitch to be checked must be used with this gage.

Squareness and concentricity of surfaces and dimensions related to threads can be accurately checked by use of standard attachments available for this gage, one of which is shown in Figure 142.

A gage for testing the lead of screw threads is shown in Figure 143A. This gage is set from a master, and the dial indicator shows any deviation in lead over a predetermined number of threads.

The measurement is made between two conical points. The point on the left is set to a master screw and locked in place; this point can be screwed into any one of four positions for various lengths. The point on the right can be screwed into any one of three positions and is connected to the indicator spindles. The work table is adjustable vertically for various diameters up to 1-1/2" by use of the two screws on the front of the base. With the master in the gage, the indicator is set to zero (Figure 143B).

This gage may be used to check the leads of screw threads of studs, thread plugs, taps, and the spacing of holes, grooves, notches, and rack teeth. The number of threads included between the two points is determined principally by the length of engagement of the thread, and the

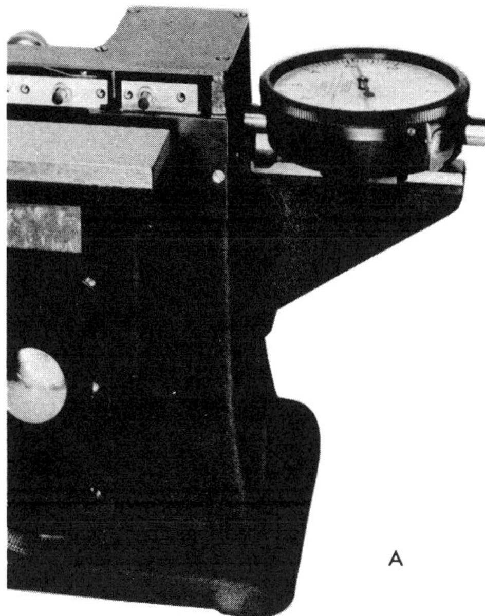

Figure 143. Dial Gage for Checking Lead

tolerance usually is given in terms of this length.

Gaging holes or bores for accuracy is probably one of the most common inspection requirements. Too often it is assumed that a simple general inspection tells all that there is to be known about the actual condition of the hole.

A general inspection to determine that a hole is finished to a specified size is not always sufficient. A hole may vary in roundness, taper, straightness, uniformity or bellmouth; and it may also have mechanical defects. Any one of the defects can cause undue vibration, weakness, wear, and other unsatisfactory performance. Therefore, gages are required which make possible determination of actual diameter variations, and which also analyze the hole for other elements of inaccuracy.

Dial-indicating hole gages are especially suitable for this type of inspection. They indicate all dimensional variations with respective fidelity and are available in a great variety of sizes for checking diameters from .122-inch up to diameters as large as may be desired.

Hole-indicating gages of the types shown require interchangeable plugs or gaging heads for different diameters and extensions for different measuring lengths. They are usually set to the desired size with ring gages or precision gage blocks.

The manner in which dial indicators are adapted to internal measurement is shown in Figure 144. Here a vertical-bore gage is being used to check holes of larger diameter (from about 1 inch to 12-1/8 inches). With the long handle and the measuring head, it is possible with this gage to explore large bored holes of considerable depth. This is not a regular three-point gage. Two studs locate the gage on a diameter so that the fixed and indicating con-

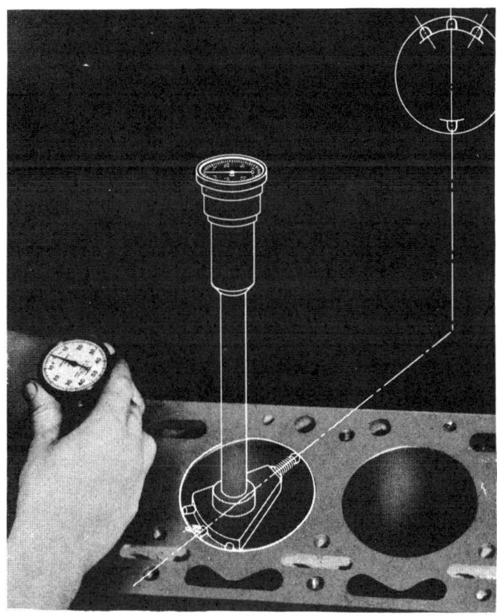

Figure 144. Dial Bore Gage

Cap interchanges for either lands or grooves. Guide shown is for lands.

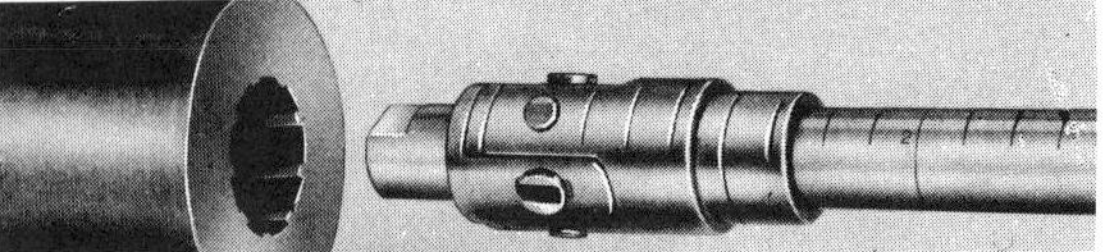

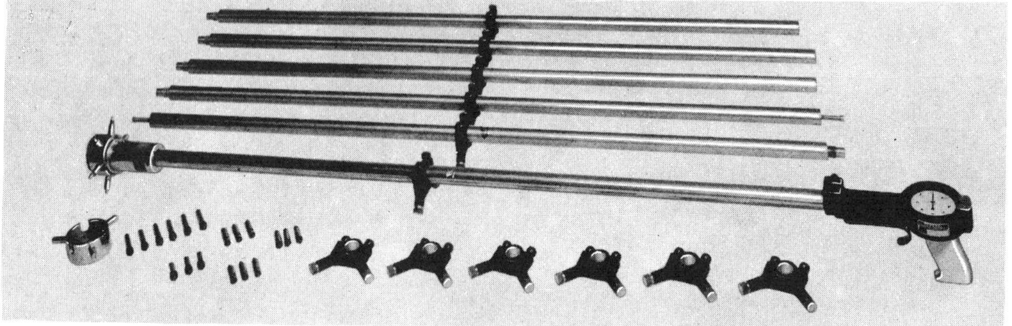

Figure 145. Gun-Bore Dial Gage

tacts read true diameter, as shown by the phantom drawing. The most common application of this gage is testing the cylinders of gasoline and steam engines. Gages of this type will indicate directly and accurately an error of as little as .0001 inch. By changing the detachable points, a variety of hole sizes can be checked with a single gage.

The gun-bore gage shown in Figure 145 is another application of the dial indicators. The complete set in Figure 145 is for five- and six-inch bores with sufficient extensions for 300 inches. Bore irregularities, out-of-roundness, taper, and diameter are checked throughout the entire length.

Guide points on the heads are interchangeable for lands and grooves, and the heads revolve independently as the gage is traversed through the length of the bore to follow the rifling. The tube is graduated every 1/4 inch and numbered each inch to indicate the exact position of the contact points.

The contact points are interchangeable to give a range of .350 inch and the dial indicator is graduated in .0005 inch. The close-up in Figure 145 shows a different style head for a 20 mm gun bore.

Figure 146 shows a dial plug gage designed to provide a highly accurate gage for use where it is desired to produce identical holes in quantities. Basically, the gage consists of two units: an amplifier and a detachable expansion plug. The amplifier contains the mechanism for operating the dial indicator and for expanding and contracting the plug. The expansion plug consists of a split plug, expanded by a tapered plunger on which there is constant spring pressure, except when the compression button is squeezed to contract the plug.

Each expansion plug is made for measuring only one size of hole within a range of $\pm$ .005 inch. For each size of expansion plug, a master reference or setting ring is required which is the exact size of the hole to be made.

Set the gage for use as follows: press the compression button, which contracts the expansion plug; insert the head of the plug in the

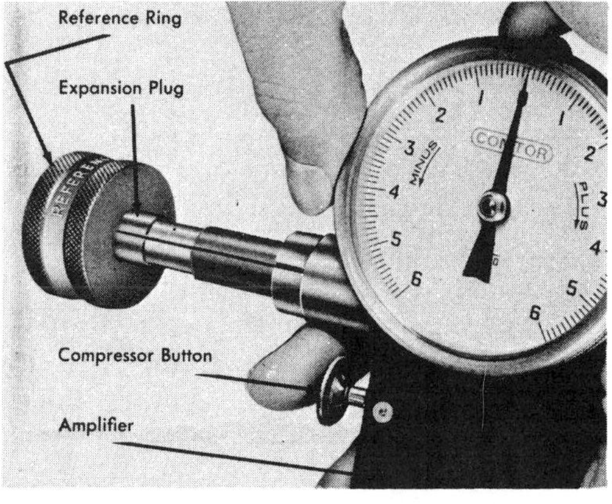

Figure 146. Dial Plug Gage

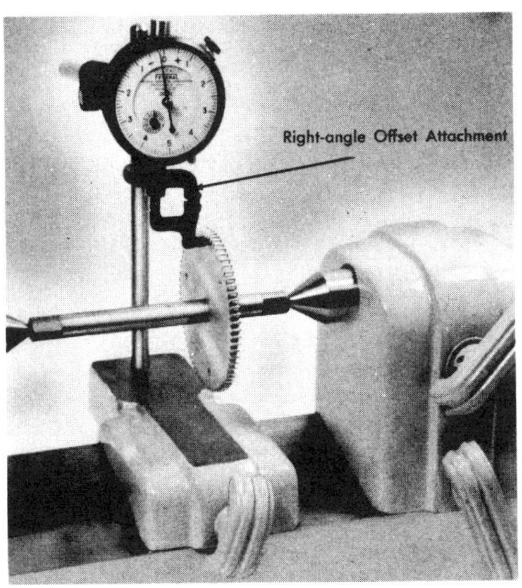

Figure 147. Indicating Side Wobble

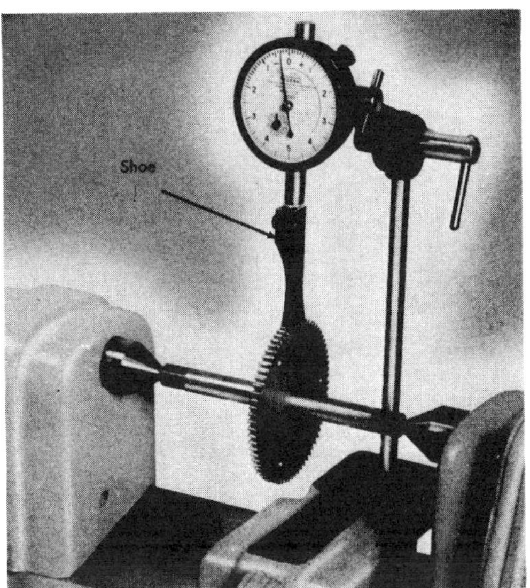

Figure 148. Checking Concentricity

master reference ring, giving the plug a slight shake in the plane of the slot to permit self-aligning and centering; adjust the dial to zero, and the gage is ready for use.

In operation, apply the plug to the work by allowing it to expand in the hole and giving it a slight shake to permit alignment and centering. Any deviation in size from the exact size of the reference ring will be shown plus or minus on the dial which is graduated in ten-thousandths of an inch.

All expansion plugs are quickly interchangeable in any amplifier, and the size range of plugs is from .125 inch to about 10 inches.

A feature of this gage is its ability to check to the bottom of blind holes or shoulders and to detect irregularities such as out-of-round, back or front taper, and barrel shape.

A dial indicator is frequently attached to a surface gage to check the location of surfaces from a surface plate or work table; this indicates the wobble or error in concentricity of work held between bench centers. Figure 147 shows a dial indicator attached to a special surface gage and being used to check the amount of wobble in the gear mounted on a mandrel between centers. As the gear is rotated, the error in squareness between the face and the axis of the hole through the center is reflected by the limits of the deflection of the indicator hand.

A dial gage should be mounted on a rigid support so that the contact pressure of the indicator does not cause deflection of the supporting means. If clamped to a long non-rigid arm, the run-out or wobble of the work may go into deflection of the support and not be indicated on the dial gage.

A similar application of the dial indicator is shown in Figure 148. In this case a gear has been pressed on a mandrel and is mounted between bench centers. This setup checks the concentricity of the gear teeth in relation to the gear shaft. A special shoe for the spindle of the indicator eliminates the jump of the hand between teeth.

The dial-indicator method is a practical way to measure the amount of wobble or eccentricity in a part. Where the variation with respect to a shaft hole is desired, a setup similar to that shown in Figures 147 and 148, where the piece is mounted on a mandrel held on centers, is generally used.

For very accurate work a comparator of the type shown in Figure 149 may be employed. It is adaptable for a simple and accurate check of working gages and instruments against standard gage blocks.

The particular comparator shown is graduated

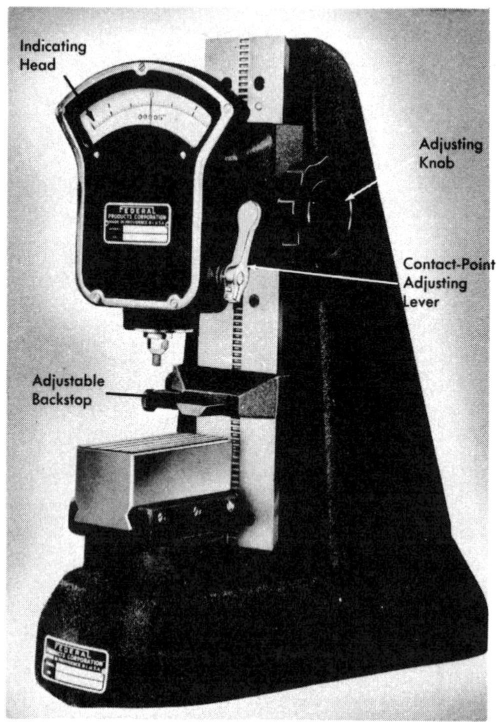

Figure 149. Supersensitive Comparator

in .00005 inches and has a range of plus or minus .0015 inch; it will check a dimension up to about 3.750 inches. The indicator head is set approximately to the proper level with the large knob; the final adjustment is made with the small lever, on the side of the head, which operates the contact point. The adjustable backstop makes it possible to set up the gage for rapid positioning of work on the table.

In the setup of a machine tool, a dial indicator is often used to check the concentricity or alignment of a part or fixture. In Figure 150, a dial indicator with a special attachment is being used to check the alignment of the table vise so that the jaws are parallel to the travel of the table. This is done simply by clamping the indicator to some stationary part on the milling machine--in this case the spindle--and running the table back and forth. The vise is shifted, until the jaws are parallel to the table travel, and clamped.

Another application of the dial indicator to machine-tool work is shown in Figure 151.

Figure 150. Checking Alignment of a Fixture

Here the outside diameter of a cutter is being checked to insure that each tooth has been ground concentric. By setting the dial at zero on the highest tooth and checking for the lowest tooth, a tolerance can be established for further grinding if necessary.

A feature of this particular gage is the magnetic base which permits positioning it either horizontally or vertically. This feature insures a steady support not subject to vibration. It is a permanent magnet not requiring any electrical connection.

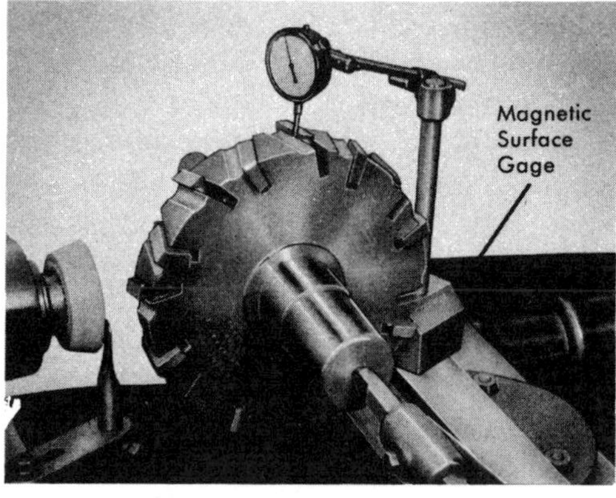

Figure 151. Checking Diameter of Cutter

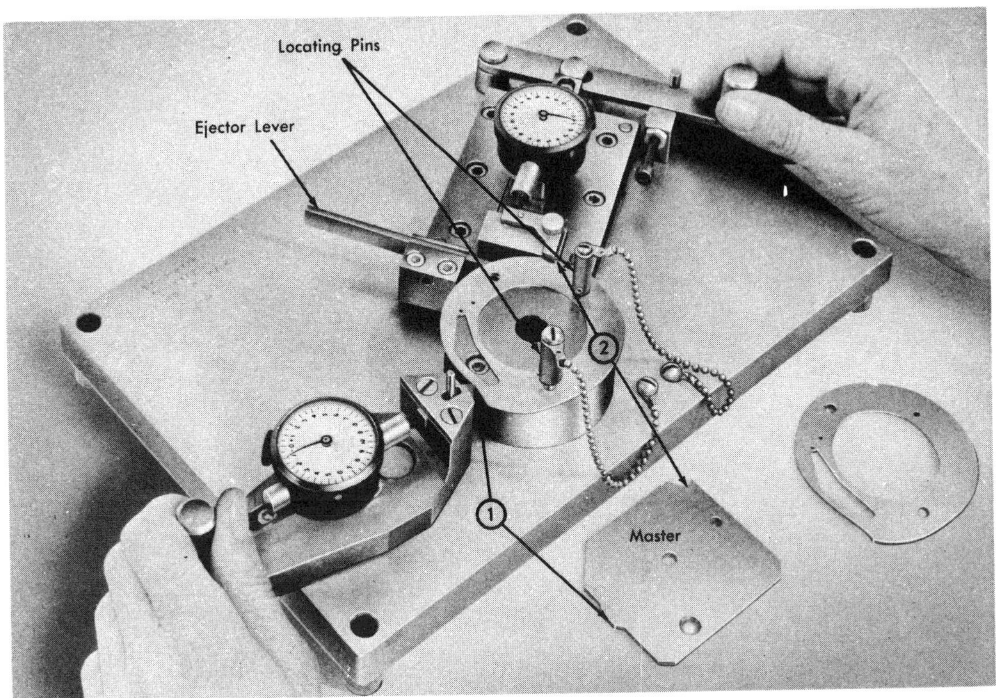

Figure 152. Special Gage with Master

SPECIAL INSPECTION GAGES

A distinction between standard and special gages is often made to classify gages used in inspection of semi-finished and finished parts. Most of the gages discussed in this book can be classified as standard gages, applicable to any number of parts. Special gages, however, are usually designed for inspection of a single part, or similar parts that differ only in some minor detail.

Because these gages must be built especially for the inspection of one or two parts, there must be sufficient production to warrant their cost. For the special gage to be worthwhile, the inspection time saved must more than make up for the difference in initial cost between a standard and a special gage.

In general, it is not practicable to provide special gages for every dimension; so the gage analyst considers: (a) Method of manufacture (is the dimension one that will be held by the tools so that only a token check is necessary?). (b) Time consumed in checking dimensions with standard equipment as compared with a special gage. (c) Importance of the dimension to the function of the part in assembly. (d) Skills of those who are to do the inspection; if a dimension is one that is checked by a machine operator, more assistance and convenience will be provided by the gage. (e) Time available for inspection; if the machine operator has a waiting period in his work cycle, the time spent in gaging work is not so important as when this is not the case. (f) Combination of gaging operations into one multi-purpose gage, such as gages equipped with several dial indicators that will check several dimensions simultaneously. (g) Making gaging part of the machining operation; for example, gages that can be attached directly to a machine and operated manually or automatically. (h) Coordination of gaging and tooling so that reference lines and locating surfaces are consistent with one another.

The setup of a special gage using dial indicators is made with a master. Figure 152 shows an inspection gage and the setting master designed to check two important dimensions on a cam. The setting master is not a duplicate of the part, but it has finished measuring surfaces which bear the same relation to one another

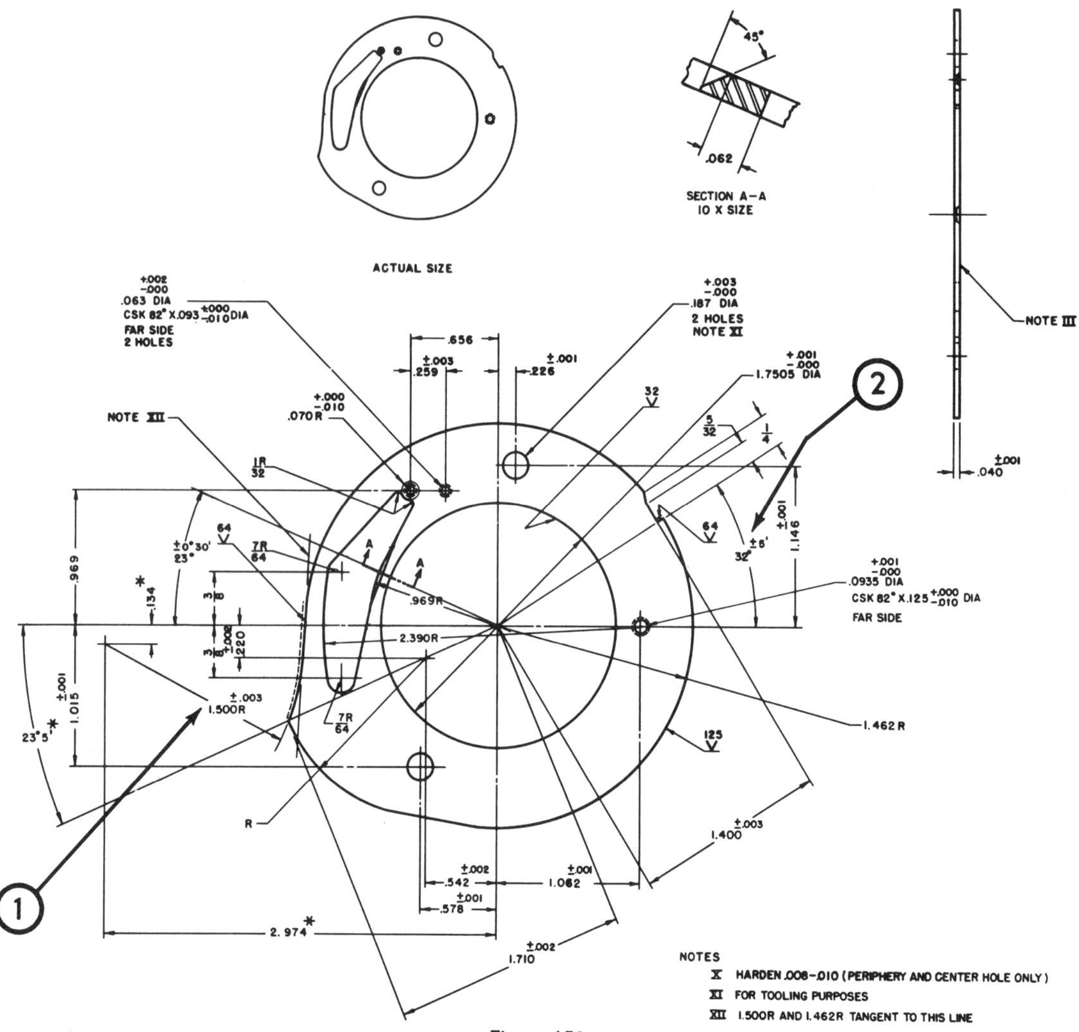

Figure 153

as the corresponding surface on the part.

Figure 153 is a print of the part indicating the dimensions to be inspected and the tolerances to be maintained in order to pass inspection.

USE OF THE GAGE

In Figure 152 the setting master is positioned in the gage by use of the two locating pins as shown. The indicators are movable: the one on the right slides in and out; the one on the left moves in an arc. With the master in place, move the indicators into contact with the setting points (1) and (2) and set the dials to zero. Move the indicators out of contact to prevent damaging them, and remove the master. This gage is equipped with an ejector to aid in removal of the master and the parts.

Place the part to be checked in the gage, using the locating pins, and move the indicators into contact with the part. Position 1 checks a 1.500-inch radius; its tolerance is ± .003 inch and is checked by moving the indicator in an arc. Position 2 checks an angle of 32 degrees; its tolerance in thousandths is ± .002 inch, as marked on the dial. The indicators will show the deviation from the zero setting.

Figure 154 shows an inspection gage for simultaneously checking six diameters on a workpiece. Each indicator, and its opposite anvil, is mounted on a separate pantograph unit; this allows it to adjust itself independently of any other measuring unit of the gage.

The pantograph mechanism, as used here, is a link-motion transfer unit consisting of two parallel steel blocks connected at each corner by flat parallel springs. In use, one of the blocks is anchored to the base of the gage; the other, which is the movable portion of the pantograph, carries the gaging contact and transmits size variations to the indicator, without friction.

Pantograph mechanisms are widely used for simultaneously checking inside and outside diameters and concentricity. They can be used to advantage for either single or multiple applications; they permit the use of dial indicators for accurately checking surfaces difficult to reach by other means.

SPECIAL GAGES--GENERAL

The variety in design of special gages is due principally to the fact that the gage is built around the part and that the gage must provide a means of locating and clamping the part. Other factors which influence the design of the gages are:

1. Size and weight of the part.

Figure 154. Special Inspection Gage

2. Production of the part.
3. Tolerance on dimensions to be checked.
4. Ease of operation in checking parts.
5. Care, maintenance and adjustment.

Applications of special gages have been curtailed to some extent by the development of the projection comparator. This measuring device, which gages a piece by optically projecting its profile on a screen, will be discussed in a later chapter; but it may be said here that for applications which permit inspection by projection of the shadow of a contour, this method of inspection is definitely superior.

Some gages, instead of using indicators, employ flush pins or sliding bars. The flush-pin gage has buttons which contact the surfaces to be checked. These buttons operate pins which are so positioned that the operator is able to sight or feel the top of the pin in relation to "go" and "not-go" steps. In this way a tolerance of about .005 inch may be held.

The sliding-bar type of gage has one or more slides with "go" and "not-go" steps so that when the bar is pushed forward over the piece, it will pass over the "go" step and strike the "not-go" step if it is within the tolerance specified.

Special gages often are equipped with ejectors or knockout pins. This device pushes the piece in the gage off the locating pins and eliminates the necessity of picking each part out of the gage with the fingers or a pick. The ejector mechanism usually is operated by a knob, button, or lever, which is struck when the piece is to be removed.

METHOD OF GEAR INSPECTION

The commonest method of determining accuracy in gears is to rotate the gear through a complete revolution in intimate contact with a master gear of known accuracy. Both gears are mounted on a variable-center-distance fixture; the resulting radial displacements or variations in center-distance during the rotation of the gear are measured by a suitable device, such as a dial indicator. Excluding the effect of backlash, this check approaches the action of the gear under operating conditions.

Permissible radial displacement will be negative in relation to one-half the sum of the mean pitch diameter of the master gear and the maximum pitch diameter of the gear being tested, as specified on the part drawing. This is called "composite check" and gives the combined effect of the following errors:

Runout (radial displacement)
Pitch error
Tooth thickness variation
Profile error
Lateral runout (sometimes called wobble)

Gears are classified in two groups: commercial, (four classes--1, 2, 3 and 4); precision (three classes--1, 2 and 3). Tables VIII and IX give the tolerances for the respective classes.

TABLE VIII

TOLERANCES FOR COMMERCIAL FINE-PITCH GEARS

Class	Total Composite Error (Inches)
Commercial 1	Minus 0.006
Commercial 2	Minus 0.004
Commercial 3	Minus 0.002
Commercial 4	Minus 0.0015

TABLE IX

TOLERANCES FOR PRECISION FINE-PITCH GEARS

Class	Total Composite Error (Inches)
Precision 1	Minus 0.001
Precision 2	Minus 0.0005
Precision 3	Minus 0.00025*

*This possibly would be the result of selection and segregation.

When checking gears on a variable center distance fixture, the amount of applied pressure is important. Excessive pressure on fine-pitch gears of narrow face width will result in incorrect readings caused by deflection of the teeth. Variable-center-distance fixtures are usually equipped with calibrated springs and scales for measuring pressure in ounces.

Recommended pressures between gear and master are based on diametral pitches. Values given in Table X represent the pressure required to maintain satisfactory contact for checking purposes:

TABLE X

RECOMMENDED CHECKING PRESSURES

Diametral Pitch	Pressure Ounces
20 to 30	28
30 to 40	24
40 to 50	20
50 to 60	16
60 to 80	12
80 to 100	8
100 to 149	4
150 and finer	2 minimum

For any gear coarser than 20 D.P., add 4 ounces pressure down to 10 D.P. and increase accordingly below 10 D.P.

RUNNING GAGES

Running gages are applicable to almost any moving part of a mechanism, but are most frequently employed to check gears, cams; worms and worm wheels.

On some gages, instead of using an indicator an attempt is made to check the part under conditions which more nearly duplicate service requirements. An example of such a gage is the running gage. By running a gear with a master at an accurately-set center-distance, it is possible to check the pitch diameter and the concentricity of the teeth with respect to the hole in the hub, as well as a combination of tooth errors which might affect the operation of the gear in the machine.

Figure 155 shows a non-adjustable running gage, which is simply two solid pins (the size of the hole in the hubs of the gears) mounted in a base plate. The master gear is placed on one stud, and the gear to be tested is placed on the other. The master gear usually is a part that has by special inspection of every tooth been designated as the master.

The gears are rotated by hand. Accuracy

Figure 155. Non-Adjustable Running Gage

of the test gear at any point may be checked by holding the master and feeling the play or backlash at that point. Any variation in the torque required to rotate the two gears also will show up as error--in concentricity, a faulty tooth, or error in tooth spacing--and will tell whether the gears, when assembled at their proper center-distance, will mesh and revolve.

The gage shown in Figure 156 is a variable center-distance running gage and is adjustable to accommodate many sizes of gears, as well as several sizes of holes in the gear hub.

On gages of this type, gears are checked in pairs, one of the pair being a master gear of known accuracy. Both gears are mounted to revolve freely on pins. The gear to be checked is mounted on the floating arm that actuates the dial indicator.

The master is mounted on the adjustable slide, which is clamped in position after setting to the correct center-distance. The base pins are used for setting up the gage. The pin in

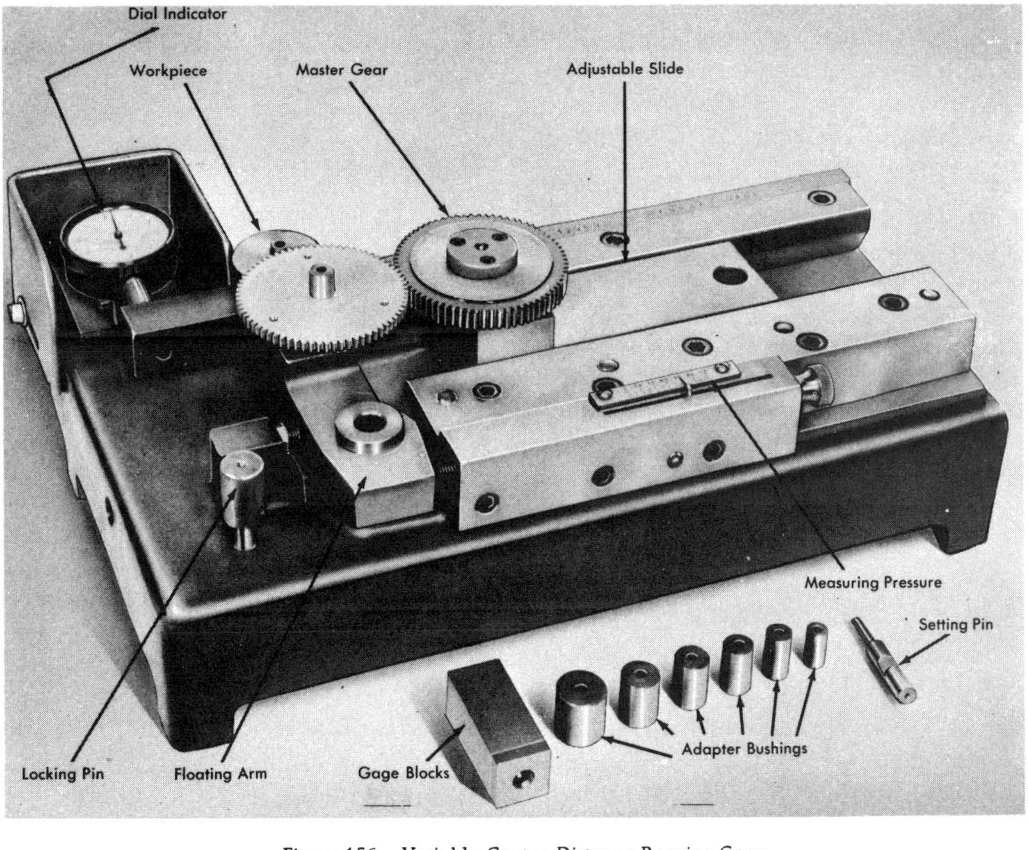

Figure 156. Variable Center-Distance Running Gage

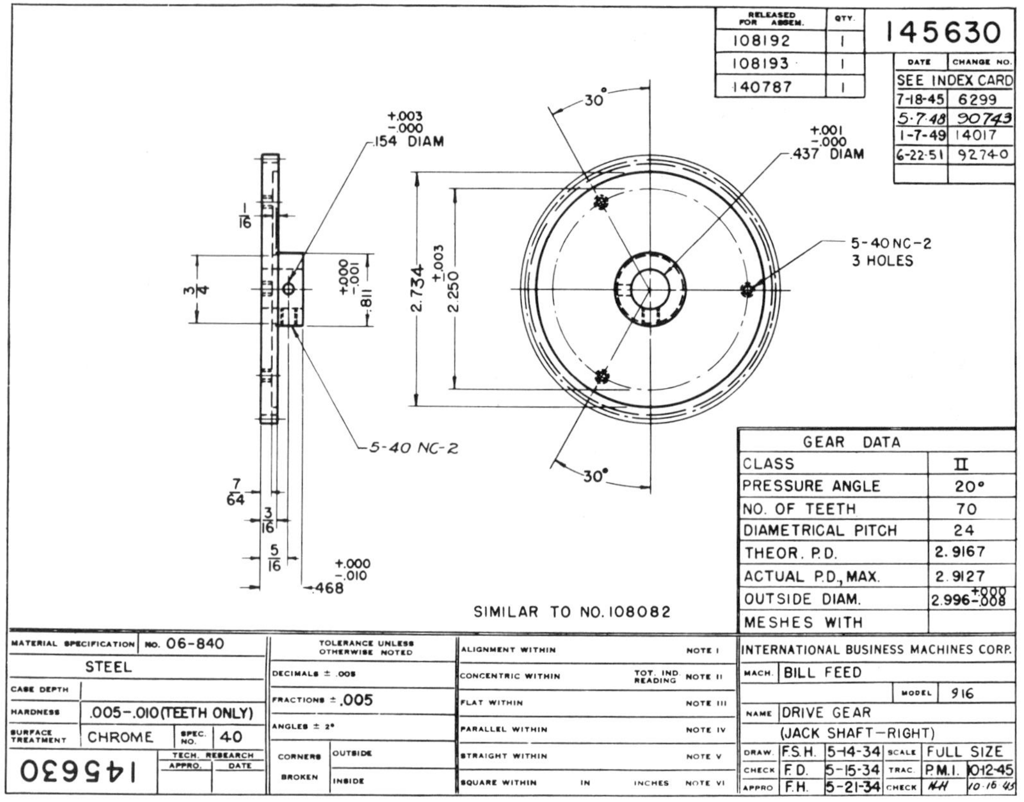

Figure 157

the floating arm is fixed. The pin in the adjustable slide is removed after setting, and the master gear inserted. A sleeve or bushing is then placed on the fixed-base pin to bring it up to the size of the hole in the gear after the gage has been set. A locking pin is used to hold the loating arm while setting the indicator to zero, and to prevent damage to it when the gage is not in use.

Figure 157 shows a print of a gear and the gear data needed to set up the gage.

The steps involved in setting up the gage are as follows:

1. Select a master gear of the same pressure angle and diametral pitch as the gear to be checked.

2. Determine the center-distance at which the gear is to run with the master by adding one-half of the actual pitch diameter shown on the print to the pitch radius as marked on the master gear; from the sum obtained, subtract one-half of the sum of the base pin diameters. This will be the distance between the two base pins.

3. Set up this distance with precision gage blocks and, with the locking pin in the floating arm, set the indicator to zero. Place the blocks between the pins; remove the locking pin and move the slide until the indicator returns to zero and lock the slide.

4. Remove the removable base pin; place the gear and master on the gage. The gears will then be in mesh; by turning the master, any deviation from zero will show on the dial indicator.

Figure 158 shows another type of adjustable running gage for checking gears. This gage is set up and operates in much the same manner as the one shown in Figure 157. Although especially adapted to checking worm wheels, it may, by use of special attachments, be used for checking helical gears or bevel gears. With the vertical head removed, it can be used as a tester for spur gears.

When used to check worm wheels, as shown, the worm wheel rotates freely on a fixed ver-

Figure 158. Gear Tester

tical arbor. The worm is mounted horizontally, either on centers or on an arbor that turns in bushings contained in the saddle which is vertically adjustable so that the height of the worm wheel may be adjusted to suit the position of the worm.

SUMMARY OF DIAL INDICATORS

In the past few years, attempts have been made to standardize the indicator and to obtain a greater degree of accuracy. As the accuracy usually decreases with increase in range, the tendency has been to limit the range to a value that will permit obtaining the accuracy desired. The indicator is intended primarily as a comparator-type gage, and as such has a high degree of accuracy. For example, a new .001-inch indicator, with a 1-1/4 to 1-3/8 inch dial of 100 divisions of .001 inch each, should be within .0004 inch for a deflection of .010 inch. For deflections of several revolutions, up to the limit of the plunger's travel, the error should not exceed ± .001 inch. A new .0001-inch indicator, with a 1-3/4 to 2-inch dial of 100 divisions of .0001 inch each, should be within .00005 inch for a deflection of .001 inch. For deflections of several revolutions, up to the limit of travel, the error should not exceed ± .0004 inch.

A dial indicator also may be expected to repeat a measurement within one-fifth of a unit division. In other words, if a reading made on a scale graduated in thousandths of an inch is repeated, it should be within .0002 inch of the former position.

Although the indicator is very accurate, it is not so sensitive as may be required for some types of work. It takes two to three ounces to start the plunger of an indicator moving, and the force increases about 3/4 of an ounce per .100-inch movement of the plunger. This point is important because the force required to operate the plunger may be sufficient to cause a deflection of its support or distortion of a frail part. As this will make the indicator show less than the actual variation, care should

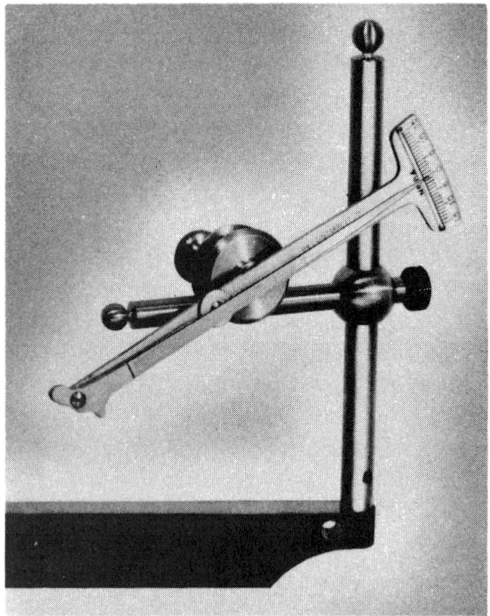

Figure 159. Lathe Indicator

Figure 160. Application of Test Indicator

be taken to support properly both the part and the indicator. In some cases it may be desirable to use a more sensitive type of indicator--the test indicator, the measuring machine or comparator with optical, electrical, pneumatic, electronic, or elastic means of magnification.

LATHE AND TEST INDICATORS

The lathe indicator is similar to the dial indicator, except that it employs the principle of the lever rather than the rack and pinion. Its principle is illustrated by the instrument shown in Figure 159: it can be attached to the spindle of any surface gage, and used to show a variation in thousandths; it may be clamped to a flat or round support, up to 3/8 inch flat or round. A holder, as shown, is designed to go in the tool post of a lathe, adapting it to show the accuracy of all types of lathe work such as turning, chucking, or locating and centering work on a face plate.

The head of the needle has three working points equally distant from its fulcrum, so that the needle will vibrate, reading in thousandths, when work is in contact with either point--in front, above, or below it. When in front, the spring operating the needle must be reversed to throw the point of the needle up instead of down, as when used above or below the work. This may be done by a slight turn of the knurled disc to which the vibrating spring is attached. In setting the indicator, bring the contact point against the work, so that the needle will point to 0; then the variation plus or minus from the setting will show.

Test indicators have their contact points constrained to move either in the arc of a circle or in a straight line. Most commercial test indicators are the former type.

The more sensitive indicators are sometimes referred to as test indicators because they are often used with height gages, as well as on machine tools. A test indicator operates on the lever principle and may have either a dial sector, known as the fan type, as in Figure 159, or a graduated dial.

Figure 160 shows how a test indicator may

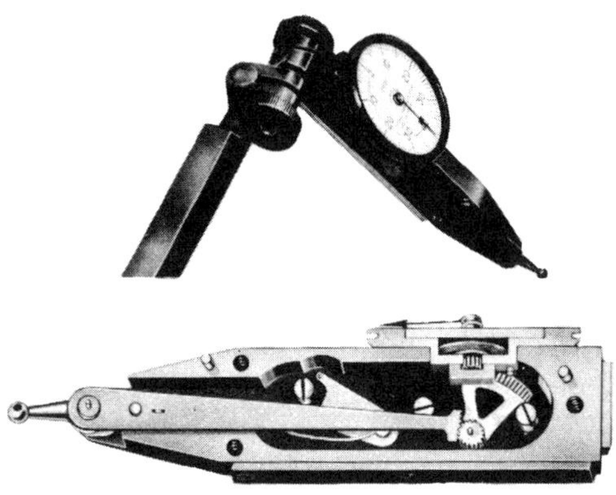

Figure 161. Dial Test Indicator

be used on work held in a chuck or between centers in a lathe. The indicator shown has a fan-type scale and a fulcrum lever-type amplifying mechanism. The indicator is mounted so that it may be turned at any angle, and the contact point is so designed that it, too, may be set at practically any angle. Means of clamping the arm to the body of the indicator must be very rigid.

The dial-test indicator is available in two types, one having a screw and lever amplifying mechanism, the other having crown gears to convert an angular deflection into a rotating movement of a dial pointer. Figure 161 shows a view of the indicator with the mounting for a tool post; another view shows the internal

Figure 162. Application of Dial Test Indicator

Figure 163. Using Indicator on Height Gage

construction. The principle of operation is self-explanatory. The lever on the side of the indicator case is to reverse the direction of movement of the contact point.

An application of a dial-test indicator is shown in Figure 162. The indicator is being used to set up for work in a vise on a horizontal milling machine.

Where applied in the inspection of parts, test indicators are used with height gages in two ways:

1. As a means of duplicating the measuring pressure at which a series of readings are taken.

2. As a comparator for determining the variation, plus or minus, from the basic dimension. This is the more important application, because an indicator can be read more precisely than the scale on a height gage.

A certain amount of spring tension is built into test indicators--from one-half to one ounce, depending on the make--to start the pointer. They are commonly used to duplicate a given measuring pressure and to eliminate feeling the height-gage measurement.

The procedure for inspecting a part with a height gage is shown by the setup in Figure 163. (This application is to establish the center location of a number of holes in the casting in relation to the side resting on the parallel.) The inspector takes a reading from the top of the parallel and adjusts the sliding head of the

Figure 164. Deming Indicator

height gage so that the test indicator shows 5. He then reads the top of the plug in the hole, setting the indicator by means of the sliding head again to 5. So he knows the measuring pressure used in both measurements was exactly the same. The distance from the center of the hole to the edge of the casting is: equal to the difference in the two readings of the height gage, minus one-half of the plug diameter.

Where a test indicator is used to measure a variation or compare a dimension with a standard, it must be more sensitive and more accurate. For this kind of work, an indicator of the type shown in Figure 164 is commonly used. This indicator is made with graduations in .0005 inch and is called a Deming. It has a compound-lever system so that a deflection at the point is actually magnified twice; it has a number of replaceable ball points --straight, bent, long, as well as a thin disc--for reaching into small holes, slots and recesses.

Because these test indicators are provided with a selection of points, some of which are longer than others, it is necessary to compensate for the accompanying change in ratio in the lever arms. For example, if a double-length pointer is used, the reading on the scale will be only half its normal value; so if the indicator is being used to measure, the reading must be multiplied by two. The angular position of the contact arm must be considered as well as the lever ratio, unless used as null point indicator. It is best to check the range used with two gage-block combinations.

Figure 165 illustrates the use of one of these indicators with a height gage. Here the tip of the indicator is resting on a long plug gage which is being used to check the size and alignment of the line-reamed holes. The indicator will be used on each end to check the parallelism of the plug gage, which is comparable to the shaft which will go through these holes.

CARE OF DIAL INDICATORS

A dial indicator is a sensitive instrument. Care should be taken that the contact point is not subjected to impact or excessive pressure. If a dial indicator does not function properly, it should be returned to the manufacturer or the gage room where it can be repaired by specially-trained men. Do not attempt to repair it yourself.

Dial indicators should be periodically inspected for accuracy by setting the indicator to zero and checking against gage blocks of various dimensions.

When using portable dial comparators, select a location where the gage will be least subject to vibration and sudden temperature changes.

CARE OF TEST INDICATORS

The sensitivity and accuracy of a test indicator is largely a matter of the condition of the lever points. Any dirt, grit, dust, rust, or gummy lubricant in the pivots will make the indicator sluggish and may cause it to bind. The lever system of the indicator should be kept clean.

Any wear in the pivots will seriously affect the accuracy of the test indicator. Some models are provided with a means for compensating for wear in the pivots and pivot seats, while on others it is necessary to replace worn parts with new ones. The indicator should be checked occasionally for any lost motion in the lever system.

Another source of trouble may be the touching of the scale or the case by the pointer. Should the pointer become bent as a result of careless handling, and in its travel rub against either the scale or the case, its action is seriously affected because of the high ratio in the lever system. When a test indicator develops a bind, the travel of the pointer should be checked to see that at no point is its free travel obstructed.

Figure 165: Deming Used with Height Gage

QUESTIONS AND PROBLEMS

1. How are dial indicators classified?
2. What is meant by a balanced scale? A continuous scale?
3. If a standard indicator has a scale of 0-25-0, what, according to the American Gage Design Committee, should the spindle travel be?
4. What is the purpose of the small inner dial on some indicators?
5. How may the limits of a dimension be indicated on a dial indicator?
6. What special features does the shock-proof indicator have?
7. What is the purpose of different contact points?
8. What is the function of the maximum hand on a dial indicator?
9. What is the function of the bell-crank attachment on a dial indicator?
10. What is a perpendicular spindle indicator?
11. How may the accuracy of a dial indicator be checked?
12. What are some of the common faults of indicators?

13. Explain how the dial comparator shown in Figure 133 is set up to check a dimension of 1.750 inches.
14. Name some advantages dial-indicator gages have over conventional fixed-type gages.
15. For what kind of work is the caliper gage shown in Figure 138 particularly adapted?
16. Explain how a gear blank can be checked for concentricity, and side wobble, with an indicator.
17. What are the advantages of a dial-plug gage over a standard plug gage?
18. Name some of the advantages that recommend the use of special gages; some of. the disadvantages that discourages their use.
19. What are some of the fundamental features of a special gage?
20. What does the running gage in Figure 156 check on a gear?
21. Explain why the matter of support is important in using a dial indicator.
22. What important difference is there between a dial and a lathe or test indicator?
23. Explain how the test indicator is used with a height gage as a device for duplicating measuring pressure.
24. Why must an indicator be more accurate when it is used as a comparator for determining a variation than when it is used as a device for duplicating measuring pressure?
25. In using the indicator in Figure 164, what must you remember when changing the point?
26. What should be checked on a test indicator to determine its condition?
27. (a) What is the dimensional difference indicated by the pointer on indicator number 1 which was set to zero before the measurement was taken?

(b) What is the tolerance indicated by the tolerance hands on indicator number 2? How much is the part over or under?

(c) What is the dimensional difference indicated by the pointer on indicator number 3 which was set to zero before the measurement was taken?

(d) What is the dimension on dial 4?

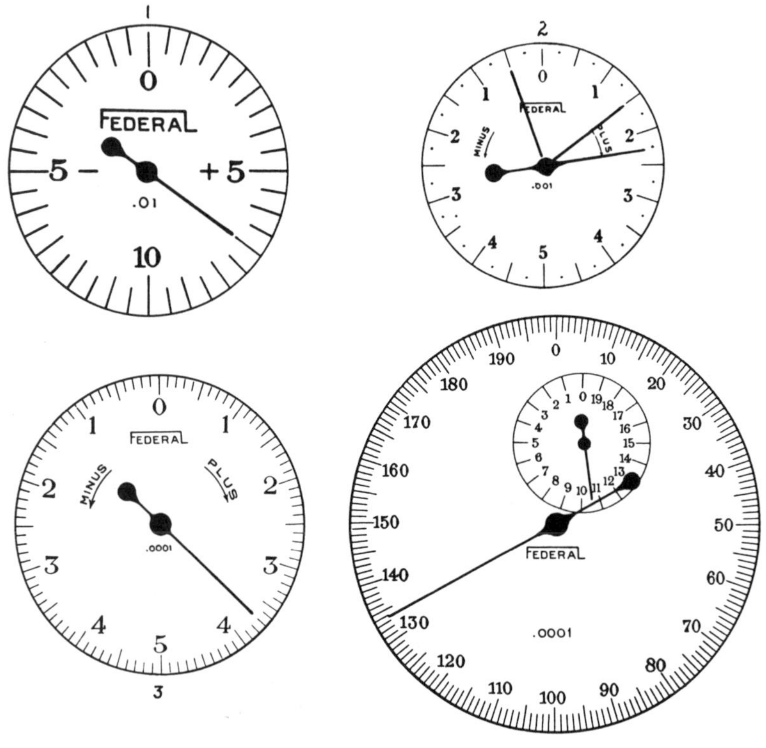

Chapter VI

MICROMETERS AND VERNIERS

THE micrometer caliper is by far the most common type of precision instrument in a machine shop. Second only to the steel rule in the extent of its use in the shop, a good micrometer is practically the starting point of the toolmaker's kit.

The micrometer's application is due to outstanding features:

1. Small, portable, sufficiently rugged to withstand normal use, which helps to make it a versatile instrument.
2. A fairly good range of measurement--usually one inch.
3. Retains its accuracy well, and has adjustments to compensate for wear.
4. Easy to handle, easy to read.
5. Relatively inexpensive.

The prototype of the present-day micrometer was an instrument which bears striking resemblance to the latest models. This instrument, known as the "Système Palmer," was the first to employ the principle of the calibrated screw thread for the purpose of measurement.

Patented in 1848, it first attained recognition at the Paris Exposition in 1867. Two pioneer tool manufacturers, J. R. Brown and Lucian Sharpe, brought the idea back with them and so offered American industries a gage of the type shown in Figure 166, for measuring the thickness of sheet metal.

Ten years later a larger micrometer, adapted to general shop use, was introduced by Brown and Sharpe. This micrometer is shown in Figure 167. In 1885 the improved micrometer caliper, having essentially the same features of construction as the present-day models, was introduced. The biggest improvement made at this time was incorporating the screw into the thimble and thus protecting the thread from abrasive wear or damage from a blow. The improved micrometer of 1885 is shown in Figure 168.

PRINCIPLE OF THE MICROMETER

The micrometer makes use of the relation of the circular movement of a screw to its axial movement. The amount of axial movement of a screw per unit of circular movement (revolution) depends on the thread, and is known as the lead. For example, if a circular nut on a screw has its circumference divided into 25 equal spaces, and if the nut advances axially 1/40 inch for each revolution, then if it is turned just one division, or 1/25 of a revolution, it will move axially only 1/25 x 1/40 or .001 inch.

In the micrometer the nut is stationary, and the screw moves forward axially a distance proportional to the amount it is turned. The screw on a micrometer has forty threads to the inch, and the thimble circumference is divided into 25 parts; so one division on the thimble represents .001 inch.

Accuracy of the micrometer is, therefore, dependent on the lead accuracy of the screw itself. It is not the practice of the micrometer manufacturers to publish guarantees of accuracy; but, in general, a new micrometer of good quality will be within .0002 inch in the range of spindle travel. The error in lead which causes the error in measurement may or may not be accumulative within this range.

The allowable error in a micrometer depends on the tolerance on the work, increasing as

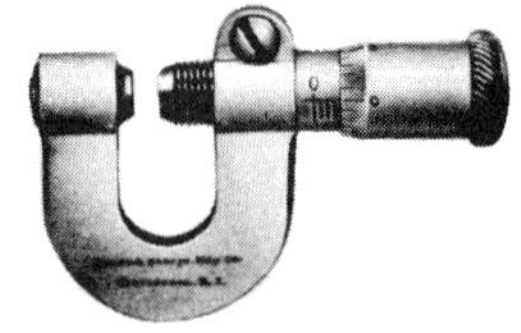

Figure 166. Pocket Sheet-Metal Gage of 1867

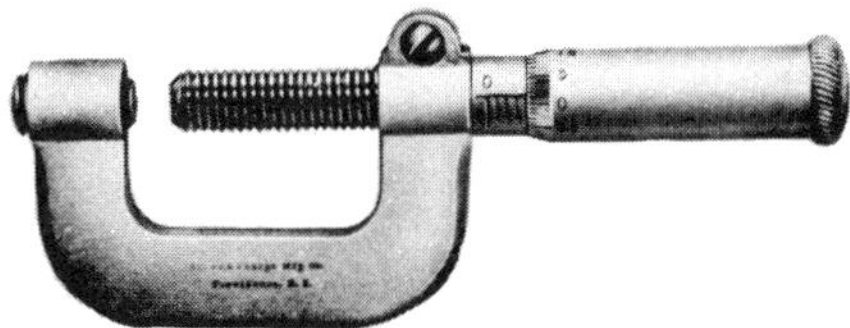

Figure 167. Micrometer Caliper of 1877

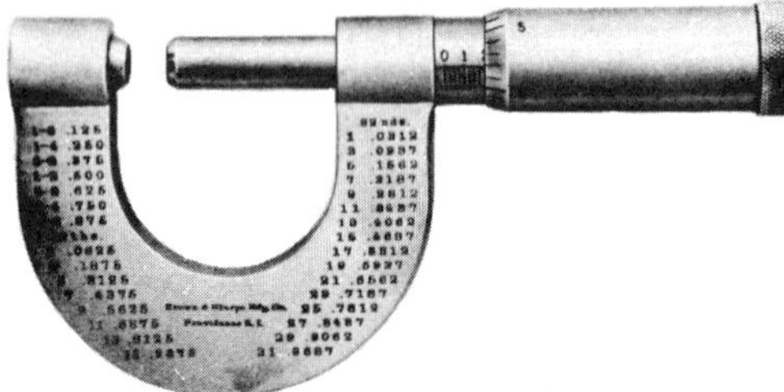

Figure 168. Micrometer Caliper of 1885

tolerances become more liberal.

DETAILS OF MICROMETER CONSTRUCTION

Figure 169 is a cutaway view of a micrometer, showing the mechanism in the barrel. The frame is U-shaped with one end holding the stationary anvil. The stationary anvil is a hardened button either pressed or screwed into the frame. On the older types the anvil was adjustable.

The spindle (actually the unthreaded part of the screw) advances or retracts to open or close the open side of the U-frame. The spindle bearing is a plain bearing and a part of the frame, and from it the hollow barrel extends. On the side of the barrel is the micrometer scale, which is graduated in tenths of an inch; these in turn are divided into sub-divisions of .025 inch. The end of the barrel supports the nut which engages the screw. This nut is slotted, and its outer surface has a taper thread which makes it possible to compensate for wear by adjusting the diameter of the nut within limits.

Attached to the screw is the thimble, a sleeve that fits over the barrel. The front edge of the thimble carries the scale by which the relatively large divisions on the barrel are broken down into 25 parts. This scale indicates parts (25ths) of a revolution, while the scale on the barrel indicates the number of revolutions.

On the particular micrometer caliper shown, the thimble is connected to the screw through a sleeve for the purpose of adjustment which permits the thimble to be slipped in relation to the screw. The under sleeve is clamped to the inner one by the thimble cap. Loosening the cap makes it possible to slip the one in relation to the other.

On the top of the thimble cap there may be a ratchet. This device consists of an overriding clutch held together by a spring; when the spindle is brought up against the work, the clutch will slip as the correct measuring pressure is reached. The purpose of the ratchet is to eliminate any difference in personal touch and thus to reduce the possibility of error which may be due to a

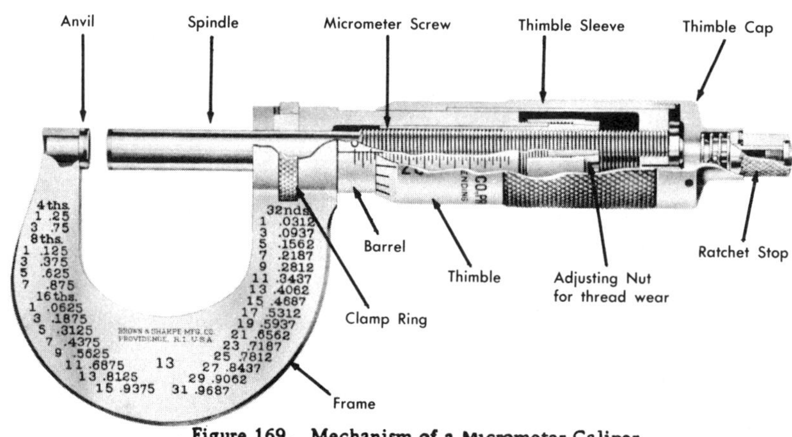

Figure 169. Mechanism of a Micrometer Caliper

difference in measuring pressure. Not all micrometers have a ratchet, and some mechanics prefer micrometer calipers without one, on the personal grounds that they can "feel" the dimension better with their fingers.

Another device, the clamp ring or lock nut, makes it possible to lock the spindle in any position to preserve a setting. This enables the machine operator or inspector to retain a reading when it is necessary to remove the micrometer from the work before reading. To clamp the spindle, the ring is turned until it is just tight, because only a little pressure is required to hold it securely.

The clamp ring is located in the middle of the spindle bearing on those micrometers equipped with it. It operates on the principle of a camming roll which works against a split ring to bind the spindle when the nut is turned. The particular micrometer caliper shown has a table of decimal equivalents stamped in the side of the frame, facilitating conversion from fractions to decimals.

Details of construction vary with the make of micrometer caliper, but in general all micrometers are constructed along similar lines. Later, when various types of micrometers are discussed, some of those with special features will be described. They are mostly design variations that adapt the instrument to the physical limitations of certain work.

HOW TO READ A MICROMETER

Reading a micrometer is only a matter of counting the revolutions of the thimble and adding to this any fraction of a revolution--because, as has been pointed out, the circular movement of the thimble is directly related to the longitudinal movement of the spindle.

Each small division on the barrel represents one revolution of the spindle; so by reading the scale it is possible to count the number of revolutions. Since the standard micrometer has 40 threads to the inch and each revolution is equivalent to 1/40 or .025 inch, each division on the scale represents .025 inch.

From 0 to 1 are four divisions; so the 1 on the scale represents 4 x .025 or .100 inch; 2 is equivalent to .200 inch; 3 to .300 inch, etc. Therefore, in Figure 170 the amount shown on the barrel scale is .200 inch plus one division, or .225 inch plus a part of a division. A part of a division is read on the circular scale on the thimble. The horizontal line on the barrel is the index. Because this scale has 25 divisions, each one is 1/25 x .025 or .001 inch. The part of a revolution indicated on the barrel is read on the thimble scale to be 16, which means .016 inch.

When this part of a division is added to the whole divisions, the total is the distance between the anvils:

.225 inch on the barrel
.016 inch on the thimble
.241 inch-distance between anvils

Check the examples of micrometer readings in Figure 171 to verify understanding of how the micrometer scale is read. It is customary to read the micrometer to the closest thousandth of an inch.

Some micrometers have a third scale for indicating a part of a division on the thimble. It will be remembered that the purpose of the thimble scale itself is to measure a part of a division on the barrel. The vernier scale is for the same purpose, but it is applied to the divisions on the thimble. In other words, a thousandths of an inch is sub-divided by means of the vernier scale so that it is possible to read tenths of a thousandth of an inch. (The theory of the vernier scale will be explained later in this chapter.) As applied to a micrometer caliper this consists of ten divisions on the barrel, which occupy the same space as nine divisions on the thimble. The difference between the width of one of the ten spaces on the barrel and one of the nine spaces on the thimble,

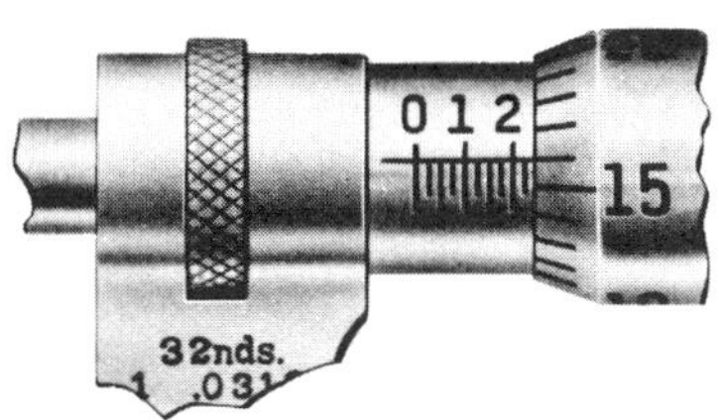

Figure 170. Micrometer Scale

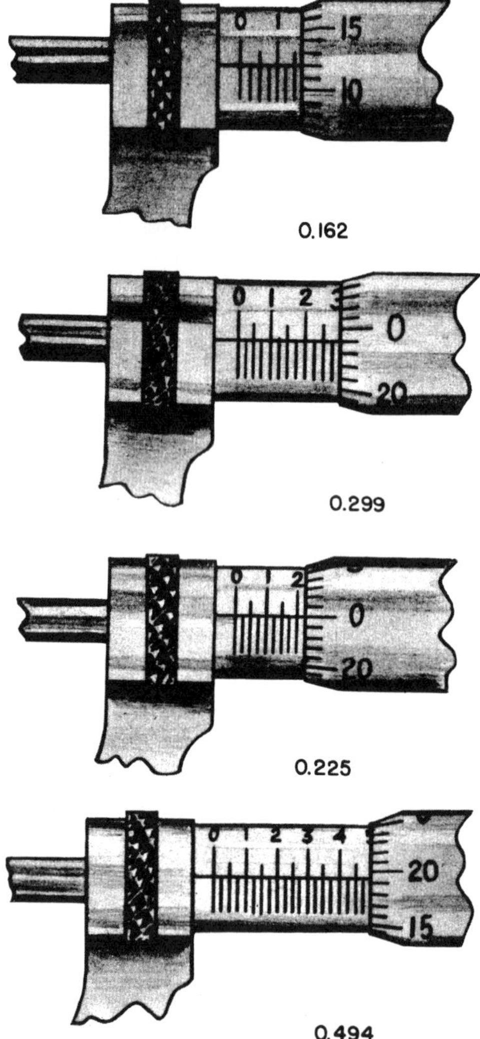

Figure 171. Micrometer Readings

therefore, is one-tenth of a space on the thimble.

The scale itself consists of ten equally-spaced lines, etched on the barrel as shown in Figure 172. The first nine divisions are numbered from 0 to 9, the tenth division being marked with another 0. With a vernier scale, the procedure is practically the same as that outlined before: (1) The reading of the whole divisions on the barrel scale is added to (2) the reading of whole divisions on the thimble scale, which is added to (3) the reading of the vernier scale; the total is the distance between the anvils.

Although before, the reading on the thimble was estimated to the closest thousandth, the

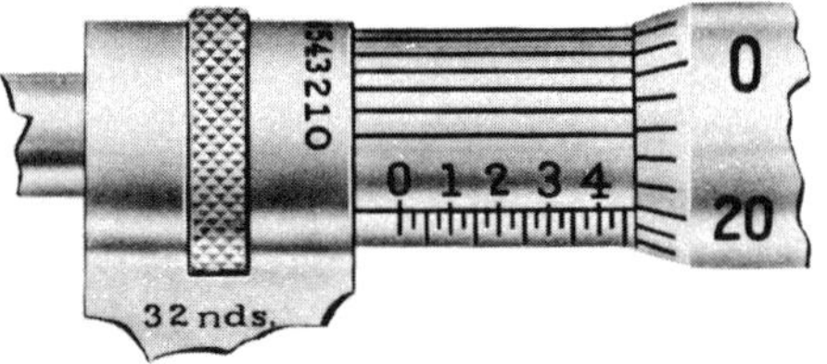

Figure 172. Micrometer Scale with Vernier

thimble reading now includes only the number of whole divisions. Decimal parts of a thimble division are read on the vernier scale. Reading of the vernier is taken by observing which line on the vernier scale coincides with some graduation on the thimble scale. The number of this line represents the number of tenths of thousandths of an inch that must be added to the number of whole thousandths to give a total micrometer reading in ten thousandths of an inch.

If, for example, the two scales were drawn flat, as shown in Figure 173, they would read as follows:

In the first example the barrel reads .250 inch, the thimble reads .019 inch, and the graduation is right on the index so that there is no decimal part of a division to be added. This is verified by the fact that both 0 lines at each end of the vernier scale coincide with lines on the thimble scale, but that none between these 0 lines does.

In the second example the barrel reads .250 inch, the thimble reads not quite .020 inch. The vernier scale coincides with the thimble scale at 7. This means that the thimble reading is .0197. When added to the .250 inch, the total is the actual distance between the anvils, .2697 inch.

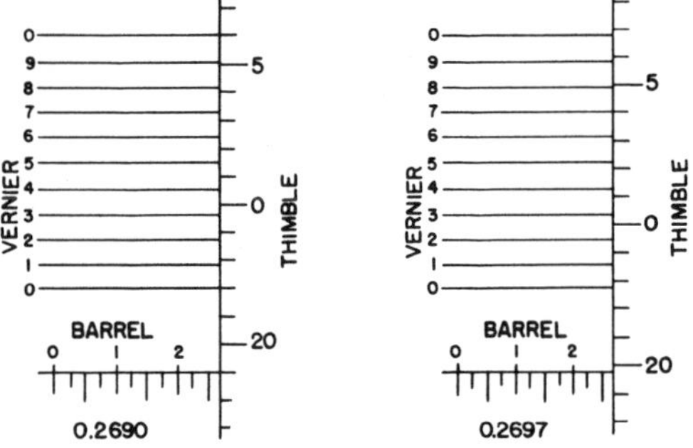

Figure 173. Micrometer Readings with Vernier

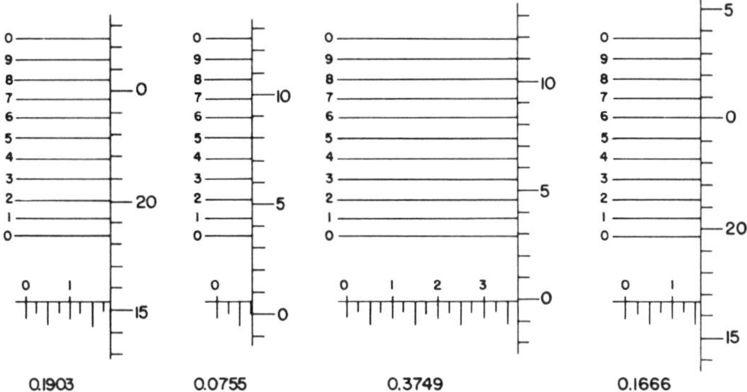

Figure 174. Vernier Micrometer Readings

The examples in Figure 174 further show how the vernier scale is applied to read a micrometer to ten-thousandths of an inch.

TYPES OF MICROMETERS AND APPLICATIONS

Many instruments employ the micrometer-screw principle, and some use only the micrometer head itself. However, this discussion will be limited to commonly-used standard forms of the micrometer: (1) micrometer caliper, (2) inside micrometer caliper, (3) inside micrometer, and (4) micrometer depth gage.

MICROMETER CALIPER

The micrometer caliper shown in Figure 175 is the commonest type. It has a range of 0 to 1 inch and is graduated to read in thousandths of an inch, or in units of the metric system from 0 to 25 millimeters by hundredths of a millimeter.

It may or may not have: (1) a stainless-steel frame to resist corrosion or tarnish, (2) a ratchet for applying a constant measuring pressure, (3) a special vernier scale for reading tenths of thousandths of an inch, (4) a clamp ring or lock nut for clamping the spindle to hold a setting, (5) cemented-carbide tips on the measuring anvils to reduce wear.

There are many variations of this more or less basic form. The frame can be smaller to the extent that the range of the caliper is only 0 to 1/2 inch or 0 to 13 millimeters; or it can be larger so that the range is 23 to 24 inches. The head has a constant range of 0 to 1 inch as the frame gets larger; so a six-inch micrometer caliper measures only from 5 to 6 inches.

The shape of the frame may be varied to adapt it to the physical requirements of some types of work. For example:

1. The frame back of the anvil may be cut away to allow the anvil to get into a narrow slot.

2. The frame may have a deep throat (sheet metal or paper gage) to permit it to reach into center sections of a sheet.

3. The frame may be in the form of a base so that the gage can be used as a bench micrometer.

4. The frame may have a wooden handle and may be of extra-heavy construction for use in a steel mill for gaging hot sheet metal.

The spindle and anvil also may vary in design to accommodate special physical requirements. For example:

1. The chamfered spindle and anvil are convenient when it is necessary that the gage slide on and off the work easily, as when gaging hot metal.

2. The ball anvil is convenient in measuring the thickness of a pipe section of small diameter.

3. The V-shaped anvil is necessary on the

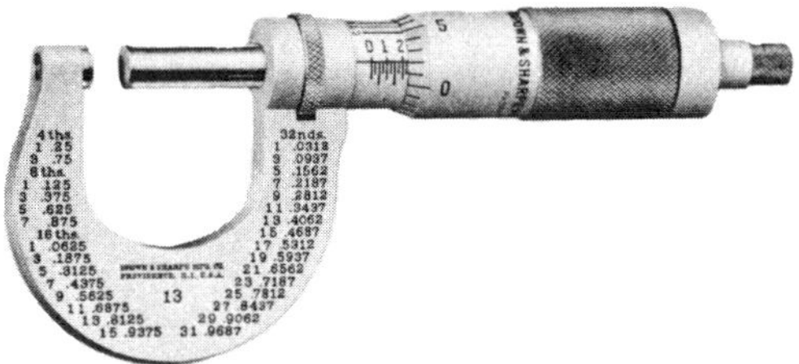

Figure 175. Micrometer Caliper

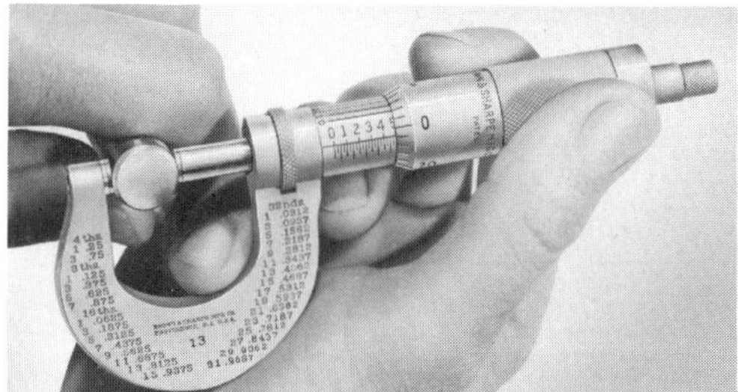

Figure 176. Checking a Small Part

screw thread micrometer caliper to mesh properly with the screw thread. On the screw thread micrometer caliper the spindle tip is cone-shaped. This gage measures the pitch diameter.

4. Interchangeable anvils of various lengths make it possible to reduce the range of a caliper. A micrometer having a range from 5 to 6 inches can be changed to one having a 4- to 5 or a 3- to 4-inch range by inserting a special anvil of the proper length.

Other optional features and attachments will be described later.

Figure 176 shows the proper way to hold a micrometer caliper in checking a small part. The left hand holds the part. The right hand holds the instrument so that the thimble rests between the thumb and the forefinger. The third or fourth finger is then in position to hold the frame against the palm of the hand. This supports the frame, making it easy to guide work over the anvil.

The thumb and forefinger are in position to turn the thimble and bring the spindle down on the piece. Small pieces can be positioned readily and moved around easily for several measurements.

On larger and bulkier work another method is used. The work is positioned, if necessary, to permit access with the micrometer, but the work remains stationary while being checked.

Figure 177 shows the proper method for holding a micrometer when checking a large part (a large part might be defined as one too large to be held in one hand). The frame is held by one hand to position it and to locate it square to the measured surface. The other hand operates the thimble either directly or through the ratchet, as illustrated. A large

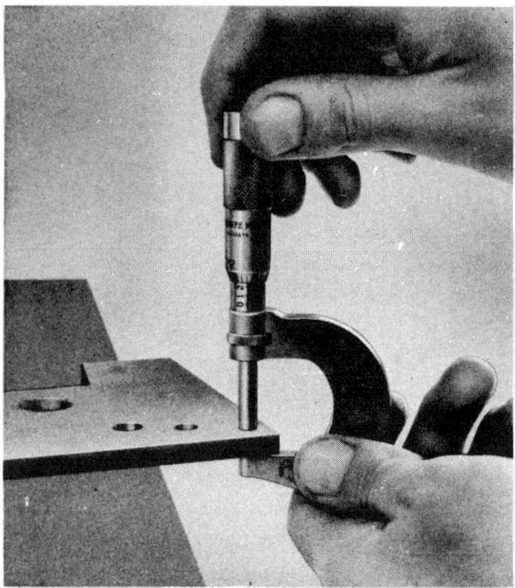

Figure 177. Checking a Large Part

flat part should be checked in several places to determine the amount of variation.

Gaging of a shaft being turned between centers is shown in Figure 178. The frame is held by one hand, while the thimble is operated by the other. In gaging a cylindrical part with a micrometer, it is necessary to "feel" the setting to be sure the spindle is on the diameter, and also to check the diameter in several places to ascertain the amount of "out-of-roundness."

To facilitate accurate measuring of outside diameters on large work, micrometer calipers are built in various sizes up to 168 inches. Figure 179 shows two styles of these calipers.

A cone pulley is being checked in Figure

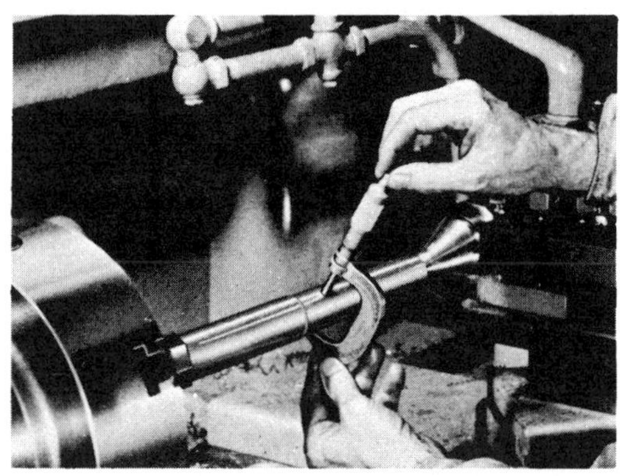

Figure 178. Measuring a Part being Turned

Figure 179A. Checking Diameter of a Pulley

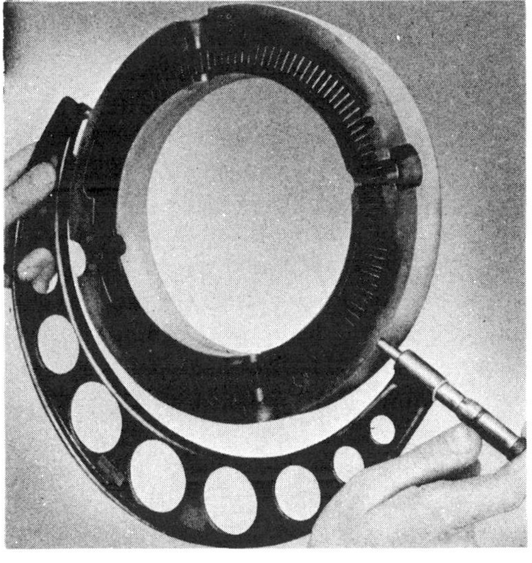

Figure 179B. Using 10-11 Inch Micrometer

179A with a caliper whose range has been reduced by a special anvil which has been screwed into the frame. A set of different-length anvils permits the use of this micrometer over a wide range of sizes; yet the spindle has only 1 inch of travel.

An assembly is being measured in Figure 179B with an 11-inch caliper of different style, This instrument has a fixed anvil; therefore, its range is confined to 10-11 inches.

To facilitate handling, these micrometers

Figure 180. Using Micrometer with Blade Anvils

have been lightened in weight by their I-section construction. Tubular frame micrometers, with or without interchangeable anvils, are available and are light in proportion to size.

Frequently the dimension to be checked is inaccessible to the measuring anvils of ordinary micrometers. When this is the case, either a special anvil or a special micrometer should be used. Figure 180 shows a micrometer with blade anvils for reaching into narrow grooves. The blades of the particular tool shown are .030 inch thick, permitting measurements in slots up to 7/32 inch deep. The spindle is of a special design that does not turn--necessary if the spindle blade is to enter a slot. The principal use of a micrometer with blade anvils is to check the grooves in circular-form tools used on screw machines.

Figure 181 shows another micrometer for the thickness of flanges on circular-form tools, which are inaccessible to the anvils of ordinary micrometers. The anvils are discs 1/2 inch in diameter, 1/16 inch thick at the center, and .020 inch thick at the edge. Flanges up to 1/8 inch deep and with as little as .020 inch clearance can be measured.

An important application of the special anvil micrometer is the screw thread micrometer shown in Figure 182. The anvil, which may or may not be free to rotate, has one truncated V groove to fit a thread. The spindle has a cone-shaped tip which also is truncated so that it does

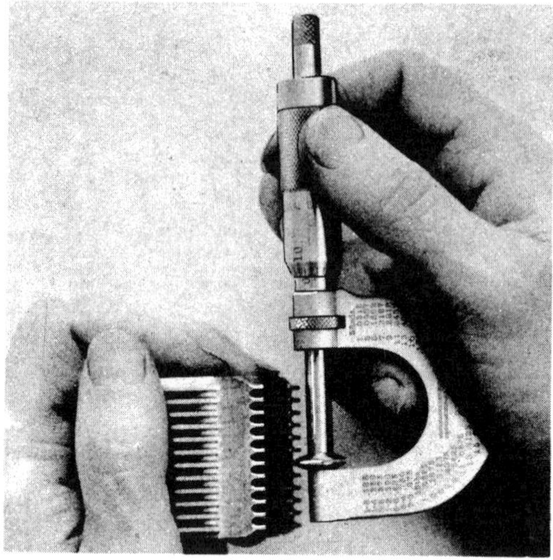

Figure 181. Using Micrometer with Disc Anvils

not contact the bottom of the thread. When set to zero, the pitch line of the spindle and the anvil coincide. When the caliper is opened, the reading represents the distance between the two pitch lines, or the pitch diameter.

Thread micrometers are subject to certain limitations. When set to read zero with the spindle and the anvil together, as is usually done, their readings over threads are always slightly distorted, the amount of distortion depending on the helix angle of the thread being measured. Readings will be more accurate when the thread micrometer is set to a standard thread plug and used for measuring threads having the same pitch diameter as that plug. When so set, however, the micrometer does not read exactly zero when the spindle and anvil are brought together.

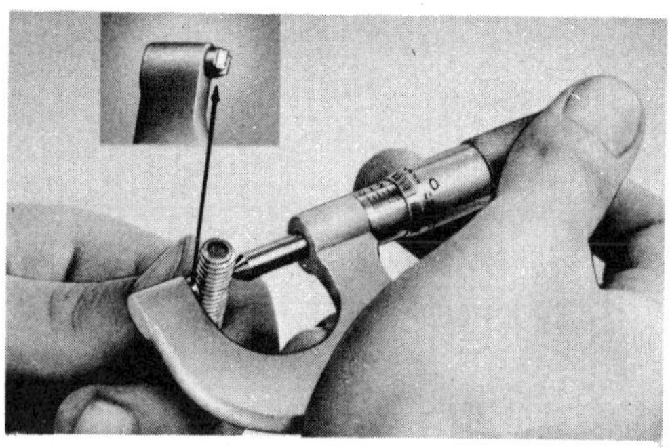

Figure 182. Screw-Thread Micrometer

Figure 183. Bench-Type Micrometer with Indicator

Because size of the thread increases as the pitch increases, or as the number of threads per inch decreases, one screw thread micrometer will not cover the entire range of screw threads. Each thread micrometer has a range for which it may be used. The limits of these ranges have been standardized by tool manufacturers, for example:

Thread Micrometer	Range in Number of Threads per Inch
A	8 to 13
B	14 to 20
C	22 to 30
D	32 to 40
E	4-1/2 to 7

Thread micrometers are graduated in either the English or metric systems and are furnished commercially for the V, American National, and Whitworth thread forms.

SPECIAL MICROMETERS

A special micrometer is one whose design adapts it to some particular type of work. Although a regular micrometer may be made to fit some application by using a special anvil or frame, a special micrometer goes a step further to satisfy specific job requirements.

The bench micrometer in Figure 183 is designed for measurements requiring greater accuracy than is obtainable with an ordinary micrometer. It differs from the regular micrometer in that it has a finer screw, having 50 threads per inch instead of 40, and a much larger sleeve (200 graduations instead of 25).

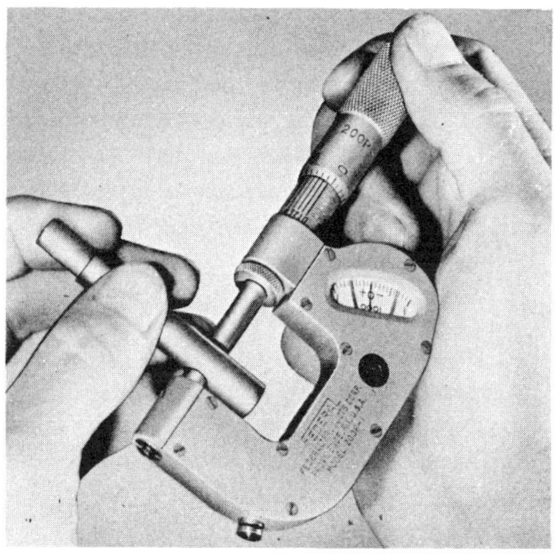

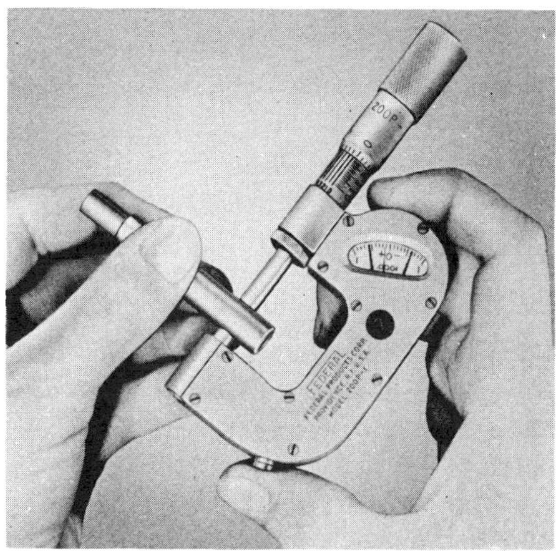

Figure 184. Indicating Micrometer

Because each revolution of the thimble is equivalent to one-fiftieth of an inch, and the circumference of the thimble is divided into two hundred parts, each division on the scale is 1/50 x 1/200 or .0001 inch. It measures 0 to 1/2 inches in .0001 inch directly from the micrometer thimble.

For multiple inspection of similar parts, it serves as a comparator and may be set without standards and read directly from the dial.

For fast setting and measuring, it is set to the nearest .001 inch, and the reading in .0001 inch taken directly from the dial.

It has adjustable measuring pressure from 8 oz. to 2 lb. This meets the needs of those who require standardized measurements at predetermined pressures.

Measuring faces are carbide-tipped for wear resistance. The knurled clamp ring locks the spindle when desired, and the retractor lever withdraws the anvil for repeat measurements.

This particular instrument is of service to wire-drawers, watchmakers, instrument makers, and others manufacturing products having small components which require very accurate machining.

INDICATING MICROMETER

Figure 184 shows another special micrometer with an indicating dial which reads up to .0001 inch. This instrument combines the precision of the dial indicator for uniformity of contact pressure, with the accuracy of the micrometer screw. It takes the element of "feel" out of the measuring operation, mechanically insuring the same pressure for each measurement, regardless of who does the measuring.

The anvil of this micrometer is movable. A light pressure on the button at the lower end of the frame draws back the anvil for inserting and removing the work. This permits use of the micrometer as a comparator without changing the position of the spindle or causing wear on the anvil and spindle faces. When the relieving button is released, the movable anvil is pressed against the work by a light pressure return spring.

This instrument may be used in the regular manner to measure size; or it may be used as a comparator, or to measure roundness.

Used as a micrometer, the spindle is brought into contact with the work. The thimble is adjusted to the nearest .001 inch and the dial then read to .0001 inch.

Used as a comparator, the spindle is clamped in position and dimensions gaged for accuracy; or the gage may be set to the desired size with gage blocks. The tolerance hands are then set to the required limits, and the reading taken from the dial.

INSIDE MICROMETER CALIPERS

The inside micrometer caliper in Figure

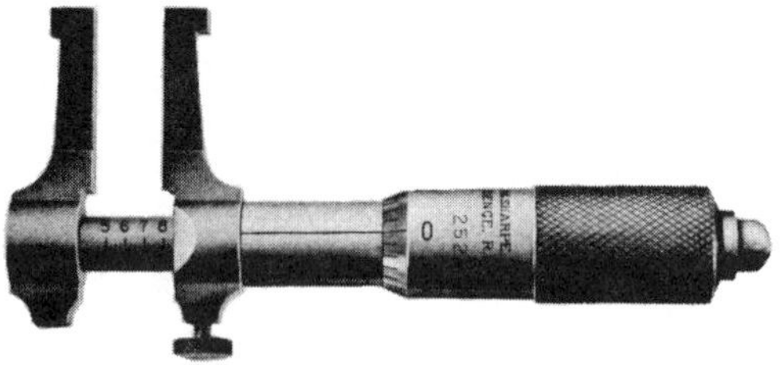

Figure 185. Inside Micrometer Caliper

185 is designed for gaging holes, narrow slots, and other small inside dimensions. The caliper shown has a range of .200 to 1 inch and measures, of course, in thousandths of an inch.

The minimum dimension is determined by the width of the tips of the measuring jaws. These tips are ground with a small radius to enable small holes to be measured at their true diameter.

Not all tools of this type have the barrel scale on the spindle. When the scale is on the barrel, it is in a more convenient position, but it runs from right to left instead of from left to right. This is sometimes confusing, and if the user is accustomed to the scale on a regular micrometer, he can easily forget and misread a reversed scale.

The spindle on an inside micrometer does not turn. It is coupled to the thimble which carries the screw through the snap ring shown on the right end of the tool in the figure.

Inside micrometer calipers are commercially available with ranges of .200 to 1 inch and 1/2 to 1-1/2 inches.

Although not so commonly used as the outside micrometer, the inside micrometer is a useful tool for checking the diameter of holes. On production work a plug gage is commonly used because it is faster and checks more of the hole. Where a plug gage of the proper size is not available, or where it is desired to know the amount of stock to be removed in machining a hole, the inside micrometer caliper has an application.

INSIDE MICROMETERS

The inside micrometer shown in Figure 186 is used to measure inside dimensions larger than those covered by the range of the inside micrometer caliper. The instrument in the figure has a range from 2 to 9-1/2 inches by thousandths.

The minimum dimension that can be checked is determined by the length of unit with its shortest anvil in place and the screw set up to zero. An especially short inside micrometer with a range from 1 to 2 inches may be obtained, and by means of extension rods, a length of a hundred inches or more can be measured.

The range of the micrometer screw itself is very short when compared to its measuring range. The smallest models have a 1/4-inch screw, the average have a 1/2-inch screw, and the largest inside micrometers have only a 1-inch screw.

The usual inside micrometer has a range extending from 2 inches to 8 or 10 inches; others are available in ranges from 2 to 42 inches, and from 32 to 107 inches. The various steps in covering this range are obtained by means of a set of extension rods which accompany the micrometer unit.

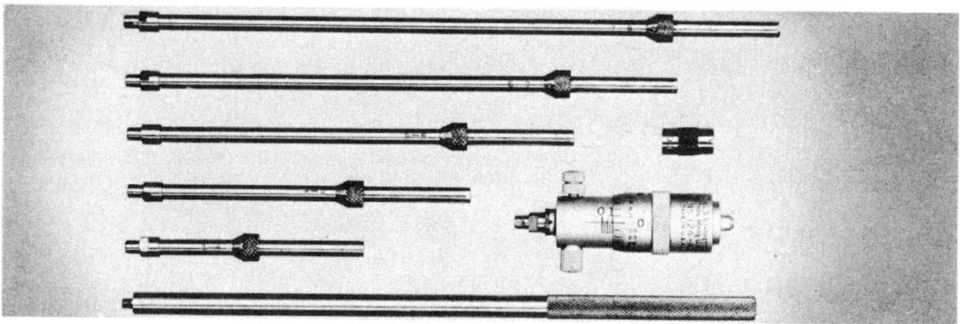

Figure 186. Inside Micrometer

Figure 187. Checking Inside Diameter of a Bored Hole

The micrometer set may also contain a collar for splitting the inch steps between two rods. The collar, which is one-half inch long, extends the rod another half inch so that the range of each step can be made to overlap the next.

The inside micrometer in Figure 187 is checking an inside diameter being bored on a vertical boring mill. The arrows indicate the directions the operator is feeling for the largest dimension horizontally and the smallest dimension vertically.

The customary procedure in using an inside micrometer is to set it across a diameter or between the inside surfaces, remove it, and then read the dimension. For this reason, the thimble on an inside micrometer is much stiffer than on a regular one, and holds the dimension well. It is good practice to verify the reading of an inside micrometer by measuring it with a micrometer caliper.

By means of an attachment, it is possible to convert the inside micrometer into a form of height gage. The attachment consists of a steel base into which the micrometer unit may be mounted, and can be used to measure the distance from a surface plate to the underside of a plug, flange, or other projection on a part.

For measuring the diameter of a deep hole or one in a small restricted place, a handle attachment of the type shown in Figure 188 may be used. This handle clamps to the body of the micrometer and is of great help in holding the micrometer.

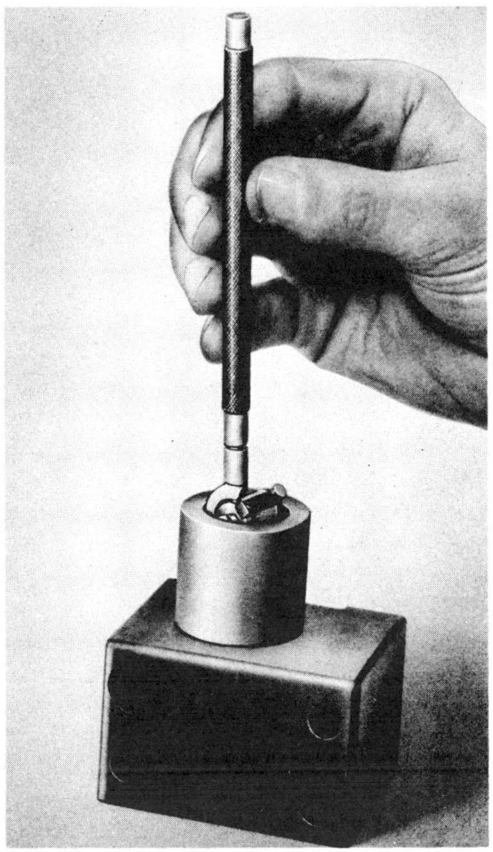

Figure 188. Inside Micrometer with Handle

Another style of inside micrometer (Figure 189) is used with a set of precision-end standards that accompany the style of jig borer on which they are being used. These standards are comparable to a working set of precision gage blocks.

Here they are being used to set up the horizontal movement of the table. Start with a zero setting from a locating bar where the part is located; set the exact dimension from the locating point to the center of the hole to be bored by (1) inserting the nearest amount of end standards in whole inches and (2) adjusting the micrometer to the fractional part of an inch.

As an example, for the position now being bored, assume there are the following rods in the machine: two 12-inch, one 3-inch, and one 2-inch--a total of 29 inches. To this add the setting on the micrometer, which is .682 inch or a total of 29.682 inches. (The dial indicator is set on zero.) It is necessary to move

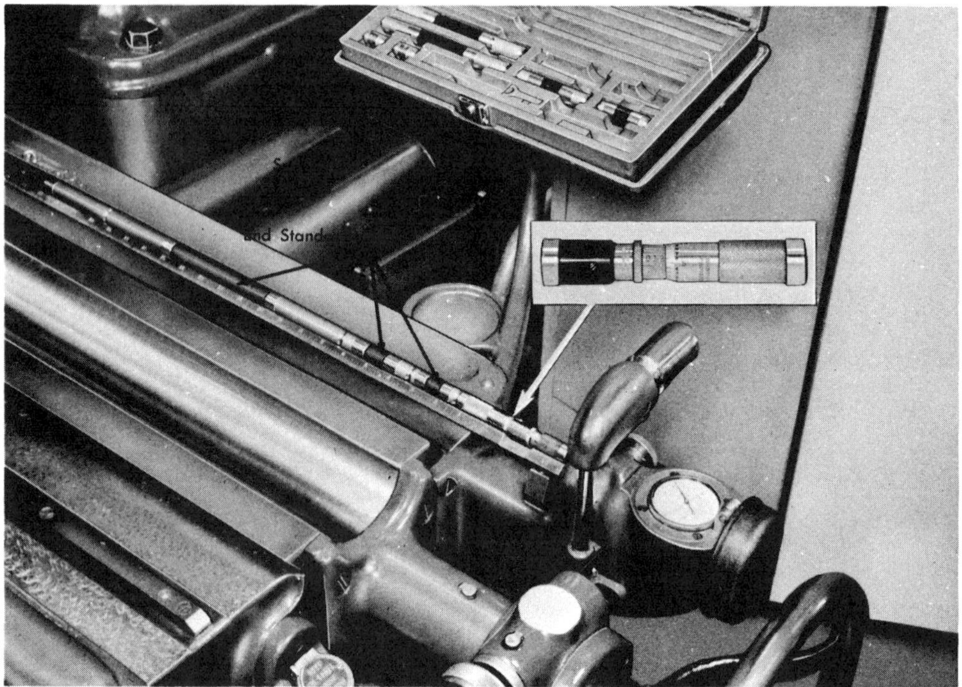

Figure 189. Using End Standards and Inside Micrometer on Jig Borer

the work 1.212 inches to the left in order to bore the next hole; therefore, this amount must be added to 29.682 inches.

Move the table to the left enough to allow for the insertion of a 1-inch standard and to change the reading on the micrometer to .894 inch. Then move the table to the right until the dial indicator reads zero. The total length of the standards plus the micrometer is now 30.894 inches.

Follow the same procedure for the vertical movement. The small white spot in front of the indicator is for marking the setting. Also, both the micrometer and the indicators are graduated in .0001 inch.

MICROMETER DEPTH GAGE

The micrometer depth gage is a simple and important instrument for measuring the depth of blind holes and slots, the distance between stepped surfaces and the distance between surfaces displaced so that they cannot be easily bridged by the measuring anvils of other types of gages.

The micrometer depth gage consists of a flat base attached to the barrel of a micrometer head with the spindle protruding from the base in the manner shown in Figure 190. These gages have a range from 0 to as much as 9 inches, using an extension rod. The micrometer screw itself has a range of either 1/2 or 1 inch.

The gage shown in Figure 190 has a range of 3 inches. This range is obtained by means of extension rods, which may be inserted by unscrewing the cap. These rods are not interchangeable in different depth micrometers. In some shops each micrometer and its extension rods are marked with the same number so that rods belonging to one micrometer will not be used in another. The rod is inserted in the hole in the micrometer screw so that the locating surface on the end of the thimble is tight against the bottom face of the nut on the extension rod. By adjusting this nut, it is possible to compensate for wear in the rod. Note that the scale on the barrel and thimble is the reverse of that on a standard micrometer (see Figure 190).

Figure 191 shows a micrometer depth gage used to measure the depth of a bored hole. The lathe operator on this job is measuring the dis-

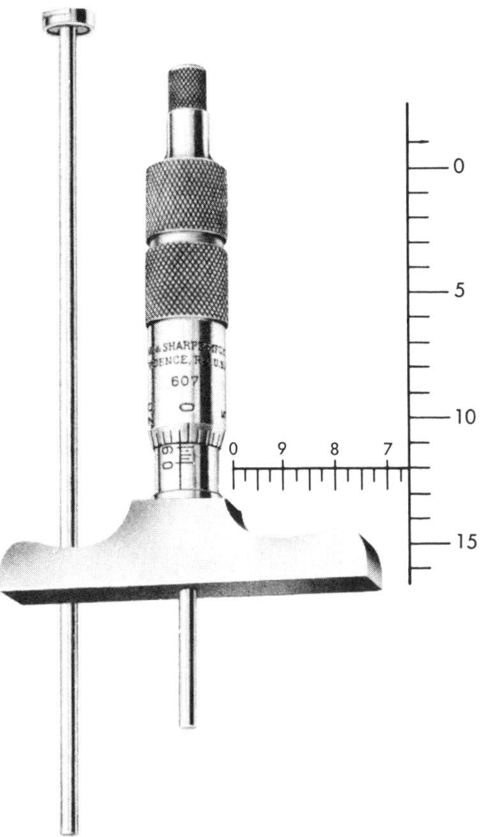

Figure 190. Micrometer Depth Gage

tance between the bottom of an inside shoulder and the front surface of the hole.

A similar application is shown in Figure 192. The depth gage is being used to check the distance from the end of the sleeve to the

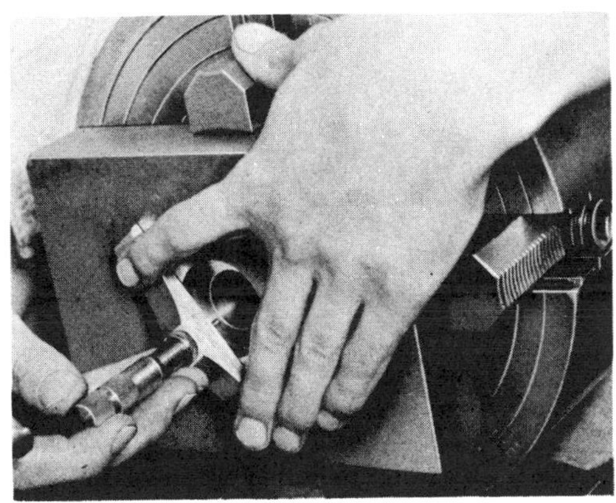

Figure 191. Measuring Depth of a Bored Hole

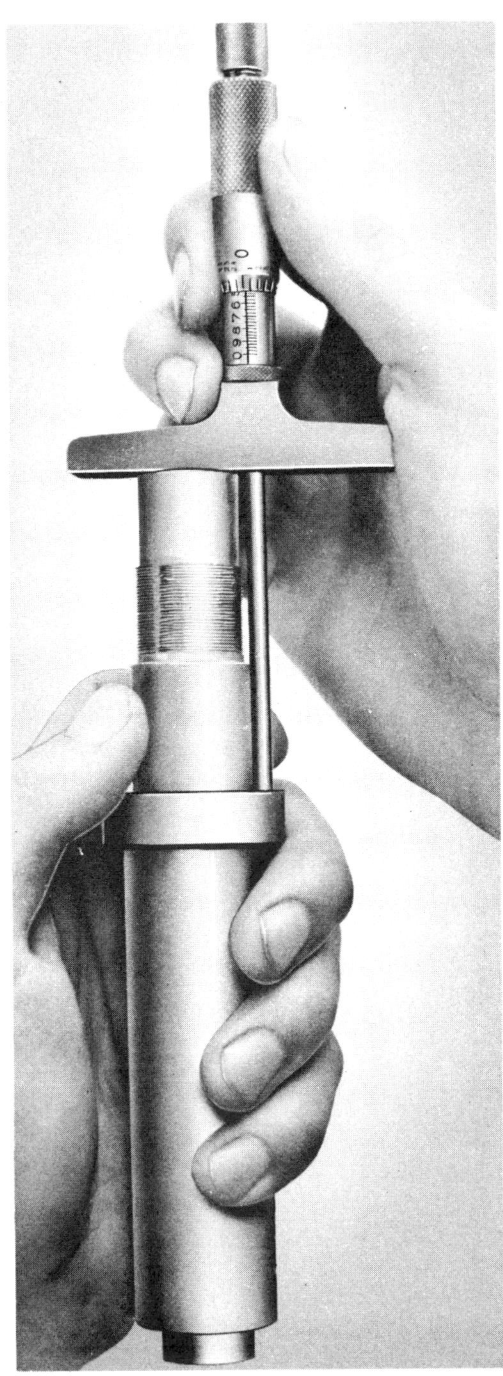

Figure 192. Measuring Location of a Flange

top surface of the flange at the base. A study of the conditions of this application indicates why the depth micrometer is particularly useful around the shop. It is capable of measuring

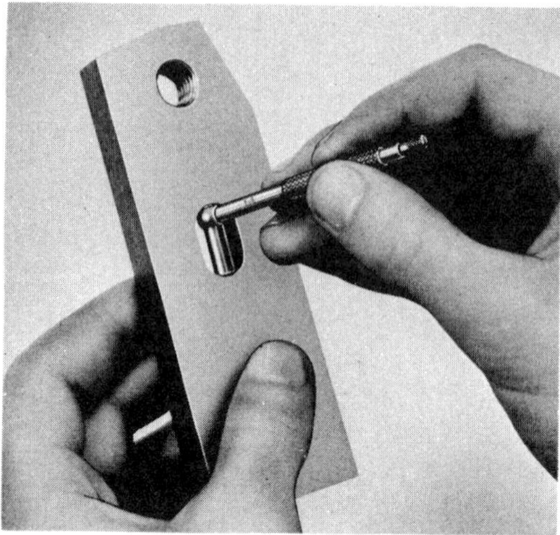

Figure 193. Using Telescope Gage

easily what might otherwise involve a very difficult setup.

TELESCOPE AND SMALL-HOLE GAGES

These gages facilitate the reading of measurements which cannot be obtained with a standard micrometer, or for which there are no plug gages available.

Telescope gages range in size from 1/2 inch to 6 inches, and are particularly adaptable for rough-bored work and odd sizes and shapes of holes. As shown in Figure 193, the gage is fitted into the hole by telescoping the plunger, which is then held against the sides by spring tension, and is locked in position by turning down the knurled screw located in the end of the handle. To obtain the measurement, the gage is then removed from the work and checked with a micrometer, as shown in Figure 194.

Small-hole gages are made in sizes ranging from 1/8 inch to 1/2 inch; they perform the same function as the telescope gages except in much smaller work. The ball point is fitted into the hole and expanded by turning in on the tapered plunger. Representative sizes of these gages are shown in Figure 195, and the measurement taken with these gages, as with the telescope gage, is obtained by using a micrometer.

CARE AND ADJUSTMENT OF MICROMETERS

The micrometer caliper and other instru-

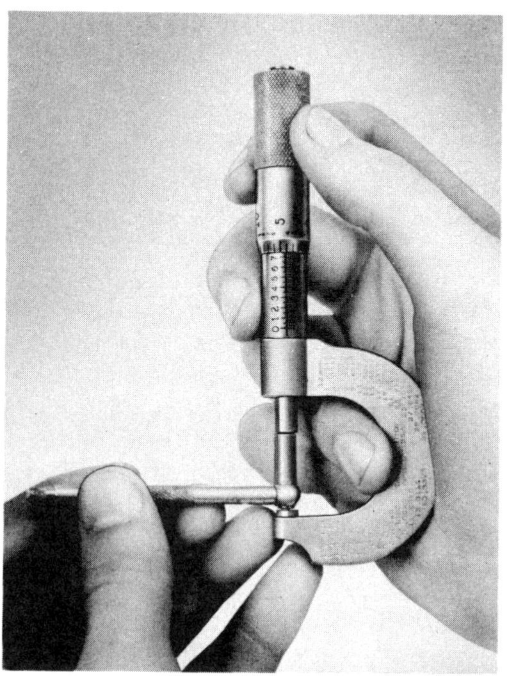

Figure 194. Taking Dimension from Telescope Gage

ments equipped with micrometer heads are vital to the control of manufacturing accuracy. Recognizing their importance, many large organizations have established in their gage department service stations for the personal micrometers of their employees. Employees are encouraged to bring their micrometers as often as once a week for checking against gage blocks; minor repairs and adjustments are made without charge.

This policy enables the inspection department to control the accuracy of gages which

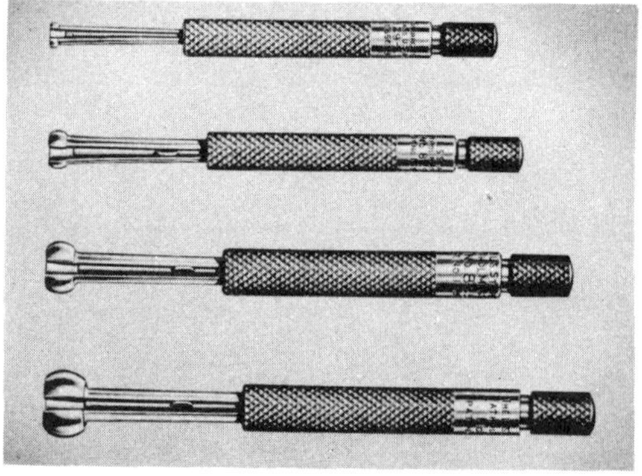

Figure 195. Small-Hole Gages

are the personal property of employees. It benefits the individual by keeping his tools in first-class condition.

Maintenance of a micrometer in good condition and assurance of measuring accuracy are possible only if the user observes the rules of good practice. Some of these rules are:

1. A micrometer should never be stored with its measuring anvils closed, because flat surfaces wrung together for any length of time may corrode. To prevent this, leave a small gap between the anvils.

2. A micrometer should be oiled in only one place, the micrometer screw, and only with light oil. Using heavy oil, which can become gummy, makes the micrometer insensitive and may cover up an error in the screw. If a micrometer is stored for any length of time, or where it is likely to rust, it should be covered with a light film of oil and wrapped in oiled paper.

3. A micrometer caliper may be traversed several hundred-thousandths by holding the frame and rolling the thimble along the hand or arm in several strokes. Holding the thimble and twirling the frame to open or close a "mike" is frowned upon by good mechanics, probably because it causes excessive wear of the screw.

4. A micrometer caliper should never be used as a snap gage. The development of many types of special-purpose gages has limited the old practice of trying to make one gage or a few general gages do all of the work. Using a micrometer as a snap gage is not practical: it does not hold its dimension well enough even with the clamp ring, and there is danger of springing the mechanism in gaging a slightly oversize part.

5. Accuracy of measurements made with a micrometer depends to some extent on the measuring pressure used. The "feel" necessary to produce the correct measuring pressure between the anvils is acquired by practice and comparing measurements with precision gage blocks or some other accurate standard. Whenever a machine operator or inspector has an opportunity, he should compare his "feel" with that of someone else to be sure that he is using the correct pressure. A gage block or accurate plug gage can be used to check any difference.

6. Before a micrometer is used, it is customary to wipe it off and to pull a piece of paper between the anvils in the manner shown in Figure 196.

7. The micrometer screw should run freely with no play at any point in its travel. This adjustment is made with the nut, which causes the threaded sleeve to tighten on the screw.

8. A micrometer screw that binds or has loose points along its travel cannot be corrected with the adjustable nut. It must be returned to the gage maker. This condition is caused by abuse or uneven wear.

9. The micrometer mechanism should be cleaned whenever it becomes gummy, contains an abrasive grit, or is to be adjusted. Any good cleaning agents, such as gasoline or benzine, are satisfactory. For most purposes a non-inflammable solvent is equally satisfactory and is safer to handle.

10. The micrometer screw should be checked with a precision gage block in at least four places other than zero to verify its accuracy. The procedure consists simply of measuring a selected group of blocks ranging between zero and one inch.

11. On all the latest micrometers the thimble can be slipped in relation to the screw. On

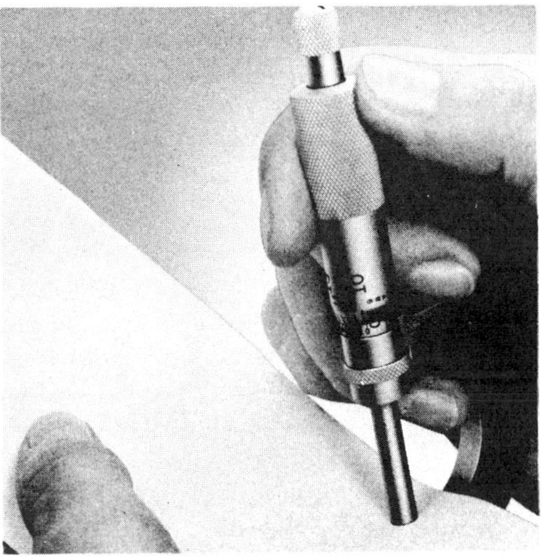

Figure 196. Cleaning Micrometer Anvils

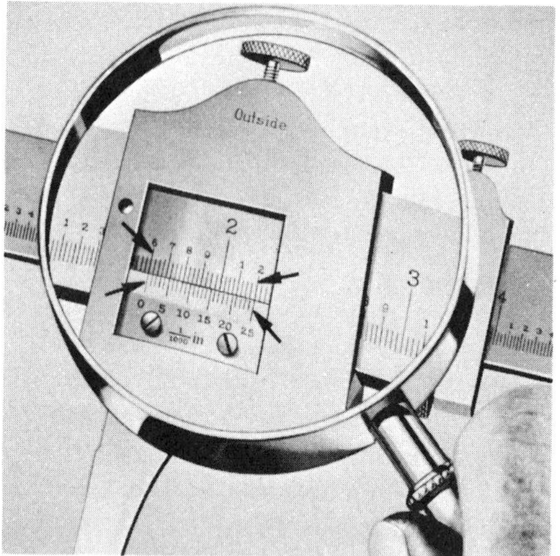

Figure 197. Vernier Scale

the earlier models the anvil was adjustable. In adjusting a micrometer to read correctly, the thimble is not set to zero when the anvils are in contact, but is set at some dimension to distribute the error. For example, if a micrometer screw had an accumulating error of .0003 inch in the length of its travel and it was set correctly at zero, it would be off .0003 inch at one inch. However, if the caliper were set correctly, in the middle of its travel it would be .00015 inch under at zero and .00015 over at one inch, which is a much better condition. It is customary to consider the size of the work measured with the micrometer and to set, if possible, the base somewhat near the mean. From this explanation, it is obvious that because a micrometer does not return exactly to zero when the anvils are in contact does not mean that it is not adjusted properly.

12. When the faces of the spindle and anvil become worn and are no longer flat and parallel to each other, the error should not exceed .0002 inch on a micrometer which is graduated to control measurements to a limit of .001 inch, and should not exceed .00005 inch on a micrometer which is graduated to control measurements to a limit of .0001 inch. Measuring a ball at several points over the surface of the anvils will show up any error in parallelism. Parallelism can be tested by means of two different-size balls mounted in an aluminum holder. If the anvils are in error more than the allowable maximum, the micrometer usually is returned to the manufacturer for repair. However, laps for correcting an error in parallelism are commercially available.

VERNIER INSTRUMENTS

The principle of the vernier has been applied to many kinds of instruments. In the foregoing discussion its use to increase the limit of micrometer measurement was described. Where the vernier is employed on an instrument as the sole agent for magnifying ordinarily imperceptible differences in length, the instrument is usually known as a vernier caliper, or a vernier protractor.

The vernier was invented by Pierre Vernier in 1631. It consists of a short auxiliary scale having usually one more graduation in the same length as the longer main scale. Therefore, the vernier scale contains one more division than the main scale over an equal length, and each division on the vernier scale is proportionally smaller than a corresponding division on the main scale. Consider in Figure 197 the vernier scale which has 25 divisions. Note that the 0 coincides with the small figure 6 and the 25 coincides with the small figure 2. Therefore, the 25 divisions on the vernier scale are equal to 24 divisions on the main scale. As each main-scale division is equal to 1/40" or .025", and each vernier-scale division is 1/25 smaller than a main-scale division, each vernier-scale division is equal to 1/25 of 1/40 (.025") or .001" less than a division on the main scale.

As we travel across the vernier scale from 0, 1/25th of a main scale division or .001 inch is added to the reading obtained just to the left of the 0 on the vernier scale for each division on the vernier scale.

In Figure 198, the 0 on the vernier scale indicates a reading on the main scale of 1.625 inch, but the 0 is past the .625 mark; therefore, we scan the vernier scale, finding that the 16 line coincides with a main scale division. (The value of the main scale division does not matter.) To 1.625 inches must be added 16/25th of a main scale division or:

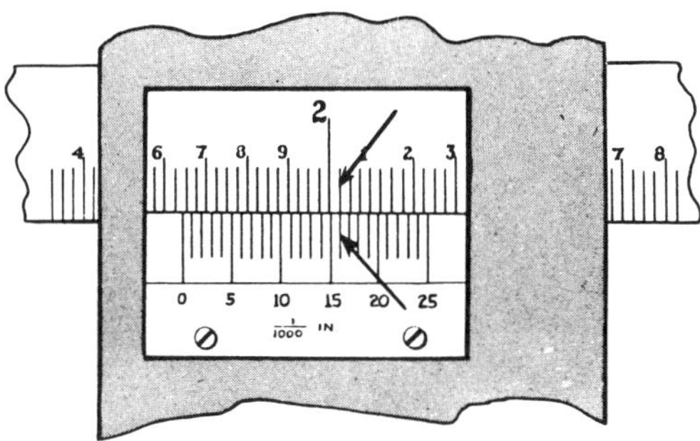

Figure 198

1.625 main scale
.016 vernier scale
1.641 reading

Figure 199A shows the zero on the vernier scale coinciding with a graduation on the main scale. In this case the vernier scale is unnecessary because there is no fractional part of a division to determine. The reading is 2.350 inches.

In Figure 199B, however, the reading shows at 0 on the vernier scale that the dimension is 2.350 inches plus a part of a division. To determine what fractional part of a whole main scale division, or how many thousandths are to be added to the 2.350-inch reading, it is necessary to find the division on the vernier scale that coincides with a division on the main scale--in this case the 18th. This indicates 18/25 of a whole division or .018 inch.

Because each division on the main scale represents .025 inch, this part of a division is equal to .018 inch, and the total dimension is equal to 2.350 plus .018, or 2.368 inches.

Vernier scales are not necessarily 25 units long, but may have any number of units. They may have only ten units, as on the vernier scale of the ten thousandths micrometer. The graduated hand wheels of a machine tool such as a jig borer often employ the vernier scale to indicate tenths or half-tenths of a thousandth of an inch table travel.

Application of the vernier to angular measure will be described later with the vernier protractor.

TYPES OF VERNIER INSTRUMENTS

The vernier scale has been applied to a

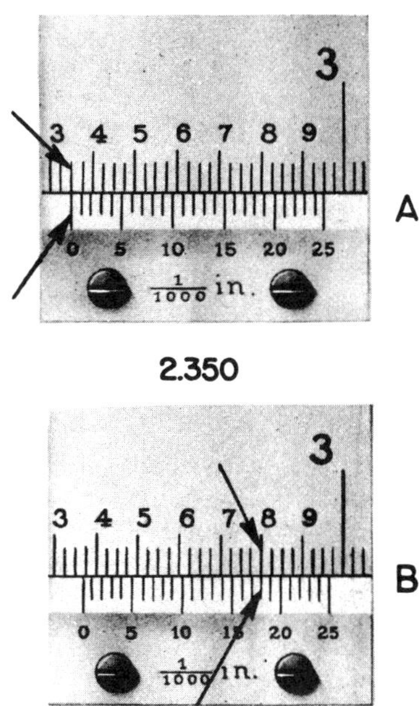

2.368

Figure 199. Reading a Vernier

variety of instruments and tools. In this chapter the discussion will be limited to the following: (1) vernier caliper, (2) vernier height gage, (3) vernier depth gage, and (4) vernier protractor.

The first vernier caliper (Figure 200) was invented in 1851 by J. R. Brown. It embodies all of the principal features of the models on the market today.

The vernier caliper in Figure 201 consists of an L-shaped frame, the end of which is one of the jaws. On the long arm of the L is scribed the main scale, which may be 6, 12, 24, 36, or 48 inches long.

The sliding jaw carries a vernier scale on either side. The scale on the front side is for outside measurements, while the scale on the back is for inside measurements. Tips of the jaws have been formed to make an inside measurement. Care should be taken to use the correct scale, because allowance is provided on the inside scale for the thickness of the jaws.

The sliding jaw assembly consists of two sections joined by a horizontal screw. By clamping the right-hand section at its approximate location, a fine adjustment of the movable jaw may be obtained by turning the adjusting nut. The

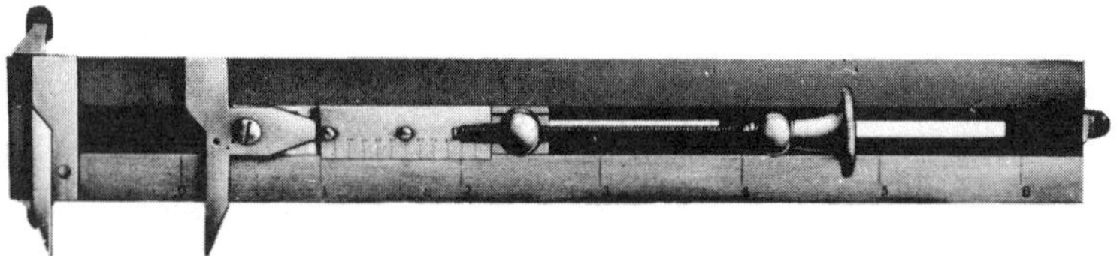

Figure 200. Original Vernier Caliper of 1851

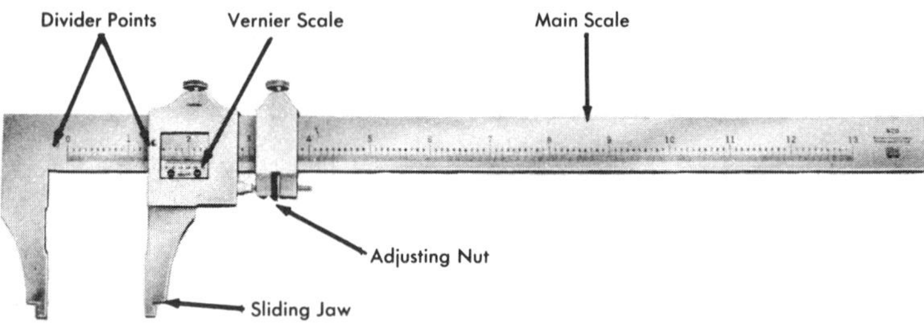

Figure 201. Vernier Caliper

sliding jaw in Figure 201 may be clamped in any position with the locking screw.

Vernier calipers are made in the standard sizes of 6, 12, 24, 36, and 48 inches; and 150, 300, 600, and 900 millimeters. The length of the jaws will range from 1-1/4 inches to 3-1/2 inches, and the minimum inside measurement with the smallest calipers is 1/4 inch and 6 millimeters.

The jaws of all vernier calipers, except the larger sizes, have two center points which are particularly useful in setting dividers to close dimensions.

Applications of the vernier caliper are shown in Figure 202A and 202B. In Figure 202A, a machinist is checking the outside diameter of a part being turned on the boring mill. One hand holds the stationary jaw to locate it, while the other operates the adjusting nut and "feels" the dimension. The same procedure is used in checking the inside dimension shown in Figure 202B.

The vernier caliper has a wide range of measurement, and the shape of the measuring anvils and their position with respect to the scale adapts this instrument to certain jobs where a micrometer, for example, could not satisfactorily be applied.

However, it does not have the accuracy of a micrometer. In any one inch of its length a vernier caliper should be accurate within .001 inch. In any 12 inches it should be accurate

Figure 202. Checking Diameter with a Vernier Caliper

within .002 inch and increase about .001 inch for every 12 inches thereafter.

CARE AND ADJUSTMENT OF VERNIER CALIPERS

Accuracy of the vernier caliper depends on the condition of fit of the sliding jaw, and the wear and distortion in the measuring surfaces.

The fit of the sliding jaw should be such that it can be moved easily and still not have any play. It may be adjusted by removing the gib in the sliding jaw assembly and bending it. The gib holds the adjustable jaw against the inside surface of the blade with just the right pressure to give it proper friction. Most industries have a department to handle such repairs.

The wear that takes place between the jaws of a vernier caliper is mostly at the tips where most measurements are made. A certain amount of this wear may be taken up by adjusting the vernier scale itself. This scale is mounted with screws in elongated holes, which permit it to be adjusted slightly to compensate for wear or distortion. When the error exceeds approximately .0002 inch, either in parallelism or flatness, the instrument should be returned to its maker for reconditioning. This figure will vary somewhat, depending on the accuracy required of its measurements.

Accuracy of measurements made with a vernier caliper depends largely on the user's ability to feel the measurement. Because the jaws are long, and because there is the possibility of some play in the adjustable jaw--especially if an excessive measuring pressure is used--it is necessary to develop an ability to use the vernier caliper properly. As with a micrometer, this feel may be acquired by measuring such known standards as gage blocks and plug gages.

GEAR TOOTH VERNIER CALIPER

The gear tooth vernier in Figure 203 is a special form of caliper for measuring the thickness of a gear tooth at the pitch line. It also may be used to measure the thickness of an acme thread at the pitch line.

This type of gage has an advantage over running gages because it measures a specific

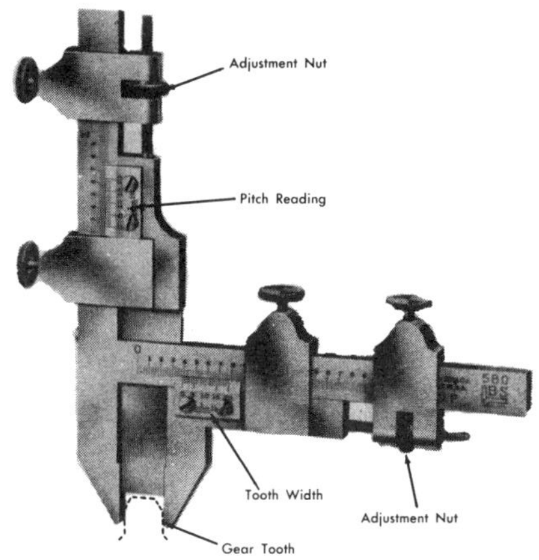

Figure 203. Gear-Tooth Vernier Caliper

dimension on the gear, while a running gage checks a composite of many dimensions. It permits an analysis of error, which is not so readily made with a running-type gage.

The projection comparator, to be discussed later, is for some inspection purposes, a good production method of checking gear teeth. The gear-tooth vernier caliper, on the other hand, has wide range; it is adaptable to the inspection of special shapes and sizes of gear teeth where the production is small, or where a portable gage is required.

It is made commercially in two sizes, and in both English and metric measure: 20 diametral to 2-diametral pitch, and 10 diametral to 1-diametral pitch, in thousands of an inch; 1-1/4 mm to 12-mm module, and 2-1/2 mm to 25-mm module in fiftieths of a millimeter.

To use the gage set the tongue for the proper addendum distance and adjust the horizontal screw to the proper width, as shown in Figure 203. Allowance may be made for variation or error in the blank diameter when setting for the distance from the top of the pitch line of the tooth.

The addendum of a gear tooth is tne distance from the pitch circle to the outside diameter of a gear; it has a definite value for each diametral pitch and number of teeth. If the calipers are adjusted to the true addendum dimension, the measurement will be taken on the tooth sides on the Line "AA," as in Figure

204B. However, the dimension actually required is at the tooth sides where the pitch circle intersects the tooth outline, as in Figure 204A, line BB. The values of these differences may be calculated from formulas, but for convenience they have been arranged in tabular form for each gear. A determined value has been added to the addendum figure, so that the caliper jaws may contact the sides of the gear tooth exactly on the pitch circle.

The new figure is known as the "corrected addendum" for each gear, and must be used instead of the true addendum when measuring gear teeth.

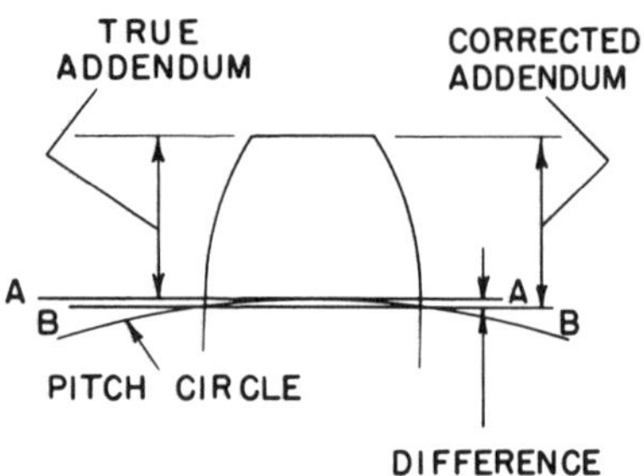

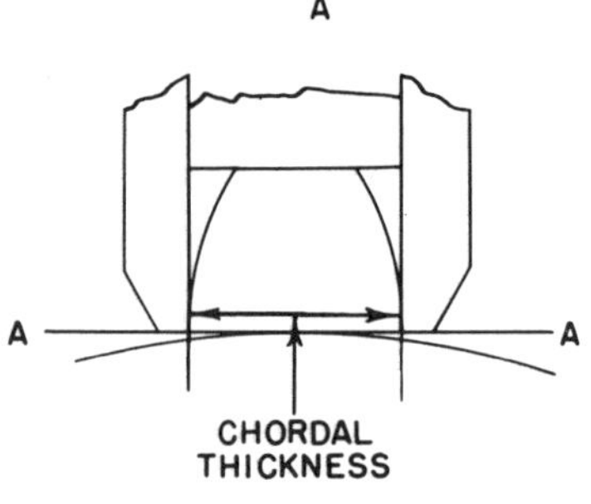

Figure 204. Gear-Tooth Measurement

VERNIER HEIGHT GAGES

The vernier height gage is a caliper with a special base block adapting it for use on a surface plate. The adjustable jaw is shaped so that a scriber or indicator can be attached with a clamp, as shown in Figure 205. A height gage is a particularly useful gage in the tool room where used in the layout of jigs and fixtures, and on the inspector's bench where used to check the location of holes and surfaces.

Height gages are commercially available in several sizes. The most commonly used sizes are the 10-, 18-, and 24-inch gages in English measure, and the 25-, 46-cm gages in metric measure. Height gages are classified by the dimension they will measure above the surface plate. A 10-inch gage, for example, will measure a point ten inches above the surface plate. The blade itself will be a little over 12 inches long.

Like the vernier caliper, height gages are graduated in divisions of .025 inch and are equipped with a vernier of 25 units for reading the scale to thousandths of an inch. Ten-inch height gages are available with a scale and a base-block design that permits them to be used as inside and outside vernier calipers.

Another height gage having some unusual features is shown in Figure 206. This type of gage (of English manufacture and known as a "Chesterman" height gage) is made in 12-, 18-, 24-, and 40-inch sizes. The 40-inch size is shown; it actually stands a little over 44 inches.

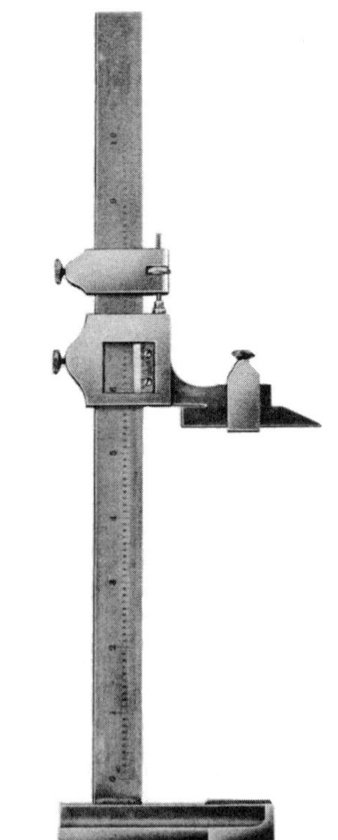

Figure 205. Vernier Height Gage

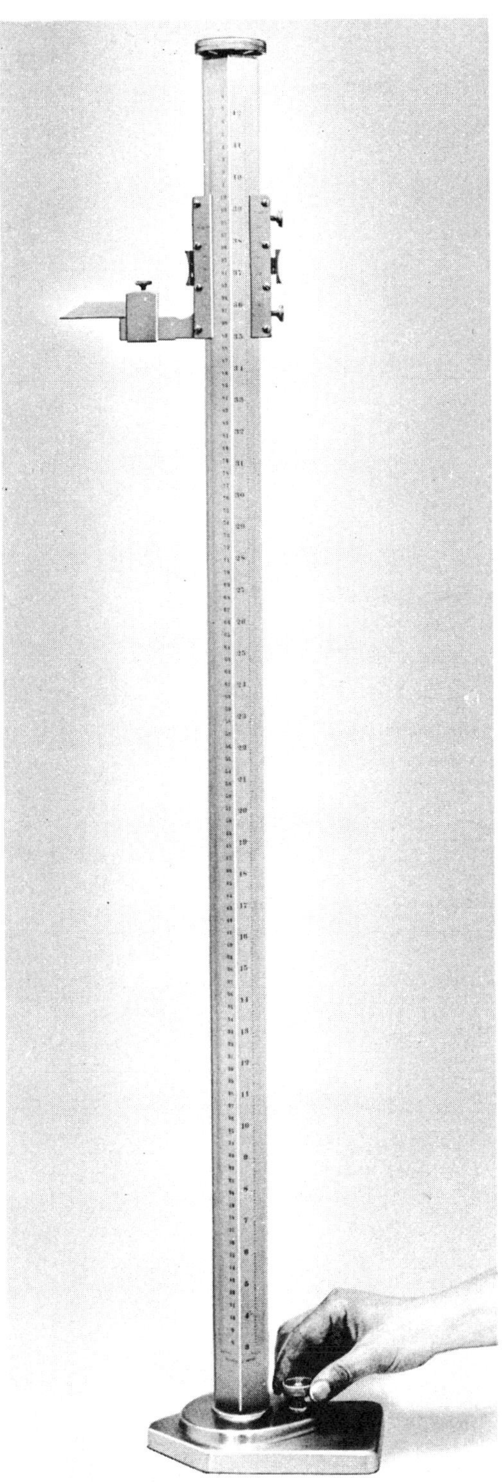

Figure 206. Chesterman Forty-Inch Height Gage

It has a large base, an L-shaped column, and is graduated in divisions of .050 inch instead of .025 inch, as is common practice. The vernier has 50 divisions on its scale and is nearly 2-1/2 inches long. This makes the spacing much wider so that the scale may be read to a thousandth of an inch without the aid of a magnifying glass.

The head may be moved rapidly to the approximate location by grasping it so that the thumb and the forefinger depress the two concave buttons. These buttons operate a split nut which engages the long screw running the whole length of the column. Final adjustment is made by turning the knurled knob either on the base or on the top of the column. Figure 206 shows how the final adjustment is made. The top knob is used when making a measurement near the upper limit, and the lower knob when working near the base.

APPLICATION OF A VERNIER HEIGHT GAGE

An application of a vernier height gage is shown in Figure 207. The gage is employed here to check the location of reamed holes in relation to the edge of the casting, and in relation to one another.

Because the base of the casting is resting directly on the surface plate, the plate is the reference plane. As all the holes in this casting are located from its base or lower edge, it is necessary to obtain a base reading. To do this, move the slide holding the indicator until the point is barely in contact with the surface plate, lock the upper slide, and with the adjusting screw, set the indicator to zero and lock the lower slide. Read the vernier (Figure 207B) and record the reading on a slip of paper. This will be the base reading. Next, take a reading from the top of a plug as shown in Figure 207A. Read the vernier and subtract half the diameter of the plug from this reading to obtain the horizontal center line of the hole. Subtract the base reading from the center-line reading to obtain the dimension as given on the blueprint. All holes in this casting can be checked in this manner. The base reading need only be taken once.

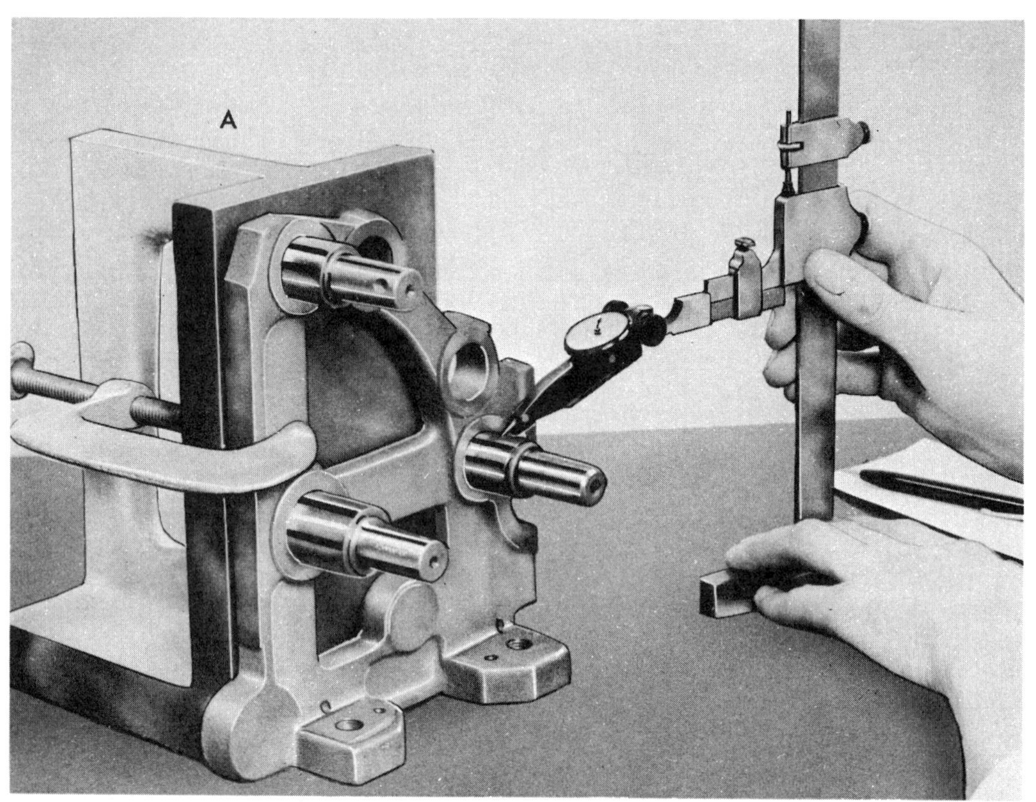

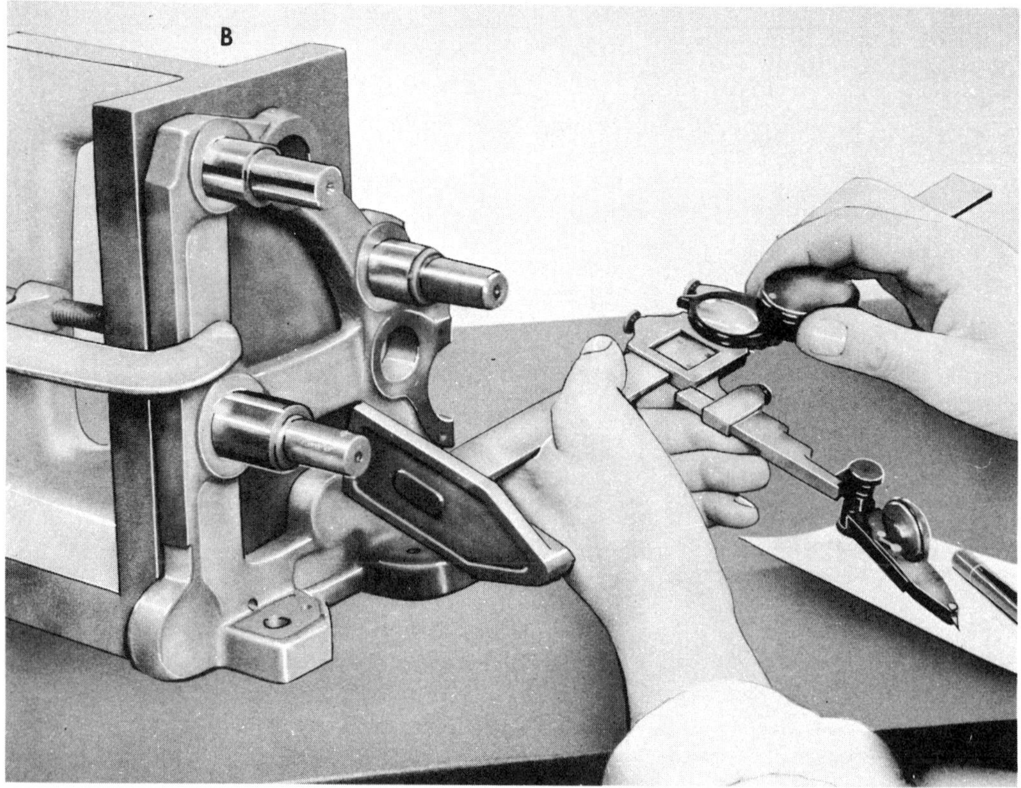

Figure 207. Checking Location of Holes

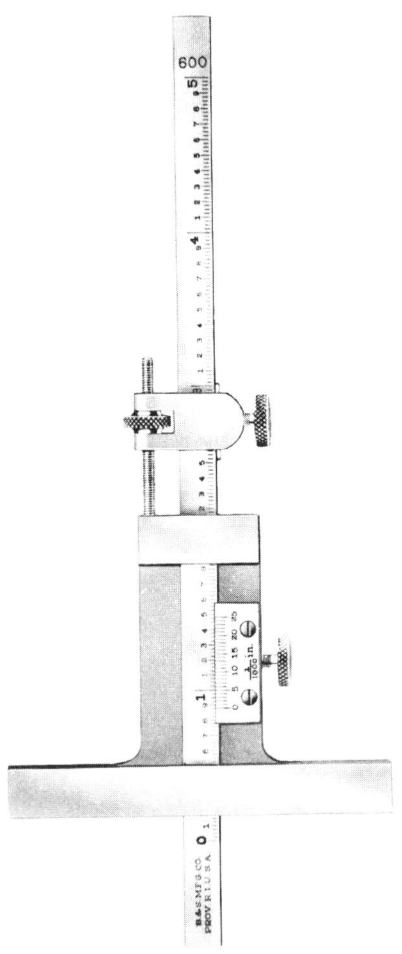

Figure 208. Vernier Depth Gage

VERNIER DEPTH GAGES

The vernier depth gage in Figure 208 is similar to the rule depth gage and the micrometer depth gage. It also is used to measure the depth of holes, slots, counterbores, recesses, and the distance from a surface to a projection.

It consists of a scale, either 6 or 12 inches long, graduated in 40ths of an inch, with a sliding head similar to the one on the vernier caliper except that it is especially designed to bridge holes and slots. The metric depth gage may have a scale 88 or 238 mm long.

The vernier depth gage is between the rule depth gage and the micrometer depth gage. It has the range of the rule depth gage and approaches the accuracy of the micrometer type, being capable of measuring to .001 inch. It

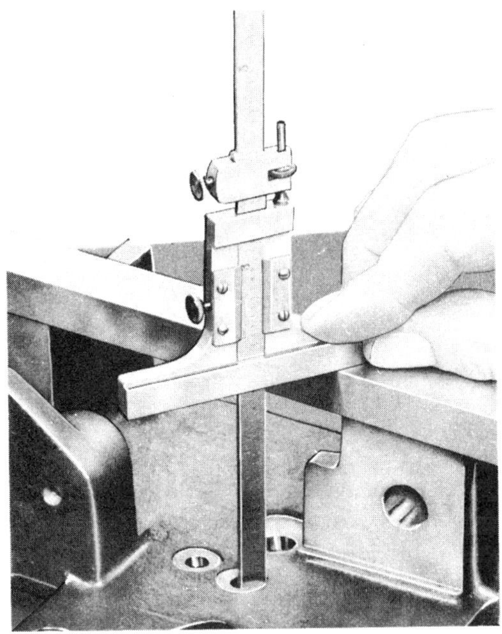

Figure 209. Checking Depth of Spot Face

will not enter holes less than 1/4 inch in diameter, but a micrometer depth gage will enter a 3/32-inch hole. On the other hand, it will enter a slot of 1/32 inch, where a micrometer depth gage will not.

The vernier scale on the depth gage is adjustable to compensate for wear. The accuracy of the gage may be checked by measuring the distance from the top of a precision gage block to a surface plate.

Figure 209 shows a vernier depth gage being used to check the depth of a spot face on a casting. A parallel is laid across the casting to serve as a flat surface from which to measure. The distance from the top of the parallel to the spot face is measured, and from this reading is subtracted the width of the parallel used. The difference is the distance from the surface of the casting to the spot face.

VERNIER BEVEL PROTRACTOR

The vernier bevel protractor in Figure 210 is a precision angle-measuring tool designed for layout and checking of angles to an accuracy closer than one degree.

It consists of two members, a base and a blade, which may be revolved in relation to each

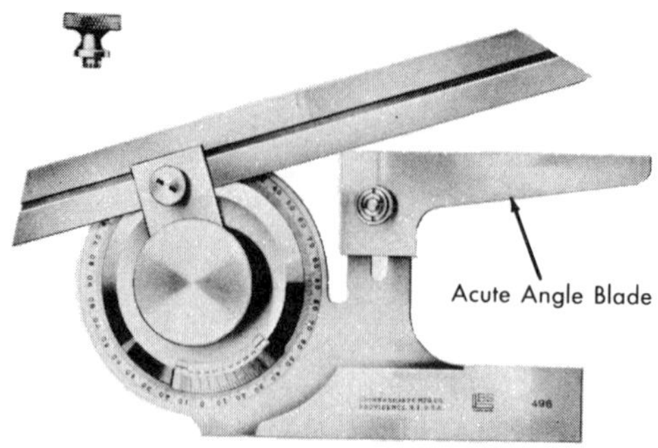

Figure 210. Vernier Bevel Protractor

other and set in any desired position. The position of one member in relation to the other is indicated on a dial scale graduated with four scales from 0 to 90 degrees. The addition of a vernier plate increases the accuracy of direct readings to a limit of 5 minutes or 1/12 of a degree.

Blades range from 6 to 12 inches long. They are mounted so that they can be moved along their axis and locked in any position. The clamping mechanism fits in the groove in the blade and is operated by the small thumb nut as shown.

The blade assembly is locked to the scale and base with the large central locking nut. The small thumb pinion shown to the upper left of the tool is for securing fine adjustment. It is inserted in the back of the protractor and engages a gear which swings the blade in either direction. On an instrument made by one firm, the fine adjustment, blade-locking, and the main clamping screw are all mounted in the center. The back of the tool is flat so that it may be laid flat on a surface plate or on the work.

The particular protractor shown in Figure 210 has what appears like an extra base. This is an acute angle attachment. This attachment permits a longer line of contact in measuring small angles and makes the tool more adaptable to the measurement of small angles (30 degrees and less). The acute angle attachment is adjustable in the slot shown in the figure.

Figure 211A shows the acute angle attachment being used to check the angle on a machine part. This angle is about 20 degrees, and it is easy to visualize how much more difficult it would be to measure it reliably using the other base.

Figure 211B shows an application using the regular base. The protractor is used here to check the angle of a surface on a die block, which is only one of many toolroom applications of this tool. It also may be used to check the taper on a collet, the angle of a dovetail, the included angle of a lathe center, the point on a drill or cutting tool, the face angle on a bevel gear and angles on parts for jigs and fixtures.

Accuracy of the vernier bevel protractor may be checked with the aid of a surface plate and a height gage and indicator. By setting the protractor index on 0, a height gage may be used to indicate the error by taking a reading

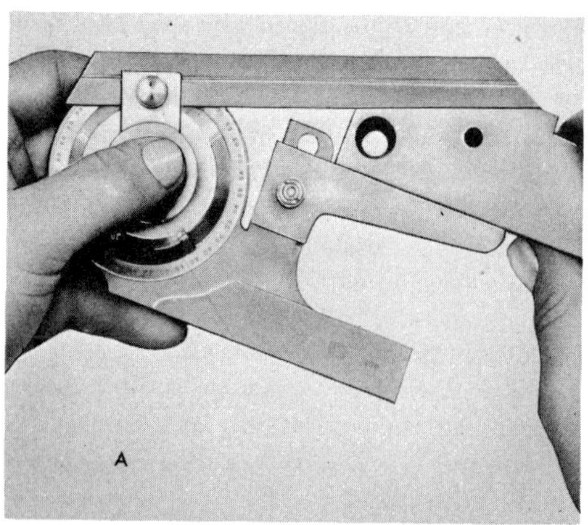

Figure 211. Applications of Vernier Bevel Protractor

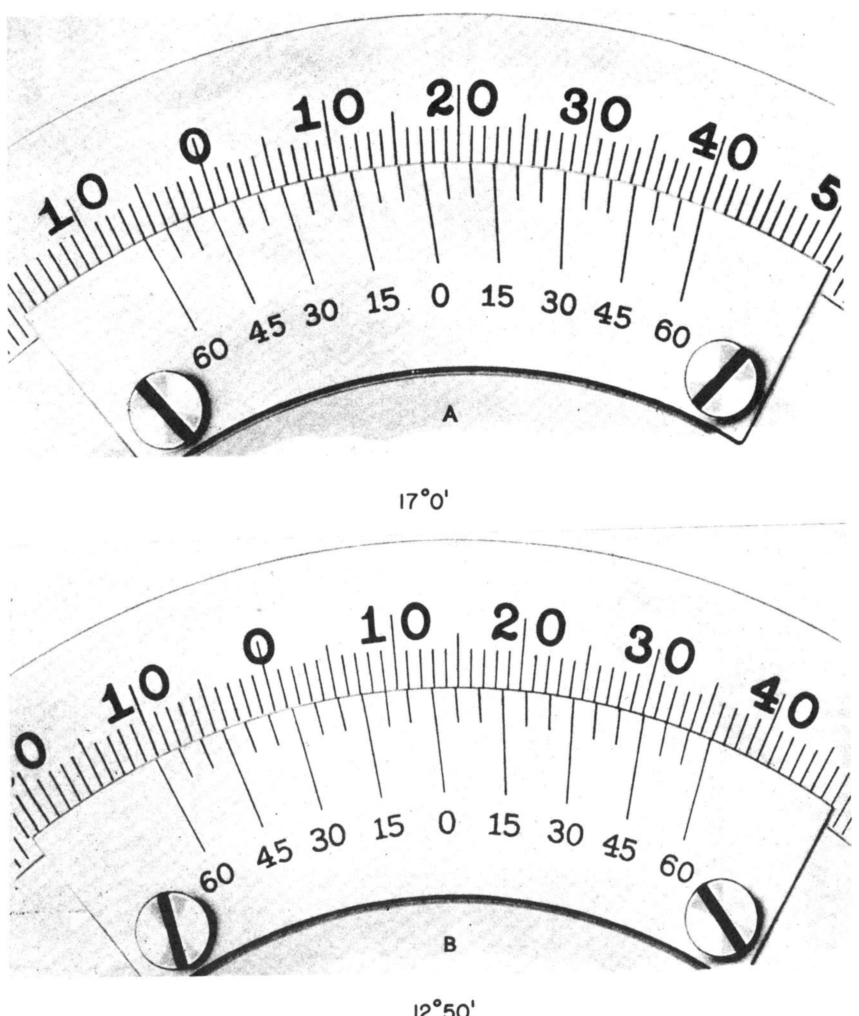

Figure 212. Vernier Scale of Protractor

at each end of the blade. Other points on the scale can be checked with the aid of a sine bar and precision gage blocks, a method of angular measurement which will be discussed later. The bevel protractor should check within 5 minutes of a true reading.

HOW TO READ THE VERNIER SCALE ON A BEVEL PROTRACTOR

The vernier on a bevel protractor is somewhat different from the one on micrometers and height gages. The main scale on a vernier protractor starts at zero and increases both to the right and to the left to 90 degrees, and then may decrease to zero again on some protractors. Because the main scale has a zero center, the vernier must also have a zero center as shown in Figure 212. When the main scale runs from right to left, the vernier scale running in the same direction must be used to obtain the correct reading of part of a main scale division and vice versa.

From the figure it will also be noted that the vernier scale is equal to 23 main scale divisions. This is done to spread out the vernier scale and make it easier to read.

The 23 main scale divisions represent an angle of 23' times 60' or 1380', and each pair of spaces represents 120'. The vernier has 12 divisions spread over the 1380', which gives each space a value of 115', and the difference between a main and a vernier scale division of 5'.

In Figure 212 it will be noted that every third division on the vernier scale is numbered 0, 15, 30, 45, 60. The reading in Figure 212A is taken at the 0 mark, which coincides with 17 degrees. In 212B the 0 mark is past 12 degrees and almost at 13 degrees. Using the vernier scale running in the same direction, note that the lines coincide at 50. So the angle is 12 degrees and 50'. With a magnifying glass it is not difficult to read the vernier on a bevel protractor correct to 5'.

CARE AND ADJUSTMENT OF BEVEL PROTRACTORS

Like all instruments equipped with a vernier, the sliding member must be kept clean and free of abrasive grit. Accuracy of any direct reading instrument, such as a vernier, depends on the user's ability to read to dimension on the scale. Abrasive grit on the scale parts of the vernier wears down the graduations, makes them indistinct and harder to read.

The vernier on the bevel protractor is adjustable to compensate for any general error. Any local wear or damage to the blade usually is corrected by replacing the blade.

QUESTIONS AND PROBLEMS

1. What are some of the outstanding features that have made the micrometer caliper such a popular tool in the shop?
2. For what type of work was the micrometer caliper developed in this country?
3. What are some of the major improvements and special features of micrometers that have been made since their invention?
4. Describe in your own words how a micrometer is read.
5. Upon what principle does the micrometer depend for its magnification?
6. Why have most organizations established in their gage department a free service station for the personal micrometers of their employees?
7. Read the following diagrams which represent micrometer scales.

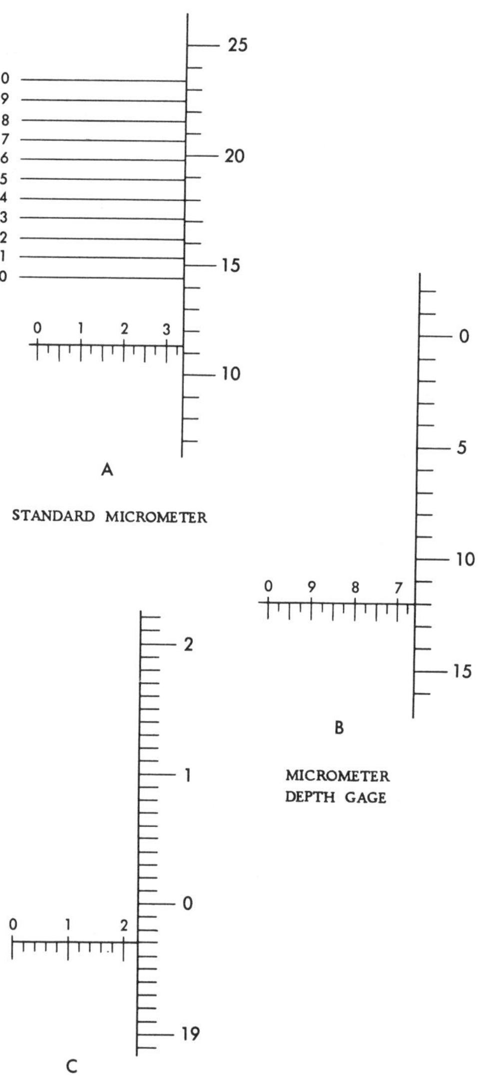

8. How can an inspector develop the proper sense of feel or touch in using micrometers?
9. How can the range of the micrometer shown in Figure 179 be changed?
10. Describe the method used when checking small parts; large parts.
11. What dimension of a thread does the thread micrometer check? Why will one thread micrometer check only a few sizes of threads?

12. What advantages and disadvantages does the inside micrometer caliper have over the inside micrometer?
13. Is the scale on depth micrometers read the same as micrometer caliper? Why?
14. Name several important precautions that should be taken when using and storing micrometers.
15. Explain how a vernier operates.
16. If a scale is graduated in divisions of .020 inch, how many graduations should the vernier scale have so that .001 of an inch can be read? What would be the over-all length of the vernier scale?
17. Why is a vernier caliper considered a particularly versatile instrument?
18. In what sizes are vernier calipers made commercially?
19. What is the purpose of a scale on both sides of the vernier caliper?
20. What points should be checked on a vernier caliper to determine the accuracy?
21. For what purpose is the gear-tooth vernier caliper used?
22. What are the two main applications of the vernier height gage?
23. What are the most commonly used sizes of height gages?
24. The base of a casting rests on a surface plate. A base reading was taken from the surface plate, which in this case was 1.000. The following readings were taken from the top of the plugs inserted in the holes of the casting as shown in Figure 207. Which of the following holes are off-location?

Plug Diameter	Inspector's Reading
A. .750	A. 5.365
B. .368	B. 8.264
C. .480	C. 9.421
D. 1.000	D. 3.100
E. 1.200	E. 7.300

Blueprint Dimensions

A. 3.950 $\pm$ .005
B. 7.084 $\pm$ .002
C. 8.180 $\pm$ .002
D. 1.600 $\pm$ .005
E. 5.695 $\pm$ .005

25. Compare the vernier depth gage with the micrometer depth gage and give the advantage of each.
26. How accurately can the measured angle be read on a vernier bevel protractor?
27. When is the acute angle attachment of the vernier protractor used?
28. Why does the vernier scale on the protractor have the zero in the middle instead of at one end as on other vernier scales?
29. Why is the vernier scale division on the protractor nearly double the main scale division instead of being somewhat smaller as on other vernier scales?
30. What is the reading of the following scale from a vernier height gage?

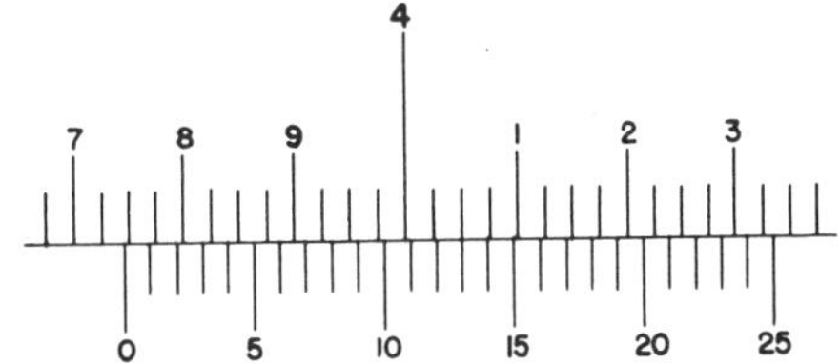

31. The diagrams below represent settings on a vernier bevel protractor. What are the angles?

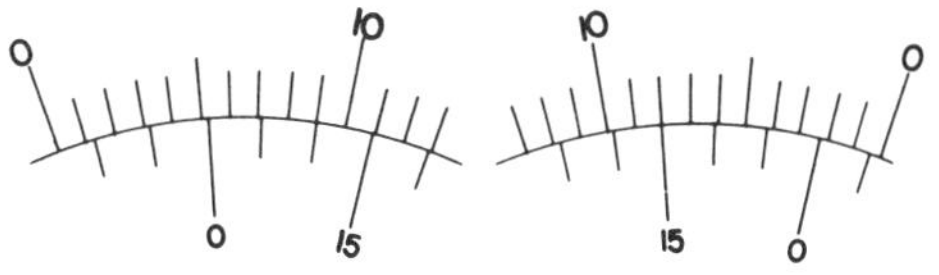

32. Where and how are telescoping gages used?
33. What are some of the advantages of indicating micrometers?
34. What are precision end standards?

Chapter VII

PRECISION GAGE BLOCKS

A PRECISION gage block is a piece of hardened alloy steel of square or rectangular cross-section with a measuring surface on each end. These measuring surfaces are lapped and polished to an optically flat surface, and the dimension between them is the length of the block and the measuring dimension. Individual gage blocks vary in size from .010" to 20".

Chromium-plated blocks are available in complete sets or as replacements for worn blocks in a standard set, and are recommended for working sets where resistance to wear is important. Solid carbide gage blocks are also manufactured in complete sets but are customarily used as wear blocks at both ends of a stack of steel blocks, because their extremely long-wear life prolongs the accuracy of the other blocks in a set. When carbide blocks are to be used as wear blocks, they may be purchased in sets of two or four in two sizes, .050" and .100".

Precision measuring instruments themselves are controlled and calibrated from gage blocks that are the industrial standards of length. In addition to their use as standards in setting gages and instruments, gage blocks are frequently used in the layout of work, in the setup of machine tools, and in the inspection of parts where very close tolerances must be maintained.

Use of gage blocks as industrial standards is not general; many industries do not require any such degree of accuracy in their measurements. For example, in such fields as pattern-making or foundry work, blacksmithing, carpentry, and rough machining, the simplest kinds of line-measuring tools adequately control the accuracy required. But where tolerances are close and where the quality of the product is largely determined by the accuracy with which it is made--such as production of firearms, automotive engines, typewriters, adding machines, mechanical and electrical insturments and meters, and so on--precision gage blocks have a most important manufacturing role.

They are usually purchased in sets of from five to as many as 85 blocks of different lengths. Two standard sets--one rectangular, one square--are shown in Figure 213. Each large standard set of these blocks may be used singly and in combinations to produce as many as 120,000 different size gages in steps of .0001". Special conversion blocks also can be obtained to produce gages for fractional and metric dimensions.

MASTER AND WORKING SETS

In the organization of an industrial gaging system of any size, primary and secondary standards are necessary. The primary standard is a master set of gage blocks used only to check the accuracy of other blocks used in the shop. The master set is carefully preserved and used only on occasions where it is necessary to check the wear, distortion, or "growth" of working standards.

Where several plants or large sections of a plant must be coordinated through a common measuring system, each department or manufacturing unit will have its master set; these masters in turn will be certified to a grand master set. The grand master set in effect controls the accuracy of manufacture throughout the organization.

Twice a year the master or grand master

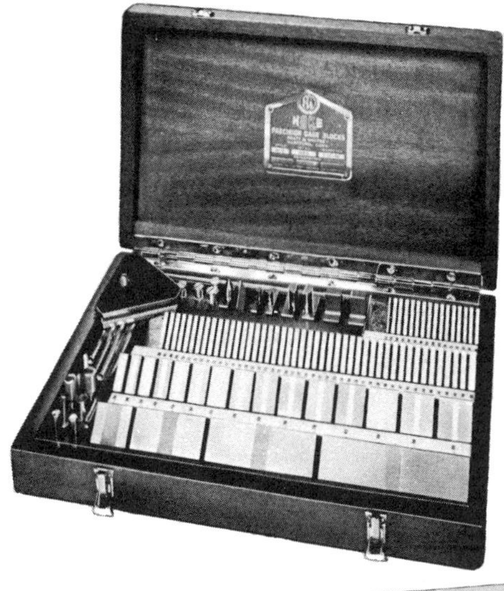

Figure 213. Sets of Gage Blocks (81 blocks)

set will be returned to the manufacturer or sent to the National Bureau of Standards in Washington for certification. There each block is checked for flatness, parallelism, and length, and in the reporting certificate the variation of error of each block is given in millionths of an inch (Figure 214). The blocks are calibrated at the international standard measuring temperature of 68 degrees F.

Although a master set may lose its accuracy, it rarely becomes a working set or is replaced, because each block is certified; and because the error is known, accurate measurements still may be made. As a working set gets older and becomes worn through use, however, it is put on less important work and may in the course of time degenerate to even a third grade of work. The average shop will have sets of precision gage blocks at several levels of accuracy, usually determined by the shop standards.

GRADES OF BLOCKS

Gage blocks are commercially manufactured in several grades or degrees of accuracy. The most accurate blocks which Brown and Sharpe Manufacturing Company lists as its quality "AA" Johansson blocks are guaranteed accurate to plus or minus two millionths of an inch in one inch. Pratt and Whitney list a "laboratory" set of Hoke blocks for similar work with a guaranteed accuracy of two millionths of an inch in one inch. Gage blocks in this classification are special; they are used mostly in exacting research and experimental work or as grand master sets in large manufacturing plants.

For general use as industrial standards, manufacturers produce gage blocks with a guaranteed accuracy of four to eight millionths of an inch in an inch. These are used to check the accuracy of other gage blocks, to calibrate measuring instruments, to set adjustable gages and tools, and in layout work where very accurate measurement is desired. Blocks of less accuracy are also made for this same application, but where the accuracy required is not so exacting.

The guaranteed accuracy of precision gage blocks is expressed in such terms as plus or minus .000002" in one inch. The tolerance is not divided for blocks less than one inch; blocks under one inch have the same tolerances as a one-inch block. This means that a precision gage block of .500" may vary between .499998" and .500002" and still be in the finest grade. It also means that a precision gage block 3.000000" long may vary 3 times .000002, or .000006", in its length, or between 2.999994" and 3.000006" and still be in this grade.

If this allowable error were permitted to be only in one direction--that is, all plus or all minus--the accumulated error in a stack of blocks would be considerable. But, because variations in blocks have a random distribution between pluses and minuses, the total error in a stack of blocks will rarely, if ever, exceed twice that of a single block. Frequently the total error is actually less than that of a single block.

UNITED STATES DEPARTMENT OF COMMERCE
WASHINGTON

National Bureau of Standards

IHF:EGH
II-9/Tw 93053

REPORT

on the test of
One Set (81) Hoke Precision Gage Blocks
Set number 567

Submitted by

International Business Machines Corporation,
Endicott, New York
Order No. 60291

Nominal size	Identification	Measured length at 68° F (20° C)	Maximum error in flatness across width of gage	Maximum error in flatness along length of gage
0.050 in.	V67L	0.049994 in.		
.100	F151L	.099998		
.1001	F65L	.100100		
.1002	F144L	.100199		
.1003	F288L	.100295		
.1004	F232L	.100394		
.1005	U32L	.100499		
.1006	U121L	.100601		
.1007	F191L	.100697		
.1008	C11L	.100797		
.1009	U90L	.100897		
.101	U107L	.100997		
.102	U118L	.102000		
.103	A123L	.103002		
.104	U156L	.103997		
.105	U215L	.104998		
.106	F120L	.106000		
.107	A12L	.107002		
.108	F258L	.107997		
.109	U119L	.108999		
.110	F55L	.110000		
.111	U41L	.111000		0.000006 in.
.112	V7L	.111998	0.000008 in.	
.113	U48L	.113000		
.114	F171L	.114003		

Nominal size	Identification	Measured length at 68° F (20° C)	Maximum error in flatness across width of gage	Maximum error in flatness along length of gage
0.650 in.	F49L	0.650001 in.		
.700	U93L	.700015		
.750	V18L	.750008		
.800	U80L	.799999		
.850	U36L	.849997		
.900	F56L	.899997		
.950	U122L	.950004		
1.000	B41L	.999997		
2.000	F280L	2.000002		
3.000	F170L	3.000005		
4.000	V22L	3.999983		

The measured length is in each case the perpendicular distance from one gaging surface to a point on the other gaging surface near the middle of and about 1/16 inch from the edge nearest the size marking.

An observational error of ±0.000005 inch on blocks up to 1 inch in length, and of ±0.000005 inch per inch of length on longer blocks is possible in the measurements of length.

The error in parallelism does not exceed ±0.000005 inch in any case. The error in parallelism indicates the correction to be applied to the measured length to obtain the length at a point about 1/16 inch in from the opposite edge.

The error in flatness does not exceed 0.000005 inch unless otherwise specified. The errors in flatness along the length and across the width are measured along lines midway between the edges, but do not include the 1/16 inch of surface nearest the edges. The flatness of the 0.050 inch gage was measured when the gage was wrung down on an optical flat.

L J Briggs
Lyman J. Briggs, Director
HW3

Date of completion of test: August 19, 1941

Lot No. L 8957

Test No. Tw 93053

Figure 214

In addition to being correct to length, the measuring surfaces of a precision gage block must be flat and parallel. Although no specific tolerances are given, blocks usually are accepted if the error in flatness and parallelism is less than one half of the total tolerance, which in the case of blocks having an allowable variation of plus or minus .000005" in length would be .000005" in flatness. But practically all new precision gage blocks are so nearly flat and their measuring surfaces are so nearly parallel, that with the finest systems of measurement the variations found, in general, are very small.

MANUFACTURE OF PRECISION GAGE BLOCKS

Precision gage blocks are made from a special alloy steel. They are hardened, ground, and then stablized over a period of time to reduce subsequent warping and growing. Despite the great care taken in seasoning the metal before it is made into blocks, there will be a certain amount of distortion and adjustment from internal stresses.

Master sets used only rarely will in time grow or shrink, becoming slightly larger or smaller than they were originally. This change is not uniform and will destroy the perfect flatness and parallelism of the measuring surfaces. Because blocks do change, although they are not used, it is periodically necessary to send them for certification.

After blocks are properly seasoned they are lapped on special lapping machines, then measured and selectively graded into sets. Although the blocks themselves seem simple, and the process for making them not complicated, it takes a great deal of care and skill to produce a precision gage block whose error is measured in a few millionths.

TESTING PRECISION GAGE BLOCKS

Before the blocks can be matched and made into sets, they must be measured for length, flatness, and parallelism. Flatness and parallelism of a gage block are checked with optical flats as shown in Figure 215. Interference of light rays reflected from the measuring surface produces the dark bands; when the bands are straight (as in Figure 215) the surface is flat within a millionth of an inch or less.

Figure 215. Checking a Block for Flatness with an Optical Flat

Length of the master precision block is also measured by light-interference bands. An interferometer used to measure master blocks is shown in Figure 216.

The hardness of a block is a measure of its resistance to wear; the more resistant it is, the longer it will retain accuracy in normal use.

CARE OF PRECISION GAGE BLOCKS

Gage blocks come from the manufacturer with a coating of rust-proofing which must be removed before they can be used. This may be done by wiping them clean with a chamois or a piece of cleansing tissue, or by cleansing them with a cleaning fluid such as kerosene, benzine, gasoline, or non-inflammable solvent. When it is especially important to remove all grease film, a final rinse in 95-percent grain alcohol is recommended. The cleaning fluid should not contain any substance that would be deposited on the measuring surfaces of the blocks when the liquid evaporates. Before blocks can be used they must be free from grease, oil, and dirt. Also, the surface plate and other tools used with gage blocks must be free from grease, oil, and dirt, to avoid a lapping action

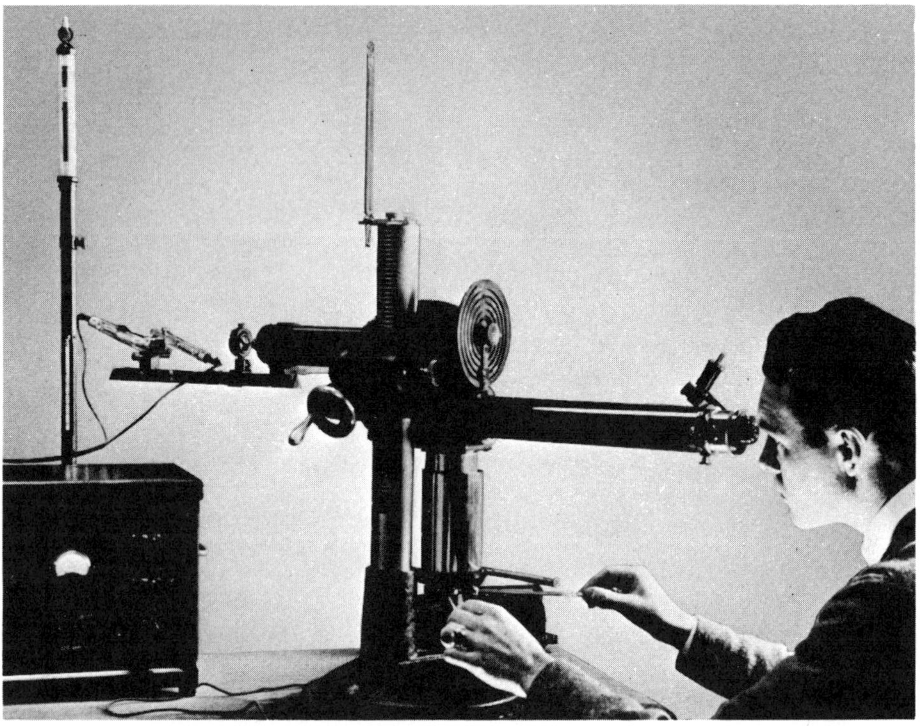

Figure 216. Checking a Gage Block by Interferometry

(between the block and the surface in contact with it) whenever the block is moved, and to assure accurate measurement.

Particular care should be observed when using gage blocks to measure hardened work. Where the object being measured with gage blocks is as hard as the block itself, the danger of scratching the measuring surfaces is greatly increased.

Gaging surfaces of precision blocks should not be touched any more than necessary. Natural moisture of the hands contains an acid which will eventually stain the blocks if it is not removed.

When through using gage blocks, wipe them with a piece of dry chamois to remove finger marks and then wipe with a chamois permeated with an acid-free oil such as white petrolatum, and return them to their case. The gage block case should be kept free from dust, filings and moisture.

When wringing two blocks together, take care not to damage them. The following illustrations indicate the best method of joining two blocks.

1. The two blocks are brought together flat as in Figure 217 and oscillated slightly. This minimizes the danger of the corner of one block scratching the other, and will detect any foreign particles between the surfaces.

2. Slide the upper block over the lower with a circular motion as shown in Figure 218. If the blocks are clean they will begin to take hold.

3. Slide the upper block half out of engagement, as shown in Figure 219, using light pressure.

4. Slide the upper block back into full engagement under light pressure as shown in Figure 220; at this point they should be wrung together ready for use.

Using heavy pressure in joining gage blocks is unnecessary; it makes them harder to separate and contributes nothing to the accuracy of the setup.

The combination for all practical purposes is as solid as a single block and may be handled as a unit. The clinging qualities of a stack of gage blocks are shown in Figures 221 and 222.

Explanation of this phenomenon is given in

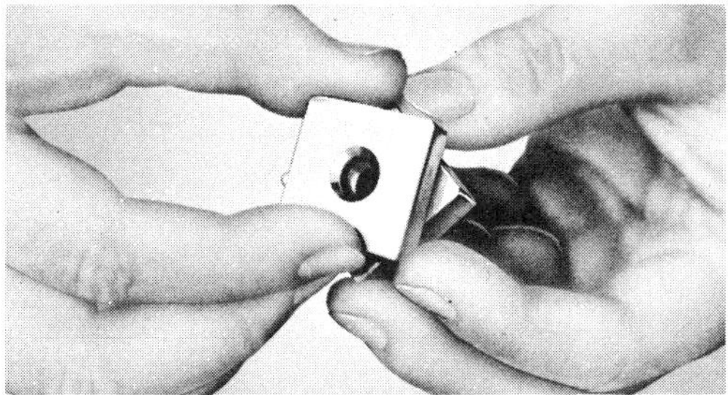

Figure 217. First Step in Wringing Gage Blocks Together

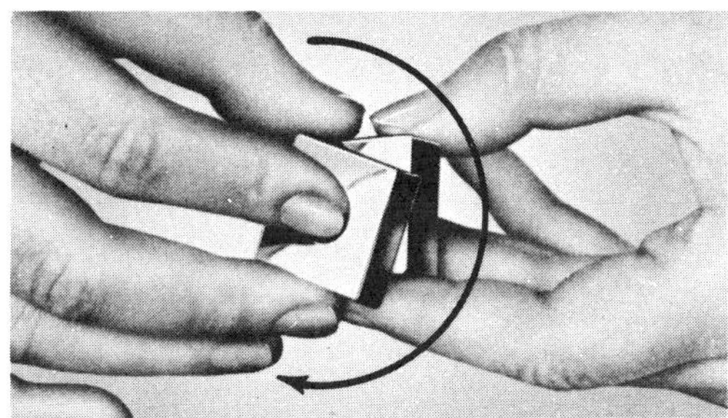

Figure 218. Second Step in Wringing Gage Blocks Together

N. K. Adam's book, "Physics and the Chemistry of Surfaces." According to this British scientist the adhesive force that binds two gage blocks wrung together is a combination of molecular attraction and the cementing action of the film of oil or moisture on the surfaces. He further points out that perfectly dry blocks do not cling together with enough force to support their own weight. So the portion of the total force contributed by a molecular attraction is small, and the cementing action of the thin film of oil or moisture may thus be the principal reason why gage blocks adhere so strongly.

Despite extraordinary clinging power, the blocks are easily separated by the same simple

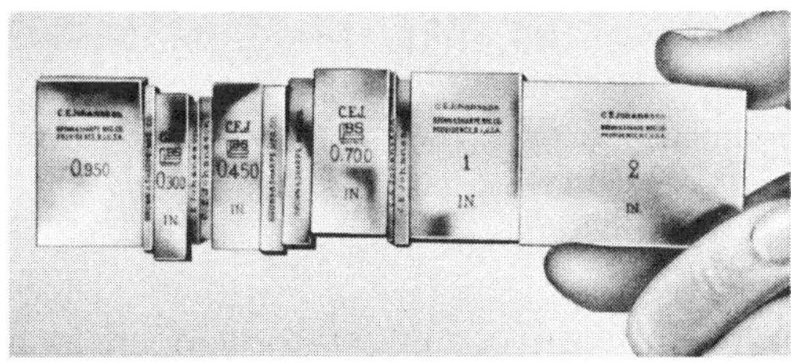

Figure 221. Clinging Power of Gage Blocks

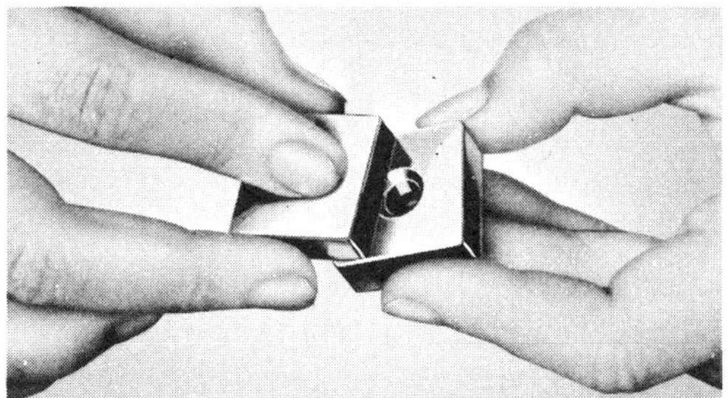

Figure 219. Third Step in Wringing Gage Blocks Together

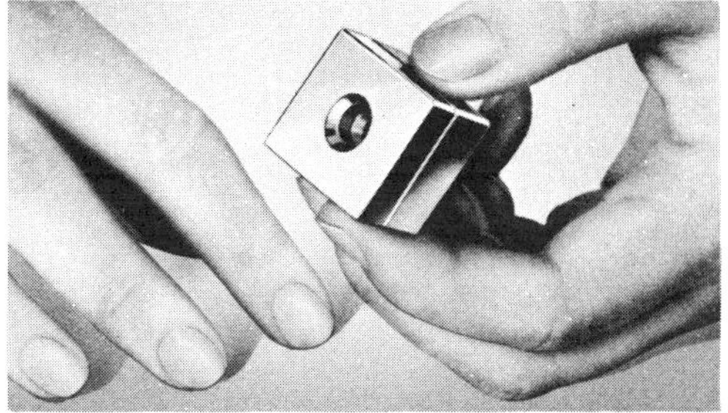

Figure 220. Fourth Step in Wringing Gage Blocks Together

sliding movement by which they are wrung together. Blocks should not be left wrung together for long intervals of time, because surfaces in contact may corrode.

EFFECT OF TEMPERATURE

The effect of temperature is negligible in the use of ordinary measuring instruments; but expansion and contraction of metal through

Figure 222. Gage Blocks Supporting Toolmaker's Flat

changes in temperature become very important when handling precision gage blocks.

The standard measuring temperature is 68 degrees F., which is just a little lower than the average temperature in most shops in this country. Since the alloy steel from which gage blocks are made has about the average coefficient of linear expansion for steel, the size of the block in relation to the work does not vary appreciably for small changes in temperature, and for anything but the finest work no correction is necessary if the part or tool is made from iron or steel. However, where the tolerances are very close, modern gage rooms have regulated temperature to insure the accuracy of all measurements made with blocks.

Since the room temperature affects the work as well as the block, the expansion in the work will be matched in most cases by a similar expansion in the block. The coefficient of linear expansion of several metals and blocks is listed below.

	Millionths of an Inch Per Inch Per Degree Fahrenheit
Steel	5.5 to 7.2
Iron	5.5 to 6.7
Phosphor Bronze	9.3
Aluminum	12.8
Copper	9.4
Gage Blocks	6.36 to 7.0

To illustrate the effect of temperature: an inspector checks an aluminum casting 2-1/2" in diameter in a room where the temperature on a hot summer afternoon is 90 degrees F. This is a practical case, and the error due to the fact that the measurement is not made at 68 degrees F. is computed as follows:

(Difference in coefficients of expansion) x (Difference between measuring temperature and international standard--68 degrees F.) x (Length of dimension) equals (Error due to the temperature difference).

(12.8 - 7.00) x (90 - 68) x 2-1/2 = 319 millionths, or about 3 ten-thousandths of an inch.

This example shows that ordinary changes in temperature do have sizeable effect on measurements made with precision gage blocks. Because the measuring temperature is so important, it is customary to have temperature-controlled gaging rooms in plants where parts are manufactured to very close tolerances.

The coefficient of linear expansion of gage blocks is practically the same as for ordinary iron and steel; so the amount of error introduced by even a large temperature difference would be much less and probably negligible if the casting had been made of iron.

In the control of the measuring temperature, it is also important to avoid conducting the heat of the body into the block by careless handling. Handling a gage block with bare hands in a manner that allows the heat of the body to raise the temperature of the block may introduce a serious error in a measurement, particularly if a long stack of blocks is being handled.

The best practice, of course, is to handle them only when they must be moved; to hold them between the tips of the fingers so that the area of contact is small; and to hold them for as short a time as possible. In checking working blocks against the master set, the possible error due to handling can be eliminated by leaving the blocks to be compared side-by-side on the same bench for several hours, by submerging them in a pan of kerosene, or by using special tongs to insulate against the heat of the hands.

Still another source of temperature error is in the work itself. Often work is heated considerably in machining. Precision measurements made on work in a machine or just taken from a machine may be in error due to the difference in temperature between the work and the gage block. Also, work frequently is cooled by the coolant used in some machining operations; this may cause a temperature error.

In using blocks with assurance that measurements made are really accurate to a tenth of a thousandth or better, the influence of temperature and temperature differences must be considered.

EFFECT OF MEASURING PRESSURE

Still another factor in the use of precision gage blocks is the measuring pressure. Where with line-graduated measuring instruments any measuring pressure is acceptable as long as it does not injure either the work or the instrument, with gage blocks it becomes a matter of accuracy; because the use of insufficient or excessive pressure does affect the measurement.

Although, theoretically, the measuring pressure should increase proportionally with the area of contact and the length of measurement, for practical purposes it is better to use a standard measuring pressure. The measuring pressure commonly used with precision gage blocks is between one-half and two pounds. The size and nature of the instrument, the work, etc., may require that a heavier or lighter measuring pressure be used. In such cases, specific reference is usually made to the measuring pressure so that uniform and accurate results will always be obtained. An example of where such specific reference is made to measuring pressure is in measuring the pitch diameter of screw threads using three wires. Instruments for this work must be capable of controlling the measuring pressure; for threads finer than twenty per inch, a pressure of 14 to 16 ounces is recommended. For threads of twenty per inch or coarser, a pressure of 2-1/4 to 2-1/2 pounds is recommended.

SELECTING BLOCKS TO OBTAIN A DESIRED DIMENSION

Part of the technique in using a set of blocks is in being able to make a selection quickly from the set that will combine to produce a gage of the desired dimension.

The set shown in Figure 213 and represented by the table in Figure 223 consists of 81 sizes of blocks, ranging from 0.050 to 4.000". The set is divided in several series--known, for example, as the one ten-thousandth series or the one-thousandth series. The larger the set, the more blocks there are in each series and the greater the number of possible combinations. With the largest sets, as many as 120,000 different gaging lengths may be obtained.

The method of combining blocks is best described by an example. From the 81-block set in Figure 223, it is desired to make a gage 1.4373" in length. The procedure:

First: One Ten-Thousandth Series--9 Blocks									
.1001"	.1002"	.1003"	.1004"	.1005"	.1006"	.1007"	.1008"	.1009"	
Second: One Thousandth Series--49 Blocks									
.101"	.102"	.103"	.104"	.105"	.106"	.107"	.108"	.109"	.110"
.111"	.112"	.113"	.114"	.115"	.116"	.117"	.118"	.119"	.120"
.121"	.122"	.123"	.124"	.125"	.126"	.127"	.128"	.129"	.130"
.131"	.132"	.133"	.134"	.135"	.136"	.137"	.138"	.139"	.140"
.141"	.142"	.143"	.144"	.145"	.146"	.147"	.148"	.149"	
Third: Fifty Thousandth Series--19 Blocks									
.050"	.100"	.150"	.200"	.250"	.300"	.350"	.400"	.450"	.500"
.550"	.600"	.650"	.700"	.750"	.800"	.850"	.900"	.950"	
Fourth: Inch Series--4 Blocks									
1.000"	2.000"	3.000"	4.000"						

Figure 223. Sizes of Blocks in an 81-block Set

1. Write the figure. 1.4373
2. Eliminate the last figure to the right by selecting a block with a 3 in the fourth decimal place; in this case this means the .1003" block; it must be subtracted from 1.4373" to obtain the amount still to be selected. The figure is written in the column on the right to prove the selection later.
   .1003 .1003
   1.3370
3. Eliminate the last two figures to the right other than zero by now selecting the .1370" block. A double elimination should be made wherever it is possible.
   .1370 .1370
   1.200
4. Eliminate the last figure to the right, other than zero, by now selecting the .200" block.
   .200 .200
   1.000
5. Eliminate the 1.000 by using the 1.000" block.
   1.000 1.000
   .000 1.4373

By reducing from right to left, the difference has progressively been reduced to zero, and the blocks selected total 1.4373".

In general, the fewest number of blocks that can be used to make up the desired gaging length the better, because each block introduces the possibility of error. It is considered good practice not to go over five blocks in making a combination, especially where the gaging length is six inches or more and the tolerances are close.

The combination selected above to make up a gage length of 1.4373" is not necessarily the only combination that will produce this dimension. Frequently two gages of the same length are required, in which case the second stack of blocks must be made up without using any of those selected for the first. For example, assume that another gage 1.4373" in length is required, using other blocks than those used before:

1. To eliminate the last figure to the right as before, a combination of two blocks is selected.	1.4373 .2013 1.2360	.1004 .1009
2. Eliminate the last two figures to the right by selecting the .136 block.	.136 1.100	.136
3. Adjust the remainder so that it may be removed with one more block.	.300 .800	.300
4. Eliminate the remainder by selecting the .800 block.	.800 .000	.800 1.4373

Using just one more block, the desired dimension is again obtained, and two gages from the one set, 1.4373" in length, are available for use.

In step 1 elimination of the last figure to the right was obtained by a combination of blocks rather than a single block. This usually is necessary on the first step in duplicating a gaging length already taken from the set.

In step 4 the customary procedure of eliminating the last figure to the right was not followed. At this point, although a .100 block remains in the set, it is not advisable to select it, because the remainder would then become 1.000" and the one-inch block has already been selected. The remainder would then have to be made up with two blocks, such as .600 and .400; and, although there is nothing wrong in selecting the blocks this way, it is always good practice to use as few blocks as possible. So, the blocks .300 and .800 were selected to make up the 1.100" remainder; the stack contains only five blocks instead of the six which would be required if the rule of eliminating the last figure to the right had been consistently applied.

EXAMPLES IN SETTING UP GAGE BLOCKS

These typical problems show further the procedure for setting up blocks to make a gage of the desired length.

EXAMPLE 1. Set up two gaging lengths of 2.1358 and .9753" for use in inspecting a lot of parts. Use the set in Figure 223.

2.1358		.9753	
2.1358		.9753	.1003
.1008	.1008	.1003	
2.0350		.8750	
.135	.135	.125	.125
1.900		.750	
.900	.900	.750	.750
1.000		.000	.9753
1.000	1.000		
.000	2.1358		

EXAMPLE 2. Set up two equal gaging lengths of 2.0142" for use in setting up a machine tool. Use the set in Figure 223.

2.0142		2.0142	
2.0142		2.0142	
.1002	.1002	.1008	.1008
1.9140		1.9134	
.114	.114	.1004	.1004
1.800		1.8130	
.800	.800	.113	.113
1.000		1.7000	
1.000	1.000	.950	.950
.000	2.0142	.750	
		.750	.750
		.000	2.0142

EXAMPLE 3. Set up the two gaging lengths for the purpose of checking a "go not-go" snap gage. The lengths required are .17510" and .17515." Use the 28-block set shown in Figure 224.

.17510		.17515	
.17510		.17515	
.0201	.0201	.02005	.02005
.1550		.15510	
.025	.025	.02040	.0204
.130		.13470	
.060	.060	.02070	.0207
.070		.11400	
.070	.070	.02400	.024
.000	.17510	.09000	
		.09000	.090
		.000	.17515

EXAMPLE 4. Set up gaging lengths of 29.990 and 29.995 mm to check the diameter of a shaft. Use the metric set shown in Figure 225.

29.990		29.995	
29.990		29.995	
1.49	1.49	1.005	1.005
28.50		28.990	
14.50	14.50	1.39	1.39
14.00		27.600	
14.00	14.00	1.38	1.38
.00	29.990	26.220	
	mm	1.22	1.22
		25.000	29.995 mm
		25.000	
		.000	

ACCESSORIES FOR PRECISION GAGE BLOCKS

The range of application of a set of precision gage blocks is greatly increased by the standard attachments (Figure 226):

First: One-Half Ten-Thousandth Series--1 Block								
.02005"								
Second: One Ten-Thousandth Series--9 Blocks								
.0201"	.0202"	.0203"	.0204"	.0205"	.0206"	.0207"	.0208"	.0209"
Third: One Thousandth Series--9 Blocks								
.021"	.022"	.023"	.024"	.025"	.026"	.027"	.028"	.029"
Fourth: Ten Thousandth Series--9 Blocks								
.010"	.020"	.030"	.040"	.050"	.060"	.070"	.080"	.090"

Figure 224. Sizes of Blocks in a 28-block Set

First: Five Thousandth Series--1 Block									
1.005									
Second: One Hundredth Series--49 Blocks									
1.01	1.02	1.03	1.04	1.05	1.06	1.07	1.08	1.09	1.10
1.11	1.12	1.13	1.14	1.15	1.16	1.17	1.18	1.19	1.20
1.21	1.22	1.23	1.24	1.25	1.26	1.27	1.28	1.29	1.30
1.31	1.32	1.33	1.34	1.35	1.36	1.37	1.38	1.39	1.40
1.41	1.42	1.43	1.44	1.45	1.46	1.47	1.48	1.49	
Third: Fifty Hundredth Series--49 Blocks									
0.50	1.00	1.50	2.00	2.50	3.00	3.50	4.00	4.50	5.00
5.50	6.00	6.50	7.00	7.50	8.00	8.50	9.00	9.50	10.00
10.50	11.00	11.50	12.00	12.50	13.00	13.50	14.00	14.50	15.00
15.50	16.00	16.50	17.00	17.50	18.00	18.50	19.00	19.50	20.00
20.50	21.00	21.50	22.00	22.50	23.00	23.50	24.00	24.50	
Fourth: Twenty-Five Millimeter Series--4 Blocks									
25	50	75	100						

Figure 225. Sizes of Blocks in a 103-block (Metric) Set

1 and 2: inside caliper jaws, used for all applications requiring an accurate inside measurement, such as the inside diameter of holes and the width of slots.

3: set of outside caliper jaws which, when put on the ends of a stack of precision gage blocks, will convert the blocks into a gage for checking the outside diameter of plug gages and the distance between outside surfaces.

4: scriber, used for scribing lines on metal

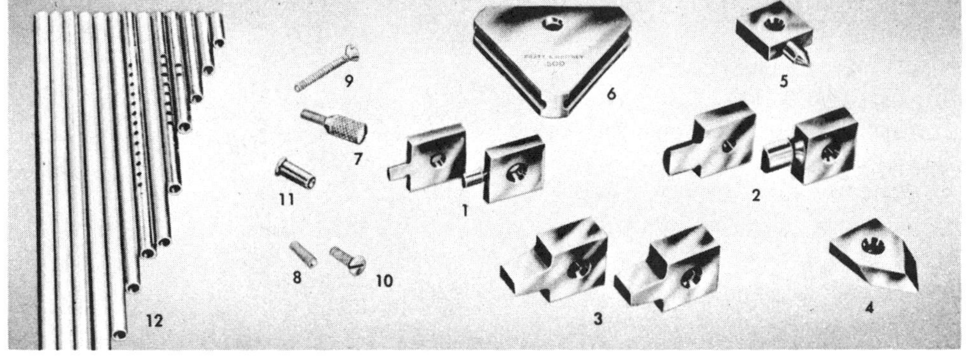

Figure 226. Standard Attachments for Hoke Precision Gage Blocks

in laying out dies, jigs and fixtures.

5: center point (a 60-degree cone), used to check inside and outside threads and to measure the distance between scribed lines.

6: base, to which blocks may be wrung and clamped. When used with the scriber, the center point, or one of the caliper jaws, it becomes an accurate height gage for the inspection of parts, the checking of other gages, and the layout of dies, jigs, and fixtures.

7, 8, 9, 10, 11, and 12: screws, studs, tie rods, etc., used to clamp a stack of blocks together. Clamping a combination of blocks makes them easier to handle and allows them to be used as a unit within all of the above-mentioned attachments.

APPLICATION OF PRECISION GAGE BLOCKS

Applications of gage blocks are so numerous and varied that only a few of the more representative ones will be discussed here. Extent of their application also is increasing very rapidly, and it is common to see them used in the machine shop today, where a few years ago they were locked in a vault and handled only by a chosen few.

This change in attitude toward gage blocks has come as a result of:

1. Reductions in their cost.
2. Need for greater accuracy in machining operations.
3. Their flexibility and field of application, particularly when used with standard attachments.
4. Education in their use, and discovery of new applications.

Some of the manufacturing fields to which precision gage blocks have been applied are:

1. Layout of tools, dies, and fixtures.
2. Checking the accuracy of tools, dies, and fixtures.
3. Setting machine-tool cutters.
4. Checking parts in the process of manufacture--process inspection.
5. Checking completed parts directly--parts' inspection.
6. Setting adjustable instruments used in the inspection of parts.
7. Verifying the accuracy of gages, and masters used in the inspection of parts.

LAYOUT OF TOOLS, DIES, AND FIXTURES

Before parts can be produced in large quantities, special tools must be built to control the accuracy of the parts and to reduce their manufacture to a series of more or less automatic operations. These special tools--such as punch press dies, milling fixtures, and drill jigs--must be machined from blocks of steel.

Before the toolmaker machines the block, it is customary for him to lay out or outline the material to be removed. Figure 227 shows a block scribed with a straight line. The height of the line is accurately determined by the gage blocks held in the adjustable holder and foot block.

CHECKING ACCURACY OF TOOLS, DIES, AND FIXTURES

Tools used in the manufacture of parts generally are held to one-tenth of the tolerance of the part. It is very important that these tools

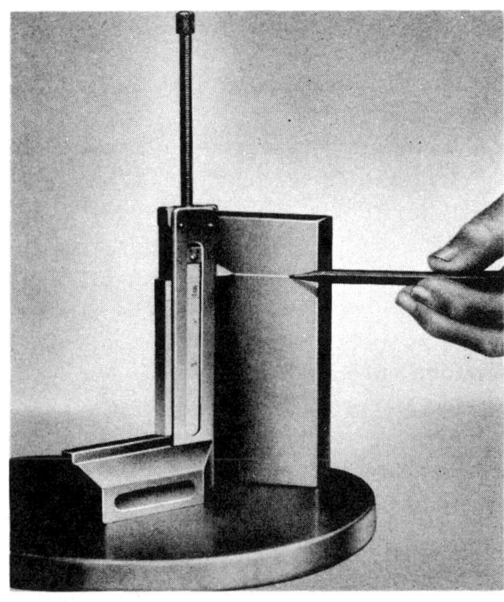

Figure 227. Scribing a Line with Gage Blocks and Attachments

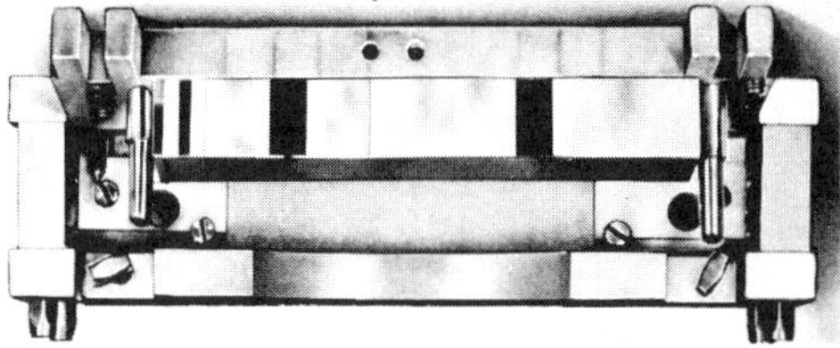

Figure 228. Checking Distance between Holes with Gage Blocks

be checked accurately; so the checking of dies, jigs, and fixtures is one of the basic applications of precision gage blocks. In most plants the finest working blocks are assigned to this work.

Figure 228 shows a stack of blocks being used to check the distance between two holes in a drill jig. Two plugs are inserted in the holes, and the distance between the plugs is measured by a stack of blocks. This is not the center-to-center distance between the holes; it is necessary to add one half of the sum of the diameters of the plugs to the length of the stack of blocks.

MACHINE SETUPS

In recent years, gage blocks have been used to facilitate and increase the accuracy of machine setups. Figure 229 shows use of gage blocks in the setup of a gear hob. This particular hob must be set off-center, with relation to the work arbor, .562". Therefore, to set the gear hob properly off-center this amount from the face, subtract one-half the diameter of the work arbor, which is .312" or .156", from .562" or .406" and insert gage blocks for this amount between the arbor and a straight-edge in line with the face of the hob.

CHECKING PARTS IN PROCESS OF MANUFACTURE

Inspection of the vital dimensions on parts in the process of manufacture has become one of the most important shop applications of gage blocks. The process inspector checks the setup by certifying the first part produced, then controls the accuracy of the lot by periodic inspection of important dimensions. Gage blocks are frequently used in process inspection, because their accuracy enables the inspector to follow

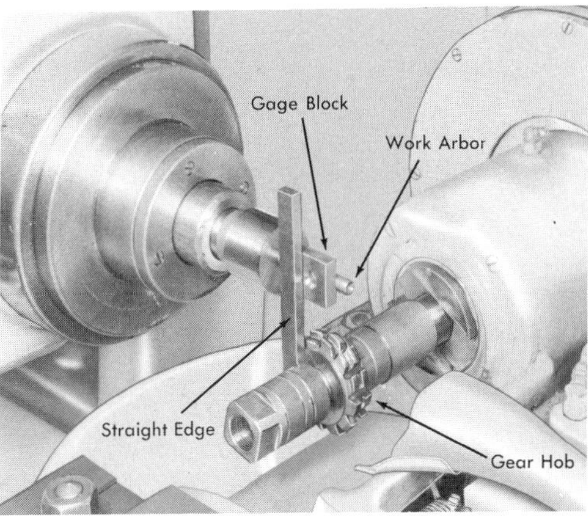

Figure 229. Setting Gear Hob with Gage Blocks

the tendency of the parts to approach either the low or the high dimension limits and correct it before they have been exceeded.

Gage blocks are used in drilling, milling, grinding, and other machining operations--particularly in grinding, which is a finishing operation in which skill of the operator is important to accuracy.

Figure 230 shows the use of gage blocks in checking the accuracy of splines in a spline gage. By inserting a stack of blocks as shown, the operator checks the width of the slots; by moving the blocks along the slot, he checks the parallelism of the sides.

CHECKING FINISHED PARTS

Finished parts are usually inspected by special gages and masters having an accuracy something less than that of gage blocks but more adaptable to the production process. However, where

Figure 230. Checking Width of Slot with Gage Blocks

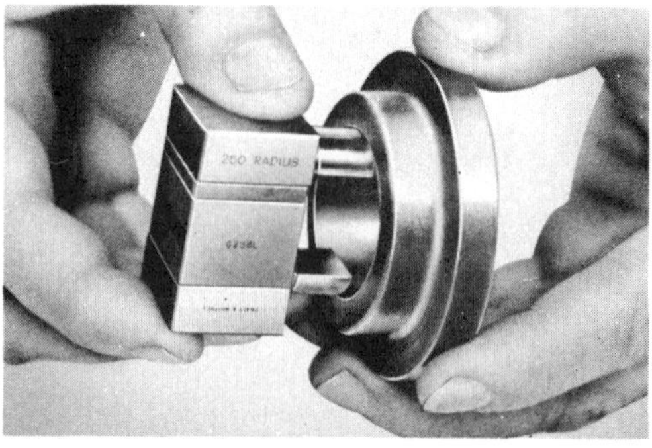

Figure 231. Checking Inside Diameter with Gage Blocks

Figure 232. Checking Spacing of Pins with Gage Blocks

the conditions are such that these methods of inspection are not practical, gage blocks are frequently used.

The gage blocks generally used for checking parts are in the third class; they are a degree lower in accuracy than the blocks used to check gages and tools. The tolerance on parts is naturally greater than that on tools and gages.

Figure 233. Checking Location of Pins on Gear Running Gage

Figure 231 shows a stack of gage blocks, with inside caliper jaws attached, being used to measure the diameter of a hole in a bushing.

Figure 232 shows a stack of rectangular gage blocks being used to check the spacing of pins.

SETTING ADJUSTABLE INSTRUMENTS AND INDICATING GAGES

Adjustable instruments and gages are set from either a master part or a stack of gage blocks. The average gage of this type measures the variation from the master in one or more dimensions and is quicker and easier to use than gage blocks.

Figure 233 shows gage blocks being used to check the mounting distances on a running gage for worm gears. On this gage the worm gear is run in contact with a master so that the smoothness of mesh and the amount of lost motion may be determined. Similar gages are employed to check spur gears, spiral gears,

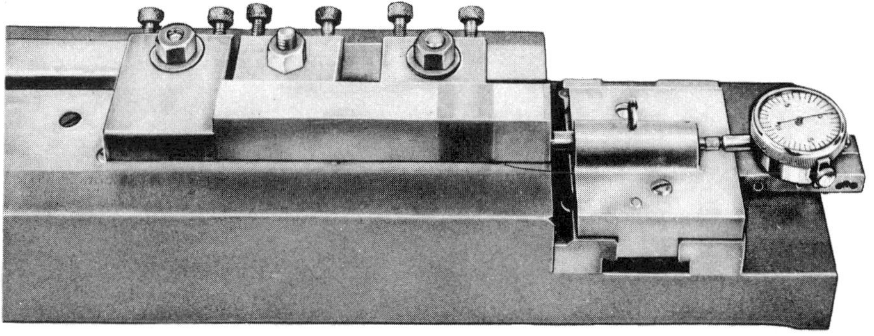

Figure 234. Setting an Adjustable-Length Gage

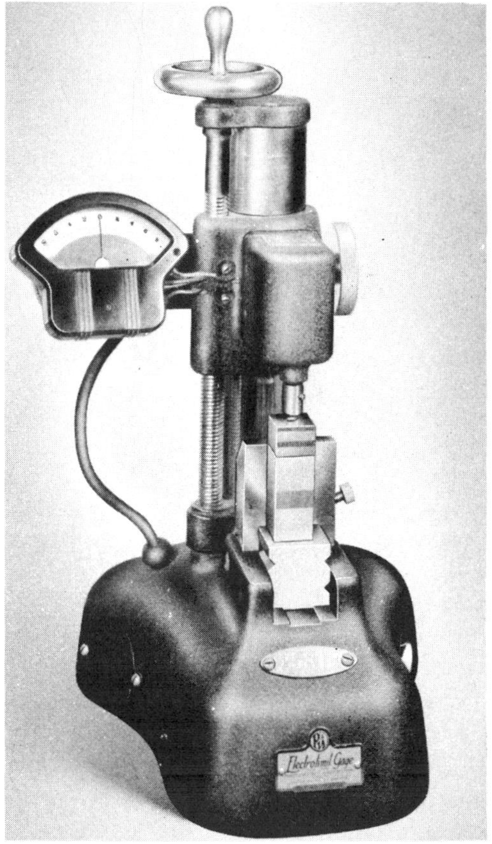

Figure 235. Setting Gage with Blocks

worms, and worm wheels.

Figure 234 shows a stack of gage blocks used to set an adjustable length gage. The dial is on zero when the gage is properly adjusted.

Shown in Figure 235 is a visual gage set up to check the pitch diameter of gears with the aid of wires. Here the gage is being adjusted to zero, with a stack of blocks corresponding to the pitch diameter of the gears to be checked, plus the constant for the wire.

Figure 236 shows the comparison between the gage blocks and the gear. Because each graduation is .00005", the difference is minus .00025", which is well within the tolerance for this gear.

VERIFYING INSPECTION GAGES

Periodic verification of the many working gages used throughout the manufacturing plant still remains the biggest job of precision gage

Figure 236. Checking a Gear on Electrolimit Gage

blocks. Through constant supervision of the accuracy of the gages used in the inspection of parts, accuracy is built into the product.

Figures 237 and 238 show two sets of precision gage blocks, a master set and a working set. The working set is being checked against the master set. This set may be one of several similar sets through a plant, all depending upon the accuracy of the master set.

The accuracy of a company's product rests in the master set. As indicated earlier, the known errors in this set are periodically determined; therefore, when checking another set against the master, these errors can be taken into consideration.

In Figure 237 a block has been set in the instrument, and the indicator is set on zero. In Figure 238 the corresponding block is checked against the master block, which shows that it is minus .000010", as the total range of the scale is .00030".

Gage blocks are used to verify plug gages, ring gages, snap gages, thread gages, running gages, and many types of special gages.

Gage blocks may be used to check threads, but in most shops a better and faster method is used. Gage blocks can be used where necessary gaging equipment is lacking.

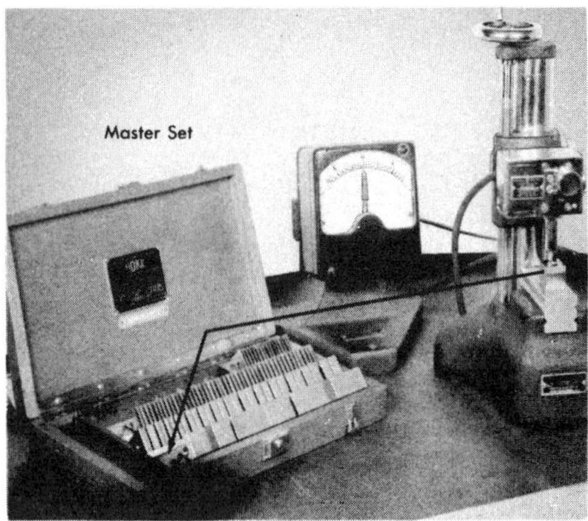

Figure 237. Setting Electrolimit Gage with Master Set

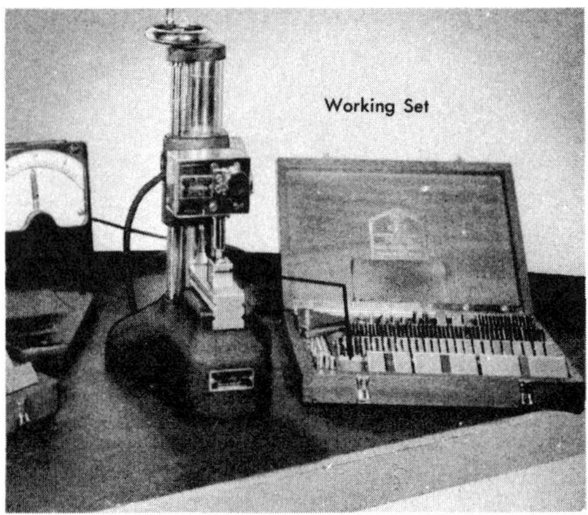

Figure 238. Checking Block from Working Set

Figures 239 and 240 show the application of gage blocks in verifying thread gages. In Figure 239 the lead of the thread at one point is being checked. The measuring points are tried in several places in making this measurement. In Figure 240 the length to the gaging point of a taper-pipe thread gage is measured with a stack of blocks. The 60 degree center point on the top of the stack of blocks meshes with the threads on the gage.

Where toolmaker's microscopes or projection comparators are available, these instruments are frequently used in checking thread gages instead of gage blocks, because at one setting they are capable of checking the pitch diameter, the lead, the major (O.D.) and minor (root) diameters, and the thread angle. The microscope and comparator will be discussed in a later chapter.

Plug and ring gages are used to inspect hole and shaft diameters, respectively. The accuracy of these gages is verified periodically with gage blocks to check the wear they receive in use. Figure 241 shows a plug gage being checked. Note use of accessories shown in Figure 226 as 3-10-12.

The setup for checking ring gages is similar to that for plug gages and is shown in Figure 242. In Figures 241 and 242, two different styles of caliper jaws are shown, both of which can be

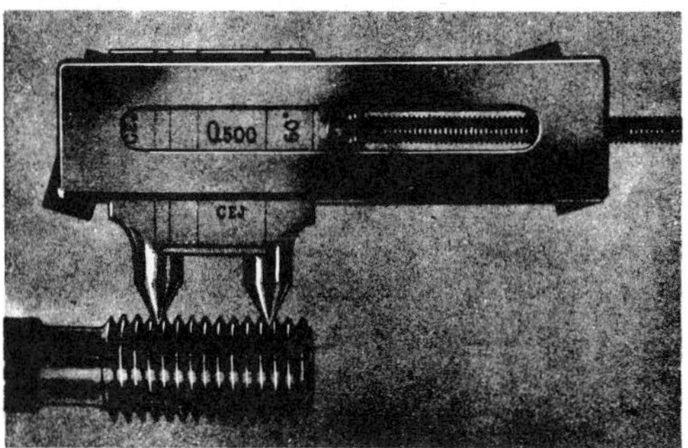

Figure 239. Checking Lead of a Thread Gage

Figure 240. Checking the Gaging Length of a Taper-Thread Gage

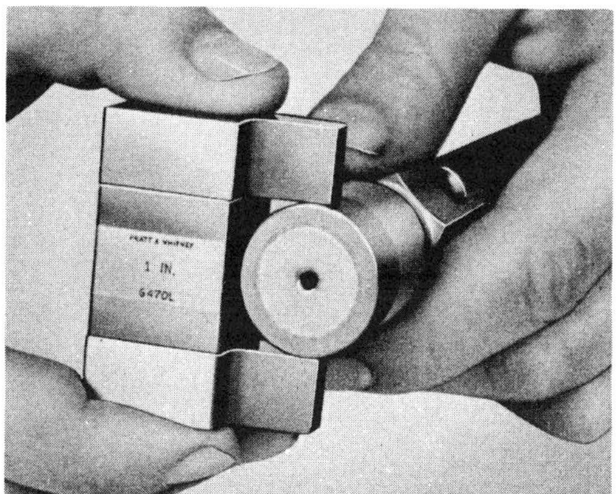

Figure 241. Checking the Diameter of a Plug Gage

used on plug gages; but only the radius style shown in Figure 242 can be used on ring gages.

Snap gages are checked as shown in Figure 243; here a set of blocks is being used to check the "not-go" jaws. In Figure 243 the blocks pass through both jaws, while in Figure 244 the second set of blocks does not pass through the "not-go" jaws.

In addition to controlling the accuracy of the standard types of gages previously shown, gage blocks are used to verify gages especially built to inspect a particular part.

GAGE BLOCKS--SUMMARY

From a study of the previously mentioned applications of gage blocks, it is not difficult

Figure 243. Checking "Not-Go" Jaws on Snap Gage

Figure 242. Checking Ring Gage

to see why they are becoming an increasingly important tool in modern manufacturing. Also, it should be evident that all gage blocks do not have to have the same accuracy; and as working sets become worn they are put on a class of work not requiring the former degree of accuracy.

It is well to remember that accuracy costs money. Wasting accuracy by verifying gages, adjusting instruments, building jigs, dies, and fixtures, setting up machines and inspecting parts to tolerances that are too close, is wasting money in the extra time and effort required, as well as in the work needlessly scrapped.

The classification of blocks in any plant will depend largely on the accuracy required to insure satisfactory performance in the product; this, of course, will vary considerably with dif-

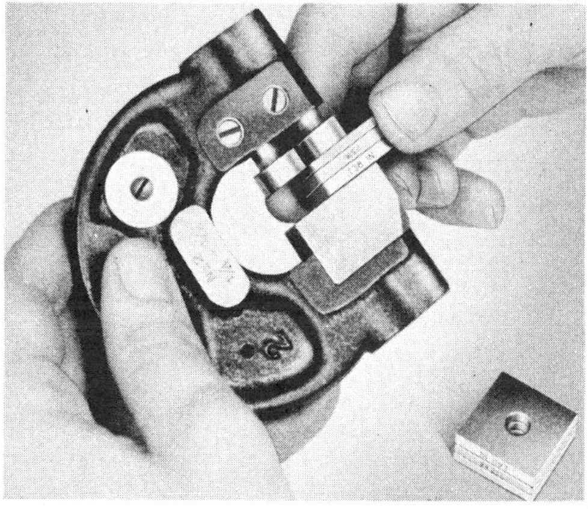

Figure 244. Checking "Go" Jaws on Snap Gage

ferent products. A typical classification follows:

Class	Class Work	Range of Error in Millionths of an Inch
1	Verifying gages, setting instruments and tool inspection.	5 to 20
2	Layout of jigs, fixtures, and dies, setting instruments, and tool inspections.	20 to 40
3	Setup of grinding, milling, and drilling machines and parts' inspection.	40 to 100

When gage blocks exceed the 100-millionth error limit, they may be either scrapped or chromium-plated and relapped to size. This process for reclaiming old blocks costs a little more than new blocks, but a chromium-plated block will last longer; so it is practical, particularly on work where the blocks are subject to hard use.

Gage blocks from sets of a low class of accuracy should not be used with blocks from sets of a higher class of accuracy, because they are likely to introduce an error in the length of a stack of blocks.

QUESTIONS AND PROBLEMS

1. Of what materials are precision gage blocks made?
2. How did development of the manufacture of interchangeable parts promote use of gage blocks?
3. Approximately how many gage lengths may be produced from a large standard set, and what is the increment of each step?
4. What is the difference between a master set and a working set? Is the master set necessarily more accurate?
5. What is the guaranteed accuracy of gage blocks made commercially?
6. If the accuracy of a gage block is said to be plus or minus .000002" in one inch, between what limits may a block .500" long vary?
7. Which is likely to be more accurate: a single four-inch block, or four blocks having a total length of four inches?
8. What is meant by "growth" in gage blocks, and how is it minimized by gage block makers?
9. How is the flatness of the measuring surface on gage blocks checked? How is the error in flatness shown?
10. What precautions must be taken in using gage blocks to avoid damage or excessive wear?
11. What effect does temperature have on all metals? Do all metals expand and contract at the same rate?
12. Why is no temperature correction necessary when using gage blocks for ordinary work?
13. Compute the expansion of a four-inch block if the temperature of the block is 85 degrees F. Use coefficient of 7 millionths.
14. Compute the error in thousandths of an inch, due to the difference in temperature between the gage block and the work. A phosphor bronze journal, two inches inside diameter, has been heated by machining so that its temperature is 150 degrees F. The temperature of the gage block is 68 degrees F.
15. What is the standard measuring pressure?
16. Make up the following gaging lengths without using the same block twice, from a large set of blocks:
    (1) 1.782 (2) 2.9826 (3) .8543
17. Name several applications of a set of precision gage blocks and its accessories.

Chapter VIII

SURFACE PLATES AND ACCESSORIES

THE STANDARD method of dimensioning mechanical drawings involves use of a base line, or reference plane, to which all dimensions are referred, either directly or indirectly. On the usual part made in the shop this line coincides with, or passes through, the portion of the part known as the locating surface or locating point; this point may be the center of a hole, the finished surface of locating feet, or, in general, a portion of the part which comes into contact or alignment with other parts in the assembly.

For example, two brackets mounted on the same base at assembly would have their important dimensions referred to the finished feet of the bracket; so, when both brackets were mounted to the base, the holes in one bracket would be in proper relation to the holes in the other bracket. Two brackets of this type are shown in Figure 245.

THE SURFACE PLATE

A surface plate is a rigid block or table whose flat or finished surface provides a reference plane for inspection and layout work. The cast-iron surface plate shown in Figure 246 has three legs so that it will not rock when placed on an uneven surface such as the top of a workbench or an inspector's bench; these are shown in Figure 246B.

Rigidity is secured by ribs connecting the legs and reinforcing every section of the plate. The plate must be rigid to insure that the working surface will not deflect when subjected to the load of heavy parts, angle irons, parallels, and other accessories to be discussed later in this chapter.

The surface plate in Figure 246 is the type usually found on work and inspection benches. Note the flat finished surface, the heavy construction and reinforcing ribs. Also on either end provision has been made for removable handles, shown in Figure 246B, to assist in handling the plate during the scraping operation to be described later.

The size of surface plates will vary from a small 3-1/2" by 4" plate, weighing about three pounds, to a plate 48" by 144" weighing as much as 6,000 pounds. The larger sizes are made to special order and are required for large work that cannot be handled on the smaller plates. The sizes most commonly used in toolrooms and on inspectors' benches are the 12" by 18" and the 18" by 24".

MACHINING AND SCRAPING

The original casting is first seasoned to remove all internal stresses. This is done either by aging or by subjecting the casting to a heat-treatment to relieve the tension between the molecules of cast iron, so that when the surface plate is machined and finished it will not warp but will retain its flat, even surface indefinitely.

After seasoning, the surface of the plate is planed in preparation for scraping. Scraping is a hand operation in which the surface of the plate is finished with hand-scraping tools to a flat plane.

It is a well-known fact that two smooth flat metal surfaces will adhere to each other. This is illustrated in the use of gage blocks where the tendency of the blocks to adhere to one another is essential to using them in stacks. The bases of the instruments used on the surface

Figure 245. Alignment of Bearing Brackets on an Assembly

plate are flat and smooth and will adhere in like manner to a smooth surface plate. Therefore, it is necessary to resort to hand scraping which leaves the surface covered with a multitude of small flat spots or "nodes" upon which the instruments will slide freely.

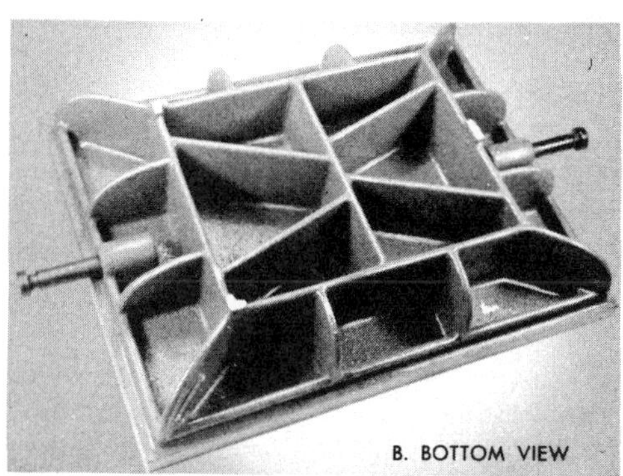

Figure 246. Cast-iron Surface Plate

Surface plates are scraped in groups of three. The planed castings are compared with one another by lightly coating the surface of one plate with Prussian blue and inverting it upon another so that the surfaces of both plates are in intimate contact. Then, by sliding one upon the other, the high spots of the under plate will become coated with blue. These high spots are then scraped and the procedure repeated until the number and distribution of the spots indicate that the two surfaces match each other.

If only two plates were matched in this way, there would be no assurance that the surfaces were flat, because the two plates might develop mating irregularities. Therefore, both of these plates are compared with a third plate; when all three plates match one another, it can be

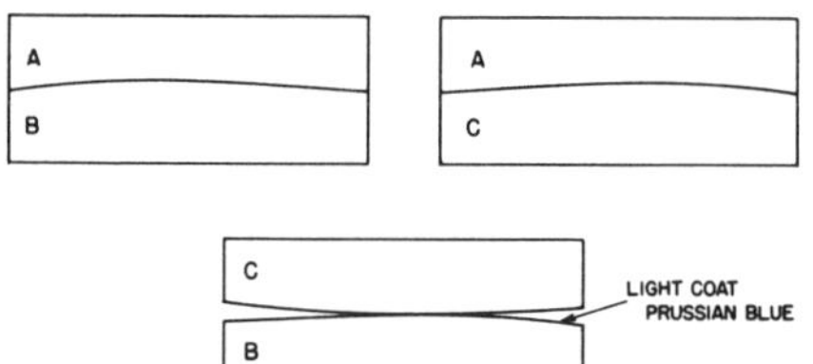

Figure 247. Comparing Scraped Surface Plates

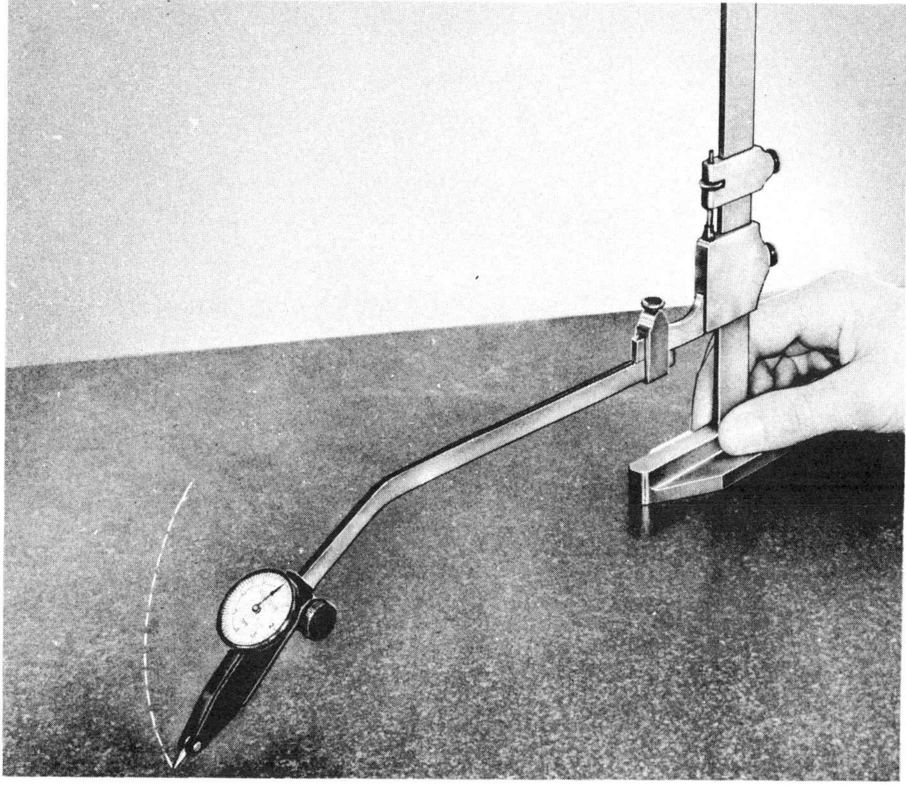

Figure 248. Indicating Flatness of Surface Plate

assumed that the three surfaces are flat. Figure 247 shows how the third plate reveals the mating irregularities of two matching plates.

As a final check, surface plates are compared to a master plate; if the number and size of contacts are large enough, the plate is ready for use. Because the spots or nodes are fairly close together and because the gages and tools used on the surface plates have bases that extend over a considerable number of these nodes, there is always sufficient bearing surface.

ACCURACY

There is no American standard for size or accuracy of surface plates. Some manufacturers regard the British standard as tentative for this country. This standard requires that "the deviation from a mean plane must not exceed ± .0003", and the deviation in 12" must not exceed ± .00015" for an 18" by 24" surface plate." This limit increases with the size of the plate.

As a surface plate is used, it begins to wear and will no longer be as accurate as it was when new. The amount of allowable error that may develop before the plate should be rescraped depends on the accuracy required of work done on the plate. For ordinary toolroom and inspection work, an 18" by 24" surface plate should be within .001" from the mean plane at any place on its surface.

TESTING FOR FLATNESS

A rough check can be made with a sensitive indicator mounted on a long arm extending from a height gage. By swinging the indicator in large arcs and shifting the base to different positions on the plate, any variation or worn spot may be detected by deflection of the pointer on the indicator. This method is shown in Figure 248.

A more accurate check may be obtained by supporting a straightedge upon two equal stacks of gage blocks (Figure 249). With a third

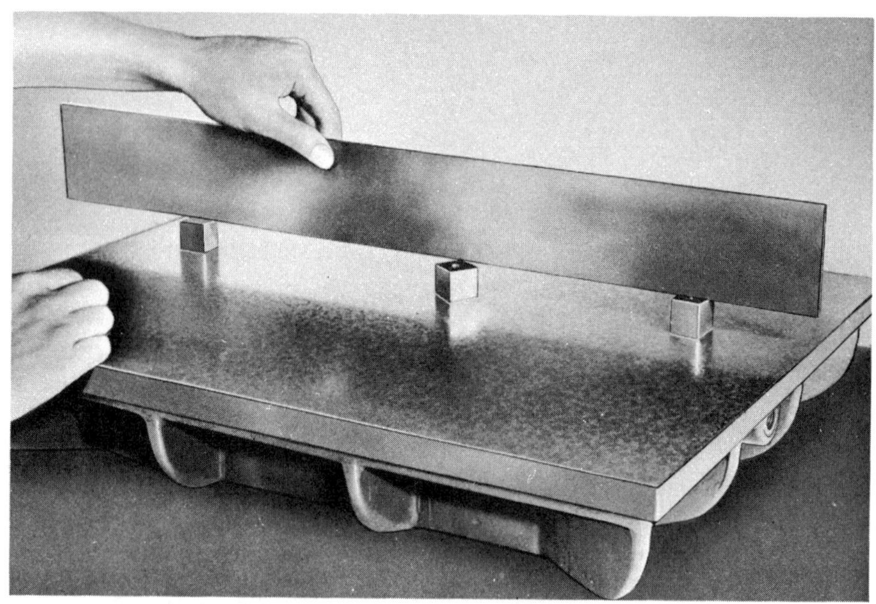

Figure 249. Checking Flatness with Straightedge

stack of gage blocks, .0005" higher than the two supporting the straightedge, check the space between the straightedge and the surface plate at various points to determine the amount of wear. Maximum wear is usually at the center of the plate, because it is used more in this area. Accuracy of the plate determines the class of work for which it will be used.

CARE OF SURFACE PLATES

Surface plates are rugged and able to withstand a certain amount of abuse, but it is easy to destroy their accuracy. For their care and preservation, observe the following:

1. Keep clean. This is the most important rule in connection with the care of surface plates. Cleanliness is necessary not only because dirt getting under the work and the instrument base, affects the accuracy of measurements made on the plate, but also because the abrasive action of a gritty substance will rapidly wear down the surface and destroy its accuracy. Occasionally the surface of a surface plate should be cleaned to remove any sticky film, rust, or tarnish. A good cleaning fluid may be used. Some machinists use "Bon Ami," which emulsifies the dirt and grease and removes rust and tarnish so that when it is dry it may be wiped off with a dry cloth. Dust on a plate may start corrosion.

2. Cover when not using. When not in use, cast iron surface plates should be protected by a wooden cover and should be wiped clean before being used. Over a weekend or holiday, a film of light oil should be spread on the surface before putting on the cover.

3. Slide heavy parts or blocks onto the plate. Heavy tools, such as angle irons and parallels, should be set down on the edge of the plate and slid into working position. It is good practice to wipe off both plate and tools first. Sliding tools into position is an additional precaution against scratching either the plate or the tool with any foreign material adhering to the surfaces. The sliding action tends to brush it away. If a heavy tool or part must be placed directly on the plate, it should never be dropped or allowed to strike on a corner and nick the plate. If a plate does become nicked, the high spot should be removed with a fine oilstone.

4. Tap the work; do not hammer. Never hammer on a surface plate. Sometimes it is necessary to tap a part into position; however, any hard pounding will injure the plate.

5. Use parallels when possible. Rough castings, forgings, and other large work likely to have imbedded grit or hard scale on surfaces,

Figure 250. Toolmaker's Flat

should be set on parallels to protect the plate.

6. Remove burrs. Parts which have burrs should be filed or stoned before setting them on a surface plate, to save the plate from being scratched, as well as to prevent any error in the measurement.

7. Remove clamps. Parallels, angle irons, or parts should not be left clamped to, or allowed to rest on a plate for a long period of time; this will cause corrosion. The surface of the plate is its most expensive part; once accuracy is destroyed, there is no easy way to restore it. A damaged plate is repaired by the manufacturer, who refinishes it by scraping. This is an expensive hand operation which may cost several hundred dollars on a large plate.

TOOLMAKER'S FLAT

The toolmaker's flat (Figure 250) is a very accurate surface plate for the finest and most exacting layout work and measurements. Where accuracy of an ordinary surface plate is measured in terms of thousandths, the toolmaker's flat is lapped and parallel to .00001" on both top and bottom. Precision gage blocks will wring to the flat just as they do to one another.

The toolmaker's flat is about 5" in diameter and 7/8" thick, and is made of an alloy steel, heat-treated to prevent distortion due to internal stresses.

The toolmaker's flat is nearly always used in conjunction with gage blocks; its field of application, in general, is limited to small work requiring the accuracy of the gage blocks themselves.

Figure 251. A Granite Surface Plate

GRANITE SURFACE PLATES

Granite surface plates were first introduced in 1942; since then they have gained in popularity to a point where they have practically replaced cast-iron plates in many shops.

Granite or natural hard-stone plates, such as the one shown in Figure 251, have several advantages when compared with the ordinary hand-scraped cast-iron surface plate.

Granite is granular in structure and composed of hard quartz, and softer materials such as feldspar and mica. In lapping operations, the abrasive cuts out the softer materials and allows the hard high points to project; this process permits instruments, gage blocks and tools used on granite plates to move easily across the surface without wringing or "freezing."

Other advantages of granite plates: their surface is harder than iron and free from shock; this prevents distortion from heavy objects dropped on the surface.

Their wear shows up as scratches or small nicks which do not burr or bulge as on metal plates, and do not impair the accuracy unless grouped densely.

They have a low heat conductivity and are not appreciably affected by ordinary temperature changes.

They are rust proof and require no protection against corrosion.

They are non-magnetic.

As abrasives will not imbed in the surface, they can be used for convenient checking near

Figure 252. Sets of Steel Parallels

Figure 253. Set of Cast-iron Parallels

grinding operations.

They are lapped to a high degree of accuracy and are available in sizes from 8" by 12" to 60" by 84", with surface accuracy of .0001" to .00005".

PARALLELS

Parallels (Figure 252) are rectangular bars of steel or iron accurately ground on four sides so that the sides are almost perfectly flat and parallel. They are used as spacers between the work and the surface plate. Cast-iron parallels are usually constructed as shown in Figure 253. This construction combines maximum strength and rigidity with least weight.

It is recommended that parallels be used, whenever possible, when checking or scribing parts on a surface plate. For example, parallels should be used:

1. To make sure the locating surface of the part is in contact with the reference plane, which when parallels are used, is transferred to the top surface of the parallels. Contact can be assured by moving the parallel slightly while under the part. If the parallel moves freely, it is not in contact; but if it resists the effort to move it, there is good contact between plate, parallel, and work.

2. To place work on a surface plate when sections of the work extend beyond or below the machined surface from which measurements are to be taken (Figure 254).

Figure 254. Use of Parallels

3. To protect the surface of the surface plate. Parallels are usually made of hardened steel and will resist abrasion or scratching, while the cast-iron surface plate, being soft, can easily be marred by a rough surface or a sharp edge. Also, the cost of a pair of parallels is considerably less than the cost of a surface plate.

COMMERCIAL SIZES OF PARALLELS

Sizes of steel parallels range from 5/16" by 1/4" by 5" to 3" by 1-1/2" by 12".

Sizes of cast-iron parallels range from 1-1/2" by 3" by 24" to 4" by 8" by 100".

Parallels are usually manufactured and sold in pairs, so that both parallels will be the same in width and thickness. Sides of parallels are not necessarily square. They are not guaranteed to be square by the manufacturer, and therefore, should never be used as a square. The accuracy of parallels obtained commercially is given in Table XI.

TABLE XI
ACCURACY OF PARALLELS

	Size Tolerance (Inches)	Parallelism Within Length (Inches)
STEEL		
Under 1-1/8" wide and up to 6" long.	± .0002	.0001
From 1-1/8" to 1-1/2" wide and from 6" to 8" long.	± .0003	.00015
From 1-1/2" to 3" wide and from 8 to 12" long.	± .0004	.0002
CAST IRON		
From 24" to 36" long.	± .0005	.0005

The accuracy of parallels decreases as the width and length increase. No tolerance is given on the squareness of the sides because all manufacturers specify that parallels should never be used as squares.

BOX PARALLELS

Box parallels such as those shown in Figure 255 are used much like ordinary parallels. However, they have an additional application: providing a flat working surface or reference plane, elevated above the surface of the surface plate, upon which height gages, surface gages, or other gages can be placed to measure dimensions beyond the range of the instrument.

For example, suppose it is required to check the location of a hole in a casting, the dimension of which is 13.127" to the base, and a 10" height gage is available. The dimension is beyond the range of the measuring instrument; but by placing the height gage upon a 6" box parallel, its range is increased to 16 inches and location of the hole can readily be checked.

Box parallels are ground flat on all surfaces to within ± .0005". Opposite sides are parallel within .00025" in 6" and adjacent sides are square within .0005" in 6". Box parallels are commercially available in the following sizes in inches: 4 by 4 by 6, 5 by 8 by 12, 4 by 6 by 6 and 6 by 14 by 16.

ANGLE IRONS

An angle iron, as the name implies, is an iron casting in the form of a right angle. The sides and surfaces are ground square or parallel with each other. When the angle iron is placed on the surface plate, it provides a perpendicular working surface to which the work may be clamped for inspection with a height gage, surface gage, or other measuring instrument.

The universal right-angle iron (Figure 256) is provided with a rectangular web or rib between the two legs; this gives it some of the features of the box parallel, because it can be placed on any one of its six sides--yet it retains the clamping advantages of any angle iron.

Sizes of universal angle irons range from 3-3/4" by 4" by 5" to 8" by 9" by 16". All

Figure 255. Box Parallels

Figure 256. Universal Right-angle Iron

Figure 257. Measuring Iron

but the largest size have a guaranteed accuracy of .0002" for parallelism and squareness of sides and ends. The largest size has a guaranteed accuracy of .0003".

The measuring iron shown in Figure 257 is a tall angle iron originally developed for machine-tool applications. The key slot and bolt hole in the base provide a means for accurately locating and clamping this iron to the machine table. The measuring iron is commonly used on the surface plate by the inspector and toolmaker for inspecting and laying out large castings and long parts beyond the clamping capacity of the ordinary angle iron. Narrow in relation to its height, it occupies a small space on the surface plate.

The front face and bottom are finished accurately and square. Edges are machined for finish only; therefore, it should not be laid on its side for accurate measurements.

Measuring irons are available in two sizes as shown in the following table:

Width	Short Leg	Long Leg	Weight
2-1/2"	6"	12-1/2"	12 lb.
4"	8"	21-1/2"	50 lb.

V-BLOCKS

A V-block is a rectangular block of steel, the top of which contains a V-shaped slot, ground to an included angle of 90 degrees, in which a cylindrical part can be held by means of a suitable clamp.

Figure 258 shows a variety of V-blocks and their relative capacities. They are used on the surface plate for holding cylindrical parts, or arbors upon which the work is mounted. The V-shaped clamp (A) makes this block adaptable to both large and small work by simply inverting the clamp.

V-blocks are usually ground and purchased in pairs so that they will be identical in dimensions. A pair of V-blocks will have a common identifying number stamped on the end of the blocks so that they may be kept together. Care should be taken when using a setup requiring two V-blocks that the blocks are mates, otherwise the accuracy of the setup may be impaired by a slight difference in the dimensions of the V-blocks.

The relation of the V to the sides is important; it should be central between the two sides and parallel to the sides. This relation can be checked as shown in Figure 259. Clamp a perfectly straight shaft, of constant diameter, in the V and indicate both ends of shaft while the block is resting, first on one side, then the other. The difference between the readings at the two ends in one position represents the amount the V is out of parallel. The difference between the two readings at one end when the block is placed--first on one side, then the other--will represent twice the amount the V is off center at that end. V-blocks are accurately ground on all sides, and the V is central between the sides within an accuracy of .0003".

A combination of the right-angle iron and

Figure 258. V-Blocks

the V-block in one unit makes this angle iron the most convenient used on the surface plate (Figure 260). It has all the features of a universal right-angle iron plus those of a V-block and, in addition, provides a means of clamping the work in position when the work is mounted on an arbor.

THE SQUARE

The square is a hardened-steel right angle, accurately ground to 90 degrees both outside and inside, made in a variety of sizes (Figure 261).

Figure 259. Checking Relation of V and Sides

Figure 260. Universal Right-angle Iron with V

Figure 261. Steel Squares

It is usually made of two pieces: the base, called the beam, and the upright, called the blade. On the standard square, the beam and blade are permanently fastened together. A small relief is provided on the inside of the square at the junction of beam and blade, so that when checking over a sharp corner, a burr on the corner of the part will not interfere with the measurement.

The use of the square is shown in Figure 262. It is used mainly to check the squareness of perpendicular surfaces, and for squaring-up parts on a surface plate before inspecting.

When using the square against a bright background, the presence of a beam of light between the square and part indicates that the surface is not square. A piece of white paper, held at an angle behind the work (as illustrated in Figure 262A), will give light reflection for visible checking; the amount the part is out of square can be estimated by the amount of light visible between the part and the square.

More definite measurement can be made by inserting the leaf of a feeler gage between part and blade (Figure 262B).

Another method commonly used by the inspector consists of inserting, between the work and the square, two pieces of thin paper or cellophane (as shown in Figure 262C). One piece of paper is inserted near the top of the blade, and the other near the bottom. If the papers are held in position by the blade, squareness is within the thickness of the paper. If, by pulling gently on the papers one at a time, an even amount of tension is observed on the two papers, squareness is within a fraction of the thickness of the paper. Cigarette paper and cellophane are about .001" thick; so by this method, it is possible to square-up the work within .0005".

A magnifying glass can be used to advantage in checking squareness, by sighting the amount of light visible between the work and the square, or examining the fit of the edge of the blade against the work (Figure 262D).

Sometimes the standard square is used at an angle to the surface being checked in order to have the corner of the blade make a line contact with the surface. This practice should be discouraged, because the side of the blade is rarely perfectly square with the beam. If line contact is desired, a bevel-edge square or universal square should be used.

Figure 263 shows a bevel-edge square which permits the blade to make a line contact with the work. This feature enables the inspector to make a more accurate check of the squareness of a surface. Against a bright background a very small error can be sighted with this square. The thin edge of the blade is susceptible to wear and damage; it should be handled with care.

The adjustable square (Figure 264) was designed for checking the squareness of surfaces inaccessible to the standard square with a fixed blade.

The square shown in Figure 264A is equipped with four blades, any one of which can be inserted in the beam, adjusted to length, and locked in position by the thumbscrew on the end of the beam. The bevel blade has an angle of 30 degrees on one end and 45 degrees on the other. The narrow blade is 1/8" wide.

Figure 264B shows an application of the bevel blade in checking the squareness of a portion of a dovetail.

PRECISION SQUARES

All steel squares used in inspection and setup work are ground to precision accuracy.

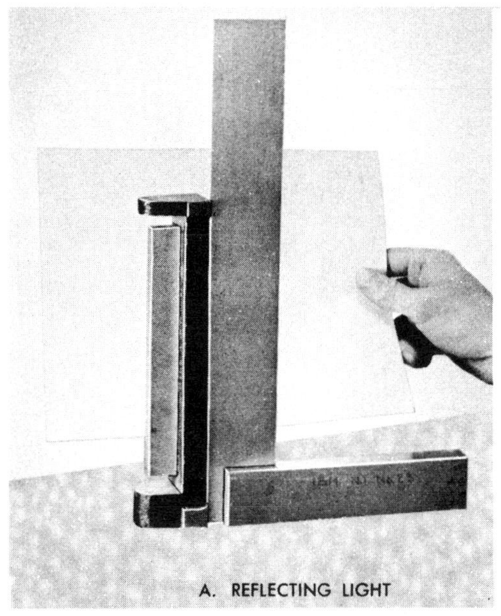

A. REFLECTING LIGHT

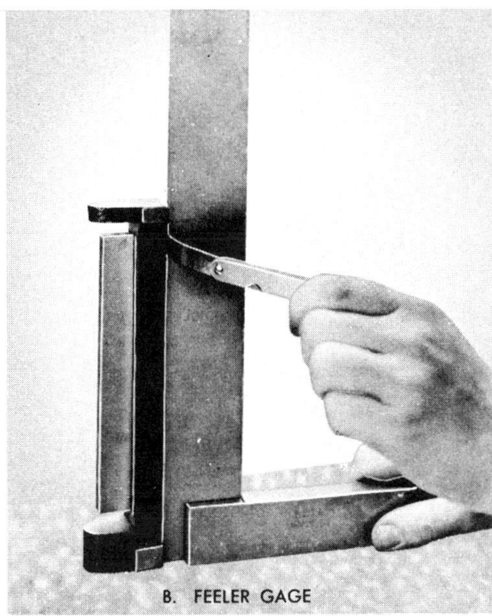

B. FEELER GAGE

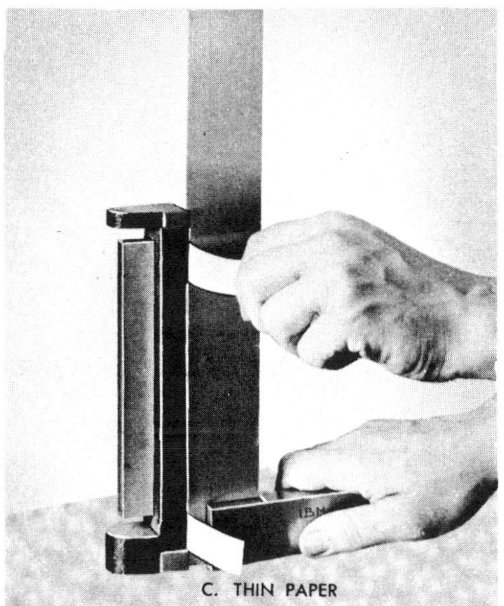

C. THIN PAPER

D. MAGNIFYING GLASS

Figure 262. Methods of Using Square

But, as with other precision gages, it is necessary to compare them with a master occasionally to determine whether the accuracy has been impaired by wear or accidental injury.

The cylindrical square (Figure 265) is commonly used as a master square. It is an alloy steel cylinder; its diameter is accurately ground and lapped to a true cylinder, and the ends are ground and lapped square with the axes of the cylinder.

Notches on the ends provide spaces for small particles of grit or dirt, present on all surfaces. When the square is placed on the surface plate, it is turned slightly on its axis so that these small particles will be picked up by the notches, and the end of the square will make close contact

Figure 263. Bevel-edge Square

with the surface of the plate.

The cylindrical square provides a perfect line contact with another surface, no matter what portion of the cylinder is making contact. In other words, the cylindrical square represents a beam with an infinite number of blades fanning out in all directions; each blade will make a perfect line contact with another surface. It is used to check the squareness of angle irons, V-blocks, steel squares, and occasionally to check jigs, fixtures, and die parts where tolerances are very small.

The large 12" cylindrical square is square within .0002" in its length or .0001" in 6". It weighs approximately 64 pounds, and is so smooth that it is difficult to handle. Particular care should be exercised in placing a cylindrical square on the surface plate. Allowing the cylindrical square to fall even a slight distance may injure both square and plate.

The universal square (Figure 266) was designed to fill the need for a high-precision square which can be used in any position. The rectangular construction with knife edges and two flat sides finished square to a high degree of accuracy makes a variety of applications possible. It is available in two sizes: 2" by 2-1/2" by 1/2" and 2-1/2" by 3-1/2" by 9/16".

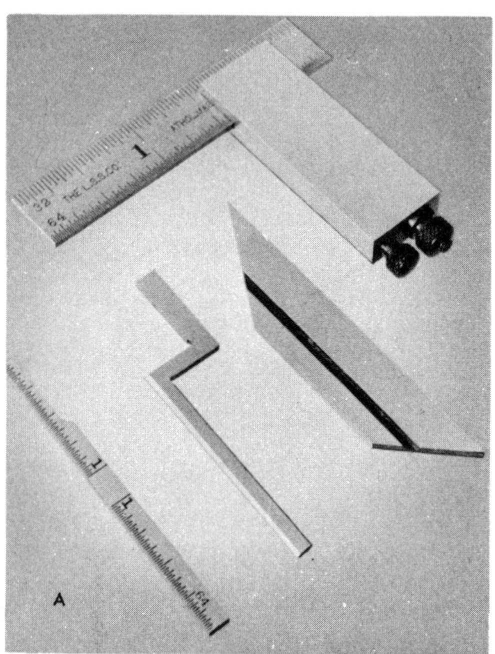

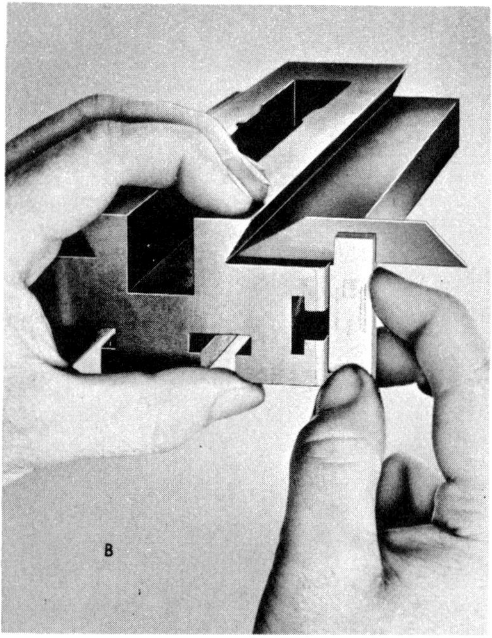

Figure 264. Adjustable Square

Figure 265. Cylindrical Square

STRAIGHTEDGES

A straightedge is a flat bar of steel or iron, whose edge is perfectly straight. It is used mainly for checking flatness of surface and for scribing straight lines in layout work. The standard steel straightedge (Figure 267) is the type commonly used by machinists and inspectors. Both edges are ground straight and parallel to each other.

Because this straightedge is comparatively thin, it will bend sidewise readily; so, it must be held square with the surface being checked--otherwise deflection of the straightedge will cause error.

Long standard steel straightedges will deflect slightly when supported near the ends, even when held square with the surface being checked. However, there are two points approximately .223 of the length from each end at which the straightedge, if supported, will have a minimum deflection either at ends or center. These points are called balancing points and are marked on long standard straightedges with an arrow and the words "support here."

Figure 266. Universal Square

The beveled-steel straightedge (Figure 268) is thicker than the standard straightedge; one edge is beveled to provide an edge approximately 1/16" wide. This narrow edge permits a more accurate inspection of surface flatness and straightness of parts. The straightedge can be held at an angle to the surface without materially affecting the accuracy of the measurement. Thickness is proportioned to the length to give it rigidity; it will not deflect sidewise when held at an angle.

Cast-iron straightedges are usually of ribbed construction to reduce weight and still retain rigidity. The cast-iron straightedge in Figure 269 was designed for leveling and setting up machinery, and for inspecting the alignment and flatness of bearing surfaces of machines, tools, and fixtures.

The knife-edge straightedge is used for

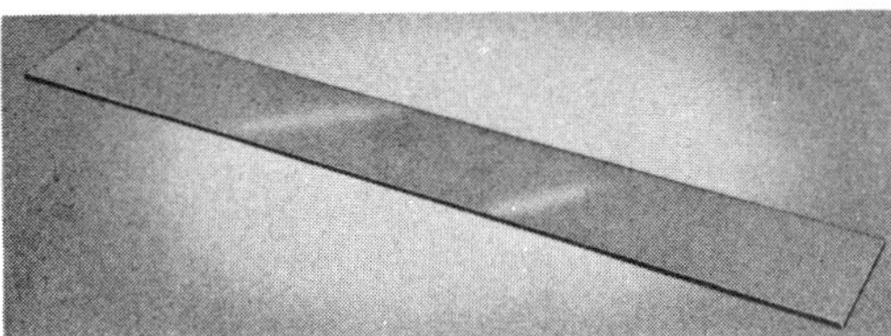

Figure 267. Standard Steel Straightedge

Figure 268. Beveled Straightedge

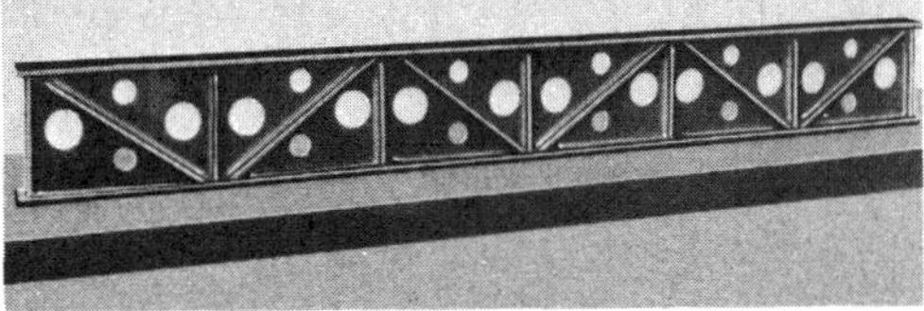

Figure 269. Cast-iron Straightedge

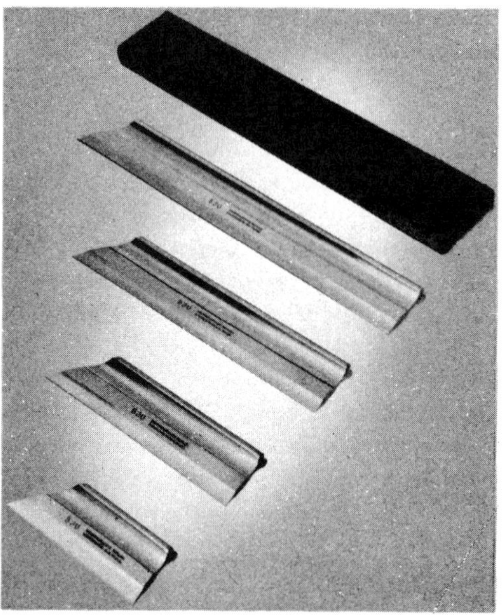

Figure 270. Knife-edge Straightedges

work requiring extreme accuracy. The edge is very narrow and is of semicircular cross-section so that line contact is maintained even when held at an angle to a surface. Figure 270 shows a set of knife-edge straightedges ranging in length from 1" to 6-1/4". With the set, is a glass test bar for checking the straightedges for accuracy.

APPLICATION OF STRAIGHTEDGES

Figure 271 shows use of the standard steel straightedge in checking a large flat surface. A feeler gage is inserted between the straightedge and the surface to determine the flatness of the surface. If a very accurate check is desired, coat the surface with Prussian blue and

Figure 271. Checking Large Surface with Straightedge

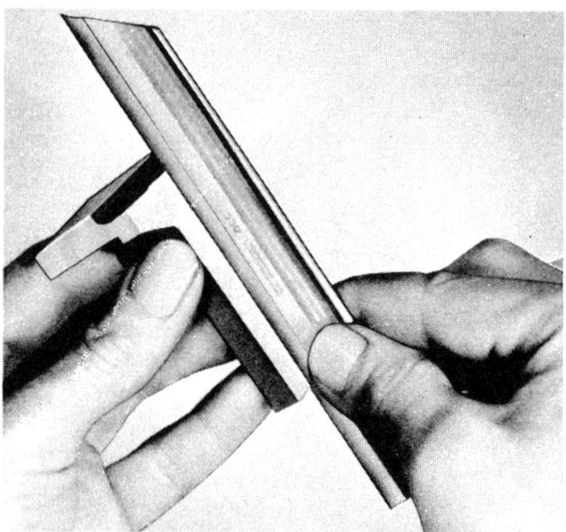

Figure 272. Using a Knife-edge Straightedge

slide the edge of the straightedge over the surface. This exposes high spots.

Figure 272 shows use of the knife-edge straightedge in checking the flatness of a surface against a bright background. The amount of light visible between the straightedge and the part indicates the error in flatness.

THE PLANER GAGE

The planer gage was originally designed for setting the cutting tool of a planer or shaper, but because of its versatility, it has a wide variety of uses.

The planer gage in Figure 273 consists of an inclined base with a slide. It is equipped with a scriber, an offset "foot," and a knurled extension. It can be accurately and quickly set to any required size within its range by use of a micrometer (Figure 274), a height gage, vernier

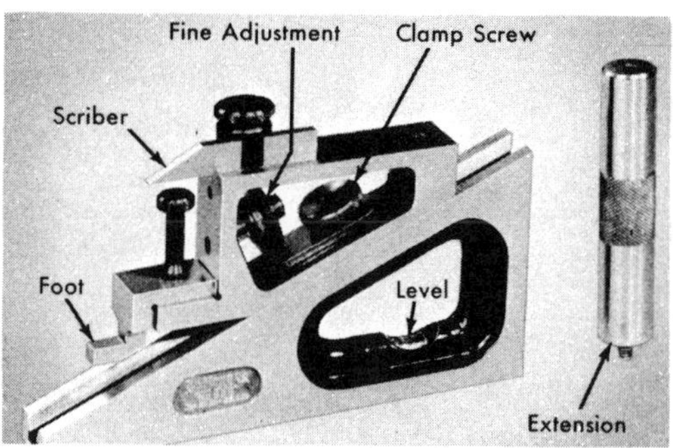

Figure 273. Planer Gage

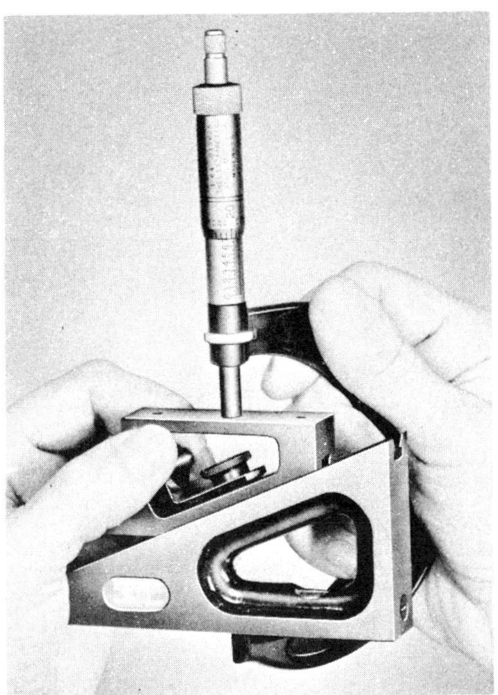

Figure 274. Setting Planer Gage with Micrometer

caliper or gage blocks, and used to transfer the setting with an indicator.

With the scriber attachment, it can be used as a height gage for laying out work. The offset attachment or "foot" permits settings in narrow areas. By use of the 3" extension, its range is increased to 9-1/4". Fine adjustment is obtained by use of the knurled thumbscrew, and a clamp nut locks the slide in place. The gage has a built-in level in the base. Figure 275 shows the planer gage being used to scribe a line on a part.

MEASURING EXAMPLES
SETTING UP ON A SURFACE PLATE

When the inspector has a part to check, he first must know what operation is to be checked and what particular dimensions on the blueprint are involved. This determines the setup on the surface plate. Referring to the operation record, and reading it carefully in relation to the print, he determines the exact nature of the operation. He must make allowance for future operations, such as leaving material in a drilled hole for future reaming, grinding, or lapping.

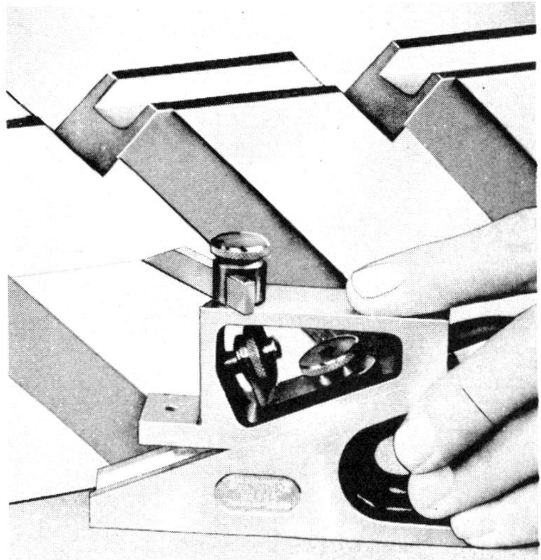

Figure 275. Scribing a Line with Planer Gage

The following important precautions must be taken in setting up work on a surface plate:

1. Surface plate, angle irons, parallels, etc., must be clean and free of burrs which might affect the accuracy of the setup.

2. The locating surface, or reference plane, must be placed parallel and square to the surface of the plate, and be in a position readily accessible to the indicator of the height gage.

3. The part must be clamped in position securely enough to prevent the part from moving during inspection, but the clamp must not be positioned so as to spring the part.

MEASURING EXAMPLE A, PART I

REFERENCE: Operation Record (Figure 276)

OPERATION RECORD-FACTORY COPY

ORDER NUMBER	DEPT. CHARGED	QUANTITY	DATE ISSUED
120993			7 30

INTERCOUNTER SHAFT
SUPPORT BEARING ASM

OPERATION	OPER. NO.	DEPT.	TOOL NO.	GROUP
ASSEMBLE 4 CAPS WITH SCREWS SCREWS TO BE VERY TIGHT 24917 8 120962 1 123600 4	5	920		
INSPECT	10	690		
MILL	15	010	37039	10
DRILL AND REAM 4 HOLES	20	060	36196	37
MILL 3125 DROP	22	010	57033	10
DRILL NO 38 HOLE CSK FOR TAP BURR AND TAP	25	030	48443	36
BURR COMPLETE AND BREAK CORNERS TO BP 120993	28	120		146

Figure 276. Operation Record for Example A

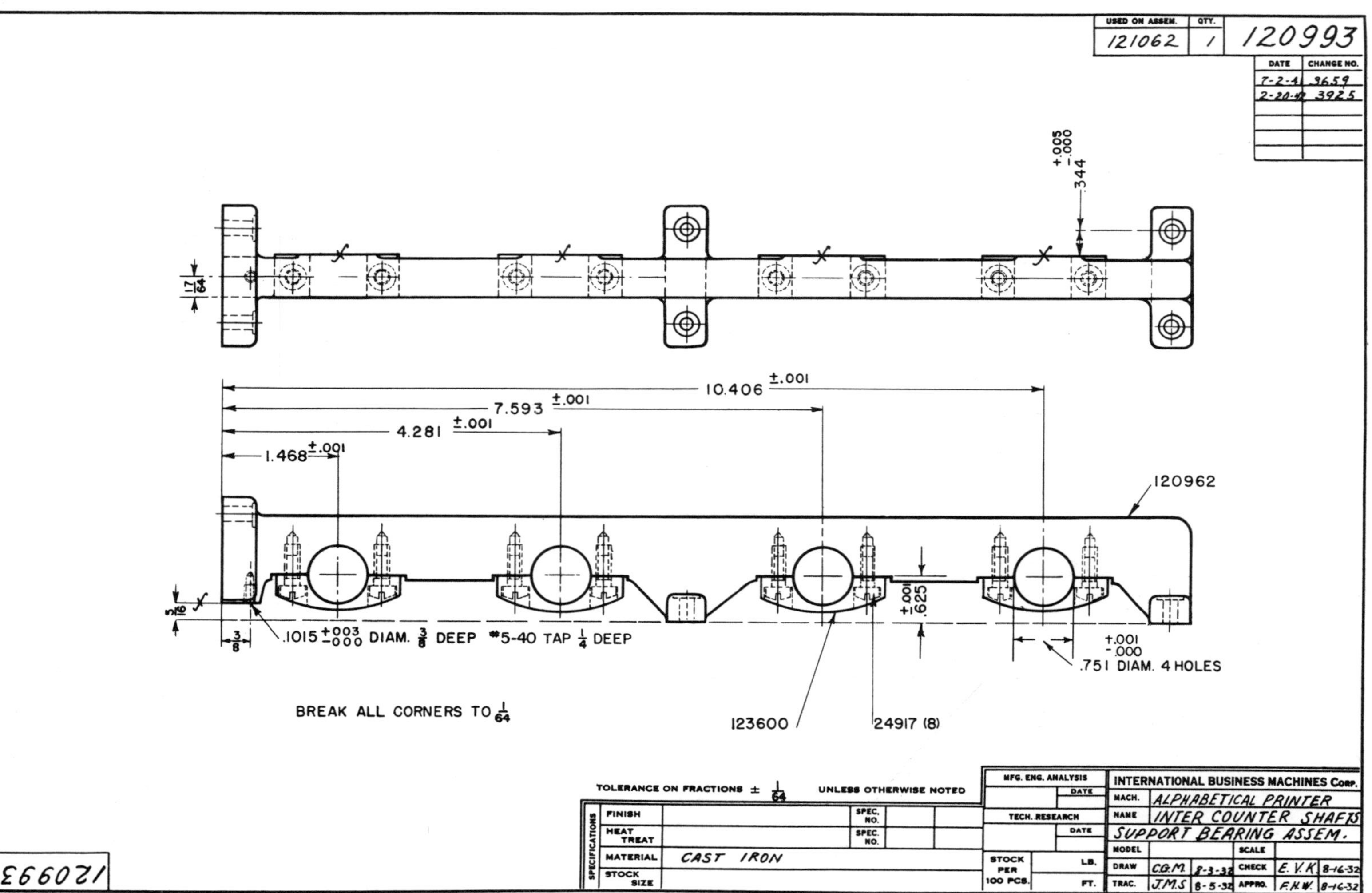

Figure 277. Blueprint of Part for Example A

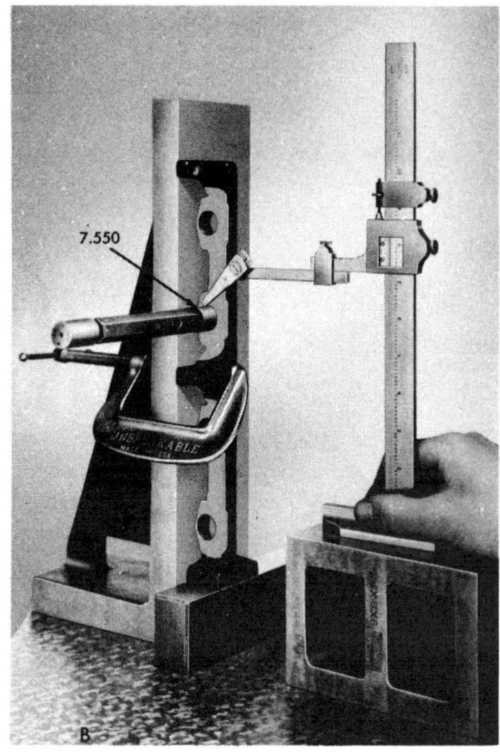

Figure 278. Measuring Example A, Part I

and blueprint of part 120993 (Figure 277).

OBJECTIVE: To check dimension 7.593" ± .001" from base of part to the center of .751" plus or minus .001" diameter hole.

PROCEDURE: Starting with a clear surface plate, select an angle iron that will accommodate the part in an upright position. Next, place a parallel at the base of the angle iron upon which the base of the part will rest. (Note that the top surface of this parallel is now the reference plane.) The locating surface of the work must coincide with the reference plane; therefore, place the part on the parallel in an upright position, with the base resting on the parallel (Figure 278A). In this position, clamp the part securely to the angle iron. (Note that the clamp is placed over one of the feet so as not to spring the part.)

Readings can now be taken with the height gage. The first reading corresponds to the reference plane or the top of the parallel. The height gage is placed on the surface plate, the slide is released, and the indicator is moved to a position in which the contact point just touches, or is very close to, the top of the parallel. Clamp the upper slide of the height gage and adjust the lower slide until the contact point of the indicator touches the parallel, and the pointer of the indicator is in the center of the scale. Clamp the lower slide, thus holding the indicator in this position. Pick up the height gage and read the vernier with a magnifying glass. The reading at the top of the parallel is 3.582". All subsequent readings are taken in the same way.

Next select a plug gage to fit the .751" hole. Insert the plug gage so as to leave a portion of the gage end extending beyond the part to provide a place to indicate, or take a reading with the height gage as shown in Figure 278B.

The distance from the surface plate to the top of the plug is now beyond the range of the height gage; so select a suitable block upon which to place the height gage to extend its range. In this example a 4" by 6" by 6" box

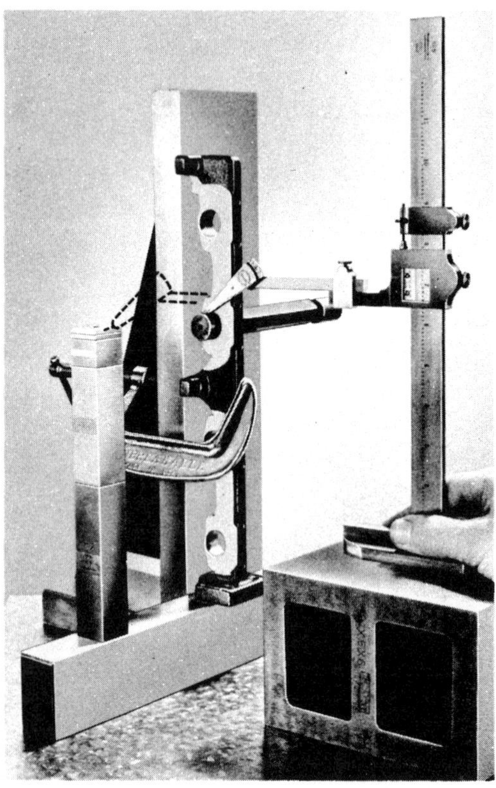

Figure 279. Measuring Example A, Part I (Alternate Method)

parallel is used. The block, in the position shown in Figure 278B increases the range of the height gage by 4".

With the height gage resting upon the box parallel, the reading over the plug end is 7.550". In this calculation 4" must be added to the reading to compensate for the height of the box parallel.

Caution: The contact point of the indicator is movable. Be careful it is not moved during a series of readings.

Information:

Dimension being checked.	7.593" ± .001
Height of box parallel.	4.000
Reading on top of parallel.	3.582
Reading over plug gage	7.550

Calculation:

Reading over plug gage	7.550
Plus height of box parallel.	4.000
Corrected reading over plug gage . .	11.550
Minus half plug diameter	.3755
Reading to center of hole	11.1745
Minus reading on top of parallel . .	3.582
Distance from base to center of hole	7.5925"

Refer to the dimension being checked (7.593" ± .001") to find that the distance arrived at by the calculations (7.5925") is within the given limits.

ALTERNATE METHOD FOR MEASURING EXAMPLE A, PART I

REFERENCE and OBJECTIVE: The same as for Measuring Example A, Part I.

PROCEDURE: An alternative and more accurate method of checking dimension 7.593" ± .001" is to set up a stack of gage blocks which will equal the distance from the base of the part to the top of a plug in the hole being checked, and indicate from the top of the blocks to top of the plug (Figure 279). Any variation in the reading of the indicator will represent the error in the location of the hole.

Information:

Dimension being checked	7.593 ± .001
Size of plug gage.	.751

Calculation:

Dimension to center of hole	7.593
Plus 1/2 diameter of plug	.3755
Height of stack of blocks	7.9685"

In this case, it is not necessary to take a reading on the height gage because the indicator is actually used to compare the height of the stack of gage blocks to the distance from the base to the top of the plug gage.

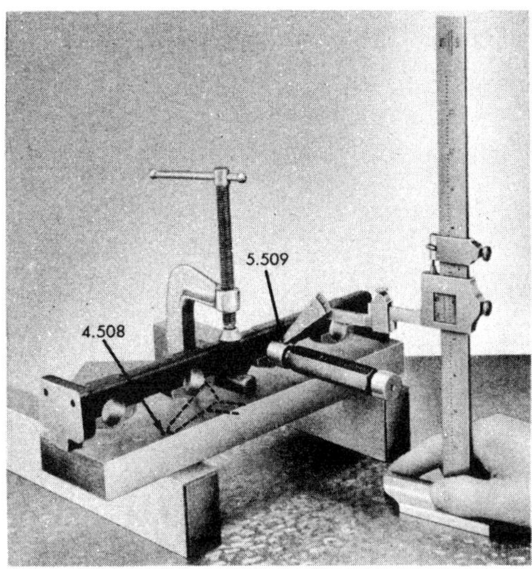

Figure 280. Measuring Example A, Part II

MEASURING EXAMPLE A, PART II

REFERENCE: Operation Record (Figure 276) and blueprint of part 120993 (Figure 277).

OBJECTIVE: To check dimension .625" ± .001" from the bottom of the feet to the center of the .751 diameter holes.

PROCEDURE: First bring the reference plane parallel to the surface plate. Place a pair of parallels on the surface plate. Clamp the part on a third parallel and place it (Figure 280) so that the clamp will clear the plate. The top surface of the third parallel, which is the reference plane, is now parallel to the surface plate.

Take the first reading on the parallel; take the second reading over the .751 plug inserted in the hole. Note that both readings are taken with the height gage resting on the surface plate. The height of the parallels does not enter into the calculations.

Information:

Dimension being checked	.625 ± .001
Size of plug gage . . .	.751
Base reading (top of parallel)	4.508
Reading over plug gage	5.509

Calculation:

Reading over plug gage	5.509
Minus 1/2 dia. of plug gage.	.3755
Reading to center of hole	5.1335
Minus base reading . .	4.508
Distance from base to center of hole . .	.6255"

OPERATION RECORD - FACTORY COPY

ORDER NUMBER		DEPT. CHARGED	QUANTITY	DATE ISSUED
108505				7 30

GEAR SUPPORT
CASTING RIGHT

OPERATION	OPER. NO.	DEPT.	TOOL NO.	GROUP
RAW STORES061108505	1	500	100	
CLEAN	10	180		
KEM ENAMEL	12	180		
GRIND BOTH SIDES TO THICKNESS	15	070		183
DRILL AND REAM	20	030	21074	36
CBORE CSK AND TAP AS BP	25	030	21127	36
PROFILE 2 BOSSES TO , 500 ,DIM	30	010	21145	12
INSPECT	40	400		
STOCK		830		
CHANGE NONE 09				

Figure 281. Operation Record for Example B

Referring to the dimension being checked (.625 ± .001) to find that the distance arrived at by the calculations (.6255") is within the given limits.

In the foregoing example a ten-inch height gage was used to illustrate how its measuring range can be increased by use of accessories, such as box parallels and angle irons. Often it is necessary to use this method when longer height gages are not available, or where the length of the part is beyond the range of the longest height gage.

MEASURING EXAMPLE B, PART I

REFERENCE: Operation Record (Figure 281) and blueprint of part 108505 (Figure 282).

OBJECTIVE: To set up part on surface plate so that one reference plane is parallel to surface plate.

PROCEDURE: The vertical reference plane passes through the centers of the two .432" + .000, - .003 diameter holes. The horizontal reference plane passes through the lower of these two holes. The outline of the part has no finished surface; therefore, it will be necessary to mount the part on an arbor and V-block.

Select an arbor to fit the center hole (.432"

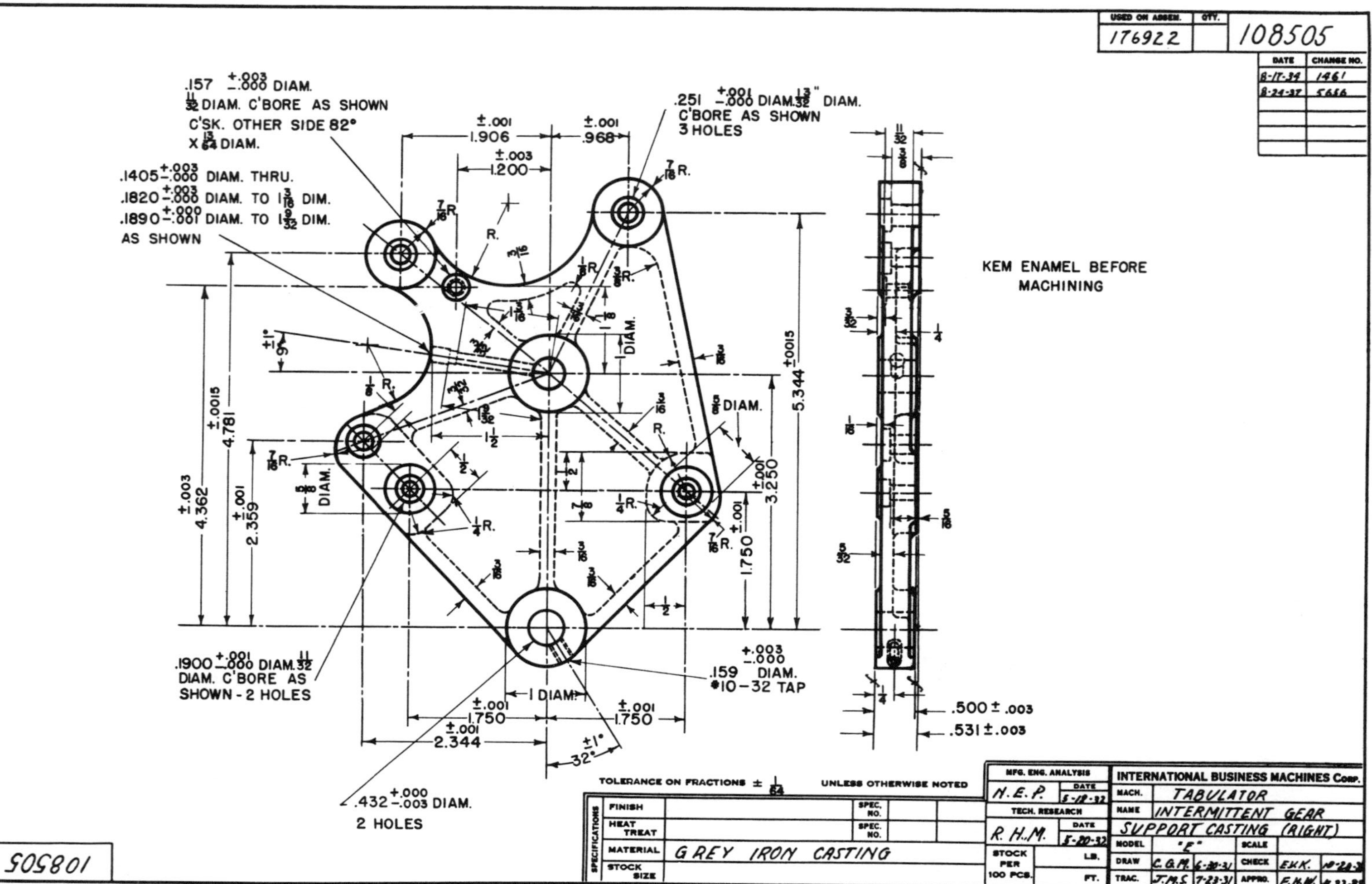

Figure 282. Blueprint of Part for Example B

diameter). The arbor, which is tapered, should be pressed into the hole until it fits the hole snugly. The longer end of the arbor, extending beyond the part, should be on the back.

Clamp the arbor in the V-slot of a universal right-angle iron with a V-slot. With the angle iron standing upright, line up the two holes vertically, by sighting, and clamp the part lightly in this position. Then turn the block 90 degrees and place it on parallels. Now the center line passing through the two .432" diameter holes is almost parallel to the surface plate (Figure 283).

Obviously, the position of the hole on the arbor is fixed and will remain in the same location if the part is revolved on the arbor. With the height gage, indicate over the top of the arbor and clamp the indicator in this position. Although it is not necessary to read the vernier, this is done for the sake of this example; the reading is 5.263".

Insert a .432" diameter plug gage in the other of the two holes; with the indicator in the same position, indicate over the plug, first tapping the casting to bring the plug within the range of the indicator. By repeated light tapping--first one way, then the other--the plug can be brought into a position where the top of the plug will indicate the same as the top of the arbor. In this position, the reference plane passing through the centers of the two .432" diameter holes is parallel to the surface of the plate, and the objective is attained (Figure 283).

Figure 283. Measuring Example B, Part I

Keep in mind at this point, that if the two holes involved were not the same diameter, the arbor, and the procedure of lining up the two holes would change as follows:

From the initial reading over the arbor subtract half the diameter of the arbor, thus obtaining a value corresponding to the center of the arbor, which is the reading of the reference plane or basic reading. But to line up the center of the plug (whose diameter in this case would differ from that of the arbor) add to the basic reading half the diameter of the plug, set the height gage to this new reading, and tap the part into a position where the top of the plug would correspond to the new setting of the height gage.

MEASURING EXAMPLE B, PART II

REFERENCE: Operation Record (Figure 281) and blueprint of part 108505 (Figure 282).

OBJECTIVE: To check dimension 2.344" ± .001 from the center line passing through the two .432" + .000, - .003 diameter holes, to the center of the .251" + .001, - .000 diameter hole.

PROCEDURE: With the part in position (Figure 284), to obtain a reading corresponding to the reference plane which passes through the center of the arbor and plug, subtract from the reading over the arbor an amount equal to half the diameter of the arbor.

The reading over the arbor was 5.263". Subtracting half the diameter of the arbor (.216) gives 5.047", the reading corresponding to the center of the arbor and the .432 diameter plug, which is the reference plane.

Insert a .251 diameter plug in the hole to be checked and take a reading over the plug:

Information:

Dimension being checked	2.344 ± .001
Reading at reference plane	5.047
Reading over .251 diameter plug	2.829

Calculation:

Reading over plug gage	2.829

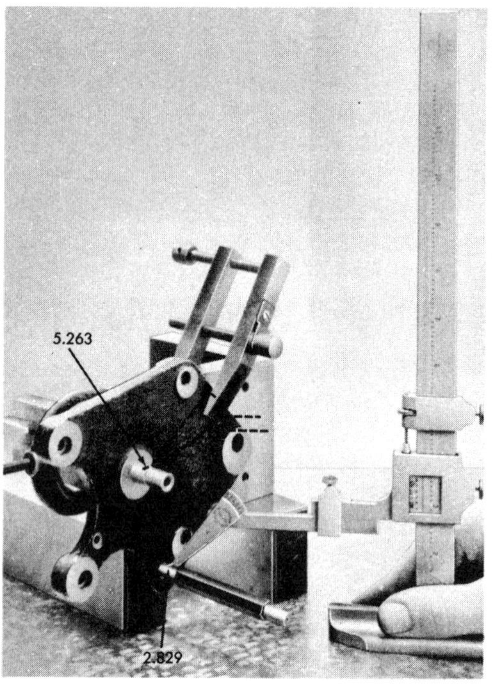

Figure 284. Measuring Example B, Part II

Minus 1/2 dia. of plug gage	.1255
Reading at center of plug gage.	2.7035"
Reading at reference plane	5.047
Minus reading at center of plug gage	2.7035
Distance from reference plane to center of hole	2.3435"

Refer to the dimension being checked (2.344" ± .001") to find that the distance arrived at by the calculations (2.3435") is within the given limits.

MEASURING EXAMPLE B, PART III

REFERENCE: Operation Record (Figure 281) and blueprint of part 108505 (Figure 282).

OBJECTIVE: To check the dimension 3.250" ± .001 between the centers of the two .432 + .001, - .003 diameter holes.

PROCEDURE: Set the angle block upright on the surface plate to bring the reference plane parallel to the surface plate (Figure 285). In

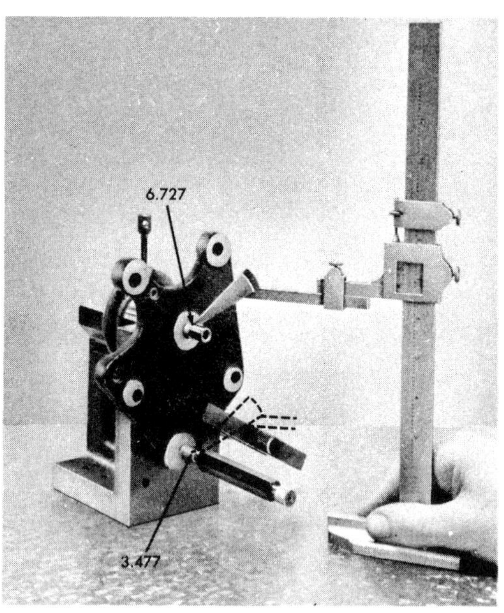

Figure 285. Measuring Example B, Part III

this position the reference plane passes through the center of the lower .432" diameter hole.

Insert a .432 diameter plug in the lower hole and indicate over top of the plug. The reading over the plug is 3.477". Now take a reading over the arbor passing through the other .432 diameter hole.

The reading over the arbor is 6.727". Because the two holes are the same diameter, it is not necessary to deduct half the diameter of the plug or the arbor before subtracting the two readings.

Information:

Dimension to be checked	3.250" ± .001
Reading over arbor . . .	6.727
Reading over plug . . .	3.477

Calculation:

Reading over arbor . . .	6.727
Minus reading over plug gage	3.477
Distance between center of holes	3.250"

Refer to the dimension being checked (3.250" ± .001) to find that the distance arrived at by the calculations (3.250") is within the given limits.

Figure 286 shows a type of gage called Pla-

Figure 286. Twenty-four-inch Pla-Chek

Chek, to be used for surface-plate work. Pla-Chek is made from a hardened steel bar with steps or measuring discs, ground and lapped to a high degree of accuracy, and spaced one inch apart. A micrometer screwthread on the lower end and a large micrometer thimble graduated .0001" on the upper end raises and lowers the measuring bar within a range of one inch. The micrometer is read in the same way as a standard micrometer, except that the graduations on the large thimble are in .0001" instead of .001".

In checking work on the surface plate, the micrometer is set to the desired dimension in thousandths and tens of thousandths of an inch, and the measurement is taken from the proper disc. On the 24- and 36-inch models, the measuring bar can be adjusted to the center line of work to be inspected and readings taken up or down in relation to a center line or other reference line on the work. The scale mounted

Figure 287. Six-inch Pla-Check with Risers

on the left side of the frame is graduated in .100" and used to identify the particular disc in use and its approximate height in relation to the actual dimension as set on the micrometer.

This gage is available in four sizes: 6, 12, 24, and 36 inches. The measuring discs on each size are one inch apart. The range of the 6" size can be increased without loss of accuracy, by the use of 6" risers (Figure 287).

Also available for use with this gage is a reverse checking plate. This is a horseshoe-shaped device with a spring. The under surface is lapped flat and will wring to the upper measuring surface of any one-inch step. This makes it easy to set a height gage indicator accurately for checking the underside of work (Figure 288). A hardened and lapped steel button in the base of the 12-, 24-, and 36-inch sizes, directly under the reverse checking plate, makes it possible to set inside micrometers or dial-type inside indicating gages. This can also be done on the

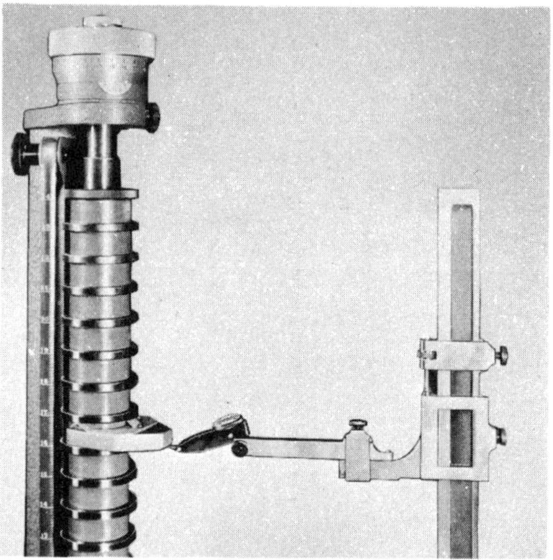

Figure 288. Use of Reverse Checking Plate on Pla-Chek

6" size by using the surface plate as the lower reference plane.

The Pla-Chek has a wide range of application. Where a number of dimensions must be checked to tolerances not closer than .0001", it eliminates the need for gage blocks. It can be used in tool inspection for checking jigs, dies, and fixtures, and in the shop for measuring the locations of holes, machined surfaces, spacing of slots, and for layout work.

Figure 286 shows the 24-inch Pla-Chek being used to check the location of holes in a casting. As the first disc on the measuring bar is 3" from the surface plate when the micrometer is set at zero, the casting is set on 3" parallels; this places its bottom edge, from which the hole locations are to be taken, on the same plane as the first disc on the lower end of the measuring bar.

Set the height gage indicator to zero on top of the plug inserted in the hole to be checked. Place the indicator point over the disc nearest the height of the plug and turn the micrometer up until the height gage indicator is at zero. The micrometer reading will be the actual distance from the bottom edge of the casting to the top of the plug. Subtract half the diameter of the plug from this reading to obtain the center of the hole; this will be the dimension for the location of the hole as given on the blueprint, plus or minus the given tolerance.

Another method is to set the Pla-Chek to the desired dimension, as given on the blueprint, plus half the diameter of the plug. Then set the height-gage indicator to zero on the proper disc and take the reading from the indicator over the plug.

All holes in the casting can be checked for location by either of these methods, because the measuring discs can be turned up or down to any desired dimension. But, it is more accurate to take the readings directly from the Pla-Chek micrometer, as in the first method, rather than from an indicator, because some indicators may not be sufficiently accurate to give correct readings where tolerances are very small.

QUESTIONS AND PROBLEMS

1. What is a reference plane? Can it pass through the center of a hole?
2. Why are important dimensions on a blueprint referred to the reference plane?
3. Why do most surface plates have three legs?
4. Why is a cast-iron surface plate aged or heat-treated before machining the surface?
5. What is the advantage of the scraped surface over a perfectly smooth surface on a cast-iron surface plate?
6. When "scraping in" a surface plate, what procedure is followed to insure a perfectly flat surface?
7. Describe two ways a surface plate may be checked for flatness. Which is the more accurate check?
8. Name seven precautions which should be taken when using a surface plate.
9. What are some of the advantages the granite surface plate has over the cast-iron surface plate?
10. How can worn surface plates be reconditioned?
11. What is a toolmaker's flat? How does it differ from a surface plate?
12. Describe parallels and give three uses for them.
13. Why are parallels purchased and used in pairs? How may pairs be identified?
14. Why should parallels never be used as a square?
15. State two advantages of a box parallel when compared with an ordinary parallel.
16. What makes the universal angle iron a more convenient tool than the ordinary angle iron?
17. What is a V-block and when is it used?
18. Why are V-blocks made in pairs?
19. How would you check the accuracy of a V-block?
20. What is a square?
21. Describe two methods for checking the squareness of a part, with a square, on a surface plate.
22. Should the side of the blade of a square be used when checking a part? Why?
23. What is the particular advantage of a bevel-edge square?
24. When is an adjustable square used?
25. Describe a cylindrical square and its principal use.
26. What is a straight edge?
27. Name four standard designs of straight edge.
28. State two ways a straight edge may be used to determine the flatness of surface.
29. How may a planer gage be used as an inspection tool?
30. What two sources of information influence the nature of the setup on a surface plate?
31. Name three important precautions that must be taken in setting up work on a surface plate.
32. If you were inspecting the drilling of a part in process and found a hole drilled .010 under the reamed size specified on the drawing, explain why you may not reject it. Can it be checked for location?
33. How can a 10-inch height gage be used to check the location of a surface 15 inches above the surface of the plate?
34. When inspecting a part which is mounted on an arbor clamped in a V-block, what preliminary steps must be taken before taking reading with a height gage?
35. How can the range of the six-inch Pla-Chek gage be increased without loss of accuracy?
36. Name three applications for the Pla-Chek gage.
37. When work being inspected is resting on a parallel, does the height of the parallel enter into the calculations?

Chapter IX

ANGULAR MEASUREMENT

ANGULAR MEASUREMENT is important in many fields. In astronomy the relation of the stars and their distances from the Earth are calculated from angular measurements. Navigation of ships and planes, location of highways and boundary lines--all involve angular measurement. In the design and manufacture of machinery, angular specifications and measurements are used so extensively that knowledge of the fundamentals of angles is essential to production and inspection.

INITIAL AND TERMINAL LINES

Geometrically speaking, a line has but one dimension--length; it has no breadth or thickness. A plane is an imaginary surface in which a straight line, joining any two points of the surface, lies wholly in that surface.

Picture a line, with one end pinned to a plane so that the other end can be moved in a circle in this plane (Figure 289): any movement of the line, from its original position (called the initial line) to a new position (called the terminal line) will represent an angle.

UNIT OF ANGULAR MEASUREMENT

If the circle, formed by a complete revolution of the line is divided into 360 equal parts, each part will represent one degree. A degree, the unit of angular measurement, is indicated by the symbol $^{\circ}$. A quadrant of that circle will represent 90 degrees or a right angle, and half of the circle will represent 180 degrees or a straight angle. Each subdivision of the circle will represent its relative proportion of 360 degrees (Figure 290).

Like an hour of time, each degree is divided into 60 minutes (60'), and each minute is divided into 60 seconds (60"). Fractions of a degree are specified in minutes; fractions of a minute are specified in seconds. If a circle were divided into 14 divisions, each division would be equal to 25-5/7 degrees, which would be specified as 25° 42' 51" (fractions of a second are not specified unless accuracy requires it).

Minutes and seconds cannot be shown on an ordinary angular scale because the graduations would be too small to be read. The vernier protractor can be used to measure to an accuracy of 5' or more; or the principle of the sine bar, where accuracy is limited only by the accuracy of the sine bar and the instrument used to measure the linear dimensions, can be used.

THE VERNIER PROTRACTOR

Principles of the vernier protractor were given in Chapter VI. These principles will now be applied to measurement of angles.

Figure 291 shows a vernier protractor upon which lines have been superimposed to indicate the initial line and the terminal line. Lines A-A and B-B are the original initial and terminal lines; but since line C-C is parallel to line A-A, line C-C is also an initial line; and since line D-D is parallel to line B-B, line D-D is also a terminal line; so it may be assumed that the base of the vernier protractor and all lines parallel to it are initial lines; and either edge of the blade, and all lines parallel to it, are terminal lines.

The angle between an initial line and a terminal line is indicated in degrees on the

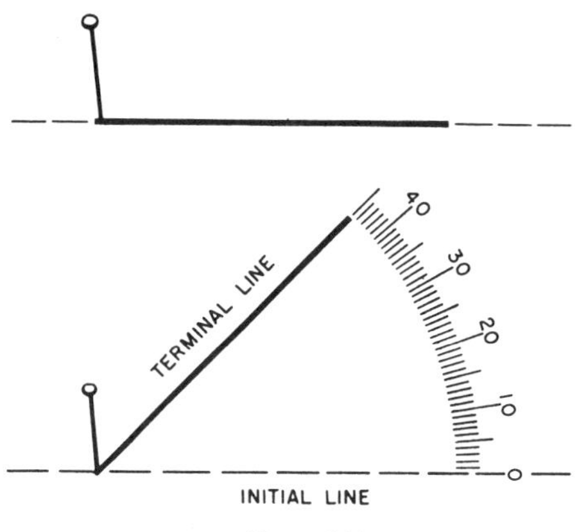

Figure 289

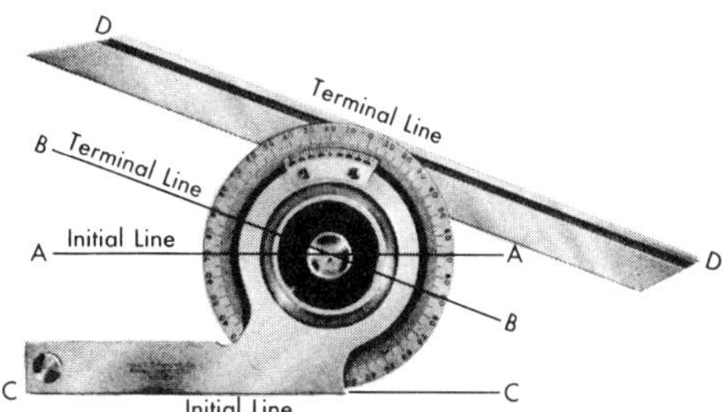

Figure 291. Vernier Protractor

scale of the protractor by the zero of the vernier. Subdivisions of degrees are read on the vernier, as described in Chapter VI.

Keep in mind that the vernier is divided by the zero into two sections, each section a vernier in itself (Figure 292). The section of the vernier to the right is used when the zero of the vernier is to the right of the zero of the scale (Figure 292A), the section of the vernier to the left is used when the zero of the vernier is to the left of the zero of the scale (Figure 292B).

READING THE VERNIER PROTRACTOR

The scale of the vernier protractor (Figure 293) is in the form of a circle divided into 360 divisions, each division representing one degree. In numbering these divisions, the scale is further divided into four quadrants of 90 degrees each. The degrees in each quadrant are numbered from zero to 90; so there are two zero positions on the scale and two 90-degree positions. The zero line, or a line passing through the two zero positions, is at a right angle to the base; a line passing through the two 90-degree positions is parallel to the base.

By definition, supplementary angles are two angles whose sum is equal to 180 degrees or a straight angle. Complementary angles are two angles whose sum is equal to 90 degrees or a right angle.

The base, being a straight line, represents

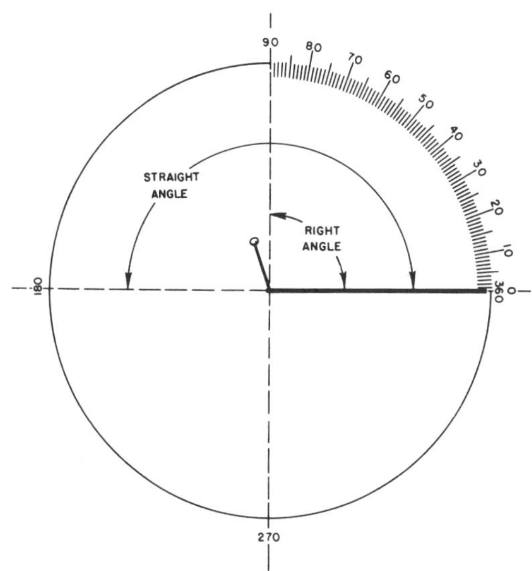

Figure 290. Angular Scale

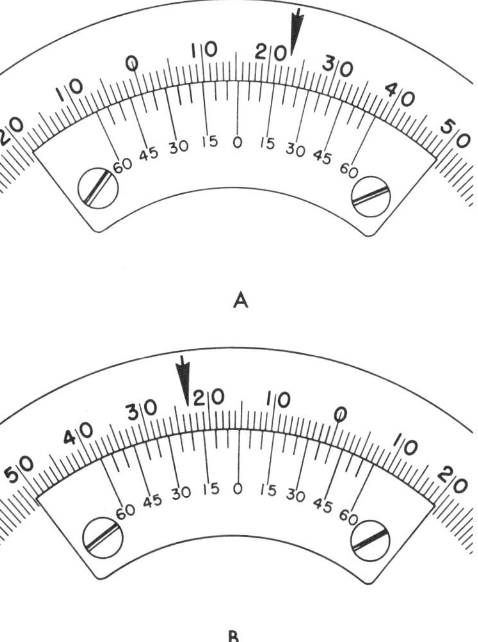

Figure 292. Right and Left Vernier Reading

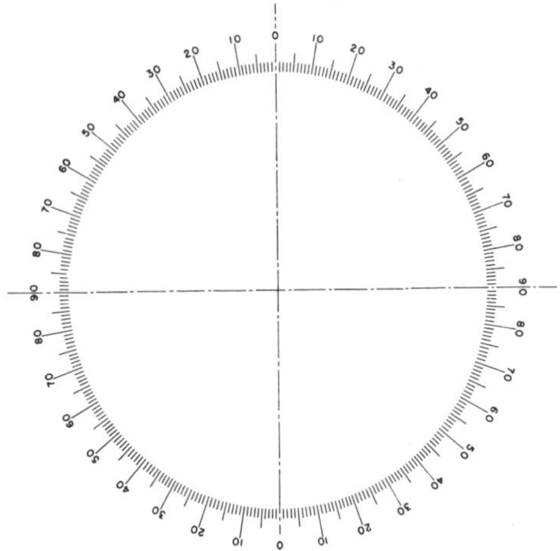

Figure 293. Scale of a Vernier Protractor

a straight angle or 180 degrees. When the blade crosses the base, of course, supplementary angles are formed; in other words, if one side of the blade makes an angle of 60 degrees with the base, the other side of the blade will make the supplement of 60 degrees, or 120 degrees, with the base. This relation of blade and reading should be considered in measuring angles. Ordinarily the side of the blade represented by the reading can be judged by noting whether the angle is acute or obtuse.

Figure 294 illustrates four different positions of the blade as applied to the measurement of the angles of a hexagonal figure. In all four cases, the reading of the vernier protractor is 60 degrees.

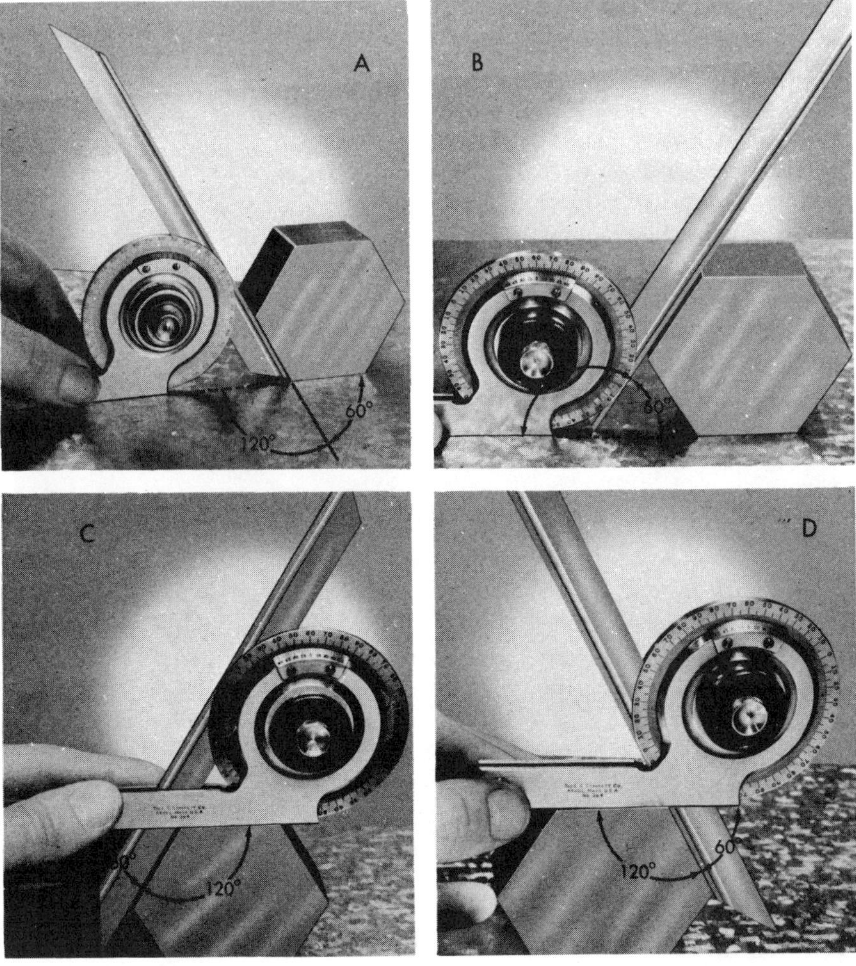

Figure 294. Applications of Vernier Protractor

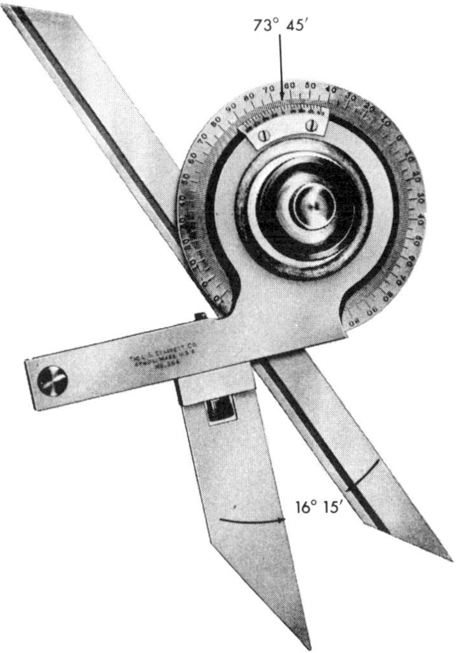

Figure 295. Acute-angle Attachment

In A and B, the reading (60°) represents the angle between the surface being measured and the surface plate. In C and D, the reading (60°) represents the supplement of the angle between two adjacent surfaces of the figure.

THE ACUTE ANGLE ATTACHMENT

The acute angle attachment lends itself to a variety of applications (one is shown in Figure 295). The objective is to set the blades to an angle of 16° 15'. As the reference plane is now the edge of the acute angle attachment, which makes an angle of 90 degrees with the base, the reading will be the complement of the angle set between the attachment and the blade. Therefore, the angle 16° 15' will be represented by a reading of 90 degrees minus 16° 15', or 73° 45'.

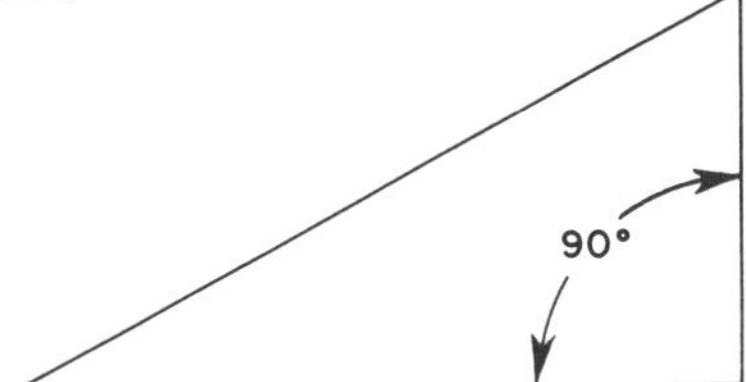

Figure 296. A Right Triangle

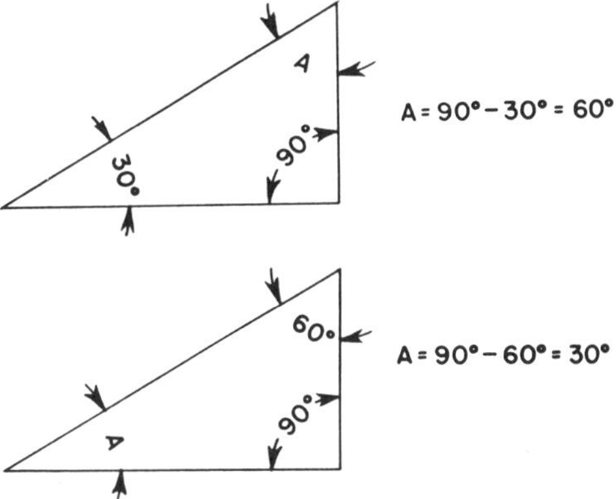

Figure 297. Complementary Angles

SHOP TRIGONOMETRY

THE RIGHT TRIANGLE

A right triangle (Figure 296) is a triangle, one angle of which is a right angle or 90 degrees. Since the sum of the angles of a triangle is equal to two right angles or 180 degrees, if one of the angles of a triangle equals 90 degrees, the two remaining angles are complementary because their sum is equal to a right angle. Therefore, if one of the two complementary angles of a right triangle is known, the other can be determined simply by subtracting the known angle from 90 degrees (Figure 297).

SIMILAR RIGHT TRIANGLES

Similar right triangles are right triangles whose corresponding complementary angles are equal. The sides of two similar right triangles are in direct proportion to each other.

Given two similar right triangles (Figure 298), if the length of the hypotenuse of the larger

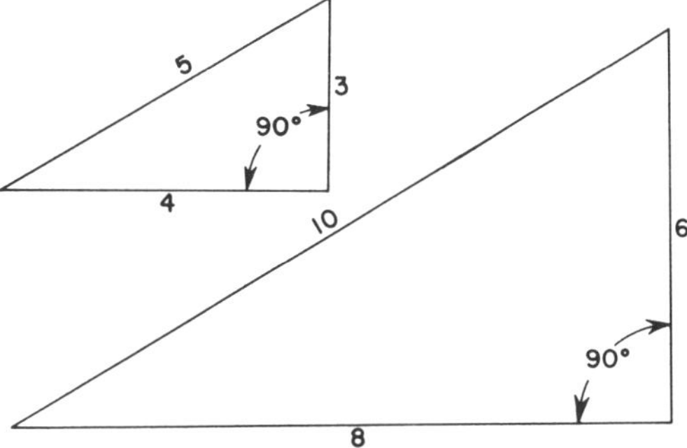

Figure 298. Similar Right Triangles

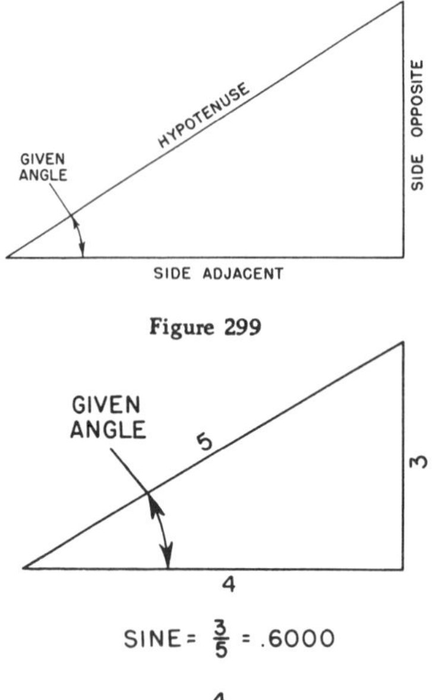

Figure 299

SINE = $\frac{3}{5}$ = .6000

COSINE = $\frac{4}{5}$ = .8000

TANGENT = $\frac{3}{4}$ = .7500

Figure 300

triangle is twice the length of the hypotenuse of the smaller triangle, then the other corresponding sides are in the same proportion. Each side of the larger triangle will be twice the length of the corresponding side of the smaller triangle.

TRIGONOMETRIC FUNCTIONS

Rules of trigonometry are based upon the known relation of the sides and angles of the right triangle. To see the relations between the sides of a right triangle, name the sides of a right triangle according to their position in relation to the given angle (Figure 299). Then, in every right triangle, the three trigonometric functions--sine, cosine, and tangent--are represented by the following relation between the sides:

1. Sine of given angle = $\frac{\text{Side opposite}}{\text{Hypotenuse}}$
2. Cosine of given angle = $\frac{\text{Side adjacent}}{\text{Hypotenuse}}$
3. Tangent of given angle = $\frac{\text{Side opposite}}{\text{Side adjacent}}$

By substituting length in inches for the names of the various sides in the foregoing fractions, the three functions--sine, cosine, and tangent--

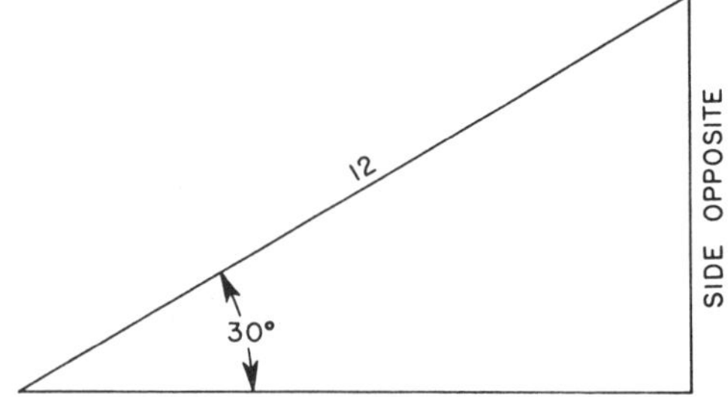

Figure 301

can be expressed in decimal form (Figure 300).

To obtain the given angle in the triangle, refer to Table XII (trigonometric functions); under the heading Sines look for .6000; or under the heading Cosines look for .8000; or under the heading Tangents, look for .7500. In each, the corresponding angle is 36° 52' 12-1/2".

Usually it is unnecessary to determine the angle to an accuracy of seconds. However, the seconds in the foregoing example were interpolated as follows:

Refer to the table of trigonometric functions to find that the sine of the given angle (.6000) falls between .59995 and .60019, the sine functions of 36° 52' and 36° 53', respectively. Therefore, a difference of 1' in the angle, at this point, makes a difference of .60019 minus .59995, or .00024, in the sine function. The sine of the given angle is .6000, which is .00005 greater than the sine of 36° 52'. Since .00024 represents a difference of 1' in the angle, .00005 is equivalent to 5/24 x 1' (60") = 12-1/2". Therefore, the given angle is 36° 52' 12-1/2".

To find the other side of the triangle when the angle and one side are given, reverse the procedure. Since the two sides involved in this problem (Figure 301) are the hypotenuse, which is given, and the side opposite, which is required, use the relation:

$$\text{Sine of given angle} = \frac{\text{Side opposite}}{\text{Hypotenuse}}$$

Refer to the table to find:

$$\text{Sine } 30^\circ = .5000$$

So the equation becomes:

$$.5000 = \frac{\text{Side opposite}}{12}$$

Side opposite = 6.000

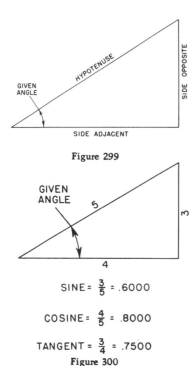

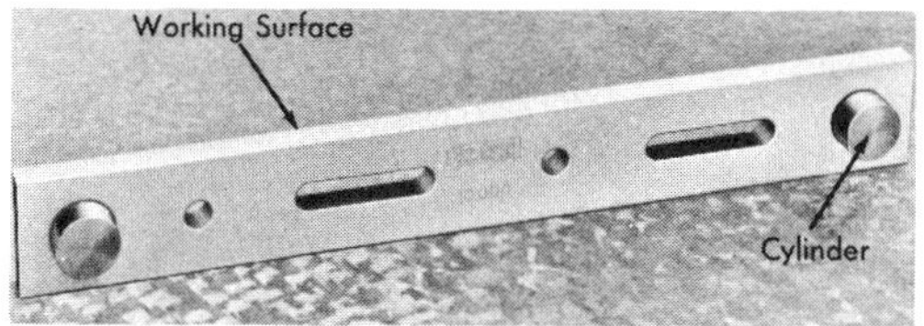

Figure 302. Ten-inch Sine Bar

THE SINE BAR

The sine bar is a device which enables one to set up, at any desired angle, a plane upon which the angle of tools and parts may be checked by indicating with a height gage or surface gage (Figure 302). It consists of a straight, rectangular bar upon which two hardened and ground cylinders are mounted so that:

1. Axes of the cylinders are at a right angle to the side of the bar.
2. Centers of the cylinders are exactly the same distance from the working edge or plane.
3. Distance--lengthwise of the bar, between the centers of the cylinders--is some fixed dimension that enters easily into calculations. This distance is usually 5 or 10 inches.

If one of the cylinders is resting on a surface, the bar can be set to any desired angle simply by raising the first cylinder above the other cylinder a distance equal to the sine of the desired angle times the distance between centers of the cylinders.

APPLICATION OF SINE BAR

For example, assume a 10-inch sine bar is being used and it is desired to set the sine bar at an angle of 30 degrees. Refer to Table XII to find that the sine of 30° equals .5000. Therefore, to set the 10-inch sine bar at an angle of 30 degrees, the one cylinder must be .5000 x 10 inches = 5.000 inches above the other cylinder (Figure 303A). If the first cylinder is placed upon a stack of gage blocks one inch high, the stack of blocks under the second cylinder must be 1 inch plus 5 inches, or 6 inches high.

If a 5-inch sine bar were used, the vertical distance between the two cylinders would be .5000 (sine of 30°) times 5.000 inches = 2.500 inches, and the height of blocks under the second cylinder would be 1 inch plus 2.5 inches = 3.5 inches (Figure 303B).

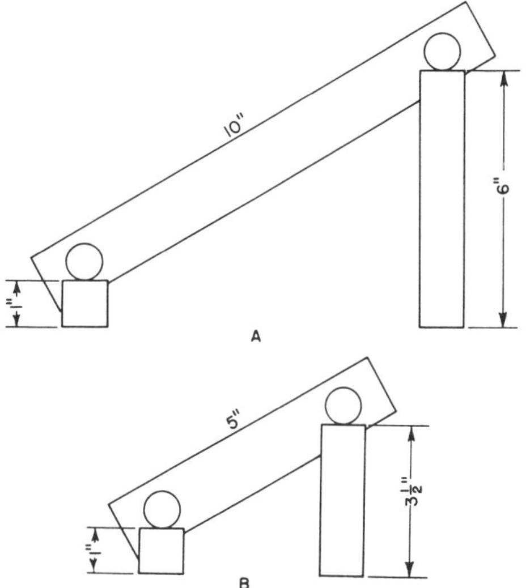

Figure 303. Application of Sine Bar

The foregoing example shows the relation of similar right triangles. Since the angle in each case was the same (30 degrees), the two sine bars, 10" and 5", respectively, formed the hypotenuses of two similar right triangles. Because the hypotenuse of the larger triangle (10-inch sine bar) was twice the hypotenuse of the smaller triangle (5-inch sine bar), the vertical distance between the two cylinders of the 10-inch sine bar would necessarily be twice the vertical distance between the cylinders of the smaller sine bar.

Table XII shows trigonometric functions based upon a hypotenuse of 1. When a 10-inch sine bar is being used, all functions must be multiplied by 10 to obtain the relative side opposite and side adjacent. This multiplication can be performed mentally simply by moving the decimal point one place to the right.

When the 5-inch sine bar is being used, reference should be made to Table XIII which gives trigonometric functions multiplied by 5. From this table, the height of blocks under the free cylinder can be read directly.

SETTING UP COMPLEMENTARY ANGLES

With the sine bar near the horizontal position, a small change in the height of the stack of gage blocks will produce a smaller change in the angle than when the sine bar is near the vertical posi-

TABLE XII. TRIGONOMETRIC FUNCTIONS (Part 1 of 9)

0°					
M	Sine	Cosine	Tan.	Cotan.	M
0	.00000	1.0000	.00000	Infinite	60
1	.00029	.0000	.00029	3437.7	59
2	.00058	.0000	.00058	1718.9	58
3	.00087	.0000	.00087	1145.9	57
4	.00116	.0000	.00116	859.44	56
5	.00145	1.0000	.00145	687.55	55
6	.00174	.0000	.00174	572.96	54
7	.00204	.0000	.00204	491.11	53
8	.00233	.0000	.00233	429.72	52
9	.00262	.0000	.00262	381.97	51
10	.00291	.99999	.00291	343.77	50
11	.00320	.99999	.00320	312.52	49
12	.00349	.99999	.00349	286.48	48
13	.00378	.99999	.00378	264.44	47
14	.00407	.99999	.00407	245.55	46
15	.00436	.99999	.00436	229.18	45
16	.00465	.99999	.00465	214.86	44
17	.00494	.99999	.00494	202.22	43
18	.00524	.99999	.00524	190.98	42
19	.00553	.99998	.00553	180.93	41
20	.00582	.99998	.00582	171.88	40
21	.00611	.99998	.00611	163.70	39
22	.00640	.99998	.00640	156.26	38
23	.00669	.99998	.00669	149.46	37
24	.00698	.99997	.00698	143.24	36
25	.00727	.99997	.00727	137.51	35
26	.00756	.99997	.00756	132.22	34
27	.00785	.99997	.00785	127.32	33
28	.00814	.99997	.00814	122.77	32
29	.00843	.99996	.00844	118.54	31
30	.00873	.99996	.00873	114.59	30
31	.00902	.99996	.00902	110.89	29
32	.00931	.99996	.00931	107.43	28
33	.00960	.99995	.00960	104.17	27
34	.00989	.99995	.00989	101.11	26
35	.01018	.99995	.01018	98.218	25
36	.01047	.99994	.01047	95.489	24
37	.01076	.99994	.01076	92.908	23
38	.01105	.99994	.01105	90.463	22
39	.01134	.99993	.01134	88.143	21
40	.01163	.99993	.01164	85.940	20
41	.01193	.99993	.01193	83.843	19
42	.01222	.99992	.01222	81.847	18
43	.01251	.99992	.01251	79.943	17
44	.01280	.99992	.01280	78.126	16
45	.01309	.99991	.01309	76.390	15
46	.01338	.99991	.01338	74.729	14
47	.01367	.99991	.01367	73.139	13
48	.01396	.99990	.01396	71.615	12
49	.01425	.99990	.01425	70.153	11
50	.01454	.99989	.01454	68.750	10
51	.01483	.99989	.01484	67.402	9
52	.01512	.99988	.01513	66.105	8
53	.01542	.99988	.01542	64.858	7
54	.01571	.99988	.01571	63.657	6
55	.01600	.99987	.01600	62.499	5
56	.01629	.99987	.01629	61.383	4
57	.01658	.99987	.01658	60.306	3
58	.01687	.99986	.01687	59.266	2
59	.01716	.99985	.01716	58.261	1
60	.01745	.99985	.01745	57.290	0
M	Cosine	Sine	Cotan.	Tan.	M
89°					

1°					
M	Sine	Cosine	Tan.	Cotan.	M
0	.01745	.99985	.01745	57.290	60
1	.01774	.99984	.01775	56.350	59
2	.01803	.99984	.01804	55.441	58
3	.01832	.99983	.01833	54.561	57
4	.01861	.99983	.01862	53.708	56
5	.01891	.99982	.01891	52.882	55
6	.01920	.99981	.01920	52.081	54
7	.01949	.99981	.01949	51.303	53
8	.01978	.99980	.01978	50.548	52
9	.02007	.99980	.02007	49.816	51
10	.02036	.99979	.02036	49.104	50
11	.02065	.99979	.02066	48.412	49
12	.02094	.99978	.02095	47.739	48
13	.02123	.99977	.02124	47.085	47
14	.02152	.99977	.02153	46.449	46
15	.02181	.99976	.02182	45.829	45
16	.02210	.99975	.02211	45.226	44
17	.02240	.99975	.02240	44.638	43
18	.02269	.99974	.02269	44.066	42
19	.02298	.99974	.02298	43.508	41
20	.02326	.99973	.02327	42.964	40
21	.02356	.99972	.02357	42.433	39
22	.02385	.99971	.02386	41.916	38
23	.02414	.99971	.02415	41.410	37
24	.02443	.99970	.02444	40.917	36
25	.02472	.99969	.02473	40.436	35
26	.02501	.99969	.02502	39.965	34
27	.02530	.99968	.02531	39.506	33
28	.02559	.99967	.02560	39.057	32
29	.02589	.99966	.02589	38.618	31
30	.02618	.99966	.02618	38.188	30
31	.02647	.99965	.02648	37.769	29
32	.02676	.99964	.02677	37.358	28
33	.02705	.99963	.02706	36.956	27
34	.02734	.99963	.02735	36.563	26
35	.02763	.99962	.02764	36.177	25
36	.02792	.99961	.02793	35.800	24
37	.02821	.99960	.02822	35.431	23
38	.02850	.99959	.02851	35.069	22
39	.02879	.99958	.02880	34.715	21
40	.02908	.99958	.02910	34.368	20
41	.02937	.99957	.02939	34.027	19
42	.02967	.99956	.02968	33.693	18
43	.02996	.99955	.02997	33.366	17
44	.03025	.99954	.03026	33.045	16
45	.03054	.99953	.03055	32.730	15
46	.03083	.99952	.03084	32.421	14
47	.03112	.99951	.03113	32.118	13
48	.03141	.99951	.03143	31.820	12
49	.03170	.99950	.03172	31.528	11
50	.03199	.99949	.03201	31.241	10
51	.03228	.99948	.03230	30.960	9
52	.03257	.99947	.03259	30.683	8
53	.03286	.99946	.03288	30.411	7
54	.03315	.99945	.03317	30.145	6
55	.03344	.99944	.03346	29.882	5
56	.03374	.99943	.03375	29.624	4
57	.03403	.99942	.03405	29.371	3
58	.03432	.99941	.03434	29.122	2
59	.03461	.99940	.03463	28.877	1
60	.03490	.99939	.03492	28.636	0
M	Cosine	Sine	Cotan.	Tan.	M
88°					

2°					
M	Sine	Cosine	Tan.	Cotan.	M
0	.03490	.99939	.03492	28.636	60
1	.03519	.99938	.03521	28.399	59
2	.03548	.99937	.03550	28.166	58
3	.03577	.99936	.03579	27.937	57
4	.03606	.99935	.03608	27.712	56
5	.03635	.99934	.03638	27.490	55
6	.03664	.99933	.03667	27.271	54
7	.03693	.99932	.03696	27.056	53
8	.03722	.99931	.03725	26.845	52
9	.03751	.99930	.03754	26.637	51
10	.03781	.99928	.03783	26.432	50
11	.03810	.99927	.03812	26.230	49
12	.03839	.99926	.03842	26.031	48
13	.03868	.99925	.03871	25.835	47
14	.03897	.99924	.03900	25.642	46
15	.03926	.99923	.03929	25.452	45
16	.03955	.99922	.03958	25.264	44
17	.03984	.99921	.03987	25.080	43
18	.04013	.99919	.04016	24.898	42
19	.04042	.99918	.04045	24.718	41
20	.04071	.99917	.04075	24.542	40
21	.04100	.99916	.04104	24.367	39
22	.04129	.99915	.04133	24.196	38
23	.04158	.99913	.04162	24.026	37
24	.04187	.99912	.04191	23.859	36
25	.04217	.99911	.04220	23.694	35
26	.04246	.99910	.04249	23.532	34
27	.04275	.99908	.04279	23.372	33
28	.04304	.99907	.04308	23.214	32
29	.04333	.99906	.04337	23.058	31
30	.04362	.99905	.04366	22.904	30
31	.04391	.99903	.04395	22.752	29
32	.04420	.99902	.04424	22.602	28
33	.04449	.99901	.04453	22.454	27
34	.04478	.99900	.04483	22.308	26
35	.04507	.99898	.04512	22.164	25
36	.04536	.99897	.04541	22.022	24
37	.04565	.99896	.04570	21.881	23
38	.04594	.99894	.04599	21.742	22
39	.04623	.99893	.04628	21.606	21
40	.04652	.99892	.04657	21.470	20
41	.04681	.99890	.04687	21.337	19
42	.04711	.99889	.04716	21.205	18
43	.04740	.99888	.04745	21.075	17
44	.04769	.99886	.04774	20.946	16
45	.04798	.99885	.04803	20.819	15
46	.04827	.99883	.04832	20.693	14
47	.04856	.99882	.04862	20.569	13
48	.04885	.99881	.04891	20.446	12
49	.04914	.99879	.04920	20.325	11
50	.04943	.99878	.04949	20.205	10
51	.04972	.99876	.04978	20.087	9
52	.05001	.99875	.05007	19.970	8
53	.05030	.99873	.05037	19.854	7
54	.05059	.99872	.05066	19.740	6
55	.05088	.99870	.05095	19.627	5
56	.05117	.99869	.05124	19.515	4
57	.05146	.99867	.05153	19.405	3
58	.05175	.99866	.05182	19.296	2
59	.05204	.99864	.05212	19.188	1
60	.05234	.99863	.05241	19.081	0
M	Cosine	Sine	Cotan.	Tan.	M
87°					

3°					
M	Sine	Cosine	Tan.	Cotan.	M
0	.05234	.99863	.05241	19.081	60
1	.05263	.99861	.05270	18.975	59
2	.05292	.99860	.05299	18.871	58
3	.05321	.99858	.05328	18.768	57
4	.05350	.99857	.05357	18.665	56
5	.05379	.99855	.05387	18.564	55
6	.05408	.99854	.05416	18.464	54
7	.05437	.99852	.05445	18.365	53
8	.05466	.99850	.05474	18.268	52
9	.05495	.99849	.05503	18.171	51
10	.05524	.99847	.05532	18.075	50
11	.05553	.99846	.05562	17.980	49
12	.05582	.99844	.05591	17.886	48
13	.05611	.99842	.05620	17.793	47
14	.05640	.99841	.05649	17.701	46
15	.05669	.99839	.05678	17.610	45
16	.05698	.99837	.05707	17.520	44
17	.05727	.99836	.05737	17.431	43
18	.05756	.99834	.05766	17.343	42
19	.05785	.99832	.05795	17.256	41
20	.05814	.99831	.05824	17.169	40
21	.05843	.99829	.05853	17.084	39
22	.05872	.99827	.05883	16.999	38
23	.05902	.99826	.05912	16.915	37
24	.05931	.99824	.05941	16.832	36
25	.05960	.99822	.05970	16.750	35
26	.05989	.99820	.05999	16.668	34
27	.06018	.99819	.06029	16.587	33
28	.06047	.99817	.06058	16.507	32
29	.06076	.99815	.06087	16.428	31
30	.06105	.99813	.06116	16.350	30
31	.06134	.99812	.06145	16.272	29
32	.06163	.99810	.06175	16.195	28
33	.06192	.99808	.06204	16.119	27
34	.06221	.99806	.06233	16.043	26
35	.06250	.99804	.06262	15.969	25
36	.06279	.99803	.06291	15.894	24
37	.06308	.99801	.06321	15.821	23
38	.06337	.99799	.06350	15.748	22
39	.06366	.99797	.06379	15.676	21
40	.06395	.99795	.06408	15.605	20
41	.06424	.99793	.06437	15.534	19
42	.06453	.99791	.06467	15.464	18
43	.06482	.99790	.06496	15.394	17
44	.06511	.99788	.06525	15.325	16
45	.06540	.99786	.06554	15.257	15
46	.06569	.99784	.06583	15.189	14
47	.06598	.99782	.06613	15.122	13
48	.06627	.99780	.06642	15.056	12
49	.06656	.99778	.06671	14.990	11
50	.06685	.99776	.06700	14.924	10
51	.06714	.99774	.06730	14.860	9
52	.06743	.99772	.06759	14.795	8
53	.06772	.99770	.06788	14.732	7
54	.06801	.99768	.06817	14.668	6
55	.06830	.99766	.06846	14.606	5
56	.06859	.99764	.06876	14.544	4
57	.06888	.99762	.06905	14.482	3
58	.06918	.99760	.06934	14.421	2
59	.06947	.99758	.06963	14.361	1
60	.06976	.99756	.06993	14.301	0
M	Cosine	Sine	Cotan.	Tan.	M
86°					

4°					
M	Sine	Cosine	Tan.	Cotan.	M
0	.06976	.99756	.06993	14.301	60
1	.07005	.99754	.07022	14.241	59
2	.07034	.99752	.07051	14.182	58
3	.07063	.99750	.07080	14.123	57
4	.07092	.99748	.07110	14.065	56
5	.07121	.99746	.07139	14.008	55
6	.07150	.99744	.07168	13.951	54
7	.07179	.99742	.07197	13.894	53
8	.07208	.99740	.07226	13.838	52
9	.07237	.99738	.07256	13.782	51
10	.07266	.99736	.07285	13.727	50
11	.07295	.99733	.07314	13.672	49
12	.07324	.99731	.07343	13.617	48
13	.07353	.99729	.07373	13.563	47
14	.07382	.99727	.07402	13.510	46
15	.07411	.99725	.07431	13.457	45
16	.07440	.99723	.07460	13.404	44
17	.07469	.99721	.07490	13.351	43
18	.07498	.99718	.07519	13.299	42
19	.07527	.99716	.07548	13.248	41
20	.07556	.99714	.07577	13.197	40
21	.07585	.99712	.07607	13.146	39
22	.07614	.99710	.07636	13.096	38
23	.07643	.99707	.07665	13.046	37
24	.07672	.99705	.07694	12.996	36
25	.07701	.99703	.07724	12.947	35
26	.07730	.99701	.07753	12.898	34
27	.07759	.99698	.07782	12.849	33
28	.07788	.99696	.07812	12.801	32
29	.07817	.99694	.07841	12.754	31
30	.07846	.99692	.07870	12.706	30
31	.07875	.99689	.07899	12.659	29
32	.07904	.99687	.07929	12.612	28
33	.07933	.99685	.07958	12.566	27
34	.07962	.99682	.07987	12.520	26
35	.07991	.99680	.08016	12.474	25
36	.08020	.99678	.08046	12.429	24
37	.08049	.99675	.08075	12.384	23
38	.08078	.99673	.08104	12.339	22
39	.08107	.99671	.08134	12.295	21
40	.08136	.99668	.08163	12.250	20
41	.08165	.99666	.08192	12.207	19
42	.08194	.99664	.08221	12.163	18
43	.08223	.99661	.08251	12.120	17
44	.08252	.99659	.08280	12.077	16
45	.08281	.99656	.08309	12.035	15
46	.08310	.99654	.08339	11.992	14
47	.08339	.99652	.08368	11.950	13
48	.08368	.99649	.08397	11.909	12
49	.08397	.99647	.08426	11.867	11
50	.08426	.99644	.08456	11.826	10
51	.08455	.99642	.08485	11.785	9
52	.08484	.99639	.08514	11.745	8
53	.08513	.99637	.08544	11.704	7
54	.08542	.99634	.08573	11.664	6
55	.08571	.99632	.08602	11.625	5
56	.08600	.99629	.08632	11.585	4
57	.08629	.99627	.08661	11.546	3
58	.08658	.99624	.08690	11.507	2
59	.08687	.99622	.08719	11.468	1
60	.08715	.99619	.08749	11.430	0
M	Cosine	Sine	Cotan.	Tan.	M
85°					

TABLE XII (Continued, part 2 of 9)

5°					
M	Sine	Cosine	Tan.	Cotan.	M
0	.08715	.99619	.08749	11.430	60
1	.08744	.99617	.08778	11.392	59
2	.08773	.99614	.08807	11.354	58
3	.08802	.99612	.08837	11.316	57
4	.08831	.99609	.08866	11.279	56
5	.08860	.99607	.08895	11.242	55
6	.08889	.99604	.08925	11.205	54
7	.08918	.99601	.08954	11.168	53
8	.08947	.99599	.08983	11.132	52
9	.08976	.99596	.09013	11.095	51
10	.09005	.99594	.09042	11.059	50
11	.09034	.99591	.09071	11.024	49
12	.09063	.99588	.09101	10.988	48
13	.09092	.99586	.09130	10.953	47
14	.09121	.99583	.09159	10.918	46
15	.09150	.99580	.09189	10.883	45
16	.09179	.99578	.09218	10.848	44
17	.09208	.99575	.09247	10.814	43
18	.09237	.99572	.09277	10.780	42
19	.09266	.99570	.09306	10.746	41
20	.09295	.99567	.09335	10.712	40
21	.09324	.99564	.09365	10.678	39
22	.09353	.99562	.09394	10.645	38
23	.09382	.99559	.09423	10.612	37
24	.09411	.99556	.09453	10.579	36
25	.09440	.99553	.09482	10.546	35
26	.09469	.99551	.09511	10.514	34
27	.09498	.99548	.09541	10.481	33
28	.09527	.99545	.09570	10.449	32
29	.09556	.99542	.09599	10.417	31
30	.09584	.99540	.09629	10.385	30
31	.09613	.99537	.09658	10.354	29
32	.09642	.99534	.09688	10.322	28
33	.09671	.99531	.09717	10.291	27
34	.09700	.99528	.09746	10.260	26
35	.09729	.99525	.09776	10.229	25
36	.09758	.99523	.09805	10.199	24
37	.09787	.99520	.09834	10.168	23
38	.09816	.99517	.09864	10.138	22
39	.09845	.99514	.09893	10.108	21
40	.09874	.99511	.09922	10.078	20
41	.09903	.99508	.09952	10.048	19
42	.09932	.99505	.09981	10.019	18
43	.09961	.99503	.10011	9.9893	17
44	.09990	.99500	.10040	9.9601	16
45	.10019	.99497	.10069	9.9310	15
46	.10048	.99494	.10099	9.9021	14
47	.10077	.99491	.10128	9.8734	13
48	.10106	.99488	.10158	9.8448	12
49	.10134	.99485	.10187	9.8164	11
50	.10163	.99482	.10216	9.7882	10
51	.10192	.99479	.10246	9.7601	9
52	.10221	.99476	.10275	9.7322	8
53	.10250	.99473	.10305	9.7044	7
54	.10279	.99470	.10334	9.6768	6
55	.10308	.99467	.10363	9.6493	5
56	.10337	.99464	.10393	9.6220	4
57	.10366	.99461	.10422	9.5949	3
58	.10395	.99458	.10452	9.5679	2
59	.10424	.99455	.10481	9.5411	1
60	.10453	.99452	.10510	9.5144	0
M	Cosine	Sine	Cotan.	Tan.	M
84°					

6°					
M	Sine	Cosine	Tan.	Cotan.	M
0	.10453	.99452	.10510	9.5144	60
1	.10482	.99449	.10540	.4878	59
2	.10511	.99446	.10569	.4614	58
3	.10540	.99443	.10599	.4351	57
4	.10568	.99440	.10628	.4090	56
5	.10597	.99437	.10657	9.3831	55
6	.10626	.99434	.10687	.3572	54
7	.10655	.99431	.10716	.3315	53
8	.10684	.99428	.10746	.3060	52
9	.10713	.99424	.10775	.2806	51
10	.10742	.99421	.10805	9.2553	50
11	.10771	.99418	.10834	.2302	49
12	.10800	.99415	.10863	.2051	48
13	.10829	.99412	.10893	.1803	47
14	.10858	.99409	.10922	.1555	46
15	.10887	.99406	.10952	9.1309	45
16	.10916	.99402	.10981	.1064	44
17	.10944	.99399	.11011	.0821	43
18	.10973	.99396	.11040	.0579	42
19	.11002	.99393	.11069	.0338	41
20	.11031	.99390	.11099	9.0098	40
21	.11060	.99386	.11128	8.9860	39
22	.11089	.99383	.11158	.9623	38
23	.11118	.99380	.11187	.9387	37
24	.11147	.99377	.11217	.9152	36
25	.11176	.99373	.11246	8.8918	35
26	.11205	.99370	.11276	.8686	34
27	.11234	.99367	.11305	.8455	33
28	.11262	.99364	.11335	.8225	32
29	.11291	.99360	.11364	.7996	31
30	.11320	.99357	.11393	8.7769	30
31	.11349	.99354	.11423	.7542	29
32	.11378	.99350	.11452	.7317	28
33	.11407	.99347	.11482	.7093	27
34	.11436	.99344	.11511	.6870	26
35	.11465	.99341	.11541	8.6648	25
36	.11494	.99337	.11570	.6427	24
37	.11523	.99334	.11600	.6208	23
38	.11551	.99330	.11629	.5989	22
39	.11580	.99327	.11659	.5772	21
40	.11609	.99324	.11688	8.5555	20
41	.11638	.99320	.11718	.5340	19
42	.11667	.99317	.11747	.5126	18
43	.11696	.99314	.11777	.4913	17
44	.11725	.99310	.11806	.4701	16
45	.11754	.99307	.11836	8.4489	15
46	.11783	.99303	.11865	.4279	14
47	.11811	.99300	.11895	.4070	13
48	.11840	.99296	.11924	.3862	12
49	.11869	.99293	.11954	.3655	11
50	.11898	.99290	.11983	8.3449	10
51	.11927	.99286	.12013	.3244	9
52	.11956	.99283	.12042	.3040	8
53	.11985	.99279	.12072	.2837	7
54	.12014	.99276	.12101	.2635	6
55	.12042	.99272	.12131	8.2434	5
56	.12071	.99269	.12160	.2234	4
57	.12100	.99265	.12190	.2035	3
58	.12129	.99262	.12219	.1837	2
59	.12158	.99258	.12249	.1640	1
60	.12187	.99255	.12278	8.1443	0
M	Cosine	Sine	Cotan.	Tan.	M
83°					

7°					
M	Sine	Cosine	Tan.	Cotan.	M
0	.12187	.99255	.12278	8.1443	60
1	.12216	.99251	.12308	.1248	59
2	.12245	.99247	.12337	.1053	58
3	.12273	.99244	.12367	.0860	57
4	.12302	.99240	.12396	.0667	56
5	.12331	.99237	.12426	8.0476	55
6	.12360	.99233	.12456	.0285	54
7	.12389	.99229	.12485	.0095	53
8	.12418	.99226	.12515	7.9906	52
9	.12447	.99222	.12544	.9717	51
10	.12476	.99219	.12574	7.9530	50
11	.12504	.99215	.12603	.9344	49
12	.12533	.99211	.12633	.9158	48
13	.12562	.99208	.12662	.8973	47
14	.12591	.99204	.12692	.8789	46
15	.12620	.99200	.12722	7.8606	45
16	.12649	.99197	.12751	.8424	44
17	.12678	.99193	.12781	.8243	43
18	.12706	.99189	.12810	.8062	42
19	.12735	.99186	.12840	.7882	41
20	.12764	.99182	.12869	7.7703	40
21	.12793	.99178	.12899	.7525	39
22	.12822	.99174	.12928	.7348	38
23	.12851	.99171	.12958	.7171	37
24	.12879	.99167	.12988	.6996	36
25	.12908	.99163	.13017	7.6821	35
26	.12937	.99160	.13047	.6646	34
27	.12966	.99156	.13076	.6473	33
28	.12995	.99152	.13106	.6300	32
29	.13024	.99148	.13136	.6129	31
30	.13053	.99144	.13165	7.5957	30
31	.13081	.99141	.13195	.5787	29
32	.13110	.99137	.13224	.5617	28
33	.13139	.99133	.13254	.5449	27
34	.13168	.99129	.13284	.5280	26
35	.13197	.99125	.13313	7.5113	25
36	.13226	.99121	.13343	.4946	24
37	.13254	.99118	.13372	.4780	23
38	.13283	.99114	.13402	.4615	22
39	.13312	.99110	.13432	.4451	21
40	.13341	.99106	.13461	7.4287	20
41	.13370	.99102	.13491	.4124	19
42	.13399	.99098	.13520	.3961	18
43	.13427	.99094	.13550	.3800	17
44	.13456	.99090	.13580	.3639	16
45	.13485	.99086	.13609	7.3479	15
46	.13514	.99083	.13639	.3319	14
47	.13543	.99079	.13669	.3160	13
48	.13571	.99075	.13698	.3002	12
49	.13600	.99071	.13728	.2844	11
50	.13629	.99067	.13757	7.2687	10
51	.13658	.99063	.13787	.2531	9
52	.13687	.99059	.13817	.2375	8
53	.13716	.99055	.13846	.2220	7
54	.13744	.99051	.13876	.2066	6
55	.13773	.99047	.13906	7.1912	5
56	.13802	.99043	.13935	.1759	4
57	.13831	.99039	.13965	.1607	3
58	.13860	.99035	.13995	.1455	2
59	.13888	.99031	.14024	.1304	1
60	.13917	.99027	.14054	7.1154	0
M	Cosine	Sine	Cotan.	Tan.	M
82°					

8°					
M	Sine	Cosine	Tan.	Cotan.	M
0	.13917	.99027	.14054	7.1154	60
1	.13946	.99023	.14084	.1004	59
2	.13975	.99019	.14113	.0854	58
3	.14004	.99015	.14143	.0706	57
4	.14032	.99010	.14173	.0558	56
5	.14061	.99006	.14202	7.0410	55
6	.14090	.99002	.14232	.0264	54
7	.14119	.98998	.14262	.0117	53
8	.14148	.98994	.14291	6.9972	52
9	.14176	.98990	.14321	.9827	51
10	.14205	.98986	.14351	6.9682	50
11	.14234	.98982	.14380	.9538	49
12	.14263	.98978	.14410	.9395	48
13	.14292	.98973	.14440	.9252	47
14	.14320	.98969	.14470	.9110	46
15	.14349	.98965	.14499	6.8969	45
16	.14378	.98961	.14529	.8828	44
17	.14407	.98957	.14559	.8687	43
18	.14436	.98952	.14588	.8547	42
19	.14464	.98948	.14618	.8408	41
20	.14493	.98944	.14648	6.8269	40
21	.14522	.98940	.14677	.8131	39
22	.14551	.98936	.14707	.7993	38
23	.14579	.98931	.14737	.7856	37
24	.14608	.98927	.14767	.7720	36
25	.14637	.98923	.14796	6.7584	35
26	.14666	.98919	.14826	.7448	34
27	.14695	.98914	.14856	.7313	33
28	.14723	.98910	.14886	.7179	32
29	.14752	.98906	.14915	.7045	31
30	.14781	.98901	.14945	6.6911	30
31	.14810	.98897	.14975	.6779	29
32	.14838	.98893	.15004	.6646	28
33	.14867	.98889	.15034	.6514	27
34	.14896	.98884	.15064	.6383	26
35	.14925	.98880	.15094	6.6252	25
36	.14953	.98876	.15123	.6122	24
37	.14982	.98871	.15153	.5992	23
38	.15011	.98867	.15183	.5863	22
39	.15040	.98862	.15213	.5734	21
40	.15068	.98858	.15243	6.5605	20
41	.15097	.98854	.15272	.5478	19
42	.15126	.98849	.15302	.5350	18
43	.15155	.98845	.15332	.5223	17
44	.15183	.98840	.15362	.5097	16
45	.15212	.98836	.15391	6.4971	15
46	.15241	.98832	.15421	.4845	14
47	.15270	.98827	.15451	.4720	13
48	.15298	.98823	.15481	.4596	12
49	.15328	.98818	.15511	.4472	11
50	.15356	.98814	.15540	6.4348	10
51	.15385	.98809	.15570	.4225	9
52	.15413	.98805	.15600	.4103	8
53	.15442	.98800	.15630	.3980	7
54	.15471	.98796	.15659	.3859	6
55	.15500	.98791	.15689	6.3737	5
56	.15528	.98787	.15719	.3616	4
57	.15557	.98782	.15749	.3496	3
58	.15586	.98778	.15779	.3376	2
59	.15615	.98773	.15809	.3257	1
60	.15643	.98769	.15838	6.3137	0
M	Cosine	Sine	Cotan.	Tan.	M
81°					

9°					
M	Sine	Cosine	Tan.	Cotan.	M
0	.15643	.98769	.15838	6.3137	60
1	.15672	.98764	.15868	.3019	59
2	.15701	.98760	.15898	.2901	58
3	.15730	.98755	.15928	.2783	57
4	.15758	.98750	.15958	.2665	56
5	.15787	.98746	.15987	6.2548	55
6	.15816	.98741	.16017	.2432	54
7	.15844	.98737	.16047	.2316	53
8	.15873	.98732	.16077	.2200	52
9	.15902	.98727	.16107	.2085	51
10	.15931	.98723	.16137	6.1970	50
11	.15959	.98718	.16167	.1856	49
12	.15988	.98714	.16196	.1742	48
13	.16017	.98709	.16226	.1628	47
14	.16045	.98704	.16256	.1515	46
15	.16074	.98700	.16286	6.1402	45
16	.16103	.98695	.16316	.1290	44
17	.16132	.98690	.16346	.1178	43
18	.16160	.98685	.16376	.1066	42
19	.16189	.98681	.16405	.0955	41
20	.16218	.98676	.16435	6.0844	40
21	.16246	.98671	.16465	.0734	39
22	.16275	.98667	.16495	.0624	38
23	.16304	.98662	.16525	.0514	37
24	.16333	.98657	.16555	.0405	36
25	.16361	.98652	.16585	6.0296	35
26	.16390	.98648	.16615	.0188	34
27	.16419	.98643	.16644	.0080	33
28	.16447	.98638	.16674	5.9972	32
29	.16476	.98633	.16704	.9865	31
30	.16505	.98628	.16734	5.9758	30
31	.16533	.98624	.16764	.9651	29
32	.16562	.98619	.16794	.9545	28
33	.16591	.98614	.16824	.9439	27
34	.16619	.98609	.16854	.9333	26
35	.16648	.98604	.16884	5.9228	25
36	.16677	.98600	.16914	.9123	24
37	.16705	.98595	.16944	.9019	23
38	.16734	.98590	.16973	.8915	22
39	.16763	.98585	.17003	.8811	21
40	.16791	.98580	.17033	5.8708	20
41	.16820	.98575	.17063	.8605	19
42	.16849	.98570	.17093	.8502	18
43	.16878	.98565	.17123	.8400	17
44	.16906	.98560	.17153	.8298	16
45	.16935	.98556	.17183	5.8196	15
46	.16964	.98551	.17213	.8095	14
47	.16992	.98546	.17243	.7994	13
48	.17021	.98541	.17273	.7894	12
49	.17050	.98536	.17303	.7794	11
50	.17078	.98531	.17333	5.7694	10
51	.17107	.98526	.17363	.7594	9
52	.17136	.98521	.17393	.7495	8
53	.17164	.98516	.17423	.7396	7
54	.17193	.98511	.17453	.7297	6
55	.17221	.98506	.17483	5.7199	5
56	.17250	.98501	.17513	.7101	4
57	.17279	.98496	.17543	.7004	3
58	.17307	.98491	.17573	.6906	2
59	.17336	.98486	.17603	.6809	1
60	.17365	.98481	.17633	5.6713	0
M	Cosine	Sine	Cotan.	Tan.	M
80°					

TABLE XII (Continued, part 3 of 9)

10°

M	Sine	Cosine	Tan.	Ootan.	M
0	.17365	.98481	.17633	5.6713	60
1	.17393	.98476	.17663	.6616	59
2	.17422	.98471	.17693	.6520	58
3	.17451	.98465	.17723	.6425	57
4	.17479	.98460	.17753	.6329	56
5	.17508	.98455	.17783	5.6234	55
6	.17537	.98450	.17813	.6140	54
7	.17565	.98445	.17843	.6045	53
8	.17594	.98440	.17873	.5951	52
9	.17622	.98435	.17903	.5857	51
10	.17651	.98430	.17933	5.5764	50
11	.17680	.98425	.17963	.5670	49
12	.17708	.98419	.17993	.5578	48
13	.17737	.98414	.18023	.5485	47
14	.17766	.98409	.18053	.5393	46
15	.17794	.98404	.18083	5.5301	45
16	.17823	.98399	.18113	.5209	44
17	.17852	.98394	.18143	.5117	43
18	.17880	.98388	.18173	.5026	42
19	.17909	.98383	.18203	.4936	41
20	.17937	.98378	.18233	5.4845	40
21	.17966	.98373	.18263	.4755	39
22	.17995	.98368	.18293	.4665	38
23	.18023	.98362	.18323	.4575	37
24	.18052	.98357	.18353	.4486	36
25	.18080	.98352	.18383	5.4396	35
26	.18109	.98347	.18413	.4308	34
27	.18138	.98341	.18444	.4219	33
28	.18166	.98336	.18474	.4131	32
29	.18195	.98331	.18504	.4043	31
30	.18223	.98325	.18534	5.3955	30
31	.18252	.98320	.18564	.3868	29
32	.18281	.98315	.18594	.3780	28
33	.18309	.98309	.18624	.3694	27
34	.18338	.98304	.18654	.3607	26
35	.18366	.98299	.18684	5.3521	25
36	.18395	.98293	.18714	.3434	24
37	.18424	.98288	.18745	.3349	23
38	.18452	.98283	.18775	.3263	22
39	.18481	.98277	.18805	.3178	21
40	.18509	.98272	.18835	5.3093	20
41	.18538	.98267	.18865	.3008	19
42	.18567	.98261	.18895	.2923	18
43	.18595	.98256	.18925	.2839	17
44	.18624	.98250	.18955	.2755	16
45	.18652	.98245	.18985	5.2671	15
46	.18681	.98240	.19016	.2588	14
47	.18709	.98234	.19046	.2505	13
48	.18738	.98229	.19076	.2422	12
49	.18767	.98223	.19106	.2339	11
50	.18795	.98218	.19136	5.2257	10
51	.18824	.98212	.19166	.2174	9
52	.18852	.98207	.19197	.2092	8
53	.18881	.98201	.19227	.2011	7
54	.18909	.98196	.19257	.1929	6
55	.18938	.98190	.19287	5.1848	5
56	.18967	.98185	.19317	.1767	4
57	.18995	.98179	.19347	.1686	3
58	.19024	.98174	.19378	.1606	2
59	.19052	.98168	.19408	.1525	1
60	.19081	.98163	.19438	5.1445	0
M	Cosine	Sine	Cotan.	Tan.	M

79°

11°

M	Sine	Cosine	Tan.	Cotan.	M
0	.19081	.98163	.19438	5.1445	60
1	.19109	.98157	.19468	.1366	59
2	.19138	.98152	.19498	.1286	58
3	.19166	.98146	.19529	.1207	57
4	.19195	.98140	.19559	.1128	56
5	.19224	.98135	.19589	5.1049	55
6	.19252	.98129	.19619	.0970	54
7	.19281	.98124	.19649	.0892	53
8	.19309	.98118	.19680	.0814	52
9	.19338	.98112	.19710	.0736	51
10	.19366	.98107	.19740	5.0658	50
11	.19395	.98101	.19770	.0581	49
12	.19423	.98095	.19800	.0504	48
13	.19452	.98090	.19831	.0427	47
14	.19480	.98084	.19861	.0350	46
15	.19509	.98078	.19891	5.0273	45
16	.19537	.98073	.19921	.0197	44
17	.19566	.98067	.19952	.0121	43
18	.19595	.98061	.19982	.0045	42
19	.19623	.98056	.20012	4.9969	41
20	.19652	.98050	.20042	4.9894	40
21	.19680	.98044	.20073	.9819	39
22	.19709	.98039	.20103	.9744	38
23	.19737	.98033	.20133	.9669	37
24	.19766	.98027	.20163	.9594	36
25	.19794	.98021	.20194	4.9520	35
26	.19823	.98016	.20224	.9446	34
27	.19851	.98010	.20254	.9372	33
28	.19880	.98004	.20285	.9298	32
29	.19908	.97998	.20315	.9225	31
30	.19937	.97992	.20345	4.9151	30
31	.19965	.97987	.20375	.9078	29
32	.19994	.97981	.20406	.9006	28
33	.20022	.97975	.20436	.8933	27
34	.20051	.97969	.20466	.8860	26
35	.20079	.97963	.20497	4.8788	25
36	.20108	.97957	.20527	.8716	24
37	.20136	.97952	.20557	.8644	23
38	.20165	.97946	.20588	.8573	22
39	.20193	.97940	.20618	.8501	21
40	.20222	.97934	.20648	4.8430	20
41	.20250	.97928	.20679	.8359	19
42	.20279	.97922	.20709	.8288	18
43	.20307	.97916	.20739	.8217	17
44	.20336	.97910	.20770	.8147	16
45	.20364	.97904	.20800	4.8077	15
46	.20393	.97899	.20830	.8007	14
47	.20421	.97893	.20861	.7937	13
48	.20450	.97887	.20891	.7867	12
49	.20478	.97881	.20921	.7798	11
50	.20506	.97875	.20952	4.7728	10
51	.20535	.97869	.20982	.7659	9
52	.20563	.97863	.21012	.7591	8
53	.20592	.97857	.21043	.7522	7
54	.20620	.97851	.21073	.7453	6
55	.20649	.97845	.21104	4.7385	5
56	.20677	.97839	.21134	.7317	4
57	.20706	.97833	.21164	.7249	3
58	.20734	.97827	.21195	.7181	2
59	.20763	.97821	.21225	.7114	1
60	.20791	.97815	.21256	4.7046	0
M	Cosine	Sine	Cotan.	Tan.	M

78°

12°

M	Sine	Cosine	Tan.	Cotan.	M
0	.20791	.97815	.21256	4.7046	60
1	.20820	.97809	.21286	.6979	59
2	.20848	.97803	.21316	.6912	58
3	.20876	.97797	.21347	.6845	57
4	.20905	.97790	.21377	.6778	56
5	.20933	.97784	.21408	4.6712	55
6	.20962	.97778	.21438	.6646	54
7	.20990	.97772	.21468	.6580	53
8	.21019	.97766	.21499	.6514	52
9	.21047	.97760	.21529	.6448	51
10	.21076	.97754	.21560	4.6382	50
11	.21104	.97748	.21590	.6317	49
12	.21132	.97741	.21621	.6252	48
13	.21161	.97735	.21651	.6187	47
14	.21189	.97729	.21682	.6122	46
15	.21218	.97723	.21712	4.6057	45
16	.21246	.97717	.21742	.5993	44
17	.21275	.97711	.21773	.5928	43
18	.21303	.97704	.21803	.5864	42
19	.21331	.97698	.21834	.5800	41
20	.21360	.97692	.21864	4.5736	40
21	.21388	.97686	.21895	.5673	39
22	.21417	.97680	.21925	.5609	38
23	.21445	.97673	.21956	.5546	37
24	.21473	.97667	.21986	.5483	36
25	.21502	.97661	.22017	4.5420	35
26	.21530	.97655	.22047	.5357	34
27	.21559	.97648	.22078	.5294	33
28	.21587	.97642	.22108	.5232	32
29	.21615	.97636	.22139	.5169	31
30	.21644	.97630	.22169	4.5107	30
31	.21672	.97623	.22200	.5045	29
32	.21701	.97617	.22230	.4983	28
33	.21729	.97611	.22261	.4921	27
34	.21757	.97604	.22291	.4860	26
35	.21786	.97598	.22322	4.4799	25
36	.21814	.97592	.22353	.4737	24
37	.21843	.97585	.22383	.4676	23
38	.21871	.97579	.22414	.4615	22
39	.21899	.97573	.22444	.4555	21
40	.21928	.97566	.22475	4.4494	20
41	.21956	.97560	.22505	.4434	19
42	.21985	.97553	.22536	.4373	18
43	.22013	.97547	.22566	.4313	17
44	.22041	.97541	.22597	.4253	16
45	.22070	.97534	.22628	4.4194	15
46	.22098	.97528	.22658	.4134	14
47	.22126	.97521	.22689	.4074	13
48	.22155	.97515	.22719	.4015	12
49	.22183	.97508	.22750	.3956	11
50	.22211	.97502	.22781	4.3897	10
51	.22240	.97495	.22811	.3838	9
52	.22268	.97489	.22842	.3779	8
53	.22297	.97483	.22872	.3721	7
54	.22325	.97476	.22903	.3662	6
55	.22353	.97470	.22934	4.3604	5
56	.22382	.97463	.22964	.3546	4
57	.22410	.97457	.22995	.3488	3
58	.22438	.97450	.23025	.3430	2
59	.22467	.97443	.23056	.3372	1
60	.22495	.97437	.23087	4.3315	0
M	Cosine	Sine	Cotan.	Tan.	M

77°

13°

M	Sine	Cosine	Tan.	Cotan.	M
0	.22495	.97437	.23087	4.3315	60
1	.22523	.97430	.23117	.3257	59
2	.22552	.97424	.23148	.3200	58
3	.22580	.97417	.23179	.3143	57
4	.22608	.97411	.23209	.3086	56
5	.22637	.97404	.23240	4.3029	55
6	.22665	.97398	.23270	.2972	54
7	.22693	.97391	.23301	.2916	53
8	.22722	.97384	.23332	.2859	52
9	.22750	.97378	.23363	.2803	51
10	.22778	.97371	.23393	4.2747	50
11	.22807	.97364	.23424	.2691	49
12	.22835	.97358	.23455	.2635	48
13	.22863	.97351	.23485	.2579	47
14	.22892	.97344	.23516	.2524	46
15	.22920	.97338	.23547	4.2468	45
16	.22948	.97331	.23577	.2413	44
17	.22977	.97324	.23608	.2358	43
18	.23005	.97318	.23639	.2303	42
19	.23033	.97311	.23670	.2248	41
20	.23061	.97304	.23700	4.2193	40
21	.23090	.97298	.23731	.2139	39
22	.23118	.97291	.23762	.2084	38
23	.23146	.97284	.23793	.2030	37
24	.23175	.97277	.23823	.1976	36
25	.23203	.97271	.23854	4.1921	35
26	.23231	.97264	.23885	.1867	34
27	.23260	.97257	.23916	.1814	33
28	.23288	.97250	.23946	.1760	32
29	.23316	.97244	.23977	.1706	31
30	.23344	.97237	.24008	4.1653	30
31	.23373	.97230	.24039	.1600	29
32	.23401	.97223	.24069	.1546	28
33	.23429	.97216	.24100	.1493	27
34	.23458	.97210	.24131	.1440	26
35	.23486	.97203	.24162	4.1388	25
36	.23514	.97196	.24192	.1335	24
37	.23542	.97189	.24223	.1282	23
38	.23571	.97182	.24254	.1230	22
39	.23599	.97175	.24285	.1178	21
40	.23627	.97169	.24316	4.1126	20
41	.23655	.97162	.24346	.1073	19
42	.23684	.97155	.24377	.1022	18
43	.23712	.97148	.24408	.0970	17
44	.23740	.97141	.24439	.0918	16
45	.23768	.97134	.24470	4.0867	15
46	.23797	.97127	.24501	.0815	14
47	.23825	.97120	.24531	.0764	13
48	.23853	.97113	.24562	.0713	12
49	.23881	.97106	.24593	.0662	11
50	.23910	.97099	.24624	4.0611	10
51	.23938	.97092	.24655	.0560	9
52	.23966	.97086	.24686	.0509	8
53	.23994	.97079	.24717	.0458	7
54	.24023	.97072	.24747	.0408	6
55	.24051	.97065	.24778	4.0358	5
56	.24079	.97058	.24809	.0307	4
57	.24107	.97051	.24840	.0257	3
58	.24136	.97044	.24871	.0207	2
59	.24164	.97037	.24902	.0157	1
60	.24192	.97029	.24933	4.0108	0
M	Cosine	Sine	Cotan.	Tan.	M

76°

14°

M	Sine	Cosine	Tan.	Cotan.	M
0	.24192	.97029	.24933	4.0108	60
1	.24220	.97022	.24964	.0058	59
2	.24249	.97015	.24995	.0009	58
3	.24277	.97008	.25025	3.9959	57
4	.24305	.97001	.25056	.9910	56
5	.24333	.96994	.25087	3.9861	55
6	.24361	.96987	.25118	.9812	54
7	.24390	.96980	.25149	.9763	53
8	.24418	.96973	.25180	.9714	52
9	.24446	.96966	.25211	.9665	51
10	.24474	.96959	.25242	3.9616	50
11	.24502	.96952	.25273	.9568	49
12	.24531	.96944	.25304	.9520	48
13	.24559	.96937	.25335	.9471	47
14	.24587	.96930	.25366	.9423	46
15	.24615	.96923	.25397	3.9375	45
16	.24643	.96916	.25428	.9327	44
17	.24672	.96909	.25459	.9279	43
18	.24700	.96901	.25490	.9231	42
19	.24728	.96894	.25521	.9184	41
20	.24756	.96887	.25552	3.9136	40
21	.24784	.96880	.25583	.9089	39
22	.24813	.96873	.25614	.9042	38
23	.24841	.96865	.25645	.8994	37
24	.24869	.96858	.25676	.8947	36
25	.24897	.96851	.25707	3.8900	35
26	.24925	.96844	.25738	.8853	34
27	.24953	.96836	.25769	.8807	33
28	.24982	.96829	.25800	.8760	32
29	.25010	.96822	.25831	.8713	31
30	.25038	.96815	.25862	3.8667	30
31	.25066	.96807	.25893	.8621	29
32	.25094	.96800	.25924	.8574	28
33	.25122	.96793	.25955	.8528	27
34	.25151	.96785	.25986	.8482	26
35	.25179	.96778	.26017	3.8436	25
36	.25207	.96771	.26048	.8390	24
37	.25235	.96763	.26079	.8345	23
38	.25263	.96756	.26110	.8299	22
39	.25291	.96749	.26141	.8254	21
40	.25319	.96741	.26172	3.8208	20
41	.25348	.96734	.26203	.8163	19
42	.25376	.96727	.26234	.8118	18
43	.25404	.96719	.26266	.8073	17
44	.25432	.96712	.26297	.8027	16
45	.25460	.96704	.26328	3.7983	15
46	.25488	.96697	.26359	.7938	14
47	.25516	.96690	.26390	.7893	13
48	.25544	.96682	.26421	.7848	12
49	.25573	.96675	.26452	.7804	11
50	.25601	.96667	.26483	3.7759	10
51	.25629	.96660	.26514	.7715	9
52	.25657	.96652	.26546	.7671	8
53	.25685	.96645	.26577	.7627	7
54	.25713	.96638	.26608	.7583	6
55	.25741	.96630	.26639	3.7539	5
56	.25769	.96623	.26670	.7495	4
57	.25798	.96615	.26701	.7451	3
58	.25826	.96608	.26732	.7407	2
59	.25854	.96600	.26764	.7364	1
60	.25882	.96592	.26795	3.7320	0
M	Cosine	Sine	Cotan.	Tan.	M

75°

TABLE XII (Continued, part 4 of 9)

15°

M	Sine	Cosine	Tan.	Cotan.	M
0	.25882	.96592	.26795	3.7320	60
1	.25910	.96585	.26826	.7277	59
2	.25938	.96577	.26857	.7234	58
3	.25966	.96570	.26888	.7191	57
4	.25994	.96562	.26920	.7147	56
5	.26022	.96555	.26951	3.7104	55
6	.26050	.96547	.26982	.7062	54
7	.26078	.96540	.27013	.7019	53
8	.26107	.96532	.27044	.6976	52
9	.26135	.96524	.27076	.6933	51
10	.26163	.96517	.27107	3.6891	50
11	.26191	.96509	.27138	.6848	49
12	.26219	.96502	.27169	.6806	48
13	.26247	.96494	.27201	.6764	47
14	.26275	.96486	.27232	.6722	46
15	.26303	.96479	.27263	3.6679	45
16	.26331	.96471	.27294	.6637	44
17	.26359	.96463	.27326	.6596	43
18	.26387	.96456	.27357	.6554	42
19	.26415	.96448	.27388	.6512	41
20	.26443	.96440	.27419	3.6470	40
21	.26471	.96433	.27451	.6429	39
22	.26499	.96425	.27482	.6387	38
23	.26527	.96417	.27513	.6346	37
24	.26556	.96409	.27544	.6305	36
25	.26584	.96402	.27576	3.6263	35
26	.26612	.96394	.27607	.6222	34
27	.26640	.96386	.27638	.6181	33
28	.26668	.96378	.27670	.6140	32
29	.26696	.96371	.27701	.6100	31
30	.26724	.96363	.27732	3.6059	30
31	.26752	.96355	.27764	.6018	29
32	.26780	.96347	.27795	.5977	28
33	.26808	.96340	.27826	.5937	27
34	.26836	.96332	.27858	.5896	26
35	.26864	.96324	.27889	3.5856	25
36	.26892	.96316	.27920	.5816	24
37	.26920	.96308	.27952	.5776	23
38	.26948	.96301	.27983	.5736	22
39	.26976	.96293	.28014	.5696	21
40	.27004	.96285	.28046	3.5656	20
41	.27032	.96277	.28077	.5616	19
42	.27060	.96269	.28109	.5576	18
43	.27088	.96261	.28140	.5536	17
44	.27116	.96253	.28171	.5497	16
45	.27144	.96245	.28203	3.5457	15
46	.27172	.96238	.28234	.5418	14
47	.27200	.96230	.28266	.5378	13
48	.27228	.96222	.28297	.5339	12
49	.27256	.96214	.28328	.5300	11
50	.27284	.96206	.28360	3.5261	10
51	.27312	.96198	.28391	.5222	9
52	.27340	.96190	.28423	.5183	8
53	.27368	.96182	.28454	.5144	7
54	.27396	.96174	.28486	.5105	6
55	.27424	.96166	.28517	3.5066	5
56	.27452	.96158	.28549	.5028	4
57	.27480	.96150	.28580	.4989	3
58	.27508	.96142	.28611	.4951	2
59	.27536	.96134	.28643	.4912	1
60	.27564	.96126	.28674	3.4874	0
M	Cosine	Sine	Cotan.	Tan.	M

74°

16°

M	Sine	Cosine	Tan.	Cotan.	M
0	.27564	.96126	.28674	3.4874	60
1	.27592	.96118	.28706	.4836	59
2	.27620	.96110	.28737	.4798	58
3	.27648	.96102	.28769	.4760	57
4	.27675	.96094	.28800	.4722	56
5	.27703	.96086	.28832	3.4684	55
6	.27731	.96078	.28863	.4646	54
7	.27759	.96070	.28895	.4608	53
8	.27787	.96062	.28926	.4570	52
9	.27815	.96054	.28958	.4533	51
10	.27843	.96045	.28990	3.4495	50
11	.27871	.96037	.29021	.4458	49
12	.27899	.96029	.29053	.4420	48
13	.27927	.96021	.29084	.4383	47
14	.27955	.96013	.29116	.4346	46
15	.27983	.96005	.29147	3.4308	45
16	.28011	.95997	.29179	.4271	44
17	.28039	.95989	.29210	.4234	43
18	.28067	.95980	.29242	.4197	42
19	.28094	.95972	.29274	.4160	41
20	.28122	.95964	.29305	3.4124	40
21	.28150	.95956	.29337	.4087	39
22	.28178	.95948	.29368	.4050	38
23	.28206	.95940	.29400	.4014	37
24	.28234	.95931	.29432	.3977	36
25	.28262	.95923	.29463	3.3941	35
26	.28290	.95915	.29495	.3904	34
27	.28318	.95907	.29526	.3868	33
28	.28346	.95898	.29558	.3832	32
29	.28374	.95890	.29590	.3795	31
30	.28401	.95882	.29621	3.3759	30
31	.28429	.95874	.29653	.3723	29
32	.28457	.95865	.29685	.3687	28
33	.28485	.95857	.29716	.3651	27
34	.28513	.95849	.29748	.3616	26
35	.28541	.95840	.29780	3.3580	25
36	.28569	.95832	.29811	.3544	24
37	.28597	.95824	.29843	.3509	23
38	.28624	.95816	.29875	.3473	22
39	.28652	.95807	.29906	.3438	21
40	.28680	.95799	.29938	3.3402	20
41	.28708	.95791	.29970	.3367	19
42	.28736	.95782	.30001	.3332	18
43	.28764	.95774	.30033	.3296	17
44	.28792	.95765	.30065	.3261	16
45	.28820	.95757	.30096	3.3226	15
46	.28847	.95749	.30128	.3191	14
47	.28875	.95740	.30160	.3156	13
48	.28903	.95732	.30192	.3121	12
49	.28931	.95723	.30223	.3087	11
50	.28959	.95715	.30255	3.3052	10
51	.28987	.95707	.30287	.3017	9
52	.29014	.95698	.30319	3.2983	8
53	.29042	.95690	.30350	.2948	7
54	.29070	.95681	.30382	.2914	6
55	.29098	.95673	.30414	3.2879	5
56	.29126	.95664	.30446	.2845	4
57	.29154	.95656	.30478	.2811	3
58	.29181	.95647	.30509	.2777	2
59	.29209	.95639	.30541	.2742	1
60	.29237	.95630	.30573	3.2708	0
M	Cosine	Sine	Cotan.	Tan.	M

73°

17°

M	Sine	Cosine	Tan.	Cotan.	M
0	.29237	.95630	.30573	3.2708	60
1	.29265	.95622	.30605	.2674	59
2	.29293	.95613	.30637	.2640	58
3	.29321	.95605	.30668	.2607	57
4	.29348	.95596	.30700	.2573	56
5	.29376	.95588	.30732	3.2539	55
6	.29404	.95579	.30764	.2505	54
7	.29432	.95571	.30796	.2472	53
8	.29460	.95562	.30828	.2438	52
9	.29487	.95554	.30859	.2405	51
10	.29515	.95545	.30891	3.2371	50
11	.29543	.95536	.30923	.2338	49
12	.29571	.95528	.30955	.2305	48
13	.29598	.95519	.30987	.2271	47
14	.29626	.95511	.31019	.2238	46
15	.29654	.95502	.31051	3.2205	45
16	.29682	.95493	.31083	.2172	44
17	.29710	.95485	.31115	.2139	43
18	.29737	.95476	.31146	.2106	42
19	.29765	.95467	.31178	.2073	41
20	.29793	.95459	.31210	3.2041	40
21	.29821	.95450	.31242	.2008	39
22	.29848	.95441	.31274	.1975	38
23	.29876	.95433	.31306	.1942	37
24	.29904	.95424	.31338	.1910	36
25	.29932	.95415	.31370	3.1877	35
26	.29959	.95407	.31402	.1845	34
27	.29987	.95398	.31434	.1813	33
28	.30015	.95389	.31466	.1780	32
29	.30043	.95380	.31498	.1748	31
30	.30070	.95372	.31530	3.1716	30
31	.30098	.95363	.31562	.1684	29
32	.30126	.95354	.31594	.1652	28
33	.30154	.95345	.31626	.1620	27
34	.30181	.95337	.31658	.1588	26
35	.30209	.95328	.31690	3.1556	25
36	.30237	.95319	.31722	.1524	24
37	.30265	.95310	.31754	.1492	23
38	.30292	.95301	.31786	.1460	22
39	.30320	.95293	.31818	.1429	21
40	.30348	.95284	.31850	3.1397	20
41	.30375	.95275	.31882	.1366	19
42	.30403	.95266	.31914	.1334	18
43	.30431	.95257	.31946	.1303	17
44	.30459	.95248	.31978	.1271	16
45	.30486	.95239	.32010	3.1240	15
46	.30514	.95231	.32042	.1209	14
47	.30542	.95222	.32074	.1177	13
48	.30569	.95213	.32106	.1146	12
49	.30597	.95204	.32138	.1115	11
50	.30625	.95195	.32171	3.1084	10
51	.30653	.95186	.32203	.1053	9
52	.30680	.95177	.32235	.1022	8
53	.30708	.95168	.32267	.0991	7
54	.30736	.95159	.32299	.0960	6
55	.30763	.95150	.32331	3.0930	5
56	.30791	.95141	.32363	.0899	4
57	.30819	.95132	.32395	.0868	3
58	.30846	.95124	.32428	.0838	2
59	.30874	.95115	.32460	.0807	1
60	.30902	.95106	.32492	3.0777	0
M	Cosine	Sine	Cotan.	Tan.	M

72°

18°

M	Sine	Cosine	Tan.	Cotan.	M
0	.30902	.95106	.32492	3.0777	60
1	.30929	.95097	.32524	.0746	59
2	.30957	.95088	.32556	.0716	58
3	.30985	.95079	.32588	.0686	57
4	.31012	.95070	.32621	.0655	56
5	.31040	.95061	.32653	3.0625	55
6	.31068	.95051	.32685	.0595	54
7	.31095	.95042	.32717	.0565	53
8	.31123	.95033	.32749	.0535	52
9	.31150	.95024	.32782	.0505	51
10	.31178	.95015	.32814	3.0475	50
11	.31206	.95006	.32846	.0445	49
12	.31233	.94997	.32878	.0415	48
13	.31261	.94988	.32910	.0385	47
14	.31289	.94979	.32943	.0356	46
15	.31316	.94970	.32975	3.0326	45
16	.31344	.94961	.33007	.0296	44
17	.31372	.94952	.33039	.0267	43
18	.31399	.94942	.33072	.0237	42
19	.31427	.94933	.33104	.0208	41
20	.31454	.94924	.33136	3.0178	40
21	.31482	.94915	.33169	.0149	39
22	.31510	.94906	.33201	.0120	38
23	.31537	.94897	.33233	.0090	37
24	.31565	.94888	.33265	.0061	36
25	.31592	.94878	.33298	3.0032	35
26	.31620	.94869	.33330	.0003	34
27	.31648	.94860	.33362	2.9974	33
28	.31675	.94851	.33395	.9945	32
29	.31703	.94841	.33427	.9916	31
30	.31730	.94832	.33459	2.9887	30
31	.31758	.94823	.33492	.9858	29
32	.31786	.94814	.33524	.9829	28
33	.31813	.94805	.33557	.9800	27
34	.31841	.94795	.33589	.9772	26
35	.31868	.94786	.33621	2.9743	25
36	.31896	.94777	.33654	.9714	24
37	.31923	.94767	.33686	.9686	23
38	.31951	.94758	.33718	.9657	22
39	.31978	.94749	.33751	.9629	21
40	.32006	.94740	.33783	2.9600	20
41	.32034	.94730	.33816	.9572	19
42	.32061	.94721	.33848	.9544	18
43	.32089	.94712	.33880	.9515	17
44	.32116	.94702	.33913	.9487	16
45	.32144	.94693	.33945	2.9459	15
46	.32171	.94684	.33978	.9431	14
47	.32199	.94674	.34010	.9403	13
48	.32226	.94665	.34043	.9375	12
49	.32254	.94655	.34075	.9347	11
50	.32282	.94646	.34108	2.9319	10
51	.32309	.94637	.34140	.9291	9
52	.32337	.94627	.34173	.9263	8
53	.32364	.94618	.34205	.9235	7
54	.32392	.94608	.34238	.9208	6
55	.32419	.94599	.34270	2.9180	5
56	.32447	.94590	.34303	.9152	4
57	.32474	.94580	.34335	.9125	3
58	.32502	.94571	.34368	.9097	2
59	.32529	.94561	.34400	.9069	1
60	.32557	.94552	.34433	2.9042	0
M	Cosine	Sine	Cotan.	Tan.	M

71°

19°

M	Sine	Cosine	Tan.	Cotan.	M
0	.32557	.94552	.34433	2.9042	60
1	.32584	.94542	.34465	.9015	59
2	.32612	.94533	.34498	.8987	58
3	.32639	.94523	.34530	.8960	57
4	.32667	.94514	.34563	.8933	56
5	.32694	.94504	.34595	2.8905	55
6	.32722	.94495	.34628	.8878	54
7	.32749	.94485	.34661	.8851	53
8	.32777	.94476	.34693	.8824	52
9	.32804	.94466	.34726	.8797	51
10	.32832	.94457	.34758	2.8770	50
11	.32859	.94447	.34791	.8743	49
12	.32887	.94438	.34824	.8716	48
13	.32914	.94428	.34856	.8689	47
14	.32942	.94418	.34889	.8662	46
15	.32969	.94409	.34921	2.8636	45
16	.32996	.94399	.34954	.8609	44
17	.33024	.94390	.34987	.8582	43
18	.33051	.94380	.35019	.8555	42
19	.33079	.94370	.35052	.8529	41
20	.33106	.94361	.35085	2.8502	40
21	.33134	.94351	.35117	.8476	39
22	.33161	.94341	.35150	.8449	38
23	.33189	.94332	.35183	.8423	37
24	.33216	.94322	.35215	.8396	36
25	.33243	.94313	.35248	2.8370	35
26	.33271	.94303	.35281	.8344	34
27	.33298	.94293	.35314	.8318	33
28	.33326	.94283	.35346	.8291	32
29	.33353	.94274	.35379	.8265	31
30	.33381	.94264	.35412	2.8239	30
31	.33408	.94254	.35445	.8213	29
32	.33435	.94245	.35477	.8187	28
33	.33463	.94235	.35510	.8161	27
34	.33490	.94225	.35543	.8135	26
35	.33518	.94215	.35576	2.8109	25
36	.33545	.94206	.35608	.8083	24
37	.33572	.94196	.35641	.8057	23
38	.33600	.94186	.35674	.8032	22
39	.33627	.94176	.35707	.8006	21
40	.33655	.94167	.35739	2.7980	20
41	.33682	.94157	.35772	.7954	19
42	.33709	.94147	.35805	.7929	18
43	.33737	.94137	.35838	.7903	17
44	.33764	.94127	.35871	.7878	16
45	.33792	.94118	.35904	2.7852	15
46	.33819	.94108	.35936	.7827	14
47	.33846	.94098	.35969	.7801	13
48	.33874	.94088	.36002	.7776	12
49	.33901	.94078	.36035	.7751	11
50	.33928	.94068	.36068	2.7725	10
51	.33956	.94058	.36101	.7700	9
52	.33983	.94049	.36134	.7675	8
53	.34011	.94039	.36167	.7650	7
54	.34038	.94029	.36199	.7625	6
55	.34065	.94019	.36232	2.7600	5
56	.34093	.94009	.36265	.7575	4
57	.34120	.93999	.36298	.7550	3
58	.34147	.93989	.36331	.7525	2
59	.34175	.93979	.36364	.7500	1
60	.34202	.93969	.36397	2.7475	0
M	Cosine	Sine	Cotan.	Tan.	M

70°

TABLE XII (Continued, part 5 of 9)

20°

M	Sine	Cosine	Tan.	Cotan.	M
0	.34202	.93969	.36397	2.7475	60
1	.34229	.93959	.36430	.7450	59
2	.34257	.93949	.36463	.7425	58
3	.34284	.93939	.36496	.7400	57
4	.34311	.93929	.36529	.7376	56
5	.34339	.93919	.36562	2.7351	55
6	.34366	.93909	.36595	.7326	54
7	.34393	.93899	.36628	.7302	53
8	.34421	.93889	.36661	.7277	52
9	.34448	.93879	.36694	.7252	51
10	.34475	.93869	.36727	2.7228	50
11	.34502	.93859	.36760	.7204	49
12	.34530	.93849	.36793	.7179	48
13	.34557	.93839	.36826	.7155	47
14	.34584	.93829	.36859	.7130	46
15	.34612	.93819	.36892	2.7106	45
16	.34639	.93809	.36925	.7082	44
17	.34666	.93799	.36958	.7058	43
18	.34693	.93789	.36991	.7033	42
19	.34721	.93779	.37024	.7009	41
20	.34748	.93769	.37057	2.6985	40
21	.34775	.93758	.37090	.6961	39
22	.34803	.93748	.37123	.6937	38
23	.34830	.93738	.37156	.6913	37
24	.34857	.93728	.37190	.6889	36
25	.34884	.93718	.37223	2.6865	35
26	.34912	.93708	.37256	.6841	34
27	.34939	.93698	.37289	.6817	33
28	.34966	.93687	.37322	.6794	32
29	.34993	.93677	.37355	.6770	31
30	.35021	.93667	.37388	2.6746	30
31	.35048	.93657	.37422	.6722	29
32	.35075	.93647	.37455	.6699	28
33	.35102	.93637	.37488	.6675	27
34	.35130	.93626	.37521	.6652	26
35	.35157	.93616	.37554	2.6628	25
36	.35184	.93606	.37587	.6604	24
37	.35211	.93596	.37621	.6581	23
38	.35239	.93585	.37654	.6558	22
39	.35266	.93575	.37687	.6534	21
40	.35293	.93565	.37720	2.6511	20
41	.35320	.93555	.37754	.6487	19
42	.35347	.93544	.37787	.6464	18
43	.35375	.93534	.37820	.6441	17
44	.35402	.93524	.37853	.6418	16
45	.35429	.93513	.37887	2.6394	15
46	.35456	.93503	.37920	.6371	14
47	.35483	.93493	.37953	.6348	13
48	.35511	.93482	.37986	.6325	12
49	.35538	.93472	.38020	.6302	11
50	.35565	.93462	.38053	2.6279	10
51	.35592	.93451	.38086	.6256	9
52	.35619	.93441	.38120	.6233	8
53	.35647	.93431	.38153	.6210	7
54	.35674	.93420	.38186	.6187	6
55	.35701	.93410	.38220	2.6164	5
56	.35728	.93400	.38253	.6142	4
57	.35755	.93389	.38286	.6119	3
58	.35782	.93379	.38320	.6096	2
59	.35810	.93368	.38353	.6073	1
60	.35837	.93358	.38386	2.6051	0
M	Cosine	Sine	Cotan.	Tan.	M

69°

21°

M	Sine	Cosine	Tan.	Cotan.	M
0	.35837	.93358	.38386	2.6051	60
1	.35864	.93348	.38420	.6028	59
2	.35891	.93337	.38453	.6006	58
3	.35918	.93327	.38486	.5983	57
4	.35945	.93316	.38520	.5960	56
5	.35972	.93306	.38553	2.5938	55
6	.36000	.93295	.38587	.5916	54
7	.36027	.93285	.38620	.5893	53
8	.36054	.93274	.38654	.5871	52
9	.36081	.93264	.38687	.5848	51
10	.36108	.93253	.38720	2.5826	50
11	.36135	.93243	.38754	.5804	49
12	.36162	.93232	.38787	.5781	48
13	.36189	.93222	.38821	.5759	47
14	.36217	.93211	.38854	.5737	46
15	.36244	.93201	.38888	2.5715	45
16	.36271	.93190	.38921	.5693	44
17	.36298	.93180	.38955	.5671	43
18	.36325	.93169	.38988	.5649	42
19	.36352	.93158	.39022	.5627	41
20	.36379	.93148	.39055	2.5605	40
21	.36406	.93137	.39089	.5583	39
22	.36433	.93127	.39122	.5561	38
23	.36460	.93116	.39156	.5539	37
24	.36488	.93105	.39189	.5517	36
25	.36515	.93095	.39223	2.5495	35
26	.36542	.93084	.39257	.5473	34
27	.36569	.93074	.39290	.5451	33
28	.36596	.93063	.39324	.5430	32
29	.36623	.93052	.39357	.5408	31
30	.36650	.93042	.39391	2.5386	30
31	.36677	.93031	.39425	.5365	29
32	.36704	.93020	.39458	.5343	28
33	.36731	.93010	.39492	.5322	27
34	.36758	.92999	.39525	.5300	26
35	.36785	.92988	.39559	2.5278	25
36	.36812	.92978	.39593	.5257	24
37	.36839	.92967	.39626	.5236	23
38	.36866	.92956	.39660	.5214	22
39	.36893	.92945	.39694	.5193	21
40	.36921	.92935	.39727	2.5171	20
41	.36948	.92924	.39761	.5150	19
42	.36975	.92913	.39795	.5129	18
43	.37002	.92902	.39828	.5108	17
44	.37029	.92892	.39862	.5086	16
45	.37056	.92881	.39896	2.5065	15
46	.37083	.92870	.39930	.5044	14
47	.37110	.92859	.39963	.5023	13
48	.37137	.92848	.39997	.5002	12
49	.37164	.92838	.40031	.4981	11
50	.37191	.92827	.40065	2.4960	10
51	.37218	.92816	.40098	.4939	9
52	.37245	.92805	.40132	.4918	8
53	.37272	.92794	.40166	.4897	7
54	.37299	.92784	.40200	.4876	6
55	.37326	.92773	.40233	2.4855	5
56	.37353	.92762	.40267	.4834	4
57	.37380	.92751	.40301	.4813	3
58	.37407	.92740	.40335	.4792	2
59	.37434	.92729	.40369	.4772	1
60	.37461	.92718	.40403	2.4751	0
M	Cosine	Sine	Cotan.	Tan.	M

68°

22°

M	Sine	Cosine	Tan.	Cotan.	M
0	.37461	.92718	.40403	2.4751	60
1	.37488	.92707	.40436	.4730	59
2	.37514	.92696	.40470	.4709	58
3	.37541	.92686	.40504	.4689	57
4	.37568	.92675	.40538	.4668	56
5	.37595	.92664	.40572	2.4647	55
6	.37622	.92653	.40606	.4627	54
7	.37649	.92642	.40640	.4606	53
8	.37676	.92631	.40673	.4586	52
9	.37703	.92620	.40707	.4565	51
10	.37730	.92609	.40741	2.4545	50
11	.37757	.92598	.40775	.4525	49
12	.37784	.92587	.40809	.4504	48
13	.37811	.92576	.40843	.4484	47
14	.37838	.92565	.40877	.4463	46
15	.37865	.92554	.40911	2.4443	45
16	.37892	.92543	.40945	.4423	44
17	.37919	.92532	.40979	.4403	43
18	.37946	.92521	.41013	.4382	42
19	.37972	.92510	.41047	.4362	41
20	.37999	.92499	.41081	2.4342	40
21	.38026	.92488	.41115	.4322	39
22	.38053	.92477	.41149	.4302	38
23	.38080	.92466	.41183	.4282	37
24	.38107	.92455	.41217	.4262	36
25	.38134	.92443	.41251	2.4242	35
26	.38161	.92432	.41285	.4222	34
27	.38188	.92421	.41319	.4202	33
28	.38214	.92410	.41353	.4182	32
29	.38241	.92399	.41387	.4162	31
30	.38268	.92388	.41421	2.4142	30
31	.38295	.92377	.41455	.4122	29
32	.38322	.92366	.41489	.4102	28
33	.38349	.92354	.41524	.4083	27
34	.38376	.92343	.41558	.4063	26
35	.38403	.92332	.41592	2.4043	25
36	.38429	.92321	.41626	.4023	24
37	.38456	.92310	.41660	.4004	23
38	.38483	.92299	.41694	.3984	22
39	.38510	.92287	.41728	.3964	21
40	.38537	.92276	.41762	2.3945	20
41	.38564	.92265	.41797	.3925	19
42	.38591	.92254	.41831	.3906	18
43	.38617	.92242	.41865	.3886	17
44	.38644	.92231	.41899	.3867	16
45	.38671	.92220	.41933	2.3847	15
46	.38698	.92209	.41968	.3828	14
47	.38725	.92197	.42002	.3808	13
48	.38751	.92186	.42036	.3789	12
49	.38778	.92175	.42070	.3770	11
50	.38805	.92164	.42105	2.3750	10
51	.38832	.92152	.42139	.3731	9
52	.38859	.92141	.42173	.3712	8
53	.38886	.92130	.42207	.3692	7
54	.38912	.92118	.42242	.3673	6
55	.38939	.92107	.42276	2.3654	5
56	.38966	.92096	.42310	.3635	4
57	.38993	.92084	.42344	.3616	3
58	.39019	.92073	.42379	.3597	2
59	.39046	.92062	.42413	.3577	1
60	.39073	.92050	.42447	2.3558	0
M	Cosine	Sine	Cotan.	Tan.	M

67°

23°

M	Sine	Cosine	Tan.	Cotan.	M
0	.39073	.92050	.42447	2.3558	60
1	.39100	.92039	.42482	.3539	59
2	.39126	.92028	.42516	.3520	58
3	.39153	.92016	.42550	.3501	57
4	.39180	.92005	.42585	.3482	56
5	.39207	.91993	.42619	2.3463	55
6	.39234	.91982	.42654	.3445	54
7	.39260	.91971	.42688	.3426	53
8	.39287	.91959	.42722	.3407	52
9	.39314	.91948	.42757	.3388	51
10	.39341	.91936	.42791	2.3369	50
11	.39367	.91925	.42826	.3350	49
12	.39394	.91913	.42860	.3332	48
13	.39421	.91902	.42894	.3313	47
14	.39448	.91891	.42929	.3294	46
15	.39474	.91879	.42963	2.3276	45
16	.39501	.91868	.42998	.3257	44
17	.39528	.91856	.43032	.3238	43
18	.39554	.91845	.43067	.3220	42
19	.39581	.91833	.43101	.3201	41
20	.39608	.91822	.43136	2.3183	40
21	.39635	.91810	.43170	.3164	39
22	.39661	.91798	.43205	.3145	38
23	.39688	.91787	.43239	.3127	37
24	.39715	.91775	.43274	.3109	36
25	.39741	.91764	.43308	2.3090	35
26	.39768	.91752	.43343	.3072	34
27	.39795	.91741	.43377	.3053	33
28	.39821	.91729	.43412	.3035	32
29	.39848	.91718	.43447	.3017	31
30	.39875	.91706	.43481	2.2998	30
31	.39901	.91694	.43516	.2980	29
32	.39928	.91683	.43550	.2962	28
33	.39955	.91671	.43585	.2944	27
34	.39981	.91659	.43620	.2925	26
35	.40008	.91648	.43654	2.2907	25
36	.40035	.91636	.43689	.2889	24
37	.40061	.91625	.43723	.2871	23
38	.40088	.91613	.43758	.2853	22
39	.40115	.91601	.43793	.2835	21
40	.40141	.91590	.43827	2.2817	20
41	.40168	.91578	.43862	.2799	19
42	.40195	.91566	.43897	.2781	18
43	.40221	.91554	.43932	.2763	17
44	.40248	.91543	.43966	.2745	16
45	.40275	.91531	.44001	2.2727	15
46	.40301	.91519	.44036	.2709	14
47	.40328	.91508	.44070	.2691	13
48	.40354	.91496	.44105	.2673	12
49	.40381	.91484	.44140	.2655	11
50	.40408	.91472	.44175	2.2637	10
51	.40434	.91461	.44209	.2619	9
52	.40461	.91449	.44244	.2602	8
53	.40487	.91437	.44279	.2584	7
54	.40514	.91425	.44314	.2566	6
55	.40541	.91414	.44349	2.2548	5
56	.40567	.91402	.44383	.2531	4
57	.40594	.91390	.44418	.2513	3
58	.40620	.91378	.44453	.2495	2
59	.40647	.91366	.44488	.2478	1
60	.40674	.91354	.44523	2.2460	0
M	Cosine	Sine	Cotan.	Tan.	M

66°

24°

M	Sine	Cosine	Tan.	Cotan.	M
0	.40674	.91354	.44523	2.2460	60
1	.40700	.91343	.44558	.2443	59
2	.40727	.91331	.44593	.2425	58
3	.40753	.91319	.44627	.2408	57
4	.40780	.91307	.44662	.2390	56
5	.40806	.91295	.44697	2.2373	55
6	.40833	.91283	.44732	.2355	54
7	.40860	.91271	.44767	.2338	53
8	.40886	.91260	.44802	.2320	52
9	.40913	.91248	.44837	.2303	51
10	.40939	.91236	.44872	2.2286	50
11	.40966	.91224	.44907	.2268	49
12	.40992	.91212	.44942	.2251	48
13	.41019	.91200	.44977	.2234	47
14	.41045	.91188	.45012	.2216	46
15	.41072	.91176	.45047	2.2199	45
16	.41098	.91164	.45082	.2182	44
17	.41125	.91152	.45117	.2165	43
18	.41151	.91140	.45152	.2147	42
19	.41178	.91128	.45187	.2130	41
20	.41204	.91116	.45222	2.2113	40
21	.41231	.91104	.45257	.2096	39
22	.41257	.91092	.45292	.2079	38
23	.41284	.91080	.45327	.2062	37
24	.41310	.91068	.45362	.2045	36
25	.41337	.91056	.45397	2.2028	35
26	.41363	.91044	.45432	.2011	34
27	.41390	.91032	.45467	.1994	33
28	.41416	.91020	.45502	.1977	32
29	.41443	.91008	.45537	.1960	31
30	.41469	.90996	.45573	2.1943	30
31	.41496	.90984	.45608	.1926	29
32	.41522	.90972	.45643	.1909	28
33	.41549	.90960	.45678	.1892	27
34	.41575	.90948	.45713	.1875	26
35	.41602	.90936	.45748	2.1859	25
36	.41628	.90924	.45783	.1842	24
37	.41654	.90911	.45819	.1825	23
38	.41681	.90899	.45854	.1808	22
39	.41707	.90887	.45889	.1792	21
40	.41734	.90875	.45924	2.1775	20
41	.41760	.90863	.45960	.1758	19
42	.41787	.90851	.45995	.1741	18
43	.41813	.90839	.46030	.1725	17
44	.41839	.90826	.46065	.1708	16
45	.41866	.90814	.46101	2.1692	15
46	.41892	.90802	.46136	.1675	14
47	.41919	.90790	.46171	.1658	13
48	.41945	.90778	.46206	.1642	12
49	.41972	.90766	.46242	.1625	11
50	.41998	.90753	.46277	2.1609	10
51	.42024	.90741	.46312	.1592	9
52	.42051	.90729	.46348	.1576	8
53	.42077	.90717	.46383	.1559	7
54	.42103	.90704	.46418	.1543	6
55	.42130	.90692	.46454	2.1527	5
56	.42156	.90680	.46489	.1510	4
57	.42183	.90668	.46524	.1494	3
58	.42209	.90655	.46560	.1478	2
59	.42235	.90643	.46595	.1461	1
60	.42262	.90631	.46631	2.1445	0
M	Cosine	Sine	Cotan.	Tan.	M

65°

TABLE XII (Continued, part 6 of 9)

25°

M	Sine	Cosine	Tan.	Cotan.	M
0	.42262	.90631	.46631	2.1445	60
1	.42288	.90618	.46666	.1429	59
2	.42314	.90606	.46702	.1412	58
3	.42341	.90594	.46737	.1396	57
4	.42367	.90581	.46772	.1380	56
5	.42394	.90569	.46808	2.1364	55
6	.42420	.90557	.46843	.1348	54
7	.42446	.90544	.46879	.1331	53
8	.42473	.90532	.46914	.1315	52
9	.42499	.90520	.46950	.1299	51
10	.42525	.90507	.46985	2.1283	50
11	.42552	.90495	.47021	.1267	49
12	.42578	.90483	.47056	.1251	48
13	.42604	.90470	.47092	.1235	47
14	.42630	.90458	.47127	.1219	46
15	.42657	.90445	.47163	2.1203	45
16	.42683	.90433	.47199	.1187	44
17	.42709	.90421	.47234	.1171	43
18	.42736	.90408	.47270	.1155	42
19	.42762	.90396	.47305	.1139	41
20	.42788	.90383	.47341	2.1123	40
21	.42815	.90371	.47376	.1107	39
22	.42841	.90358	.47412	.1092	38
23	.42867	.90346	.47448	.1076	37
24	.42893	.90333	.47483	.1060	36
25	.42920	.90321	.47519	2.1044	35
26	.42946	.90308	.47555	.1028	34
27	.42972	.90296	.47590	.1013	33
28	.42998	.90283	.47626	.0997	32
29	.43025	.90271	.47662	.0981	31
30	.43051	.90258	.47697	2.0965	30
31	.43077	.90246	.47733	.0950	29
32	.43104	.90233	.47769	.0934	28
33	.43130	.90221	.47805	.0918	27
34	.43156	.90208	.47840	.0903	26
35	.43182	.90196	.47876	2.0887	25
36	.43208	.90183	.47912	.0872	24
37	.43235	.90171	.47948	.0856	23
38	.43261	.90158	.47983	.0840	22
39	.43287	.90145	.48019	.0825	21
40	.43313	.90133	.48055	2.0809	20
41	.43340	.90120	.48091	.0794	19
42	.43366	.90108	.48127	.0778	18
43	.43392	.90095	.48162	.0763	17
44	.43418	.90082	.48198	.0747	16
45	.43444	.90070	.48234	2.0732	15
46	.43471	.90057	.48270	.0717	14
47	.43497	.90044	.48306	.0701	13
48	.43523	.90032	.48342	.0686	12
49	.43549	.90019	.48378	.0671	11
50	.43575	.90006	.48414	2.0655	10
51	.43602	.89994	.48449	.0640	9
52	.43628	.89981	.48485	.0625	8
53	.43654	.89968	.48521	.0609	7
54	.43680	.89956	.48557	.0594	6
55	.43706	.89943	.48593	2.0579	5
56	.43732	.89930	.48629	.0564	4
57	.43759	.89918	.48665	.0548	3
58	.43785	.89905	.48701	.0533	2
59	.43811	.89892	.48737	.0518	1
60	.43837	.89879	.48773	2.0503	0
M	Cosine	Sine	Cotan	Tan	M

64°

26°

M	Sine	Cosine	Tan.	Cotan.	M
0	.43837	.89879	.48773	2.0503	60
1	.43863	.89867	.48809	.0488	59
2	.43889	.89854	.48845	.0473	58
3	.43915	.89841	.48881	.0458	57
4	.43942	.89828	.48917	.0443	56
5	.43968	.89815	.48953	2.0427	55
6	.43994	.89803	.48989	.0412	54
7	.44020	.89790	.49025	.0397	53
8	.44046	.89777	.49062	2.0382	52
9	.44072	.89764	.49098	.0367	51
10	.44098	.89751	.49134	2.0352	50
11	.44124	.89739	.49170	.0338	49
12	.44150	.89726	.49206	.0323	48
13	.44177	.89713	.49242	.0308	47
14	.44203	.89700	.49278	.0293	46
15	.44229	.89687	.49314	2.0278	45
16	.44255	.89674	.49351	.0263	44
17	.44281	.89661	.49387	.0248	43
18	.44307	.89649	.49423	.0233	42
19	.44333	.89636	.49459	.0219	41
20	.44359	.89623	.49495	2.0204	40
21	.44385	.89610	.49532	.0189	39
22	.44411	.89597	.49568	.0174	38
23	.44437	.89584	.49604	.0159	37
24	.44463	.89571	.49640	.0145	36
25	.44489	.89558	.49677	2.0130	35
26	.44516	.89545	.49713	.0115	34
27	.44542	.89532	.49749	.0101	33
28	.44568	.89519	.49785	.0086	32
29	.44594	.89506	.49822	.0071	31
30	.44620	.89493	.49858	2.0057	30
31	.44646	.89480	.49894	.0042	29
32	.44672	.89467	.49931	.0028	28
33	.44698	.89454	.49967	.0013	27
34	.44724	.89441	.50003	1.9998	26
35	.44750	.89428	.50040	1.9984	25
36	.44776	.89415	.50076	.9969	24
37	.44802	.89402	.50113	.9955	23
38	.44828	.89389	.50149	.9940	22
39	.44854	.89376	.50185	.9926	21
40	.44880	.89363	.50222	1.9912	20
41	.44906	.89350	.50258	.9897	19
42	.44932	.89337	.50295	.9883	18
43	.44958	.89324	.50331	.9868	17
44	.44984	.89311	.50368	.9854	16
45	.45010	.89298	.50404	1.9840	15
46	.45036	.89285	.50441	.9825	14
47	.45062	.89272	.50477	.9811	13
48	.45088	.89258	.50514	.9797	12
49	.45114	.89245	.50550	.9782	11
50	.45140	.89232	.50587	1.9768	10
51	.45166	.89219	.50623	.9754	9
52	.45191	.89206	.50660	.9739	8
53	.45217	.89193	.50696	.9725	7
54	.45243	.89180	.50733	.9711	6
55	.45269	.89166	.50769	1.9697	5
56	.45295	.89153	.50806	.9683	4
57	.45321	.89140	.50843	.9668	3
58	.45347	.89127	.50879	.9654	2
59	.45373	.89114	.50916	.9640	1
60	.45399	.89101	.50952	1.9626	0
M	Cosine	Sine	Cotan.	Tan.	M

63°

27°

M	Sine	Cosine	Tan.	Cotan.	M
0	.45399	.89101	.50952	1.9626	60
1	.45425	.89087	.50989	.9612	59
2	.45451	.89074	.51026	.9598	58
3	.45477	.89061	.51062	.9584	57
4	.45503	.89048	.51099	.9570	56
5	.45528	.89034	.51136	1.9556	55
6	.45554	.89021	.51172	.9542	54
7	.45580	.89008	.51209	.9528	53
8	.45606	.88995	.51246	.9514	52
9	.45632	.88981	.51283	.9500	51
10	.45658	.88968	.51319	1.9486	50
11	.45684	.88955	.51356	.9472	49
12	.45710	.88942	.51393	.9458	48
13	.45736	.88928	.51430	.9444	47
14	.45761	.88915	.51466	.9430	46
15	.45787	.88902	.51503	1.9416	45
16	.45813	.88888	.51540	.9402	44
17	.45839	.88875	.51577	.9388	43
18	.45865	.88862	.51614	.9375	42
19	.45891	.88848	.51651	.9361	41
20	.45917	.88835	.51687	1.9347	40
21	.45942	.88822	.51724	.9333	39
22	.45968	.88808	.51761	.9319	38
23	.45994	.88795	.51798	.9306	37
24	.46020	.88781	.51835	.9292	36
25	.46046	.88768	.51872	1.9278	35
26	.46072	.88755	.51909	.9264	34
27	.46097	.88741	.51946	.9251	33
28	.46123	.88728	.51983	.9237	32
29	.46149	.88714	.52020	.9223	31
30	.46175	.88701	.52057	1.9210	30
31	.46201	.88688	.52094	.9196	29
32	.46226	.88674	.52131	.9182	28
33	.46252	.88661	.52168	.9169	27
34	.46278	.88647	.52205	.9155	26
35	.46304	.88634	.52242	1.9142	25
36	.46330	.88620	.52279	.9128	24
37	.46355	.88607	.52316	.9115	23
38	.46381	.88593	.52353	.9101	22
39	.46407	.88580	.52390	.9088	21
40	.46433	.88566	.52427	1.9074	20
41	.46458	.88553	.52464	.9061	19
42	.46484	.88539	.52501	.9047	18
43	.46510	.88526	.52538	.9034	17
44	.46536	.88512	.52575	.9020	16
45	.46561	.88499	.52612	1.9007	15
46	.46587	.88485	.52650	.8993	14
47	.46613	.88472	.52687	.8980	13
48	.46639	.88458	.52724	.8967	12
49	.46664	.88444	.52761	.8953	11
50	.46690	.88431	.52798	1.8940	10
51	.46716	.88417	.52836	.8927	9
52	.46741	.88404	.52873	.8913	8
53	.46767	.88390	.52910	.8900	7
54	.46793	.88376	.52947	.8887	6
55	.46819	.88363	.52984	1.8873	5
56	.46844	.88349	.53022	.8860	4
57	.46870	.88336	.53059	.8847	3
58	.46896	.88322	.53096	.8834	2
59	.46921	.88308	.53134	.8820	1
60	.46947	.88295	.53171	1.8807	0
M	Cosine	Sine	Cotan.	Tan.	M

62°

28°

M	Sine	Cosine	Tan.	Cotan.	M
0	.46947	.88295	.53171	1.8807	60
1	.46973	.88281	.53208	.8794	59
2	.46998	.88267	.53245	.8781	58
3	.47024	.88254	.53283	.8768	57
4	.47050	.88240	.53320	.8754	56
5	.47075	.88226	.53358	1.8741	55
6	.47101	.88213	.53395	.8728	54
7	.47127	.88199	.53432	.8715	53
8	.47152	.88185	.53470	.8702	52
9	.47178	.88171	.53507	.8689	51
10	.47204	.88158	.53545	1.8676	50
11	.47229	.88144	.53582	.8663	49
12	.47255	.88130	.53619	.8650	48
13	.47281	.88117	.53657	.8637	47
14	.47306	.88103	.53694	.8624	46
15	.47332	.88089	.53732	1.8611	45
16	.47357	.88075	.53769	.8598	44
17	.47383	.88061	.53807	.8585	43
18	.47409	.88048	.53844	.8572	42
19	.47434	.88034	.53882	.8559	41
20	.47460	.88020	.53919	1.8546	40
21	.47486	.88006	.53957	.8533	39
22	.47511	.87992	.53995	.8520	38
23	.47537	.87979	.54032	.8507	37
24	.47562	.87965	.54070	.8495	36
25	.47588	.87951	.54107	1.8482	35
26	.47613	.87937	.54145	.8469	34
27	.47639	.87923	.54183	.8456	33
28	.47665	.87909	.54220	.8443	32
29	.47690	.87895	.54258	.8430	31
30	.47716	.87882	.54295	1.8418	30
31	.47741	.87868	.54333	.8405	29
32	.47767	.87854	.54371	.8392	28
33	.47792	.87840	.54409	.8379	27
34	.47818	.87826	.54446	.8367	26
35	.47844	.87812	.54484	1.8354	25
36	.47869	.87798	.54522	.8341	24
37	.47895	.87784	.54559	.8329	23
38	.47920	.87770	.54597	.8316	22
39	.47946	.87756	.54635	.8303	21
40	.47971	.87742	.54673	1.8291	20
41	.47997	.87728	.54711	.8278	19
42	.48022	.87715	.54748	.8265	18
43	.48048	.87701	.54786	.8253	17
44	.48073	.87687	.54824	.8240	16
45	.48099	.87673	.54862	1.8227	15
46	.48124	.87659	.54900	.8215	14
47	.48150	.87645	.54937	.8202	13
48	.48175	.87631	.54975	.8190	12
49	.48201	.87617	.55013	.8177	11
50	.48226	.87603	.55051	1.8165	10
51	.48252	.87588	.55089	.8152	9
52	.48277	.87574	.55127	.8140	8
53	.48303	.87560	.55165	.8127	7
54	.48328	.87546	.55203	.8115	6
55	.48354	.87532	.55241	1.8102	5
56	.48379	.87518	.55279	.8090	4
57	.48405	.87504	.55317	.8078	3
58	.48430	.87490	.55355	.8065	2
59	.48455	.87476	.55393	.8053	1
60	.48481	.87462	.55431	1.8040	0
M	Cosine	Sine	Cotan.	Tan.	M

61°

29°

M	Sine	Cosine	Tan.	Cotan.	M
0	.48481	.87462	.55431	1.8040	60
1	.48506	.87448	.55469	.8028	59
2	.48532	.87434	.55507	.8016	58
3	.48557	.87420	.55545	.8003	57
4	.48583	.87405	.55583	.7991	56
5	.48608	.87391	.55621	1.7979	55
6	.48633	.87377	.55659	.7966	54
7	.48659	.87363	.55697	.7954	53
8	.48684	.87349	.55735	.7942	52
9	.48710	.87335	.55774	.7930	51
10	.48735	.87320	.55812	1.7917	50
11	.48760	.87306	.55850	.7905	49
12	.48786	.87292	.55888	.7893	48
13	.48811	.87278	.55926	.7881	47
14	.48837	.87264	.55964	.7868	46
15	.48862	.87250	.56003	1.7856	45
16	.48887	.87235	.56041	.7844	44
17	.48913	.87221	.56079	.7832	43
18	.48938	.87207	.56117	.7820	42
19	.48964	.87193	.56156	.7808	41
20	.48989	.87178	.56194	1.7795	40
21	.49014	.87164	.56232	.7783	39
22	.49040	.87150	.56270	.7771	38
23	.49065	.87136	.56309	.7759	37
24	.49090	.87121	.56347	.7747	36
25	.49116	.87107	.56385	1.7735	35
26	.49141	.87093	.56424	.7723	34
27	.49166	.87078	.56462	.7711	33
28	.49192	.87064	.56500	.7699	32
29	.49217	.87050	.56539	.7687	31
30	.49242	.87035	.56577	1.7675	30
31	.49268	.87021	.56616	.7663	29
32	.49293	.87007	.56654	.7651	28
33	.49318	.86992	.56692	.7639	27
34	.49343	.86978	.56731	.7627	26
35	.49369	.86964	.56769	1.7615	25
36	.49394	.86949	.56808	.7603	24
37	.49419	.86935	.56846	.7591	23
38	.49445	.86921	.56885	.7579	22
39	.49470	.86906	.56923	.7567	21
40	.49495	.86892	.56962	1.7555	20
41	.49521	.86877	.57000	.7544	19
42	.49546	.86863	.57039	.7532	18
43	.49571	.86849	.57077	.7520	17
44	.49596	.86834	.57116	.7508	16
45	.49622	.86820	.57155	1.7496	15
46	.49647	.86805	.57193	.7484	14
47	.49672	.86791	.57232	.7473	13
48	.49697	.86776	.57270	.7461	12
49	.49723	.86762	.57309	.7449	11
50	.49748	.86748	.57348	1.7437	10
51	.49773	.86733	.57386	.7426	9
52	.49798	.86719	.57425	.7414	8
53	.49823	.86704	.57464	.7402	7
54	.49849	.86690	.57502	.7390	6
55	.49874	.86675	.57541	1.7379	5
56	.49899	.86661	.57580	.7367	4
57	.49924	.86646	.57619	.7355	3
58	.49950	.86632	.57657	.7344	2
59	.49975	.86617	.57696	.7332	1
60	.50000	.86603	.57735	1.7320	0
M	Cosine	Sine	Cotan.	Tan.	M

60°

TABLE XII (Continued, part 7 of 9)

30°

M	Sine	Cosine	Tan.	Cotan.	M
0	.50000	.86603	.57735	1.7320	60
1	.50025	.86588	.57774	.7309	59
2	.50050	.86573	.57813	.7297	58
3	.50075	.86559	.57851	.7286	57
4	.50101	.86544	.57890	.7274	56
5	.50126	.86530	.57929	1.7262	55
6	.50151	.86515	.57968	.7251	54
7	.50176	.86500	.58007	.7239	53
8	.50201	.86486	.58046	.7228	52
9	.50226	.86471	.58085	.7216	51
10	.50252	.86457	.58123	1.7205	50
11	.50277	.86442	.58162	.7193	49
12	.50302	.86427	.58201	.7182	48
13	.50327	.86413	.58240	.7170	47
14	.50352	.86398	.58279	.7159	46
15	.50377	.86383	.58318	1.7147	45
16	.50402	.86369	.58357	.7136	44
17	.50428	.86354	.58396	.7124	43
18	.50453	.86339	.58435	.7113	42
19	.50478	.86325	.58474	.7101	41
20	.50503	.86310	.58513	1.7090	40
21	.50528	.86295	.58552	.7079	39
22	.50553	.86281	.58591	.7067	38
23	.50578	.86266	.58630	.7056	37
24	.50603	.86251	.58670	.7044	36
25	.50628	.86237	.58709	1.7033	35
26	.50653	.86222	.58748	.7022	34
27	.50679	.86207	.58787	.7010	33
28	.50704	.86192	.58826	.6999	32
29	.50729	.86178	.58865	.6988	31
30	.50754	.86163	.58904	1.6977	30
31	.50779	.86148	.58944	.6965	29
32	.50804	.86133	.58983	.6954	28
33	.50829	.86118	.59022	.6943	27
34	.50854	.86104	.59061	.6931	26
35	.50879	.86089	.59100	1.6920	25
36	.50904	.86074	.59140	.6909	24
37	.50929	.86059	.59179	.6898	23
38	.50954	.86044	.59218	.6887	22
39	.50979	.86030	.59258	.6875	21
40	.51004	.86015	.59297	1.6864	20
41	.51029	.86000	.59336	.6853	19
42	.51054	.85985	.59376	.6842	18
43	.51079	.85970	.59415	.6831	17
44	.51104	.85955	.59454	.6820	16
45	.51129	.85941	.59494	1.6808	15
46	.51154	.85926	.59533	.6797	14
47	.51179	.85911	.59572	.6786	13
48	.51204	.85896	.59612	.6775	12
49	.51229	.85881	.59651	.6764	11
50	.51254	.85866	.59691	1.6753	10
51	.51279	.85851	.59730	.6742	9
52	.51304	.85836	.59770	.6731	8
53	.51329	.85821	.59809	.6720	7
54	.51354	.85806	.59849	.6709	6
55	.51379	.85791	.59888	1.6698	5
56	.51404	.85777	.59928	.6687	4
57	.51429	.85762	.59967	.6676	3
58	.51454	.85747	.60007	.6665	2
59	.51479	.85732	.60046	.6654	1
60	.51504	.85717	.60086	1.6643	0
M	Cosine	Sine	Cotan.	Tan.	M

59°

31°

M	Sine	Cosine	Tan.	Cotan.	M
0	.51504	.85717	.60086	1.6643	60
1	.51529	.85702	.60126	.6632	59
2	.51554	.85687	.60165	.6621	58
3	.51578	.85672	.60205	.6610	57
4	.51603	.85657	.60244	.6599	56
5	.51628	.85642	.60284	1.6588	55
6	.51653	.85627	.60324	.6577	54
7	.51678	.85612	.60363	.6566	53
8	.51703	.85597	.60403	.6555	52
9	.51728	.85582	.60443	.6544	51
10	.51753	.85566	.60483	1.6534	50
11	.51778	.85551	.60522	.6523	49
12	.51803	.85536	.60562	.6512	48
13	.51827	.85521	.60602	.6501	47
14	.51852	.85506	.60642	.6490	46
15	.51877	.85491	.60681	1.6479	45
16	.51902	.85476	.60721	.6469	44
17	.51927	.85461	.60761	.6458	43
18	.51952	.85446	.60801	.6447	42
19	.51977	.85431	.60841	.6436	41
20	.52002	.85416	.60881	1.6425	40
21	.52026	.85400	.60920	.6415	39
22	.52051	.85385	.60960	.6404	38
23	.52076	.85370	.61000	.6393	37
24	.52101	.85355	.61040	.6383	36
25	.52126	.85340	.61080	1.6372	35
26	.52151	.85325	.61120	.6361	34
27	.52175	.85309	.61160	.6350	33
28	.52200	.85294	.61200	.6340	32
29	.52225	.85279	.61240	.6329	31
30	.52250	.85264	.61280	1.6318	30
31	.52275	.85249	.61320	.6308	29
32	.52299	.85234	.61360	.6297	28
33	.52324	.85218	.61400	.6286	27
34	.52349	.85203	.61440	.6276	26
35	.52374	.85188	.61480	1.6265	25
36	.52398	.85173	.61520	.6255	24
37	.52423	.85157	.61560	.6244	23
38	.52448	.85142	.61601	.6233	22
39	.52473	.85127	.61641	.6223	21
40	.52498	.85112	.61681	1.6212	20
41	.52522	.85096	.61721	.6202	19
42	.52547	.85081	.61761	.6191	18
43	.52572	.85066	.61801	.6181	17
44	.52597	.85050	.61842	.6170	16
45	.52621	.85035	.61882	1.6160	15
46	.52646	.85020	.61922	.6149	14
47	.52671	.85004	.61962	.6139	13
48	.52695	.84989	.62003	.6128	12
49	.52720	.84974	.62043	.6118	11
50	.52745	.84959	.62083	1.6107	10
51	.52770	.84943	.62123	.6097	9
52	.52794	.84928	.62164	.6086	8
53	.52819	.84912	.62204	.6076	7
54	.52844	.84897	.62244	.6066	6
55	.52868	.84882	.62285	1.6055	5
56	.52893	.84866	.62325	.6045	4
57	.52918	.84851	.62366	.6034	3
58	.52942	.84836	.62406	.6024	2
59	.52967	.84820	.62446	.6014	1
60	.52992	.84805	.62487	1.6003	0
M	Cosine	Sine	Cotan.	Tan.	M

58°

32°

M	Sine	Cosine	Tan.	Cotan.	M
0	.52992	.84805	.62487	1.6003	60
1	.53016	.84789	.62527	.5993	59
2	.53041	.84774	.62568	.5983	58
3	.53066	.84758	.62608	.5972	57
4	.53090	.84743	.62649	.5962	56
5	.53115	.84728	.62689	1.5952	55
6	.53140	.84712	.62730	.5941	54
7	.53164	.84697	.62770	.5931	53
8	.53189	.84681	.62811	.5921	52
9	.53214	.84666	.62851	.5910	51
10	.53238	.84650	.62892	1.5900	50
11	.53263	.84635	.62933	.5890	49
12	.53288	.84619	.62973	.5880	48
13	.53312	.84604	.63014	.5869	47
14	.53337	.84588	.63055	.5859	46
15	.53361	.84573	.63095	1.5849	45
16	.53386	.84557	.63136	.5839	44
17	.53411	.84542	.63177	.5829	43
18	.53435	.84526	.63217	.5818	42
19	.53460	.84511	.63258	.5808	41
20	.53484	.84495	.63299	1.5798	40
21	.53509	.84479	.63339	.5788	39
22	.53533	.84464	.63380	.5778	38
23	.53558	.84448	.63421	.5768	37
24	.53583	.84433	.63462	.5757	36
25	.53607	.84417	.63503	1.5747	35
26	.53632	.84402	.63543	.5737	34
27	.53656	.84386	.63584	.5727	33
28	.53681	.84370	.63625	.5717	32
29	.53705	.84355	.63666	.5707	31
30	.53730	.84339	.63707	1.5697	30
31	.53754	.84323	.63748	.5687	29
32	.53779	.84308	.63789	.5677	28
33	.53803	.84292	.63830	.5667	27
34	.53828	.84276	.63871	.5657	26
35	.53852	.84261	.63912	1.5646	25
36	.53877	.84245	.63953	.5636	24
37	.53901	.84229	.63994	.5626	23
38	.53926	.84214	.64035	.5616	22
39	.53950	.84198	.64076	.5606	21
40	.53975	.84182	.64117	1.5596	20
41	.53999	.84167	.64158	.5586	19
42	.54024	.84151	.64199	.5577	18
43	.54048	.84135	.64240	.5567	17
44	.54073	.84120	.64281	.5557	16
45	.54097	.84104	.64322	1.5547	15
46	.54122	.84088	.64363	.5537	14
47	.54146	.84072	.64404	.5527	13
48	.54171	.84057	.64446	.5517	12
49	.54195	.84041	.64487	.5507	11
50	.54220	.84025	.64528	1.5497	10
51	.54244	.84009	.64569	.5487	9
52	.54268	.83993	.64610	.5477	8
53	.54293	.83978	.64652	.5467	7
54	.54317	.83962	.64693	.5458	6
55	.54342	.83946	.64734	1.5448	5
56	.54366	.83930	.64775	.5438	4
57	.54391	.83914	.64817	.5428	3
58	.54415	.83899	.64858	.5418	2
59	.54439	.83883	.64899	.5408	1
60	.54464	.83867	.64941	1.5399	0
M	Cosine	Sine	Cotan.	Tan.	M

57°

33°

M	Sine	Cosine	Tan.	Cotan.	M
0	.54464	.83867	.64941	1.5399	60
1	.54488	.83851	.64982	.5389	59
2	.54513	.83835	.65023	.5379	58
3	.54537	.83819	.65065	.5369	57
4	.54561	.83804	.65106	.5359	56
5	.54586	.83788	.65148	1.5350	55
6	.54610	.83772	.65189	.5340	54
7	.54634	.83756	.65231	.5330	53
8	.54659	.83740	.65272	.5320	52
9	.54683	.83724	.65314	.5311	51
10	.54708	.83708	.65355	1.5301	50
11	.54732	.83692	.65397	.5291	49
12	.54756	.83676	.65438	.5282	48
13	.54781	.83660	.65480	.5272	47
14	.54805	.83644	.65521	.5262	46
15	.54829	.83629	.65663	1.5252	45
16	.54854	.83613	.65604	.5243	44
17	.54878	.83597	.65646	.5233	43
18	.54902	.83581	.65688	.5223	42
19	.54926	.83565	.65729	.5214	41
20	.54951	.83549	.65771	1.5204	40
21	.54975	.83533	.65813	.5195	39
22	.54999	.83517	.65854	.5185	38
23	.55024	.83501	.65896	.5175	37
24	.55048	.83485	.65938	.5166	36
25	.55072	.83469	.65980	1.5156	35
26	.55097	.83453	.66021	.5147	34
27	.55121	.83437	.66063	.5137	33
28	.55145	.83421	.66105	.5127	32
29	.55169	.83405	.66147	.5118	31
30	.55194	.83388	.66188	1.5108	30
31	.55218	.83372	.66230	.5099	29
32	.55242	.83356	.66272	.5089	28
33	.55266	.83340	.66314	.5080	27
34	.55291	.83324	.66356	.5070	26
35	.55315	.83308	.66398	1.5061	25
36	.55339	.83292	.66440	.5051	24
37	.55363	.83276	.66482	.5042	23
38	.55388	.83260	.66524	.5032	22
39	.55412	.83244	.66566	.5023	21
40	.55436	.83228	.66608	1.5013	20
41	.55460	.83211	.66650	.5004	19
42	.55484	.83195	.66692	.4994	18
43	.55509	.83179	.66734	.4985	17
44	.55533	.83163	.66776	.4975	16
45	.55557	.83147	.66818	1.4966	15
46	.55581	.83131	.66860	.4957	14
47	.55605	.83115	.66902	.4947	13
48	.55629	.83098	.66944	.4938	12
49	.55654	.83082	.66986	.4928	11
50	.55678	.83066	.67028	1.4919	10
51	.55702	.83050	.67071	.4910	9
52	.55726	.83034	.67113	.4900	8
53	.55750	.83017	.67155	.4891	7
54	.55774	.83001	.67197	.4881	6
55	.55799	.82985	.67239	1.4872	5
56	.55823	.82969	.67282	.4863	4
57	.55847	.82952	.67324	.4853	3
58	.55871	.82936	.67366	.4844	2
59	.55895	.82920	.67408	.4835	1
60	.55919	.82904	.67451	1.4826	0
M	Cosine	Sine	Cotan.	Tan.	M

56°

34°

M	Sine	Cosine	Tan.	Cotan.	M
0	.55919	.82904	.67451	1.4826	60
1	.55943	.82887	.67493	.4816	59
2	.55967	.82871	.67535	.4807	58
3	.55992	.82855	.67578	.4798	57
4	.56016	.82839	.67620	.4788	56
5	.56040	.82822	.67663	1.4779	55
6	.56064	.82806	.67705	.4770	54
7	.56088	.82790	.67747	.4761	53
8	.56112	.82773	.67790	.4751	52
9	.56136	.82757	.67832	.4742	51
10	.56160	.82741	.67875	1.4733	50
11	.56184	.82724	.67917	.4724	49
12	.56208	.82708	.67160	.4714	48
13	.56232	.82692	.68002	.4705	47
14	.56256	.82675	.68045	.4696	46
15	.56280	.82659	.68087	1.4687	45
16	.56304	.82643	.68130	.4678	44
17	.56328	.82626	.68173	.4669	43
18	.56353	.82610	.68215	.4659	42
19	.56377	.82593	.68258	.4650	41
20	.56401	.82577	.68301	1.4641	40
21	.56425	.82561	.68343	.4632	39
22	.56449	.82544	.68386	.4623	38
23	.56473	.82528	.68429	.4614	37
24	.56497	.82511	.68471	.4605	36
25	.56521	.82495	.68514	1.4595	35
26	.56545	.82478	.68557	.4586	34
27	.56569	.82462	.68600	.4577	33
28	.56593	.82445	.68642	.4568	32
29	.56617	.82429	.68685	.4559	31
30	.56641	.82413	.68728	1.4550	30
31	.56664	.82396	.68771	.4541	29
32	.56688	.82380	.68814	.4532	28
33	.56712	.82363	.68857	.4523	27
34	.56736	.82347	.68899	.4514	26
35	.56760	.82330	.68942	1.4505	25
36	.56784	.82314	.68985	.4496	24
37	.56808	.82297	.69028	.4487	23
38	.56832	.82280	.69071	.4478	22
39	.56856	.82264	.69114	.4469	21
40	.56880	.82247	.69157	1.4460	20
41	.56904	.82231	.69200	.4451	19
42	.56928	.82214	.69243	.4442	18
43	.56952	.82198	.69286	.4433	17
44	.56976	.82181	.69329	.4424	16
45	.57000	.82165	.69372	1.4415	15
46	.57023	.82148	.69415	.4406	14
47	.57047	.82131	.69459	.4397	13
48	.57071	.82115	.69502	.4388	12
49	.57095	.82098	.69545	.4379	11
50	.57119	.82082	.69588	1.4370	10
51	.57143	.82065	.69631	.4361	9
52	.57167	.82048	.69674	.4352	8
53	.57191	.82032	.69718	.4343	7
54	.57214	.82015	.69761	.4335	6
55	.57238	.81998	.69804	1.4326	5
56	.57262	.81982	.69847	.4317	4
57	.57286	.81965	.69891	.4308	3
58	.57310	.81948	.69934	.4299	2
59	.57334	.81932	.69977	.4290	1
60	.57358	.81915	.70021	1.4281	0
M	Cosine	Sine	Cotan.	Tan.	M

55°

TABLE XII (Continued, part 8 of 9)

35°					
M	Sine	Cosine	Tan.	Cotan.	M
0	.57358	.81915	.70021	1.4281	60
1	.57381	.81898	.70064	.4273	59
2	.57405	.81882	.70107	.4264	58
3	.57429	.81865	.70151	.4255	57
4	.57453	.81848	.70194	.4246	56
5	.57477	.81832	.70238	1.4237	55
6	.57500	.81815	.70281	.4228	54
7	.57524	.81798	.70325	.4220	53
8	.57548	.81781	.70368	.4211	52
9	.57572	.81765	.70412	.4202	51
10	.57596	.81748	.70455	1.4193	50
11	.57619	.81731	.70499	.4185	49
12	.57643	.81714	.70542	.4176	48
13	.57667	.81698	.70686	.4167	47
14	.57691	.81681	.70629	.4158	46
15	.57714	.81664	.70573	1.4150	45
16	.57738	.81647	.70717	.4141	44
17	.57762	.81630	.70760	.4132	43
18	.57786	.81614	.70804	.4123	42
19	.57809	.81597	.70848	.4115	41
20	.57833	.81580	.70891	1.4106	40
21	.57857	.81563	.70935	.4097	39
22	.57881	.81546	.70979	.4089	38
23	.57904	.81530	.71022	.4080	37
24	.57928	.81513	.71066	.4071	36
25	.57952	.81496	.71110	1.4063	35
26	.57975	.81479	.71154	.4054	34
27	.57999	.81462	.71198	.4045	33
28	.58023	.81445	.71241	1.4037	32
29	.58047	.81428	.71285	.4028	31
30	.58070	.81411	.71329	1.4019	30
31	.58094	.81395	.71373	.4011	29
32	.58118	.81378	.71417	.4002	28
33	.58141	.81361	.71461	.3994	27
34	.58165	.81344	.71505	.3985	26
35	.58189	.81327	.71549	1.3976	25
36	.58212	.81310	.71593	.3968	24
37	.58236	.81293	.71637	.3959	23
38	.58259	.81276	.71681	.3951	22
39	.58283	.81259	.71725	.3942	21
40	.58307	.81242	.71769	1.3933	20
41	.58330	.81225	.71813	.3925	19
42	.58354	.81208	.71857	.3916	18
43	.58378	.81191	.71901	.3908	17
44	.58401	.81174	.71945	.3899	16
45	.58425	.81157	.71990	1.3891	15
46	.58448	.81140	.72034	.3882	14
47	.58472	.81123	.72078	.3874	13
48	.58496	.81106	.72122	.3865	12
49	.58519	.81089	.72166	.3857	11
50	.58543	.81072	.72211	1.3848	10
51	.58566	.81055	.72255	.3840	9
52	.58590	.81038	.72299	1.3831	8
53	.58614	.81021	.72344	.3823	7
54	.58637	.81004	.72388	.3814	6
55	.58661	.80987	.72432	1.3806	5
56	.58684	.80970	.72477	.3797	4
57	.58708	.80953	.72521	.3789	3
58	.58731	.80936	.72565	.3781	2
59	.58755	.80919	.72610	.3772	1
60	.58778	.80902	.72654	1.3764	0
M	Cosine	Sine	Cotan.	Tan.	M
54°					

36°					
M	Sine	Cosine	Tan.	Cotan.	M
0	.58778	.80902	.72654	1.3764	60
1	.58802	.80885	.72699	.3755	59
2	.58825	.80867	.72743	.3747	58
3	.58849	.80850	.72788	.3738	57
4	.58873	.80833	.72832	1.3730	56
5	.58896	.80816	.72877	1.3722	55
6	.58920	.80799	.72921	.3713	54
7	.58943	.80782	.72966	.3705	53
8	.58967	.80765	.73010	.3697	52
9	.58990	.80747	.73055	.3688	51
10	.59014	.80730	.73100	1.3680	50
11	.59037	.80713	.73144	.3672	49
12	.59060	.80696	.73189	.3663	48
13	.59084	.80679	.73234	.3655	47
14	.59107	.80662	.73278	.3647	46
15	.59131	.80644	.73323	1.3638	45
16	.59154	.80627	.73368	.3630	44
17	.59178	.80610	.73412	.3622	43
18	.59201	.80593	.73457	.3613	42
19	.59225	.80576	.73502	.3605	41
20	.59248	.80558	.73547	1.3597	40
21	.59272	.80541	.73592	.3588	39
22	.59295	.80524	.73637	.3580	38
23	.59318	.80507	.73681	.3572	37
24	.59342	.80489	.73726	.3564	36
25	.59365	.80472	.73771	1.3555	35
26	.59389	.80455	.73816	.3547	34
27	.59412	.80437	.73861	.3539	33
28	.59435	.80420	.73906	.3531	32
29	.59459	.80403	.73951	.3522	31
30	.59482	.80386	.73996	1.3514	30
31	.59506	.80368	.74041	.3506	29
32	.59529	.80351	.74086	.3498	28
33	.59552	.80334	.74131	.3489	27
34	.59576	.80316	.74176	.3481	26
35	.59599	.80299	.74221	1.3473	25
36	.59622	.80282	.74266	.3465	24
37	.59646	.80264	.74312	.3457	23
38	.59669	.80247	.74357	.3449	22
39	.59692	.80230	.74402	.3440	21
40	.59716	.80212	.74447	1.3432	20
41	.59739	.80195	.74492	.3424	19
42	.59762	.80177	.74538	.3416	18
43	.59786	.80160	.74583	.3408	17
44	.59809	.80143	.74628	.3400	16
45	.59832	.80125	.74673	1.3392	15
46	.59856	.80108	.74719	1.3383	14
47	.59879	.80090	.74764	.3375	13
48	.59902	.80073	.74809	.3367	12
49	.59926	.80056	.74855	.3359	11
50	.59949	.80038	.74900	1.3351	10
51	.59972	.80021	.74946	.3343	9
52	.59995	.80003	.74991	.3335	8
53	.60019	.79986	.75037	.3327	7
54	.60042	.79968	.75082	.3319	6
55	.60065	.79951	.75128	1.3311	5
56	.60088	.79933	.75173	.3303	4
57	.60112	.79916	.75219	.3294	3
58	.60135	.79898	.75264	.3286	2
59	.60158	.79881	.75310	.3278	1
60	.60181	.79863	.75355	1.3270	0
M	Cosine	Sine	Cotan.	Tan.	M
53°					

37°					
M	Sine	Cosine	Tan.	Cotan.	M
0	.60181	.79863	.75355	1.3270	60
1	.60205	.79846	.75401	.3262	59
2	.60228	.79828	.75447	.3254	58
3	.60251	.79811	.75492	.3246	57
4	.60274	.79793	.75538	.3238	56
5	.60298	.79776	.75584	1.3230	55
6	.60320	.79758	.75629	.3222	54
7	.60344	.79741	.75675	.3214	53
8	.60367	.79723	.75721	.3206	52
9	.60390	.79706	.75767	.3198	51
10	.60413	.79688	.75812	1.3190	50
11	.60437	.79670	.75858	.3182	49
12	.60460	.79653	.75904	.3174	48
13	.60483	.79635	.75950	.3166	47
14	.60506	.79618	.75996	.3159	46
15	.60529	.79600	.76042	1.3151	45
16	.60552	.79582	.76088	.3143	44
17	.60576	.79565	.76134	.3135	43
18	.60599	.79547	.76179	.3127	42
19	.60622	.79530	.76225	.3119	41
20	.60645	.79512	.76271	1.3111	40
21	.60668	.79494	.76317	.3103	39
22	.60691	.79477	.76364	.3095	38
23	.60714	.79459	.76410	.3087	37
24	.60737	.79441	.76456	.3079	36
25	.60761	.79424	.76502	1.3071	35
26	.60784	.79406	.76548	.3064	34
27	.60807	.79388	.76594	.3056	33
28	.60830	.79371	.76640	.3048	32
29	.60853	.79353	.76686	.3040	31
30	.60876	.79335	.76733	1.3032	30
31	.60899	.79318	.76779	.3024	29
32	.60922	.79300	.76825	.3016	28
33	.60945	.79282	.76871	.3009	27
34	.60968	.79264	.76918	.3001	26
35	.60991	.79247	.76964	1.2993	25
36	.61014	.79229	.77010	.2985	24
37	.61037	.79211	.77057	.2977	23
38	.61061	.79193	.77103	.2970	22
39	.61084	.79176	.77149	.2962	21
40	.61107	.79158	.77196	1.2954	20
41	.61130	.79140	.77242	.2946	19
42	.61153	.79122	.77289	.2938	18
43	.61176	.79104	.77335	.2931	17
44	.61199	.79087	.77382	.2923	16
45	.61222	.79069	.77428	1.2915	15
46	.61245	.79051	.77475	.2907	14
47	.61268	.79033	.77521	.2900	13
48	.61290	.79015	.77568	.2892	12
49	.61314	.78998	.77614	.2884	11
50	.61337	.78980	.77661	1.2876	10
51	.61360	.78962	.77708	.2869	9
52	.61383	.78944	.77754	.2861	8
53	.61405	.78926	.77801	.2853	7
54	.61428	.78908	.77848	.2845	6
55	.61451	.78890	.77895	1.2838	5
56	.61474	.78873	.77941	.2830	4
57	.61497	.78855	.77988	.2822	3
58	.61520	.78837	.78035	.2815	2
59	.61543	.78819	.78082	.2807	1
60	.61566	.78801	.78128	1.2799	0
M	Cosine	Sine	Cotan.	Tan.	M
52°					

38°					
M	Sine	Cosine	Tan.	Cotan.	M
0	.61566	.78801	.78128	1.2799	60
1	.61589	.78783	.78175	.2792	59
2	.61612	.78765	.78222	.2784	58
3	.61635	.78747	.78269	.2776	57
4	.61658	.78729	.78316	.2769	56
5	.61681	.78711	.78363	1.2761	55
6	.61703	.78693	.78410	.2753	54
7	.61726	.78675	.78457	.2746	53
8	.61749	.78657	.78504	.2738	52
9	.61772	.78640	.78551	.2730	51
10	.61795	.78622	.78598	1.2723	50
11	.61818	.78604	.78645	.2715	49
12	.61841	.78586	.78692	.2708	48
13	.61864	.78568	.78739	.2700	47
14	.61886	.78550	.78786	.2692	46
15	.61909	.78532	.78834	1.2685	45
16	.61932	.78514	.78881	.2677	44
17	.61955	.78496	.78928	.2670	43
18	.61978	.78478	.78975	.2662	42
19	.62001	.78460	.79022	.2655	41
20	.62023	.78441	.79070	1.2647	40
21	.62046	.78423	.79117	.2639	39
22	.62069	.78405	.79164	.2632	38
23	.62092	.78387	.79212	.2624	37
24	.62115	.78369	.79259	.2617	36
25	.62137	.78351	.79306	1.2609	35
26	.62160	.78333	.79354	.2602	34
27	.62183	.78315	.79401	.2594	33
28	.62206	.78297	.79449	.2587	32
29	.62229	.78279	.79496	.2579	31
30	.62251	.78261	.79543	1.2572	30
31	.62274	.78243	.79591	.2564	29
32	.62297	.78224	.79639	.2557	28
33	.62320	.78206	.79686	.2549	27
34	.62342	.78188	.79734	.2542	26
35	.62365	.78170	.79781	1.2534	25
36	.62388	.78152	.79829	.2527	24
37	.62411	.78134	.79876	.2519	23
38	.62433	.78116	.79924	.2512	22
39	.62456	.78097	.79972	.2504	21
40	.62479	.78079	.80020	1.2497	20
41	.62501	.78061	.80067	.2489	19
42	.62524	.78043	.80115	.2482	18
43	.62547	.78025	.80163	.2475	17
44	.62570	.78007	.80211	.2467	16
45	.62592	.77988	.80258	1.2460	15
46	.62615	.77970	.80306	.2452	14
47	.62638	.77952	.80354	.2445	13
48	.62660	.77934	.80402	.2437	12
49	.62683	.77915	.80450	.2430	11
50	.62706	.77897	.80498	1.2423	10
51	.62728	.77879	.80546	.2415	9
52	.62751	.77861	.80594	.2408	8
53	.62774	.77842	.80642	.2400	7
54	.62796	.77824	.80690	.2393	6
55	.62819	.77806	.80738	1.2386	5
56	.62841	.77788	.80786	.2378	4
57	.62864	.77769	.80834	.2371	3
58	.62887	.77751	.80882	.2364	2
59	.62909	.77733	.80930	.2356	1
60	.62932	.77715	.80978	1.2342	0
M	Cosine	Sine	Cotan.	Tan.	M
51°					

39°					
M	Sine	Cosine	Tan.	Cotan.	M
0	.62932	.77715	.80978	1.2349	60
1	.62955	.77696	.81026	.2342	59
2	.62977	.77678	.81075	.2334	58
3	.63000	.77660	.81123	.2327	57
4	.63022	.77641	.81171	.2320	56
5	.63045	.77623	.81219	1.2312	55
6	.63067	.77605	.81268	.2305	54
7	.63090	.77586	.81316	.2297	53
8	.63113	.77568	.81364	.2290	52
9	.63135	.77549	.81413	.2283	51
10	.63158	.77531	.81461	1.2276	50
11	.63180	.77513	.81509	.2268	49
12	.63203	.77494	.81558	.2261	48
13	.63225	.77476	.81606	.2254	47
14	.63248	.77458	.81655	.2247	46
15	.63270	.77439	.81703	1.2239	45
16	.63293	.77421	.81752	.2232	44
17	.63315	.77402	.81800	.2225	43
18	.63338	.77384	.81849	.2218	42
19	.63360	.77365	.81898	.2210	41
20	.63383	.77347	.81946	1.2203	40
21	.63405	.77329	.81995	.2196	39
22	.63428	.77310	.82043	.2189	38
23	.63450	.77292	.82092	.2181	37
24	.63473	.77273	.82141	.2174	36
25	.63495	.77255	.82190	1.2167	35
26	.63518	.77236	.82238	.2160	34
27	.63540	.77218	.82287	.2152	33
28	.63563	.77199	.82336	.2145	32
29	.63585	.77181	.82385	.2138	31
30	.63608	.77162	.82434	1.2131	30
31	.63630	.77144	.82482	.2124	29
32	.63653	.77125	.82531	.2117	28
33	.63675	.77107	.82580	.2109	27
34	.63697	.77088	.82629	.2102	26
35	.63720	.77070	.82678	1.2095	25
36	.63742	.77051	.82727	.2088	24
37	.63765	.77033	.82776	.2081	23
38	.63787	.77014	.82825	.2074	22
39	.63810	.76996	.82874	.2066	21
40	.63832	.76977	.82923	1.2059	20
41	.63854	.76958	.82972	.2052	19
42	.63877	.76940	.83022	.2045	18
43	.63899	.76921	.83071	.2038	17
44	.63921	.76903	.83120	.2031	16
45	.63944	.76884	.83169	1.2024	15
46	.63966	.76865	.83218	.2016	14
47	.63989	.76847	.83267	.2009	13
48	.64011	.76828	.83317	.2002	12
49	.64033	.76810	.83366	.1995	11
50	.64056	.76791	.83415	1.1988	10
51	.64078	.76772	.83465	.1981	9
52	.64100	.76754	.83514	.1974	8
53	.64123	.76735	.83563	.1967	7
54	.64145	.76716	.83613	.1960	6
55	.64167	.76698	.83662	1.1953	5
56	.64189	.76679	.83712	.1946	4
57	.64212	.76660	.83761	.1939	3
58	.64234	.76642	.83811	.1932	2
59	.64256	.76623	.83860	.1924	1
60	.64279	.76604	.83910	1.1917	0
M	Cosine	Sine	Cotan.	Tan.	M
50°					

TABLE XII (Continued, part 9 of 9)

40°

M	Sine	Cosine	Tan.	Cotan.	M
0	.64279	.76604	.83910	1.1917	60
1	.64301	.76586	.83959	.1910	59
2	.64323	.76567	.84009	.1903	58
3	.64345	.76548	.84059	.1896	57
4	.64368	.76530	.84108	.1889	56
5	.64390	.76511	.84158	1.1882	55
6	.64412	.76492	.84208	.1875	54
7	.64435	.76473	.84257	.1868	53
8	.64457	.76455	.84307	.1861	52
9	.64479	.76436	.84357	.1854	51
10	.64501	.76417	.84407	1.1847	50
11	.64523	.76398	.84457	.1840	49
12	.64546	.76380	.84506	.1833	48
13	.64568	.76361	.84556	.1826	47
14	.64590	.76342	.84606	.1819	46
15	.64612	.76323	.84656	1.1812	45
16	.64635	.76304	.84706	.1805	44
17	.64657	.76286	.84756	.1798	43
18	.64679	.76267	.84806	.1791	42
19	.64701	.76248	.84856	.1785	41
20	.64723	.76229	.84906	1.1778	40
21	.64745	.76210	.84956	.1771	39
22	.64768	.76191	.85006	.1764	38
23	.64790	.76173	.85056	.1757	37
24	.64812	.76154	.85107	.1750	36
25	.64834	.76135	.85157	1.1743	35
26	.64856	.76116	.85207	.1736	34
27	.64878	.76097	.85257	.1729	33
28	.64900	.76078	.85307	.1722	32
29	.64923	.76059	.85358	.1715	31
30	.64945	.76041	.85408	1.1708	30
31	.64967	.76022	.85458	.1702	29
32	.64989	.76003	.85509	.1695	28
33	.65011	.75984	.85559	.1688	27
34	.65033	.75965	.85609	.1681	26
35	.65055	.75946	.85660	1.1674	25
36	.65077	.75927	.85710	.1667	24
37	.65100	.75908	.85761	.1660	23
38	.65121	.75889	.85811	.1653	22
39	.65144	.75870	.85862	.1647	21
40	.65166	.75851	.85912	1.1640	20
41	.65188	.75832	.85963	.1633	19
42	.65210	.75813	.86013	.1626	18
43	.65232	.75794	.86064	.1619	17
44	.65254	.75775	.86115	.1612	16
45	.65276	.75756	.86165	1.1605	15
46	.65298	.75737	.86216	.1599	14
47	.65320	.75718	.86267	.1592	13
48	.65342	.75700	.86318	.1585	12
49	.65364	.75680	.86368	.1578	11
50	.65386	.75661	.86419	1.1571	10
51	.65408	.75642	.86470	.1565	9
52	.65430	.75623	.86521	.1558	8
53	.65452	.75604	.86572	.1551	7
54	.65474	.75585	.86623	.1544	6
55	.65496	.75566	.86674	1.1537	5
56	.65518	.75547	.86725	.1531	4
57	.65540	.75528	.86775	.1524	3
58	.65562	.75509	.86826	.1517	2
59	.65584	.75490	.86878	.1510	1
60	.65606	.75471	.86929	1.1504	0
M	Cosine	Sine	Cotan.	Tan.	M

49°

41°

M	Sine	Cosine	Tan.	Cotan.	M
0	.65606	.75471	.86929	1.1504	60
1	.65628	.75452	.86980	.1497	59
2	.65650	.75433	.87031	.1490	58
3	.65672	.75414	.87082	.1483	57
4	.65694	.75394	.87133	.1477	56
5	.65716	.75375	.87184	1.1470	55
6	.65737	.75356	.87235	.1463	54
7	.65759	.75337	.87287	.1456	53
8	.65781	.75318	.87338	.1450	52
9	.65803	.75299	.87389	.1443	51
10	.65825	.75280	.87441	1.1436	50
11	.65847	.75261	.87492	.1430	49
12	.65869	.75241	.87543	.1423	48
13	.65891	.75222	.87595	.1416	47
14	.65913	.75203	.87646	.1409	46
15	.65934	.75184	.87698	1.1403	45
16	.65956	.75165	.87749	.1396	44
17	.65978	.75146	.87801	.1389	43
18	.66000	.75126	.87852	.1383	42
19	.66022	.75107	.87904	.1376	41
20	.66044	.75088	.87955	1.1369	40
21	.66066	.75069	.88007	.1363	39
22	.66087	.75049	.88058	.1356	38
23	.66109	.75030	.88110	.1349	37
24	.66131	.75011	.88162	.1343	36
25	.66153	.74992	.88213	1.1336	35
26	.66175	.74973	.88265	.1329	34
27	.66197	.74953	.88317	.1323	33
28	.66218	.74934	.88369	.1316	32
29	.66240	.74915	.88421	.1309	31
30	.66262	.74895	.88472	1.1303	30
31	.66284	.74876	.88524	.1296	29
32	.66305	.74857	.88576	.1290	28
33	.66327	.74838	.88628	.1283	27
34	.66349	.74818	.88680	.1276	26
35	.66371	.74799	.88732	1.1270	25
36	.66393	.74780	.88784	.1263	24
37	.66414	.74760	.88836	.1257	23
38	.66436	.74741	.88888	.1250	22
39	.66458	.74722	.88940	.1243	21
40	.66479	.74702	.88992	1.1237	20
41	.66501	.74683	.89044	.1230	19
42	.66523	.74664	.89097	.1224	18
43	.66545	.74644	.89149	.1217	17
44	.66566	.74625	.89201	.1211	16
45	.66588	.74606	.89253	1.1204	15
46	.66610	.74586	.89306	.1197	14
47	.66631	.74567	.89358	.1191	13
48	.66653	.74548	.89410	.1184	12
49	.66675	.74528	.89463	.1178	11
50	.66697	.74509	.89515	1.1171	10
51	.66718	.74489	.89567	.1165	9
52	.66740	.74470	.89620	.1158	8
53	.66762	.74450	.89672	.1152	7
54	.66783	.74431	.89725	.1145	6
55	.66805	.74412	.89777	1.1139	5
56	.66826	.74392	.89830	.1132	4
57	.66848	.74373	.89882	.1126	3
58	.66870	.74353	.89935	.1119	2
59	.66891	.74334	.89988	.1113	1
60	.66913	.74314	.90040	1.1106	0
M	Cosine	Sine	Cotan.	Tan.	M

48°

42°

M	Sine	Cosine	Tan.	Cotan.	M
0	.66913	.74314	.90040	1.1106	60
1	.66935	.74295	.90093	.1100	59
2	.66956	.74275	.90146	.1093	58
3	.66978	.74256	.90198	.1086	57
4	.66999	.74236	.90251	.1080	56
5	.67021	.74217	.90304	1.1074	55
6	.67043	.74197	.90357	.1067	54
7	.67064	.74178	.90410	.1061	53
8	.67086	.74158	.90463	.1054	52
9	.67107	.74139	.90515	.1048	51
10	.67129	.74119	.90568	1.1041	50
11	.67150	.74100	.90621	.1035	49
12	.67172	.74080	.90674	.1028	48
13	.67194	.74061	.90727	.1022	47
14	.67215	.74041	.90780	.1015	46
15	.67237	.74022	.90834	1.1009	45
16	.67258	.74002	.90887	.1003	44
17	.67280	.73983	.90940	.0996	43
18	.67301	.73963	.90993	.0990	42
19	.67323	.73943	.91046	.0983	41
20	.67344	.73924	.91099	1.0977	40
21	.67366	.73904	.91153	.0971	39
22	.67387	.73885	.91206	.0964	38
23	.67409	.73865	.91259	.0958	37
24	.67430	.73845	.91312	.0951	36
25	.67452	.73826	.91366	1.0945	35
26	.67473	.73806	.91419	.0939	34
27	.67495	.73787	.91473	.0932	33
28	.67516	.73767	.91526	.0926	32
29	.67537	.73747	.91580	.0919	31
30	.67559	.73728	.91633	1.0913	30
31	.67580	.73708	.91687	.0907	29
32	.67602	.73688	.91740	.0900	28
33	.67623	.73669	.91794	.0894	27
34	.67645	.73649	.91847	.0888	26
35	.67666	.73629	.91901	1.0881	25
36	.67688	.73610	.91955	.0875	24
37	.67709	.73590	.92008	.0868	23
38	.67730	.73570	.92062	.0862	22
39	.67752	.73551	.92116	.0856	21
40	.67773	.73531	.92170	1.0849	20
41	.67794	.73511	.92223	.0843	19
42	.67816	.73491	.92277	.0837	18
43	.67837	.73472	.92331	.0830	17
44	.67859	.73452	.92385	.0824	16
45	.67880	.73432	.92439	1.0818	15
46	.67901	.73412	.92493	.0812	14
47	.67923	.73393	.92547	.0805	13
48	.67944	.73373	.92601	.0799	12
49	.67965	.73353	.92655	.0793	11
50	.67987	.73333	.92709	1.0786	10
51	.68008	.73314	.92763	.0780	9
52	.68029	.73294	.92817	.0774	8
53	.68051	.73274	.92871	.0767	7
54	.68072	.73254	.92926	.0761	6
55	.68093	.73234	.92980	1.0755	5
56	.68115	.73215	.93034	.0749	4
57	.68136	.73195	.93088	.0742	3
58	.68157	.73175	.93143	.0736	2
59	.68178	.73155	.93197	.0730	1
60	.68200	.73135	.93251	1.0724	0
M	Cosine	Sine	Cotan.	Tan.	M

47°

43°

M	Sine	Cosine	Tan.	Cotan.	M
0	.68200	.73135	.93251	1.0724	60
1	.68221	.73115	.93306	.0717	59
2	.68242	.73096	.93360	.0711	58
3	.68264	.73076	.93415	.0705	57
4	.68285	.73056	.93469	.0699	56
5	.68306	.73036	.93524	1.0692	55
6	.68327	.73016	.93578	.0686	54
7	.68349	.72996	.93633	.0680	53
8	.68370	.72976	.93687	.0674	52
9	.68391	.72956	.93742	.0667	51
10	.68412	.72937	.93797	1.0661	50
11	.68433	.72917	.93851	.0655	49
12	.68455	.72897	.93906	.0649	48
13	.68476	.72877	.93961	.0643	47
14	.68497	.72857	.94016	.0636	46
15	.68518	.72837	.94071	1.0630	45
16	68539	.72817	.94125	.0624	44
17	.68561	.72797	.94180	.0618	43
18	.68582	.72777	.94235	.0612	42
19	.68603	.72757	.94290	.0605	41
20	.68624	.72737	.94345	1.0599	40
21	.68645	.72717	.94400	.0593	39
22	.68666	.72697	.94455	.0587	38
23	.68688	.72677	.94510	.0581	37
24	.68709	.72657	.94565	.0575	36
25	.68730	.72637	.94620	1.0568	35
26	.68751	.72617	.94675	.0562	34
27	.68772	.72597	.94731	.0556	33
28	.68793	.72577	.94786	.0550	32
29	.68814	.72557	.94841	.0544	31
30	.68835	.72537	.94896	1.0538	30
31	.68856	.72517	.94952	.0532	29
32	.68878	.72497	.95007	.0525	28
33	.68899	.72477	.95062	.0519	27
34	.68920	.72457	.95118	.0513	26
35	.68941	.72437	.95173	1.0507	25
36	.68962	.72417	.95229	.0501	24
37	.68983	.72397	.95284	.0495	23
38	.69004	.72377	.95340	.0489	22
39	.69025	.72357	.95395	.0483	21
40	.69046	.72337	.95451	1.0476	20
41	.69067	.72317	.95506	.0470	19
42	.69088	.72297	.95562	.0464	18
43	.69109	.72277	.95618	.0458	17
44	.69130	.72256	.95673	.0452	16
45	.69151	.72236	.95729	1.0446	15
46	.69172	.72216	.95785	.0440	14
47	.69193	.72196	.95841	.0434	13
48	.69214	.72176	.95896	.0428	12
49	.69235	.72156	.95952	.0422	11
50	.69256	.72136	.96008	1.0416	10
51	.69277	.72115	.96064	.0410	9
52	.69298	.72095	.96120	.0404	8
53	.69319	.72075	.96176	.0397	7
54	.69340	.72055	.96232	.0391	6
55	.69361	.72035	.96288	1.0385	5
56	.69382	.72015	.96344	.0379	4
57	.69403	.71994	.96400	.0373	3
58	.69424	.71974	.96456	.0367	2
59	.69445	.71954	.96513	.0361	1
60	.69466	.71934	.96569	1.0355	0
M	Cosine	Sine	Cotan.	Tan.	M

46°

44°

M	Sine	Cosine	Tan.	Cotan.	M
0	.69466	.71934	.96569	1.0355	60
1	.69487	.71914	.96625	.0349	59
2	.69508	.71893	.96681	.0343	58
3	.69528	.71873	.96738	.0337	57
4	.69549	.71853	96794	.0331	56
5	.69570	.71833	.96850	1.0325	55
6	.69591	.71813	.96907	.0319	54
7	.69612	.71792	.96963	.0313	53
8	.69633	.71772	.97020	.0307	52
9	.69654	.71752	.97076	.0301	51
10	.69675	.71732	.97133	1.0295	50
11	.69696	.71711	.97189	.0289	49
12	.69716	.71691	.97246	.0283	48
13	.69737	.71671	.97302	.0277	47
14	.69758	.71650	.97359	.0271	46
15	.69779	.71630	.97416	1.0265	45
16	.69800	.71610	.97472	.0259	44
17	.69821	.71589	.97529	.0253	43
18	.69841	.71569	.97586	.0247	42
19	.69862	.71549	.97643	.0241	41
20	.69883	.71529	.97700	1.0235	40
21	.69904	.71508	.97756	.0229	39
22	.69925	.71488	.97813	.0223	38
23	.69945	.71468	.97870	.0218	37
24	.69966	.71447	.97927	.0212	36
25	.69987	.71427	.97984	1.0206	35
26	.70008	.71406	.98041	.0200	34
27	.70029	.71386	.98098	.0194	33
28	.70049	.71366	.98155	.0188	32
29	.70070	.71345	.98212	.0182	31
30	.70091	.71325	.98270	1.0176	30
31	.70112	.71305	.98327	.0170	29
32	.70132	.71284	.98384	.0164	28
33	.70153	.71264	.98441	.0158	27
34	.70174	.71243	.98499	.0152	26
35	.70194	.71223	.98556	1.0146	25
36	.70215	.71203	.98613	.0141	24
37	.70236	.71182	.98671	.0135	23
38	.70257	.71162	.98728	.0129	22
39	.70277	.71141	.98786	.0123	21
40	.70298	.71121	.98843	1.0117	20
41	.70319	.71100	.98901	.0111	19
42	.70339	.71080	.98958	.0105	18
43	.70360	.71059	.99016	.0099	17
44	.70381	.71039	.99073	.0093	16
45	.70401	.71018	.99131	1.0088	15
46	.70422	.70998	.99189	.0082	14
47	.70443	.70977	.99246	.0076	13
48	.70463	.70957	.99304	.0070	12
49	.70484	.70936	.99362	.0064	11
50	.70505	.70916	.99420	1.0058	10
51	.70525	.70895	.99478	.0052	9
52	.70546	.70875	.99536	.0047	8
53	.70566	.70854	.99593	.0041	7
54	.70587	.70834	.99651	.0035	6
55	.70608	.70813	.99709	1.0029	5
56	.70628	.70793	.99767	.0023	4
57	.70649	.70772	.99826	.0017	3
58	.70669	.70752	.99884	.0012	2
59	.70690	.70731	.99942	.0006	1
60	.70711	.70711	1.0000	1.0000	0
M	Cosine	Sine	Cotan.	Tan.	M

45°

TABLE XIII. TRIGONOMETRIC CONSTANTS FOR FIVE-INCH SINE BAR (Part 1 of 5)

M	0°	1°	2°	3°	4°	5°	6°	7°	8°	9°	10°	11°
0	0.00000	0.08725	0.17450	0.26170	0.34880	0.43580	0.52265	0.60935	0.69585	0.78215	0.86825	0.95405
1	.00145	.08870	.17595	.26315	.35025	.43725	.52410	.61080	.69730	.78360	.86965	.95545
2	.00290	.09015	.17740	.26460	.35170	.43870	.52555	.61225	.69875	.78505	.87110	.95690
3	.00435	.09160	.17885	.26605	.35315	.44015	.52700	.61370	.70020	.78650	.87255	.95835
4	.00580	.09310	.18030	.26750	.35460	.44155	.52845	.61510	.70165	.78790	.87395	.95975
5	0.00725	0.09455	0.18175	0.26895	0.35605	0.44300	0.52985	0.61655	0.70305	0.78935	0.87540	0.96120
6	.00875	.09600	.18320	.27040	.35750	.44445	.53130	.61800	.70450	.79080	.87685	.96260
7	.01020	.09745	.18465	.27185	.35895	.44590	.53275	.61945	.70595	.79225	.87825	.96405
8	.01165	.09890	.18615	.27330	.36040	.44735	.53420	.62090	.70740	.79365	.87970	.96545
9	.01310	.10035	.18760	.27475	.36185	.44880	.53565	.62235	.70885	.79510	.88115	.96690
10	0.01455	0.10180	0.18905	0.27620	0.36330	0.45025	0.53710	0.62380	0.71025	0.79655	0.88255	0.96830
11	.01600	.10325	.19050	.27765	.36475	.45170	.53855	.62520	.71170	.79795	.88400	.96975
12	.01745	.10470	.19195	.27910	.36620	.45315	.54000	.62665	.71315	.79940	.88540	.97115
13	.01890	.10615	.19340	.28055	.36765	.45460	.54145	.62810	.71460	.80085	.88685	.97260
14	.02035	.10760	.19485	.28200	.36910	.45605	.54290	.62955	.71600	.80230	.88830	.97405
15	0.02180	0.10905	0.19630	0.28345	0.37055	0.45750	0.54435	0.63100	0.71745	0.80370	0.88970	0.97545
16	.02325	.11055	.19775	.28490	.37200	.45895	.54580	.63245	.71890	.80515	.89115	.97690
17	.02475	.11200	.19920	.28635	.37345	.46040	.54725	.63390	.72035	.80660	.89260	.97830
18	.02620	.11345	.20065	.28780	.37490	.46185	.54865	.63530	.72180	.80800	.89400	.97975
19	.02765	.11490	.20210	.28925	.37635	.46330	.55010	.63675	.72320	.80945	.89545	.98115
20	0.02910	0.11635	0.20355	0.29070	0.37780	0.46475	0.55155	0.63820	0.72465	0.81090	0.89685	0.98260
21	.03055	.11780	.20500	.29220	.37925	.46620	.55300	.63965	.72610	.81230	.89830	.98400
22	.03200	.11925	.20645	.29365	.38070	.46765	.55445	.64110	.72755	.81375	.89975	.98545
23	.03345	.12070	.20795	.29510	.38215	.46910	.55590	.64255	.72900	.81520	.90115	.98685
24	.03490	.12215	.20940	.29655	.38360	.47055	.55735	.64400	.73040	.81665	.90260	.98830
25	0.03635	0.12360	0.21085	0.29800	0.38505	0.47200	0.55880	0.64540	0.73185	0.81805	0.90405	0.98970
26	.03780	.12505	.21230	.29945	.38650	.47345	.56025	.64685	.73330	.81950	.90545	.99115
27	.03925	.12650	.21375	.30090	.38795	.47490	.56170	.64830	.73475	.82095	.90690	.99255
28	.04070	.12800	.21520	.30235	.38940	.47635	.56315	.64975	.73615	.82235	.90830	.99400
29	.04220	.12945	.21665	.30380	.39085	.47780	.56455	.65120	.73760	.82380	.90975	.99540
30	0.04365	0.13090	0.21810	0.30525	0.39230	0.47925	0.56600	0.65265	0.73905	0.82525	0.91120	0.99685
31	.04510	.13235	.21955	.30670	.39375	.48070	.56745	.65405	.74050	.82665	.91260	.99825
32	.04655	.13380	.22100	.30815	.39520	.48210	.56890	.65550	.74190	.82810	.91405	.99970
33	.04800	.13525	.22245	.30960	.39665	.48355	.57035	.65695	.74335	.82955	.91545	1.0011
34	.04945	.13670	.22390	.31105	.39810	.48500	.57180	.65840	.74480	.83100	.91690	.0016
35	0.05090	0.13815	0.22535	0.31250	0.39955	0.48645	0.57325	0.65985	0.74625	0.83240	0.91835	1.0039
36	.05235	.13960	.22680	.31395	.40100	.48790	.57470	.66130	.74770	.83385	.91975	.0054
37	.05380	.14105	.22825	.31540	.40245	.48935	.57615	.66270	.74910	.83530	.92120	.0068
38	.05525	.14250	.22970	.31685	.40390	.49080	.57760	.66415	.75055	.83670	.92260	.0082
39	.05670	.14395	.23115	.31830	.40535	.49225	.57900	.66560	.75200	.83815	.92405	.0096
40	0.05820	0.14540	0.23265	0.31975	0.40680	0.49370	0.58045	0.66705	0.75345	0.83960	0.92545	1.0110
41	.05965	.14690	.23410	.32120	.40825	.49515	.58190	.66850	.75485	.84100	.92690	.0125
42	.06110	.14835	.23555	.32265	.40970	.49660	.58335	.66995	.75630	.84245	.92835	.0139
43	.06255	.14980	.23700	.32410	.41115	.49805	.58480	.67135	.75775	.84390	.92975	.0153
44	.06400	.15125	.23845	.32555	.41260	.49950	.58625	.67280	.75920	.84530	.93120	.0168
45	0.06545	0.15270	0.23990	0.32700	0.41405	0.50095	0.58770	0.67425	0.76060	0.84675	0.93260	1.0182
46	.06690	.15415	.24135	.32845	.41550	.50240	.58915	.67570	.76205	.84820	.93405	.0196
47	.06835	.15560	.24280	.32990	.41695	.50385	.59060	.67715	.76350	.84960	.93550	.0210
48	.06980	.15705	.24425	.33135	.41840	.50530	.59200	.67860	.76495	.85105	.93690	.0225
49	.07125	.15850	.24570	.33280	.41985	.50675	.59345	.68000	.76635	.85250	.93835	.0239
50	0.07270	0.15995	0.24715	0.33425	0.42130	0.50820	0.59490	0.68145	0.76780	0.85390	0.93975	1.0253
51	.07415	.16140	.24860	.33570	.42275	.50960	.59635	.68290	.76925	.85535	.94120	.0267
52	.07565	.16285	.25005	.33715	.42420	.51105	.59 80	.68435	.77070	.85680	.94260	.0281
53	.07710	.16430	.25150	.33865	.42565	.51250	.59925	.68580	.77210	.85820	.94405	.0296
54	.07855	.16580	.25295	.34010	.42710	.51395	.60070	.68720	.77355	.85965	.94550	.0310
55	0.08000	0.16725	0.25440	0.34155	0.42855	0.51540	0.60215	0.68865	0.77500	0.86110	0.94690	1.0324
56	.08145	.16870	.25585	.34300	.43000	.51685	.60355	.69010	.77645	.86250	.94835	.0338
57	.08290	.17015	.25730	.34445	.43145	.51830	.60500	.69155	.77785	.86395	.94975	.0353
58	.08435	.17160	.25875	.34590	.43290	.51975	.60645	.69300	.77930	.86540	.95120	.0367
59	.08580	.17305	.26028	.34735	.43435	.52120	.60790	.69445	.78075	.86680	.95260	.0381
60	0.08725	0.17450	0.26170	0.34880	0.43580	0.52265	0.60935	0.69585	0.78215	0.86825	0.95405	1.0395

TABLE XIII (Continued, part 2 of 5)

M	12°	13°	14°	15°	16°	17°	18°	19°	20°	21°	22°	23°
0	1.0395	1.1247	1.2096	1.2941	1.3782	1.4618	1.5451	1.6278	1.7101	1.7918	1.8730	1.9536
1	.0410	.1261	.2110	.2955	.3796	.4632	.5464	.6292	.7114	.7932	.8744	.9550
2	.0424	.1276	.2124	.2969	.3810	.4646	.5478	.6306	.7128	.7945	.8757	.9563
3	.0438	.1290	.2138	.2983	.3824	.4660	.5492	.6319	.7142	.7959	.8771	.9576
4	.0452	.1304	.2152	.2997	.3838	.4674	.5506	.6333	.7155	.7972	.8784	.9590
5	1.0466	1.1318	1.2166	1.3011	1.3852	1.4588	1.5520	1.6347	1.7169	1.7986	1.8797	1.9603
6	.0481	.1332	.2181	.3025	.3865	.4702	.5534	.6361	.7183	.8000	.8811	.9617
7	.0495	.1346	.2195	.3039	.3879	.4716	.5547	.6374	.7196	.8013	.8824	.9630
8	.0509	.1361	.2209	.3053	.3893	.4730	.5561	.6388	.7210	.8027	.8838	.9643
9	.0523	.1375	.2223	.3067	.3907	.4743	.5575	.6402	.7224	.8040	.8851	.9657
10	1.0538	1.1389	1.2237	1.3081	1.3921	1.4757	1.5589	1.6416	1.7237	1.8054	1.8865	1.9670
11	.0552	.1403	.2251	.3095	.3935	.4771	.5603	.6429	.7251	.8067	.8878	.9683
12	.0566	.1417	.2265	.3109	.3949	.4785	.5616	.6443	.7265	.8081	.8892	.9697
13	.0580	.1431	.2279	.3123	.3963	.4799	.5630	.6457	.7292	.8095	.8906	.9711
14	.0594	.1446	.2293	.3137	.3977	.4813	.5644	.6471	.7242	.8108	.8919	.9724
15	1.0609	1.1460	1.2307	1.3151	1.3991	1.4827	1.5658	1.6484	1.7306	1.8122	1.8932	1.9737
16	.0623	.1474	.2322	.3165	.4005	.4841	.5672	.6498	.7319	.8135	.8946	.9750
17	.0637	.1488	.2336	.3179	.4019	.4855	.5686	.6512	.7333	.8149	.8959	.9764
18	.0651	.1502	.2350	.3193	.4033	.4868	.5699	.6525	.7347	.8162	.8973	.9777
19	.0665	.1516	.2364	.3207	.4047	.4882	.5713	.6539	.7360	.8176	.8986	.9790
20	1.0680	1.1531	1.2378	1.3221	1.4061	1.4896	1.5727	1.6553	1.7374	1.8189	1.8999	1.9804
21	.0694	.1545	.2392	.3235	.4075	.4910	.5741	.6567	.7387	.8203	.9013	.9817
22	.0708	.1559	.2406	.3250	.4089	.4924	.5755	.6580	.7401	.8217	.9026	.9830
23	.0722	.1573	.2420	.3264	.4103	.4938	.5768	.6594	.7415	.8230	.9040	.9844
24	.0737	.1587	.2434	.3278	.4117	.4952	.5782	.6608	.7428	.8244	.9053	.9857
25	1.0751	1.1601	1.2448	1.3292	1.4131	1.4966	1.5796	1.6622	1.7442	1.8257	1.9067	1.9870
26	.0765	.1615	.2462	.3306	.4145	.4980	.5810	.6635	.7456	.8271	.9080	.9884
27	.0779	.1630	.2477	.3320	.4159	.4993	.5824	.6649	.7469	.8284	.9094	.9897
28	.0793	.1644	.2491	.3334	.4173	.5007	.5837	.6663	.7483	.8298	.9107	.9911
29	.0808	.1658	.2505	.3348	.4187	.5021	.5851	.6676	.7496	.8311	.9120	.9924
30	1.0822	1.1672	1.2519	1.3362	1.4201	1.5035	1.5865	1.6690	1.7510	1.8325	1.9134	1.9937
31	.0836	.1686	.2533	.3376	.4214	.5049	.5879	.6704	.7524	.8338	.9147	.9951
32	.0850	.1700	.2547	.3390	.4228	.5063	.5893	.6718	.7537	.8352	.9161	.9964
33	.0864	.1714	.2561	.3404	.4242	.5077	.5906	.6731	.7551	.8365	.9174	.9977
34	.0879	.1729	.2575	.3418	.4256	.5091	.5920	.6745	.7565	.8379	.9188	.9991
35	1.0893	1.1743	1.2589	1.3432	1.4270	1.5104	1.5934	1.6759	1.7578	1.8392	1.9201	2.0004
36	.0907	.1757	.2603	.3446	.4284	.5118	.5948	.6772	.7592	.8406	.9215	.0017
37	.0921	.1771	.2617	.3460	.4298	.5132	.5961	.6786	.7605	.8419	.9228	.0031
38	.0935	.1785	.2631	.3474	.4312	.5146	.5975	.6800	.7619	.8433	.9241	.0044
39	.0949	.1799	.2645	.3488	.4326	.5160	.5989	.6813	.7633	.8447	.9255	.0057
40	1.0964	1.1813	1.2660	1.3502	1.4340	1.5174	1.6003	1.6827	1.7646	1.8460	1.9268	2.0070
41	.0978	.1828	.2674	.3516	.4354	.5188	.6017	.6841	.7660	.8474	.9282	.0084
42	.0992	.1842	.2688	.3530	.4368	.5201	.6030	.6855	.7673	.8487	.9295	.0097
43	.1006	.1856	.2702	.3544	.4382	.5215	.6044	.6868	.7687	.8501	.9308	.0110
44	.1020	.1870	.2716	.3558	.4396	.5229	.6058	.6882	.7701	.8514	.9322	.0124
45	1.1035	1.1884	1.2730	1.3572	1.4410	1.5243	1.6072	1.6896	1.7714	1.8528	1.9335	2.0137
46	1049	.1898	.2744	.3586	.4423	.5257	.6085	.6909	.7728	.8541	.9349	.0150
47	.1063	.1912	.2758	.3600	.4437	.5271	.6099	.6923	.7742	.8555	.9362	.0164
48	.1077	.1926	.2772	.3614	.4451	.5285	.6113	.6937	.7755	.8568	.9376	.0177
49	.1091	.1941	.2786	.3628	.4465	.5298	.6127	.6950	.7769	.8582	.9389	.0190
50	1.1106	1.1955	1.2800	1.3642	1.4479	1.5312	1.6141	1.6964	1.7782	1.8595	1.9402	2.0204
51	.1120	.1969	.2814	.3656	.4493	.5326	.6154	.6978	.7796	.8609	.9416	.0217
52	.1134	.1983	.2828	.3670	.4507	.5340	.6168	.6991	.7809	.8622	.9429	.0230
53	.1148	.1997	.2842	.3684	.4521	.5354	.6182	.7005	.7823	.8636	.9443	.0244
54	.1162	.2011	.2856	.3698	.4535	.5368	.6196	.7019	.7837	.8649	.9456	.0257
55	1.1176	1.2025	1.2870	1.3712	1.4549	1.5381	1.6209	1.7032	1.7850	1.8663	1.9469	2.0270
56	.1191	.2039	.2884	.3726	.4563	.5395	.6223	.7046	.7864	.8676	.9483	.0283
57	.1205	.2054	.2899	.3740	.4577	.5409	.6237	.7060	.7877	.8690	.9496	.0297
58	.1219	.2068	.2913	.3754	.4591	.5423	.6251	.7073	.7891	.8703	.9510	.0310
59	.1233	.2082	.2927	.3768	.4604	.5437	.6264	.7087	.7905	.8717	.9523	.0323
60	1.1247	1.2096	1.2941	1.3782	1.4618	1.5451	1.6278	1.7101	1.7918	1.8730	1.9536	2.0337

TABLE XIII (Continued, part 3 of 5)

M	24°	25°	26°	27°	28°	29°	30°	31°	32°	33°	34°	35°
0	2.0337	2.1131	2.1918	2.2699	2.3473	2.4240	2.5000	2.5752	2.6496	2.7232	2.7959	2.8679
1	.0350	.1144	.1931	.2712	.3486	.4253	.5012	.5764	.6508	.7244	.7971	.8690
2	.0363	.1157	.1944	.2725	.3499	.4266	.5025	.5777	.6520	.7256	.7984	.8702
3	.0376	.1170	.1958	.2738	.3512	.4278	.5038	.5789	.6533	.7268	.7996	.8714
4	.0390	.1183	.1971	.2751	.3525	.4291	.5050	.5802	.6545	.7280	.8008	.8726
5	2.0403	2.1197	2.1984	2.2764	2.3538	2.4304	2.5063	2.5814	2.6557	2.7293	2.8020	2.8738
6	.0416	.1210	.1997	.2777	.3550	.4317	.5075	.5826	.6570	.7305	.8032	.8750
7	.0430	.1223	.2010	.2790	.3563	.4329	.5088	.5839	.6582	.7317	.8044	.8762
8	.0443	.1236	.2023	.2803	.3576	.4342	.5100	.5851	.6594	.7329	.8056	.8774
9	.0456	.1249	.2036	.2816	.3589	.4355	.5113	.5864	.6607	.7341	.8068	.8786
10	2.0469	2.1262	2.2049	2.2829	2.3602	2.4367	2.5126	2.5876	2.6619	2.7354	2.8080	2.8798
11	.048[illegible]	.1276	.2062	.2842	.3614	.4380	.5138	.5889	.6631	.7366	.8092	.8809
12	.0496	.1289	.2075	.2855	.3627	.4393	.5151	.5901	.6644	.7378	.8104	.8821
13	.0509	.1302	.2088	.2868	.3640	.4405	.5163	.5914	.6656	.7390	.8116	.8833
14	.0522	.1315	.2101	.2881	.3653	.4418	.5176	.5926	.6668	.7402	.8128	.8845
15	2.0536	2.1328	2.2114	2.2893	2.3666	2.4431	2.5188	2.5938	2.6680	2.7414	2.8140	2.8857
16	.0549	.1341	.2127	.2906	.3679	.4444	.5201	.5951	.6693	.7427	.8152	.8869
17	.0562	.1354	.2140	.2919	.3691	.4456	.5214	.5963	.6705	.7439	.8164	.8881
18	.0575	.1368	.2153	.2932	.3704	.4469	.5226	.5976	.6717	.7451	.8176	.8893
19	.0589	.1381	.2166	.2945	.3717	.4482	.5239	.5988	.6730	.7463	.8188	.8905
20	2.0602	2.1394	2.2179	2.2958	2.3730	2.4494	2.5251	2.6001	2.6742	2.7475	2.8200	2.8916
21	.0615	.1407	.2192	.2971	.3743	.4507	.5264	.6013	.6754	.7487	.8212	.8928
22	.0628	.1420	.2205	.2984	.3755	.4520	.5276	.6025	.6767	.7499	.8224	.8940
23	.0642	.1433	.2218	.2997	.3768	.4532	.5289	.6038	.6779	.7512	.8236	.8952
24	.0655	.1447	.2232	.3010	.3781	.4545	.5301	.6050	.6791	.7524	.8248	.8964
25	2.0668	2.1460	2.2245	2.3023	2.3794	2.4558	2.5314	2.6063	2.6803	2.7536	2.8260	2.8976
26	.0681	.1473	.2258	.3036	.3807	.4570	.5327	.6075	.6816	.7548	.8272	.8988
27	.0695	.1486	.2271	.3048	.3819	.4583	.5339	.6087	.6828	.7560	.8284	.8999
28	.0708	.1499	.2284	.3061	.3832	.4596	.5352	.6100	.6840	.7572	.8296	.9011
29	.0721	.1512	.2297	.3074	.3845	.4608	.5364	.6112	.6852	.7584	.8308	.9023
30	2.0734	2.1525	2.2310	2.3087	2.3858	2.4621	2.5377	2.6125	2.6865	2.7597	2.8320	2.9035
31	.0748	.1538	.2323	.3100	.3870	.4634	.5389	.6137	.6877	.7609	.8332	.9047
32	.0761	.1552	.2336	.3113	.3883	.4646	.5402	.6149	.6889	.7621	.8344	.9059
33	.0774	.1565	.2349	.3126	.3896	.4659	.5414	.6162	.6902	.7633	.8356	.9070
34	.0787	.1578	.2362	.3139	.3909	.4672	.5427	.6174	.6914	.7645	.8368	.9082
35	2.0801	2.1591	2.2375	2.3152	2.3922	2.4684	2.5439	2.6187	2.6926	2.7657	2.8380	2.9094
36	.0814	.1604	.2388	.3165	.3934	.4697	.5452	.6199	.6938	.7669	.8392	.9106
37	.0827	.1617	.2401	.3177	.3947	.4709	.5464	.6211	.6951	.7681	.8404	.9118
38	.0840	.1630	.2414	.3190	.3960	.4722	.5477	.6224	.6963	.7694	.8416	.9130
39	.0853	.1643	.2427	.3203	.3973	.4735	.5489	.6236	.6975	.7706	.8428	.9141
40	2.0867	2.1656	2.2440	2.3216	2.3985	2.4747	2.5502	2.6249	2.6987	2.7718	2.8440	2.9153
41	.0880	.1670	.2453	.3229	.3998	.4760	.5514	.6261	.7000	.7730	.8452	.9165
42	.0893	.1683	.2466	.3242	.4011	.4773	.5527	.6273	.7012	.7742	.8464	.9177
43	.0906	.1696	.2479	.3255	.4024	.4785	.5539	.6286	.7024	.7754	.8476	.9189
44	.0920	.1709	.2492	.3268	.4036	.4798	.5552	.6298	.7036	.7766	.8488	.9200
45	2.0933	2.1722	2.2505	2.3280	2.4049	2.4811	2.5564	2.6310	2.7048	2.7778	2.8500	2.9212
46	.0946	.1735	.2518	.3293	.4062	.4823	.5577	.6323	.7061	.7790	.8512	.9224
47	.0959	.1748	.2531	.3306	.4075	.4836	.5589	.6335	.7073	.7802	.8523	.9236
48	.0972	.1761	.2544	.3319	.4087	.4848	.5602	.6348	.7085	.7815	.8535	.9248
49	.0986	.1774	.2557	.3332	.4100	.4861	.5614	.6360	.7097	.7827	.8547	.9259
50	2.0999	2.1787	2.2570	2.3345	2.4113	2.4874	2.5627	2.6372	2.7110	2.7839	2.8559	2.9271
51	.1012	.1801	.2583	.3358	.4126	.4886	.5639	.6385	.7122	.7851	.8571	.9283
52	.1025	.1814	.2596	.3371	.4138	.4899	.5652	.6397	.7134	.7863	.8583	.9295
53	.1038	.1827	.2609	.3383	.4151	.4912	.5664	.6409	.7146	.7875	.8595	.9307
54	.1052	.1840	.2621	.3396	.4164	.4924	.5677	.6422	.7158	.7887	.8607	.9318
55	2.1065	2.1853	2.2634	2.3409	2.4177	2.4937	2.5689	2.6434	2.7171	2.7899	2.8619	2.9330
56	.1078	.1866	.2647	.3422	.4189	.4949	.5702	.6446	.7183	.7911	.8631	.9342
57	.1091	.1879	.2660	.3435	.4202	.4962	.5714	.6459	.7195	.7923	.8643	.9354
58	.1104	.1892	.2673	.3448	.4215	.4975	.5727	.6471	.7207	.7935	.8655	.9365
59	.1117	.1905	.2686	.3460	.4228	.4987	.5739	.6483	.7220	.7947	.8667	.9377
60	2.1131	2.1918	2.2699	2.3473	2.4240	2.5000	2.5752	2.6496	2.7232	2.7959	2.8679	2.9389

TABLE XIII (Continued, part 4 of 5)

M	36°	37°	38°	39°	40°	41°	42°	43°	44°	45°	46°	47°
0	2.9389	3.0091	3.0783	3.1466	3.2139	3.2803	3.3456	3.4100	3.4733	3.5355	3.5967	3.6567
1	.9401	.0102	.0794	.1477	.2150	.2814	.3467	.4110	.4743	.5365	.5977	.6577
2	.9413	.0114	.0806	.1488	.2161	.2825	.3478	.4121	.4754	.5376	.5987	.6587
3	.9424	.0125	.0817	.1500	.2173	.2836	.3489	.4132	.4764	.5386	.5997	.6597
4	.9436	.0137	.0829	.1511	.2184	.2847	.3499	.4142	.4774	.5396	.6007	.6607
5	2.9448	3.0149	3.0840	3.1522	3.2195	3.2858	3.3510	3.4153	3.4785	3.5406	3.6017	3.6617
6	.9460	.0160	.0852	.1534	.2206	.2869	.3521	.4163	.4795	.5417	.6027	.6627
7	.9471	.0172	.0863	.1545	.2217	.2879	.3532	.4174	.4806	.5427	.6037	.6637
8	.9483	.018[illegible]	.0874	.1556	.2228	.2890	.3543	.4185	.4816	.5437	.6047	.6647
9	.9495	.0195	.0886	.1567	.2239	.2901	.3553	.4195	.4827	.5448	.6058	.6657
10	2.9507	3.0207	3.0897	3.1579	3.2250	3.2912	3.3564	3.4206	3.4837	3.5458	3.6068	3.6666
11	.9518	.0218	.0909	.1590	.2262	.2923	.3575	.4217	.4848	.5468	.6078	.6676
12	.9530	.0230	.0920	.1601	.2273	.2934	.3586	.4227	.4858	.5478	.6088	.6686
13	.9542	.0241	.0932	.1612	.2284	.2945	.3597	.4238	.4868	.5489	.6098	.6696
14	.9554	.0253	.0943	.1624	.2295	.2956	.3607	.4248	.4879	.5499	.6108	.6706
15	2.9565	3.0264	3.0954	3.1635	3.2306	3.2967	3.3618	3.4259	3.4889	3.5509	3.6118	3.6716
16	.9577	.0276	.0966	.1646	.2317	.2978	3629	.4269	.4900	.5519	.6128	.6726
17	.9589	.0288	.0977	.1658	.2328	.2989	.3640	.4280	.4910	.5529	.6138	.6736
18	.9600	.0299	.0989	.1669	.2339	.3000	.3651	.4291	.4921	.5540	.6148	.6745
19	.9612	.0311	.1000	.1680	.2350	.3011	.3661	.4301	.4931	.5550	.6158	.6755
20	2.9624	3.0322	3.1012	3.1691	3.2361	3.3022	3.3672	3.4312	3.4941	3.5560	3.6168	3.6765
21	.9636	.0334	.1023	.1703	.2373	.3033	.3683	.4322	.4952	.5570	.6178	.6775
22	.9647	.0345	.1034	.1714	.2384	.3044	.3693	.4333	.4962	.5581	.6188	.6785
23	.9659	.0357	.1046	.1725	.2395	.3054	.3704	.4344	.4973	.5591	.6198	.6795
24	.9671	.0369	.1057	.1736	.2406	.3065	.3715	.4354	.4983	.5601	.6208	.6805
25	2.9682	3.0380	3.1069	3.1748	3.2417	3.3076	3.3726	3.4365	3.4993	3.5611	3.6218	3.6814
26	.9694	.0392	.1080	.1759	.2428	3087	.3736	.4375	.5004	.5621	.6228	.6824
27	.9706	.0403	.1091	.1770	.2439	3098	.3747	.4386	.5014	.5632	.6238	.6834
28	.9718	.0415	.1103	.1781	2450	.3109	.3758	.4396	.5024	.5642	.6248	.6844
29	.9729	.0426	.1114	.1792	.2461	.3120	.3769	.4407	.5035	.5652	.6258	.6854
30	2.9741	3.0438	3.1125	3.1804	3.2472	3.3131	3.3779	3.4417	3.5045	3.5662	3.6268	3.6864
31	.9753	.0449	.1137	.1815	.2483	.3142	.3790	.4428	.5056	.5672	.6278	.6873
32	.9764	.0461	.1148	.1826	.2494	3153	.3801	.4439	.5066	.5683	.6288	.6883
33	.9776	.0472	.1160	.1837	.2505	.3163	.3811	.4449	.5076	.5693	.6298	.6893
34	.9788	.0484	.1171	.1849	.2516	.3174	.3822	.4460	.5087	.5703	.6308	.6903
35	2.9799	3.0495	3.1182	3.1860	3.2527	3.3185	3.3833	3.4470	3.5097	3.5713	3.6318	3.6913
36	.9811	.0507	.1194	.1871	.2538	.3196	.3844	.4481	.5107	.5723	.6328	.6923
37	.9823	.0519	.1205	.1882	.2550	.3207	.3854	.4491	.5118	.5734	.6338	.6932
38	.9834	.0530	.1216	.1893	.2561	.3218	.3865	.4502	.5128	.5744	.6348	.6942
39	.9846	.0542	.1228	.1905	.2572	.3229	.3876	.4512	.5138	.5754	.6358	.6952
40	2.9858	3.0553	3.1239	3.1916	3.2583	3.3240	3.3886	3.4523	3.5149	3.5764	3.6368	3.6962
41	.9869	.0565	.1251	.1927	.2594	.3250	.3897	.4533	.5159	.5774	.6378	.6972
42	.9881	.0576	.1262	.1938	.2605	.3261	.3908	.4544	.5169	.5784	.6388	.6981
43	.9893	.0588	.1273	.1949	.2616	.3272	.3918	.4554	.5180	.5795	.6398	.6991
44	.9904	.0599	.1285	.1961	.2627	.3283	.3929	.4565	.5190	.5805	.6408	.7001
45	2.9916	3.0611	3.1296	3.1972	3.2638	3.3294	3.3940	3.4575	3.5200	3.5815	3.6418	3.7011
46	.9928	.0622	.1307	.1983	.2649	.3305	.3950	.4586	.5211	.5825	.6428	.7020
47	.9939	.0634	.1319	.1994	.2650	.3316	.3961	.4596	.5221	.5835	.6438	.7030
48	.9951	.0645	.1330	.2005	.2671	.3326	.3972	.4607	.5231	.5845	.6448	.7040
49	.9963	.0657	1341	.2016	.2682	.3337	.3982	.4617	.5242	.5855	.6458	.7050
50	2.9974	3.0668	3.1353	3.2028	3.2693	3.3348	3.3993	3.4628	3.5252	3.5866	3.6468	3.7060
51	.9986	.0680	.1364	.2039	.2704	.3359	.4004	.4638	.5262	.5876	.6478	.7069
52	.9997	.0691	.1375	.2050	.2715	.3370	.4014	.4649	.5273	.5886	.6488	.7079
53	3.0009	.0703	.1387	.2061	.2726	.3381	.4025	.4659	.5283	.5896	.6498	.7089
54	.0021	.0714	.1398	.2072	.2737	.3391	.4036	.4670	.5293	.5906	.6508	.7099
55	3.0032	3.0725	3.1409	3.2083	3.2748	3.3402	3.4046	3.4680	3.5304	3.5916	3.6518	3.7108
56	.0044	.0737	.1421	.2095	.2759	.3413	.4057	.4691	.5314	.5926	.6528	.7118
57	.0056	.0748	.1432	.2106	.2770	.3424	.4068	.4701	.5324	.5936	.6538	.7128
58	.0067	.0760	.1443	.2117	.2781	.3435	.4078	.4712	.5335	.5947	.6548	.7138
59	.0079	.0771	.1454	.2128	.2792	.3445	.4089	.4722	.5345	.5957	.6558	.7147
60	3.0091	3.0783	3 1466	3.2139	3.2803	3.3456	3.4100	3.4733	3.5355	3.5967	3.6567	3.7157

TABLE XIII (Continued, part 5 of 5)

M	48°	49°	50°	51°	52°	53°	54°	55°	56°	57°	58°	59°
0	3.7157	3.7735	3.8302	3.8857	3.9400	3.9932	4.0451	4.0957	4.1452	4.1933	4.2402	4.2858
1	.7167	.7745	.8311	.8866	.9409	.9940	.0459	.0966	.1460	.1941	.2410	.2866
2	.7176	.7754	.8321	.8875	.9418	.9949	.0468	.0974	.1468	.1949	.2418	.2873
3	.7186	.7764	.8330	.8884	.9427	.9958	.0476	.0982	.1476	.1957	.2425	.2881
4	.7196	.7773	.8339	.8894	.9436	.9967	.0485	.0991	.1484	.1965	.2433	.2888
5	3.7206	3.7783	3.8349	3.8903	3.9445	3.9975	4.0493	4.0999	4.1492	4.1973	4.2441	4.2896
6	.7215	.7792	.8358	.8912	.9454	.9984	.0502	.1007	.1500	.1981	.2448	.2903
7	.7225	.7802	.8367	.8921	.9463	.9993	.0510	.1016	.1508	.1989	.2456	.2910
8	.7235	.7811	.8377	.8930	.9472	4.0001	.0519	.1024	.1517	.1997	.2464	.2913
9	.7244	.7821	.8386	.8939	.9481	.0010	.0527	.1032	.1525	.2004	.2471	.2925
10	3.7254	3.7830	3.8395	3.8948	3.9490	4.0019	4.0536	4.1041	4.1533	4.2012	4.2479	4.2933
11	.7264	.7840	.8405	.8958	.9499	.0028	.0544	.1049	.1541	.2020	.2487	.2940
12	.7274	.7850	.8414	.8967	.9508	.0036	.0553	.1057	.1549	.2028	.2494	.2948
13	.7283	.7859	.8423	.8976	.9516	.0045	.0561	.1066	.1557	.2036	.2502	.2955
14	.7293	.7869	.8433	.8985	.9525	.0054	.0570	.1074	.1565	.2044	.2510	.2963
15	3.7303	3.7878	3.8442	3.8994	3.9534	4.0062	4.0578	4.1082	4.1573	4.2052	4.2517	4.2970
16	.7312	.7887	.8451	.9003	.9543	.0071	.0587	.1090	.1581	.2060	.2525	.2978
17	.7322	.7897	.8460	.9012	.9552	.0080	.0595	.1099	.1589	.2067	.2533	.2985
18	.7332	.7906	.8470	.9021	.9561	.0089	.0604	.1107	.1597	.2075	.2540	.2992
19	.7341	.7916	.8479	.9030	.9570	.0097	.0612	.1115	.1606	.2083	.2548	.3000
20	3.7351	3.7925	3.8488	3.9039	3.9579	4.0106	4.0621	4.1124	4.1614	4.2091	4.2556	4.3007
21	.7361	.7935	.8498	.9049	.9588	.0115	.0629	.1132	.1622	.2099	.2563	.3015
22	.7370	.7944	.8507	.9058	.9596	.0123	.0638	.1140	.1630	.2107	.2571	.3022
23	.7380	.7954	.8516	.9067	.9605	.0132	.0646	.1148	.1638	.2115	.2578	.3029
24	.7390	.7963	.8525	.9076	.9614	.0141	.0655	.1157	.1646	.2122	.2586	.3037
25	3.7399	3.7973	3.8535	3.9085	3.9623	4.0149	4.0663	4.1165	4.1654	4.2130	4.2594	4.3044
26	.7409	.7982	.8544	.9094	.9632	.0158	.0672	.1173	.1662	.2138	.2601	.3052
27	.7419	.7992	.8553	.9103	.9641	.0167	.0680	.1181	.1670	.2146	.2609	.3059
28	.7428	.8001	.8562	.9112	.9650	.0175	.0689	.1190	.1678	.2154	.2617	.3066
29	.7438	.8011	.8572	.9121	.9659	.0184	.0697	.1198	.1686	.2162	.2624	.3074
30	3.7448	3.8020	3.8581	3.9130	3.9667	4.0193	4.0706	4.1206	4.1694	4.2169	4.2632	4.3081
31	.7457	.8029	.8590	.9139	.9676	.0201	.0714	.1214	.1702	.2177	.2639	.3089
32	.7467	.8039	.8599	.9148	.9685	.0210	.0722	.1223	.1710	.2185	.2647	.3096
33	.7476	.8048	.8609	.9157	.9694	.0219	.0731	.1231	.1718	.2193	.2655	.3103
34	.7486	.8058	.8618	.9166	.9703	.0227	.0739	.1239	.1726	.2201	.2662	3111
35	3.7496	3.8067	3.8627	3.9175	3.9712	4.0236	4.0748	4.1247	4.1734	4.2208	4.2670	4.3118
36	.7505	.8077	.8636	.9184	.9720	.0244	.0756	.1255	.1742	.2216	.2677	.3125
37	.7515	.8086	.8646	.9193	.9729	.0253	.0765	.1264	.1750	.2224	.2685	.3133
38	.7525	.8096	.8655	.9202	.9738	.0252	.0773	.1272	.1758	.2232	.2692	.3140
39	.7534	.8105	.8664	.9212	.9747	.0270	.0781	.1280	.1766	.2240	.2700	.3147
40	3.7544	3.8114	3.8673	3.9221	3.9756	4.0279	4.0790	4.1288	4.1774	4.2247	4.2708	4.3155
41	.7553	.8124	.8683	.9230	.9765	.0288	.0798	.1296	.1782	.2255	.2715	.3162
42	.7563	.8133	.8692	.9239	.9773	.0296	.0807	.1305	.1790	.2263	.2723	.3170
43	.7573	.8143	.8701	.9248	.9782	.0305	.0815	.1313	.1798	.2271	.2730	.3177
44	.7582	.8152	.8710	.9257	.9791	.0313	.0823	.1321	.1806	.2278	.2738	.3184
45	3.7592	3.8161	3.8719	3.9266	3.9800	4.0322	4.0832	4.1329	4.1814	4.2286	4.2745	4.3192
46	.7601	.8171	.8729	.9275	.9809	.0331	.0840	.1337	.1822	.2294	.2753	.3199
47	.7611	.8180	.8738	.9284	.9817	.0339	.0849	.1346	.1830	.2302	.2760	.3206
48	.7620	.8190	.8747	.9293	.9826	.0348	.0857	.1354	.1838	.2309	.2768	.3213
49	.7630	.8199	.8756	.9302	.9835	.0356	.0865	.1362	.1846	.2317	.2775	.3221
50	3.7640	3.8208	3.8765	3.9311	3.9844	4.0365	4.0874	4.1370	4.1854	4.2325	4.2783	4.3228
51	.7649	.8218	.8775	.9320	.9853	.0374	.0882	.1378	.1862	.2333	.2791	.3235
52	.7659	.8227	.8784	.9329	.9861	.0382	.0891	.1386	.1870	.2340	.2798	.3243
53	.7668	.8236	.8793	.9338	.9870	.0391	.0899	.1395	.1878	.2348	.2806	.3250
54	.7678	.8246	.8802	.9347	.9879	.0399	.0907	.1403	.1886	.2356	.2813	.3257
55	3.7687	3.8255	3.8811	3.9355	3.9899	4.0408	4.0916	4.1411	4.1894	4.2364	4.2821	4.3265
56	.7697	.8265	.8820	.9364	.9896	.0416	.0924	.1419	.1902	.2371	.2828	.3272
57	.7707	.8274	.8830	.9373	.9905	.0425	.0932	.1427	.1909	.2379	.2836	.3279
58	.7716	.8283	.8839	.9382	.9914	.0433	.0941	.1435	.1917	.2387	.2843	.3286
59	.7726	.8293	.8848	.9391	.9923	.0442	.0949	.1443	.1925	.2394	.2851	.3294
60	3.7735	3.8302	3.8857	3.9400	3.9932	4.0451	4.0957	4.1452	4.1933	4.2402	4.2858	4.3301

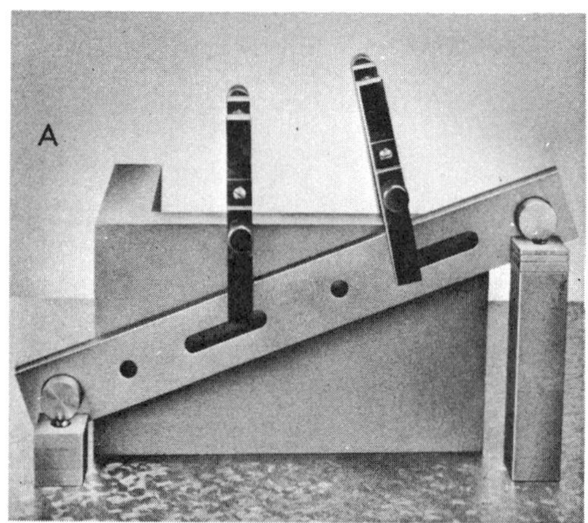

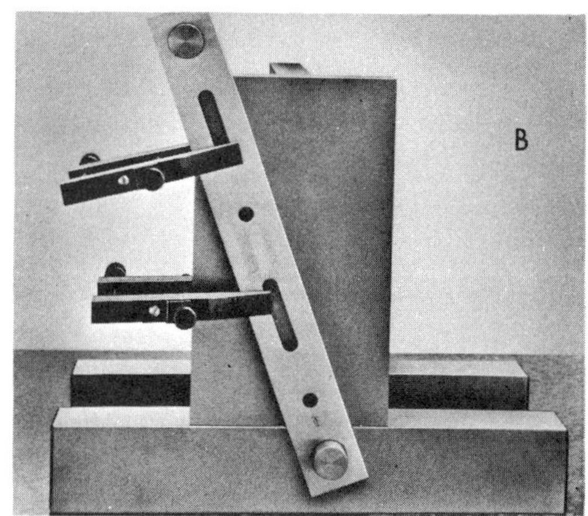

Figure 304. Setting up Complementary Angle

tion. Therefore, angles can be measured more accurately when the sine bar is nearly horizontal.

When a 10-inch sine bar makes an angle of 80 degrees with the horizontal, a one-minute change in the angle will change the height of blocks only .0005 inch. When the sine bar is at an angle of 10 degrees to the horizontal, a one-minute change will change height of blocks .0028 inch--more than five times the change when in the 80-degree position. Obviously, a slight inaccuracy in instruments or setup in the 10-degree position would affect the measurement only 1/5 as much as it would in the 80-degree position. Therefore, when angles greater than 60 degrees are being measured, the complement of the angle is usually set up with the sine bar.

By definition, the complement of an angle is the difference between the angle and 90 degrees. The complement of 80 degrees is 10 degrees; the complement of 70 degrees is 20 degrees. Therefore, for an 80-degree angle the sine bar would be set up to 10 degrees, and for a 70-degree angle the sine bar would be set up to 20 degrees. By clamping the sine bar to an angle iron and turning the angle iron 90 degrees, the sine bar assumes the angle which is complementary to the angle of its original position (Figure 304A and 304B).

MEASURING EXAMPLE

REFERENCE: Blueprint 128178 (Figure 305).

OBJECTIVE: To check the location of the keyway specified on the drawing as 7 degrees $\pm$ 1 degree between the center of the part and the center of the keyway.

PROCEDURE: First set the part at an angle of 7 degrees so that the center line passing through the center of the keyway and the center of the bore is horizontal and parallel to the surface of the surface plate (Figure 306). To do this, place an angle iron upon the surface plate to provide a perpendicular surface against which the sine bar can rest and upon which the part can be clamped when set at the proper angle.

Next, determine the vertical distance between the cylinders of the sine bar when the sine bar is set at an angle of 7 degrees. Refer to the table of sines to find that the sine of 7 degrees is .12187; therefore, the vertical distance between the cylinders of a 10-inch sine bar at an angle of 7 degrees, will be 10 x .12187, or 1.2187 inch. Now, if one cylinder is placed on a one-inch gage block, so that the other cylinder will be 1.2187 inch higher, the other cylinder must be placed on a stack of blocks 2.2187 inches high.

Place the sine bar on the two sets of blocks and against the surface of the angle iron. Place the part on the sine bar, being careful to clean the locating surface and remove all burrs. Clamp the part to the angle iron, and the setup is ready for inspection. (Note that the side of the part resting on the sine bar is the locating surface according to the drawing, Figure 305.)

In this position, the sides of the keyway should be parallel to the surface plate. This

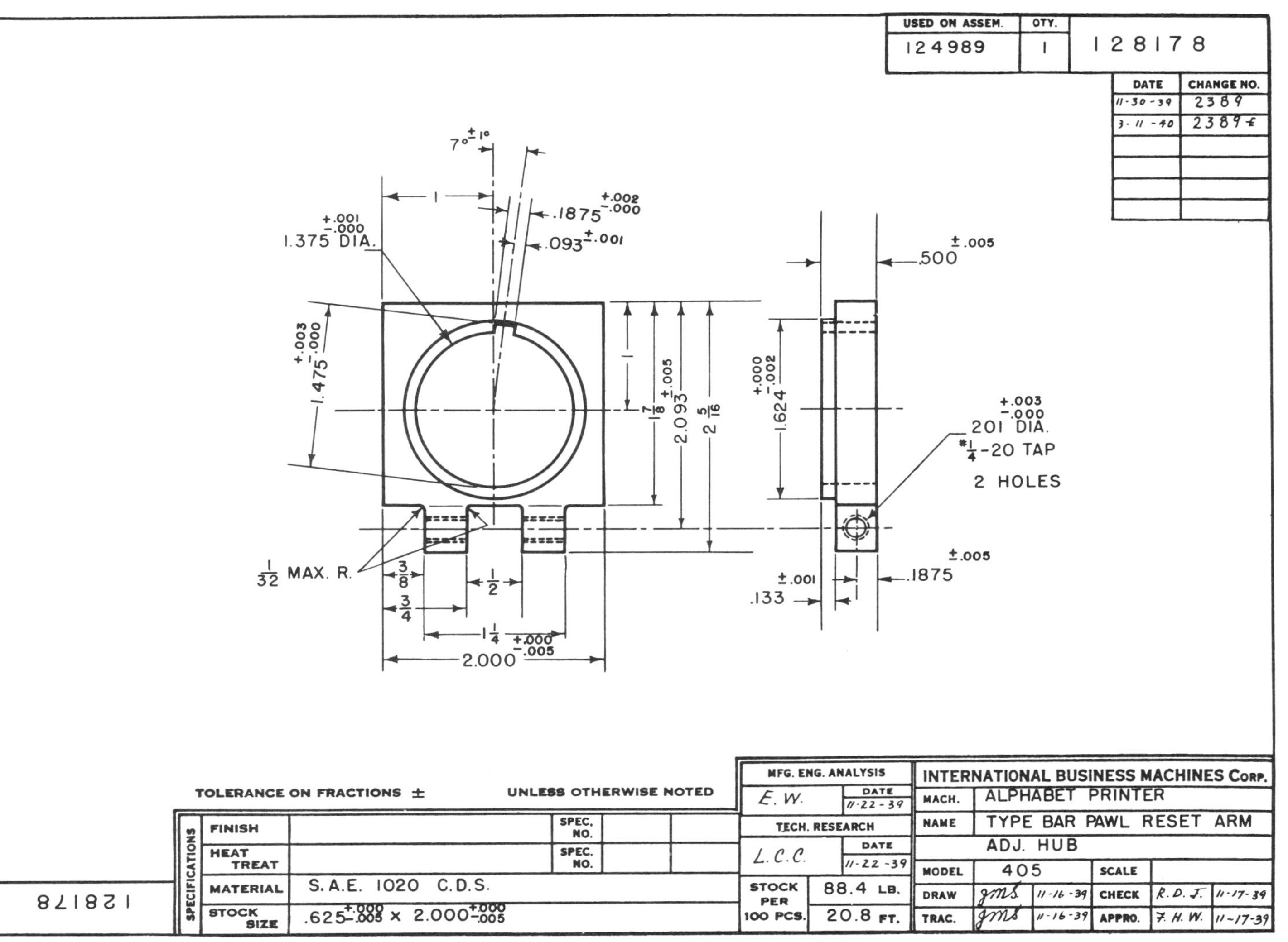

Figure 305. Blueprint for Measuring Example

Figure 306. Checking Angle of a Keyway

can be checked with a flat gage the exact size of the keyway as shown in Figure 306. If the sides of the keyway are not parallel to the surface plate in this position, the angle of the keyway is not correct. If the sides are parallel to the surface plate, the angle is correct.

At this point, the angle has been checked but not the location of the keyway, the center line of which should pass through the center of the bore, with the part in this position. To check the location of the keyway, remove the flat gage and take a reading on the bottom of the bore (Figure 307). The lowest point in the bore can be determined by moving the indicator sidewise and noting the point at which the indicator point reaches its lowest position.

Take a reading at this point, add half the diameter of the bore minus half the width of the keyway; this gives the reading corresponding to the lower side of the keyway. If a reading on the lower side of the keyway corresponds to this figure, the keyway is in location.

Figure 307. Checking Location of Keyway

SINE TABLES

SIMPLE SINE TABLE

The sine table (Figure 308) incorporates the principle of the sine bar into a hinged fixture which can be set with gage blocks to any position from 0 to 60 degrees. The top of the base is ground and lapped perfectly flat so that gage blocks can be put between it and the cylinder of the sine table to set the table at any desired angle. The wide, flat working surface will accommodate V-blocks or angle irons upon which the work is mounted, or the work can be mounted

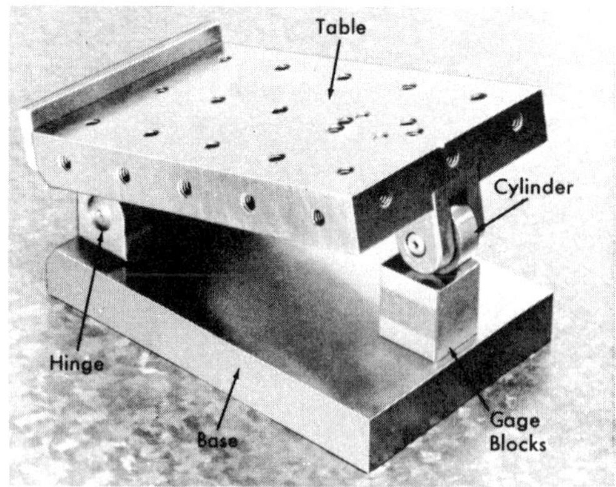

Figure 308. Simple Sine Table

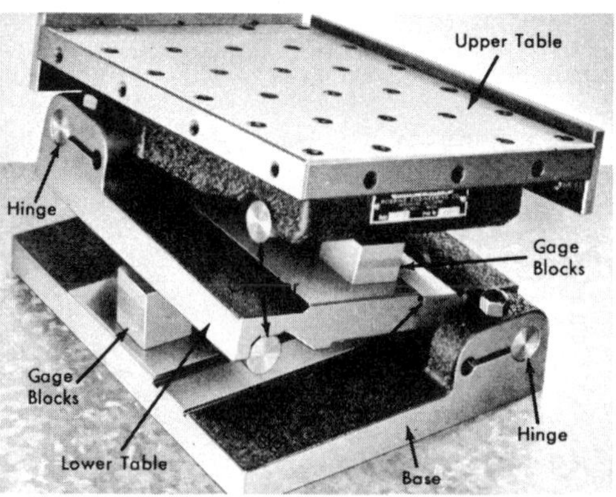

Figure 309. Compound Sine Table

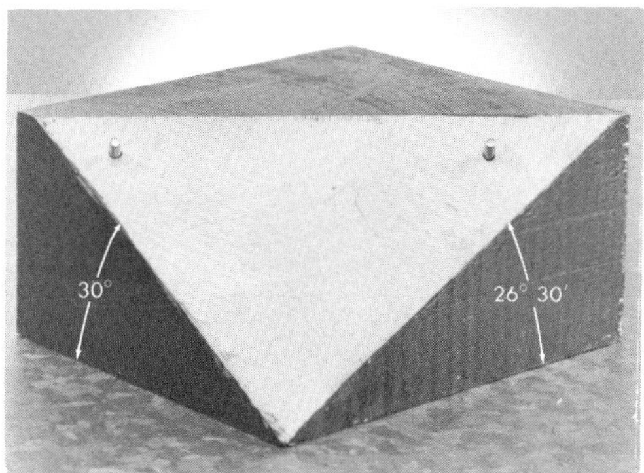

Figure 310. Compound Angle

directly upon the table. Side plates facilitate setting the work square with the edge of the table.

COMPOUND SINE TABLE

The compound sine table (Figure 309) is a combination of two tables, one superimposed upon the other. The top of the base and the top of the lower table are ground and lapped perfectly flat so that gage blocks can be inserted between these surfaces and the cylinders of the tables. A slot .100 inch deep is provided for using blocks starting at .1001 inch for very small angles.

When the lower table is closed, the fixture can be used as a simple sine table by setting the upper table to the required angle with gage blocks. Also with the upper table closed, the fixture can be used in like manner by inserting gage blocks between the lower table and the base. In this respect, the compound table has an advantage over the simple sine table in that it can be set in either of two directions without moving the fixture which, when used for machining operations, is fastened to the machine table.

The compound sine table was designed, principally, to simplify setting up compound angles. A compound angle is represented by a plane which makes two angles with the base or reference plane of the work (Figure 310). To set up a compound angle, each table is set to a predetermined angle (Figure 311). These angles are calculated mathematically by a system too lengthy to be described here. The subject is covered in any complete text on shop mathematics.

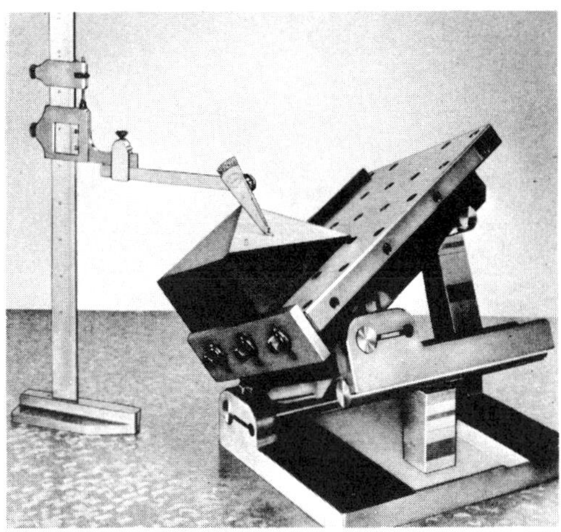

Figure 311. Checking a Compound Angle

THE ANGLE COMPUTER

The angle computer (Figure 312) incorporates all the features of a compound sine table and, in addition, can be rotated in a horizontal plane. Its main advantage is in the simplicity and speed of setting the instrument to required angles.

The angles are set by means of circular dial scales and verniers, as in setting a vernier protractor. The scales, which are graduated in degrees, can be read to an accuracy of 5' by aid of the vernier.

The angle computer was designed to simplify layout of die work or machined parts, and for inspection of tools and parts. It has three individual directions of rotation--horizontally, and vertically, in two planes at right angles to each other. Actuating screws provide for close adjustment of the angles; all three angles can be set in relation to one another. So, the instrument is ideal for setting up compound angles.

DIVIDING OR INDEXING HEADS

An index head is a fixture upon which the periphery of a part can be divided into any number of equal divisions. It was originally designed for machine operations, such as milling teeth in cutters, and flutes in reamers and drills, but it has since become a valuable aid to the inspector in measuring angles, and in checking spacing of teeth, slots, and grooves. Index

Figure 312. Angle Computer

Figure 313. Dial Index Head

heads can be divided into three classes: dial, plain, and universal.

DIAL INDEX HEAD

The dial index head (Figure 313) consists of a cast steel housing in which is mounted a spindle that can be turned on its axis and locked in any position. The angular position of the spindle is read on a dial scale graduated in degrees.

Mounted on the spindle is an index plate containing 24 holes; mounted in the housing is a plunger that engages the holes, thus providing a quick way of indexing the spindle in any one of 24 positions 15 degrees apart.

The base of the housing is finished and contains a keyway so that it can be mounted on the machine table; for inspection work, however, the head is usually placed on a surface plate.

The work is located on a special face plate (Figure 313) having an accurately ground pin in the center, upon which parts with various size bores can be located by means of suitable bushings. Small clamps hold the work.

In checking an angle, the initial line is established by setting the index line of the scale to zero and mounting the part so that the locating surface or reference plane is parallel to the surface plate. From this position, the part can be set to any desired angle by a direct reading on the dial scale.

A typical example of the use of the dial index head is shown in Figure 314. Here a number of equally-spaced slots are being checked for relation with one another and with the center bore.

The spindle is first set to zero and locked in position. The part is then mounted on the face plate so that a line through the centers of the two opposite slots is parallel to the surface plate and in line with the center of the bore. Select a suitable bushing to adapt the bore of the part to the pin on the face plate. Then mount the part and lightly clamp it in a position in which the center line through two opposite slots is approximately parallel to the surface plate.

Take a reading over the pin and subtract half the diameter of the pin. The resultant figure will be the reading corresponding to the center of the pin and the center of the slots when they are in alignment. Subtract half the width of the slot from this reading to obtain the reading corresponding to the bottom of the slot.

Set the height gage to this final reading, tap the part until the bottom of the slot indicates zero, and clamp securely to the face plate.

To check angular location of slots, indicate each successive slot while indexing them into position with the plunger. Because there are 12 slots, the angular spacing is 30 degrees, which is a multiple of 15 degrees and which can be indexed with the plunger.

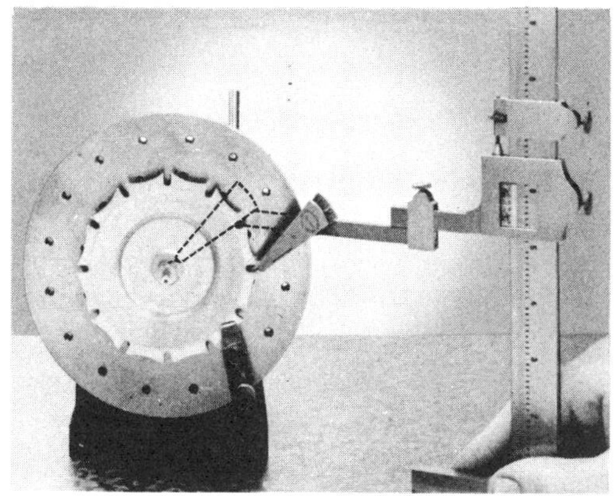

Figure 314. Checking Relation of Slots

Radial location of the slots can be checked by taking a reading on the arbor, calculating the reading at the bottom of a slot directly above the bore, and checking each slot successively by indexing as before (Figure 315).

PLAIN INDEX HEAD

The plain index head (Figure 316) differs from the dial index head in that the spindle is turned by means of a crank, a worm, and a worm gear (Figure 317). Ratio of the gear and the worm is 40:1; therefore, one turn of the crank will turn the spindle 1/40 of a revolution, or

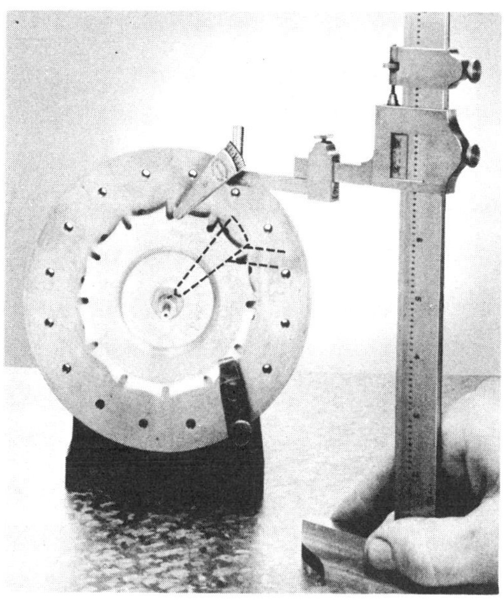

Figure 315. Checking Location of Slots

Figure 316. Plain Index Head

9 degrees. Subdivisions of a turn of the crank are obtained by means of an index plate attached to the housing under the crank.

The index plate contains several circles of holes equally spaced, each circle having a different number of holes. By means of a pin in the handle of the crank, the crank can be located in a hole of one of these circles as an

Figure 317. Worm and Gear

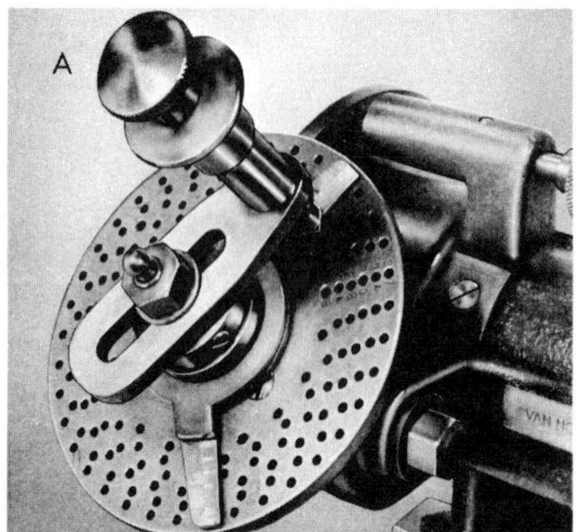

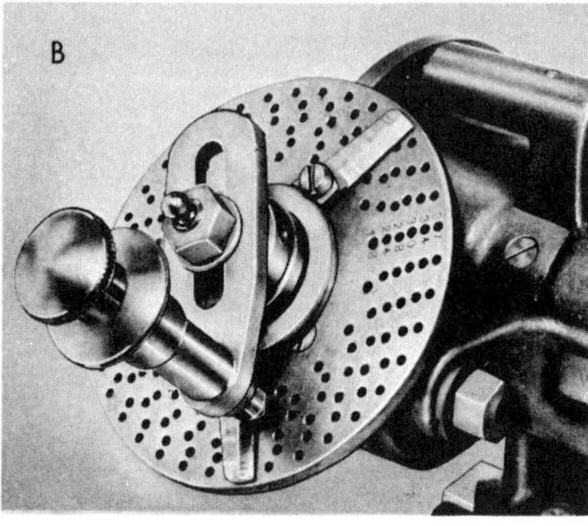

Figure 318. Indexing--Using Selector Arms

initial position, and, by counting off the holes in that particular circle, can be turned to a new or terminal position which will be a definite number of holes from the initial position. The number of holes between the initial and terminal position will represent a fraction of a turn of the crank. The denominator of the fraction will be the number of holes in the circle, and the numerator will be the number of holes between the initial and terminal position of the crank.

A sector with adjustable arms is mounted on the index plate for convenience in marking off the number of holes required to make a fractional part of a turn of the crank. The sector arms are spread to include the number of holes required and locked together so that they will move as a solid sector.

For example, suppose the index plate contained six rows of holes--18, 24, 29, 30, 34, 37--and the spindle is to be turned 30 degrees. Because one turn of the crank turns the spindle 9 degrees, three turns of the crank will turn it 27 degrees. So, 30 degrees minus 27 degrees equals 3 degrees, which is one-third turn of the crank. One-third turn will equal eight holes in the 24 circle, or ten holes in the 30 circle. Therefore, if the crank is turned three complete turns and, in addition, eight holes in the 24 circle, the spindle will turn 30 degrees.

With the crank in a hole of the 24-hole circle (as shown in Figure 318A), the sector arms are spread to include nine holes and clamped to each other so that they will move as one unit. Disregard the initial hole when counting the number of holes for a fractional turn of the crank. Since the crank is actually moved eight spaces, it is obvious that it must move to the eighth hole from the initial hole (count them); and since the initial hole is within the sector, eight spaces will be represented by nine holes between the sector arms.

Move the sector so that the left arm is against the pin as shown. This is the initial or starting position. Now withdraw the pin and turn the crank three turns plus eight holes; this brings the crank pin to the hole at the inside edge of the right sector arm (Figure 318B), and the spindle is moved 30 degrees. If the sector is now moved until the left arm is against the pin, the foregoing operation can be repeated. Should the angle be specified in minutes, the procedure is the same after the angle has been reduced to the number of turns of the crank.

Required: To index the angle 10° 12'. Since each revolution of the crank turns the spindle 9°, the angle 10° 12' will require one complete turn and a fraction of a turn represented by the remainder, or 1° 12' (72'); a complete turn, or 9°, is 540'. Therefore, the fraction of a turn is 72/540 = 4/30, or four holes in the 30-hole circle.

UNIVERSAL INDEX HEAD

The universal index head (Figure 319) although having much the same field of application as the plain index head, is radically different in design. The head is a hollow casting in which the spindle is mounted similar to the mounting of the spindle in the plain index head, except that provision has been made for rotating the index plate in relation to the movement of the crank. By this arrangement a great variety of divisions that are beyond the range of the plain index head can be indexed. The head is mounted on trunnions upon which the spindle can be set at any angle in a vertical plane, from 10 degrees below the horizontal to 5 degrees beyond the vertical.

Figure 319. Universal Index Head

PLAIN INDEXING

Plain indexing on the universal index centers is identical to that on the plain index centers. It depends upon the number of turns of the crank to produce one turn of the spindle. When this ratio is known, it is easy to calculate the number of turns of the crank to divide the periphery of the work into a particular number of divisions. It requires 40 turns of the crank to produce one turn of the spindle. To find how many turns of the crank are required for a certain division of the work, 40 is divided by the number of divisions; the quotient will be the number of turns of the crank.

When the quotient contains a fraction, it will be necessary to move the crank a fraction of a turn, and, as explained before, the denominator of the fraction will be the number of holes in the circle selected; the numerator will be the number of holes between the initial and terminal position of the crank. If the fraction is such that the denominator does not correspond to the number of holes in the circles available but is a factor of or multiple of one of the circles, then both numerator and denominator must be multiplied or divided by a common factor so that the denominator will correspond to the number of holes in one of the circles.

	15, 16, 17, 18, 19, 20
Hole Circles	21, 23, 27, 29, 31, 33
	37, 39, 41, 43, 47, 49

Example:

Objective: to divide one revolution of the spindle into nine divisions.

Procedure: 40/9 - 4-4/9 turns of the crank.

Refer to the three plates (Figure 320); no nine-hole circle is available, but there is an 18-hole circle. Multiply both numerator and denominator by 2 to change the fraction to 8/18. which can be indexed by moving the crank eight holes on the 18 circle. So for one division, the crank is given four turns and, in addition, is moved eight holes in the 18-hole circle.

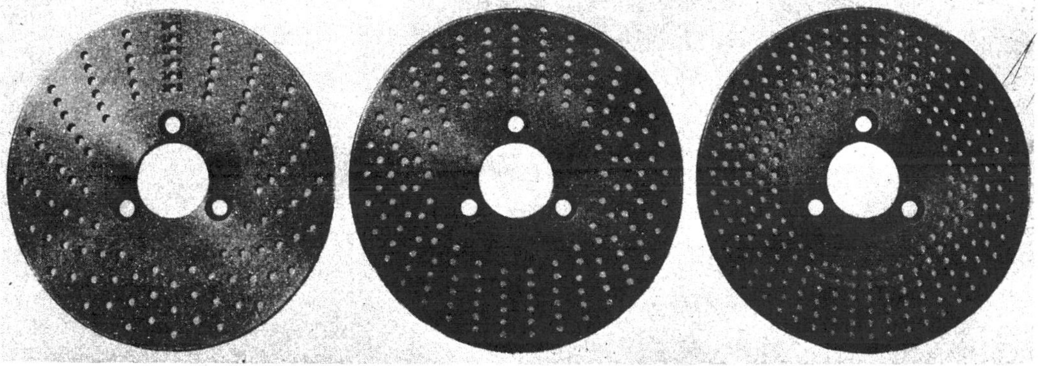

Figure 320. Index Plates

DIFFERENTIAL INDEXING

If the fraction is such that the numerator is not a factor or multiple of the number of holes in one of the circles available, the fractional turn of the crank must be obtained by means of a train of gears connecting the spindle to the index plate, so that the index plate will move in relation to the movement of the crank. This method of indexing is called differential indexing.

As example of how the movement of the index plate affects the movement of the crank, assume that the crank is to be moved 1/9 turn, and an index plate having an eight-hole circle (Figure 321) is available.

Starting in any hole of the eight-hole circle: if the crank were moved 1/9 turn in a positive direction, it would be $1/8 \times 1/9 = 1/72$ of a turn short of the next hole in the circle. Therefore, to use this next hole, the index plate will have to move 1/72 turn in the opposite direction, while the crank is moving 1/9 turn forward.

Start with the crank in the vertical position: if, while the crank is moving 1/9 turn in a positive direction, the index plate moves 1/72 turn in a negative direction, the next hole in the eight-hole circle will be directly under the crank and can be used for indexing 1/9 turn.

CHANGE GEARS

The train of gears connecting the spindle and the index plate is called change gears.

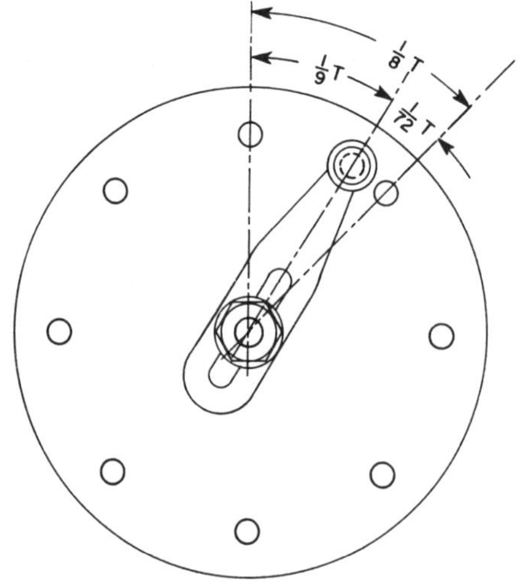

Figure 321

Twelve standard change gears are furnished with each index head. They have the following number of teeth: 24 (2 gears), 28, 32, 40, 44, 48, 56, 64, 72, 86, and 100. Besides these standard gears, eight special gears can be furnished with the following teeth: 46, 47, 52, 58, 68, 70, 76, and 84.

GEAR RATIO

When the required number of divisions cannot be indexed by plain indexing, select an approximate number of divisions which can be indexed by plain indexing. The difference between the movement of the spindle for the approximate and required divisions is corrected by the change gears. The gear ratio is determined with the following formula:

$$\text{Gear ratio} = (n - N) \times \frac{40}{n}$$

in which N = required divisions

n = approximate divisions

When the approximate number of divisions is greater than the required number, the resulting fraction is plus, and the index plate must move in the same direction as the crank. Conversely, when the approximate number of divisions is less than the required number, the resulting fraction is minus, and the index plate must move in a direction opposite to that of the crank.

The numerator of the fraction indicates the driving gear or gears, and the denominator indicates the driven gears. Idler gears control the

Figure 322. Simple Gearing

rotating direction of the index plate. The gearing can be either simple or compound.

Simple gearing:

1 idler for positive rotation of index plate.

2 idlers for negative rotation of index plate.

Compound gearing:

1 idler for negative rotation of index plate.

No idlers for positive rotation of index plate.

Simple gearing consists of one driving gear, one driven gear and one or two idler gears, depending upon direction of rotation of index plate. Compound gearing consists of two driving gears, two driven gears, and one or no idlers, depending upon direction of rotation. Example:

Objective: to index 57 divisions

Procedure: number of turns of crank =

$$\frac{40}{N} = \frac{40}{57}$$

As 57 is not a multiple or factor of number of holes in a circle available, select an approximate number of divisions.

By selecting n = 56 divisions, the number of turns of the crank will be:

$\frac{40}{n} = \frac{40}{56} = \frac{5}{7} \times \frac{3}{3} = \frac{15}{21}$ (15 holes in the 21-hole circle)

The gear ratio will be:

$$(n - N) \times \frac{40}{n} =$$

$$(56 - 57 \times \frac{40}{56} = - 5/7$$

In the foregoing example, the selected division is less than the required division; therefore, the fraction is negative (- 5/7), and the index plate must turn in a negative direction, or counterclockwise. Since the denominator (7) is a factor of the number of teeth in one of the gears, the gearing will be simple gearing, and the gears will be 5/7 x 8/8 = 40/56--40 teeth in driving gear, 56 in driven gear. Since the index plate moves in a negative direction, two idlers will be used (Figure 322).

The manufacturer of the dividing head provides a table for obtaining the hole circle, number of holes, gear ratio, gears to be used, and location of the gears on the dividing head, for any desired number of divisions. These tables are usually available in the department to which the dividing head has been assigned.

THE "VINCO" DIVIDING HEAD

This dividing head, although a radical departure from the dividing head previously described, incorporates the same principles and has the same application. It is used mainly for the inspection of machine parts, involving angular spacing, angular surfaces, and angular locations.

The head (Figure 323) consists of an iron housing (13), containing a spindle (14) upon which is mounted an adjustable driver plate (1) and a graduated dial and vernier (3) (4) on which the initial setting is made by a handwheel (15). The spindle is locked in the initial position by spindle lock (5); this automatically engages the micrometer unit (9) (10). The secondary setting is made by observing the relation of master discs (6) (7) through the telescope (8). Final setting is made by means of a micrometer thimble and barrel (9) (10).

The complete unit comprises the dividing head and tailstock mounted on a surface plate which in itself is supported by a cabinet pedestal (Figure 324). The tailstock is adjustable both vertically and horizontally. The surface plate is accurately scraped to a flat surface and ribbed for strength and rigidity. It contains a beveled keyway to insure accurate alignment of tailstock with the dividing head.

Figure 325 shows a setup on the Vinco dividing head for checking the periphery of a cam. A wedge-shaped locator mounted on a steel gallows replaces the follower which rides upon the periphery of the cam in final assembly. A height gage, resting upon the surface plate, is used to check the rise or fall of the cam surfaces at each of the critical points dimensioned on the cam drawing.

The cam, mounted on an arbor between centers of the machine, is driven by a dog connected to the driver plate. With primary and secondary scales set to zero, the zero position of the cam is set directly under the locator. Final adjustment of the cam is done by use of adjusting screws in the driver plate.

From this position, successive points on the cam are indexed as explained under operating instructions. Height gage readings, taken from

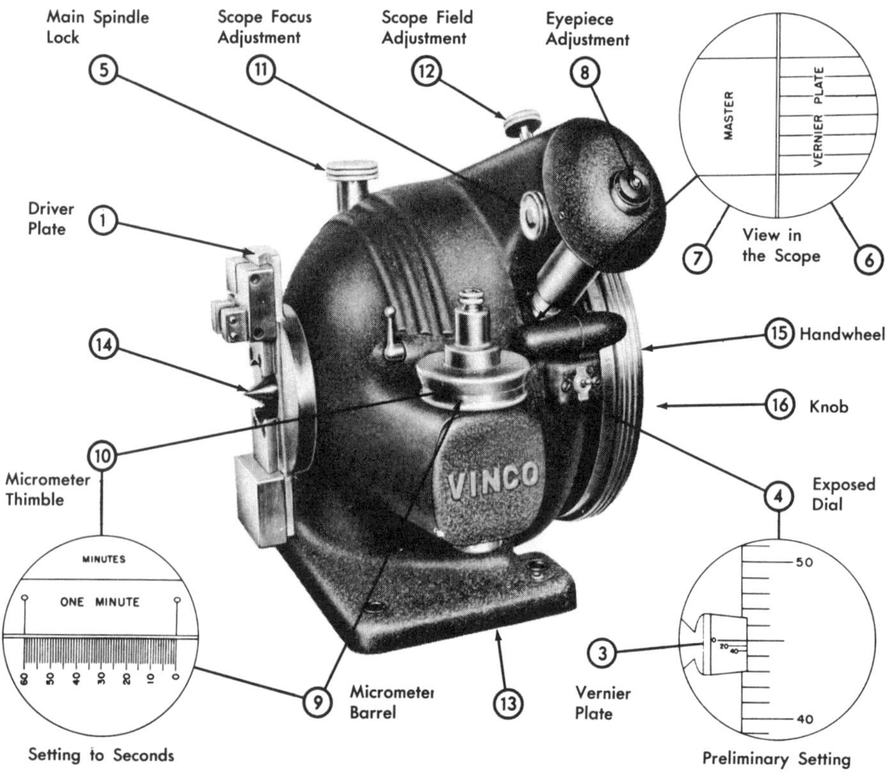

Figure 323. Vinco Dividing Head

the center of the arbor to the center of the wedge-shaped locator, should coincide with dimensions on the cam drawing.

OPERATING INSTRUCTIONS

The work is usually mounted on an arbor, between the centers of the head and the tailstock. Position a dog on the arbor into the driver plate; make final adjustment by adjusting four screws against the tail of this dog, after the head is set to zero.

The initial setting depends on the operation being checked. Zero position is usually set directly above the centers when radial dimensions are being checked, and in a plane passing through the centers horizontally when work is scribed, or when scribing is inspected.

Setting the head to zero requires adjustment of three individual scales. The first scale is on the handwheel at (3) and (4) in Figure 323. With the main spindle lock (5) closed, the initial setting is made on the scale (4) by releasing the handwheel through knob (16) and setting the scale to zero. This adjustment does not move the spindle.

The second adjustment is by means of the micrometer thimble (10) and the scale, (6) and (7), which is observed through the telescope (8).

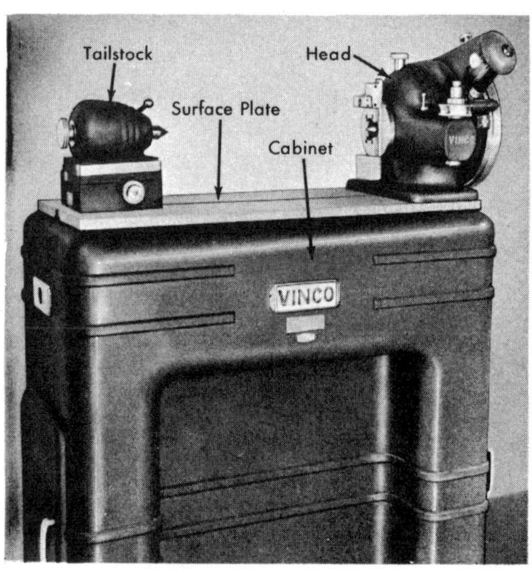

Figure 324. Vinco--Complete Unit

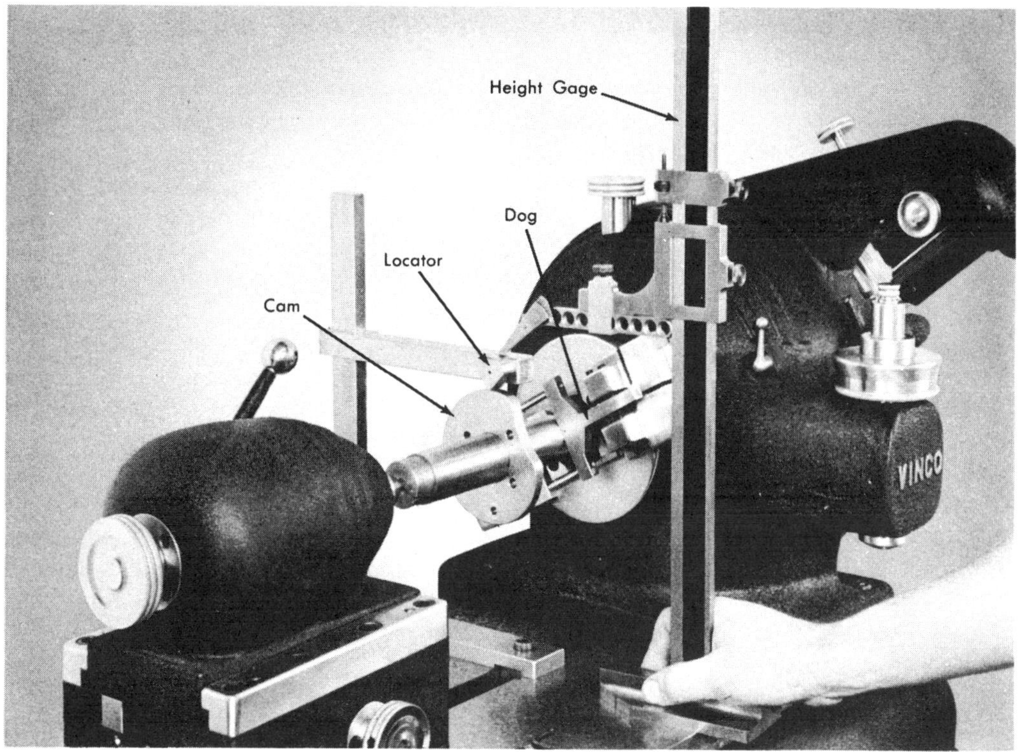

Figure 325. Checking a Cam with the Vinco Dividing Head

Looking into the telescope, observe that a graduation of the master disc (7) almost lines up with the lower line of the vernier (6). Turn the thimble (10) until these two lines coincide. This adjustment will move the spindle slightly; this movement is corrected and the work set back to its original position by means of the adjusting screws in the driver plate (1).

The third adjustment is made with the micrometer barrel (9), which is turned until the vernier lines up with two zero lines on the thimble as shown. This adjustment does not move the work. The head is now set to zero, and the work is in the initial position for indexing.

SETTING HEAD TO REQUIRED ANGLE

With the work and the head in the zero position, the required angle--say 8° 11' 28"--is set as follows:

1. Release spindle lock (5) and set scale (4) on handwheel to 8 degrees. Secure spindle lock (5). This setting has advanced the work 8 degrees.

2. Looking into the telescope, turn micrometer thimble (10) until the line on the master disc (7) exactly coincides with the first line above the bottom line of the vernier plate (6). Since each line on the vernier represents 10', the work has advanced 10' and is now set at 8° 10'.

3. Refer to micrometer scale (9) and (10) and set barrel (9) so that the vernier lines up with two zero lines of the thimble. Now move the thimble one whole division from 0 to 0, in the direction of the numbers on the vernier. This movement of the thimble advances the work one minute, making the setting 8° 11'. Move the thimble an additional 28" on the vernier, and the work will be exactly 8° 11' 28" from its initial position.

CARE OF ANGLE-MEASURING INSTRUMENTS

The most important rule for the care and preservation of all precision tools and instruments, of course, is: keep them clean. Everyone using them should be familiar enough with mechanical devices to recognize their delicacy, and should handle the instruments with proper care, so that accuracy will not be impaired by abuse or accidental injury. All persons using precision measuring instruments should observe these precautions:

1. Clean before and after using. All dirt contains abrasive, and the continual use of an instrument or tool which has not been thoroughly cleaned will eventually impair accuracy and make it unfit for use.

Perspiration contains corrosive agents which prove very injurious to steel surfaces. In some cases, although the surface is covered with a thin film of oil, rust appears on the surface within ten or fifteen minutes after prolonged contact with arm and hand. An experienced mechanic or inspector will have a chamois or clean cloth available and will wipe all fine tools before storing them in the tool box or case.

2. Oil. Cover all exposed parts with a film of oil; oil all bearing surfaces before putting the instrument in a case and setting it aside. Care should be exercised in the selection of oil. Certain cutting oils contain chemicals which are corrosive agents.

3. Cover. Keep all instruments and tools covered when not in use. The air contains dirt and abrasive which will be deposited upon an exposed surface, and eventually will work into hidden surfaces which are difficult to clean, thus starting corrosion and wear.

4. Avoid impact with other metallic parts. Any bump against hard material will leave a burr, which, although invisible, will make the instrument inaccurate; also if impact is severe enough, the instrument will be distorted and unfit for use.

In general, the life of a tool, gage, or instrument depends upon the care with which it is used. If unfamiliar with a tool, do not handle it until thoroughly instructed in its use. A good slogan for all operators, inspectors, and others who use measuring instruments is: "When in doubt, ask someone who knows."

QUESTIONS AND PROBLEMS

1. Define the following: right angle; straight angle; complementary angle; supplementary angle.
2. When the blade crosses the base of the vernier protractor, what is the usual guide to determine which of the two resultant supplementary angles is represented by the reading on the scale?
3. When using the acute angle attachment, why does the reading on the scale represent the complement of the angle being measured?
4. The given angle is 25°. What is its complementary angle? its supplementary angle?
5. What is a right triangle?
6. Define the following trigonometric functions in terms of the hypotenuse, side opposite, and side adjacent of a right triangle: sine; cosine; tangent.
7. Define similar right triangles.
8. The hypotenuse of a right angle is 7 inches. The side opposite is 3.500 inches. What is the given angle?
9. The given angle of a right triangle is 25°. The side opposite is 3 inches. What is the hypotenuse?
10. The hypotenuse of a right triangle is 8 inches. The side opposite is 5 inches. What is the cosine of the given angle?
11. The sine of 26° 20' is .44359. The sine of 26° 21' is .44385. What is the sine of 26° 20' 30"?
12. What is a sine bar? Describe how it is used.
13. If the distance between the centers of the two cylinders of a sine bar is 6 inches, what would be the height of blocks under the upper cylinder to set the sine bar at an angle of 30°, the lower cylinder resting on a one-inch stack of blocks?
14. When using the sine bar to measure an angle, why is a measurement more accurate when set at 10° than at 80°?
15. Using a 10-inch sine bar with one cylinder resting on a one-inch gage block, what angles will be subtended by placing the

following stack of blocks under the other cylinder? Specify the angle to the nearest minute.

A. 2.7500
B. 4.0680
C. 3.9044
D. 7.6912

16. Describe the simple sine table and explain how it is used.
17. State the advantages of the compound sine table, and the purpose of the .100" deep slot in the surface upon which the gage blocks are placed.
18. What is the accuracy of the angle computer? How is the accuracy obtained?
19. Define the term "indexing."
20. How does the plain index head differ from the dial index head?
21. What is the ratio of the worm and gear in the plain index head?
22. One turn of the crank will turn the spindle how many degrees?
23. Give the difference between plain and differential indexing.
24. Give three features found on the universal index head which are not on the plain index head.
25. What are change gears? When are they used?
26. Explain what is meant by simple gearing; compound gearing.
27. The gear ratio for differential indexing is determined by what formula?
28. How is the number of turns of the crank determined to produce a required number of divisions?
29. In differential indexing, what determines the direction of rotation of the index plate? How is the direction of rotation controlled?
30. Does the number of teeth on the idler gear affect the movement of the index plate?
31. Calculate the number of turns of crank, hole circle, and number of holes to index the following divisions:

Number Divisions	Full Turns of Crank	Number of Holes	Hole Circle
24			
50			
62			

32. Using a plain index head and three index plates (Figure 320), calculate the number of turns of the crank, fractional turn of crank number of holes and hole circle to index the following angles:

Angle	Full Turns of Crank	Fractional Turn of Crank	Number of Holes	Hole Circle
2° 15'				
16° 12'				
28° 30'				
55° 48'				

33. Name three uses of the Vinco dividing head.
34. Describe the three individual settings involved in checking an angle on a Vinco dividing head.
35. Why is the tailstock of the Vinco adjustable?
36. Give three important precautions which should be observed when using precision measuring instruments.

Chapter X

COMPARISON MEASUREMENT

VOLUME PRODUCTION is based on interchangeable parts, machined and inspected in large quantities and entirely independent of other parts with which they are finally assembled. Frequently parts manufacture is done by different companies and the parts brought together for final assembly; so all parts must assemble and function properly without further fitting, according to the requirements of the assembly in which they are used. To meet these requirements, accurate, rapid and economical methods of inspection are necessary.

Fundamentally, comparison measurement consists of determining the unknown dimensional variations of a part by comparing it with a master or standard having the basic dimension of the part being checked. The term "comparator" has been generally accepted as the name of an instrument or device which incorporates within itself a means of holding the part being inspected, a master with which the part is compared, and a means of amplification whereby small variations from basic dimensions can be observed easily.

Numerous types of comparison instruments for amplifying and recording dimensional variations are available from manufacturers of gaging equipment. Some of the most common types are:

Mechanical comparators having dial indicators of various types, as discussed in Chapter V.

Mechanical-optical comparators having a mechanical "reed-type" mechanism, and a light beam which casts a shadow on a scale to indicate dimensional variations.

Electrical and electronic comparators with a variety of electrical circuits.

Optical projection and reflection comparators which measure by casting a magnified shadow of contours and surfaces of parts on a screen (see Chapter XI).

Air comparators which utilize the flow or pressure of air to indicate variation of dimensions.

Multiple gaging comparators that use combinations of mechanical electronic, electrical and pneumatic principles for gaging a number of dimensions simultaneously.

REED COMPARATOR

The reed comparator, similar in construction to the dial comparator, consists of a base, column, and gaging head (Figure 326).

The base houses a transformer for reducing the line voltage of 110 volts to 8 volts, required by the lamp (which is part of the magnification system), and a switch for controlling the lamp. The base supports the anvil and a column upon which the gaging head is raised and lowered by means of a rack and pinion. By releasing a thumbscrew in the collar at the base of the column, the gaging head and rack can be swung clear of the base for checking parts larger than the working height of the comparator.

The gaging head contains the complete gaging unit, including the reed mechanism, a light-beam lever, and the scale.

THE REED MECHANISM. Figure 327 shows the principle of the reed mechanism. Two steel blocks, one stationary (D) and the other movable (E), are attached to two vertical strips of spring steel, called reeds. The upper ends of the reeds are joined, and a pointer is attached

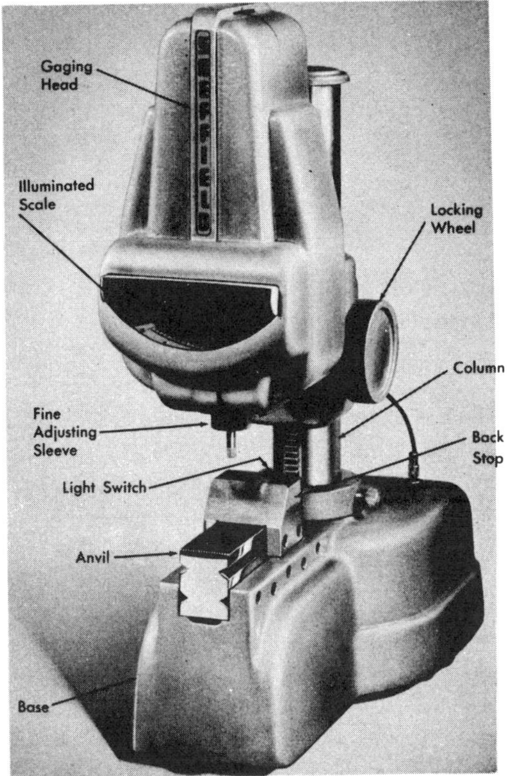

Figure 326. Reed-type Comparator

at this junction. It is obvious that if block (D) remains stationary, any movement of block (E) will force the reeds to deflect the pointer either right or left, depending upon whether block (E) is moved up or down (Figures 327B and 327C). By attaching a spindle to block (E) and mounting this simple mechanism in fixed relation to a reference plane (Figure 327A), elements of a simple but very sensitive comparator are obtained.

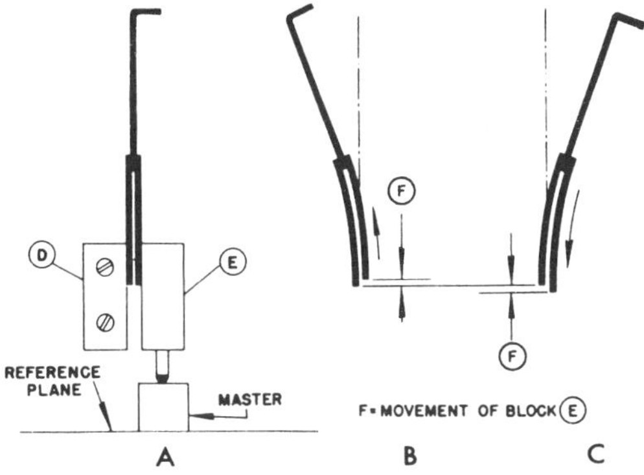

Figure 327. Principle of Reed Mechanism

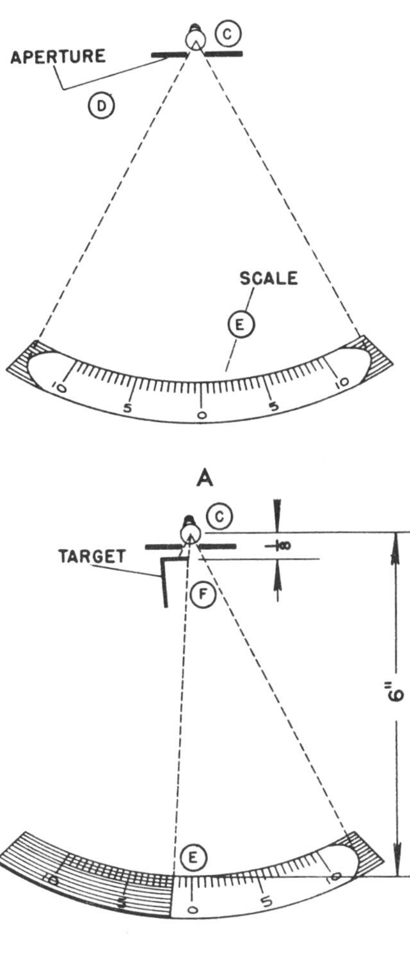

Figure 328. Light-beam Lever

LIGHT-BEAM LEVER. A simple light-beam lever is shown in Figure 328. A beam from a source of light at (C) passes through an aperture (D) and floods the full length of a scale (E) (Figure 328A). If an opaque object (F)--called a target--is moved into this beam of light, close to the source (Figure 328B), the shadow of the target will move across the scale much farther than the movement of the target at (F). The relation of the distance traversed by the shadow to that of the target will represent the magnification of the light-beam lever and will be in proportion to their relative distances from the source of light. For example: if in Figure 328B, the distance from (C) to (F) is 1/8", and the distance from (C) to (E) is 6", the mag-

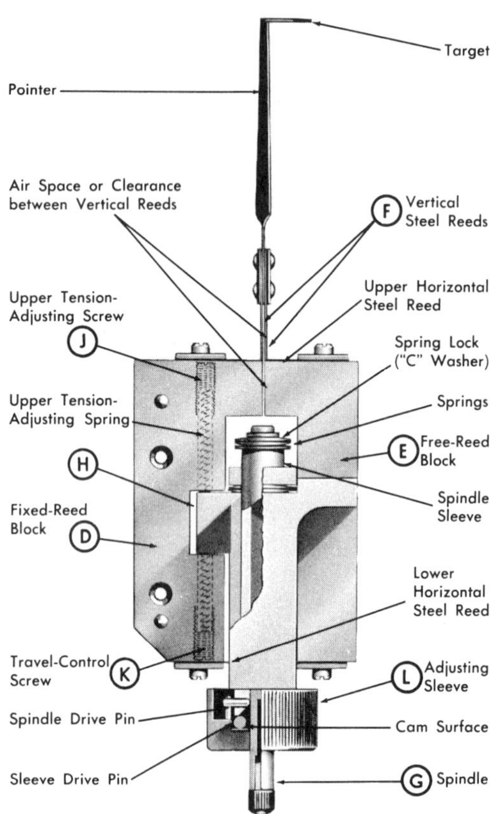

Figure 329. Reed Mechanism

nification of the light-beam lever will be 6 divided by 1/8, or 48 times.

THE GAGING HEAD. The complete reed mechanism is shown in Figure 329. Two metal blocks, one fixed (D) and one floating (E), are joined by two horizontal steel reeds whose function is to hold the blocks in relation to each other. These reeds should not be confused with the vertical reeds which actuate the pointer. The fixed block is rigidly anchored to the gage-head case. The floating block carries the spindle.

Two vertical reeds, at (F), one attached to each block, are joined together at their upper ends; beyond this extends a pointer, the tip of which is in the form of a small flat plate called a target.

When the floating block is moved by pressure upon the contact point (G) of the spindle, the vertical reeds tend to slip by each other; but since they are joined at their upper end, instead of slipping, the upper ends of the reeds swing

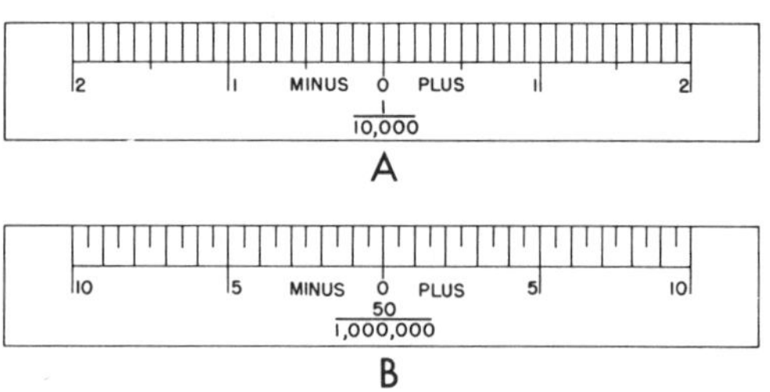

Figure 330. Scales for Reed Comparator

through an arc, and as the target is an extension of the reeds, it swings through a much wider arc.

The amount of the target swing is proportional to the movement of the floating block. The two blocks interlock at (H) in such a way that excessive impact or pressure on the spindle will be absorbed by the blocks themselves and will not damage the sensitive mechanism above.

Two adjusting screws, (J) and (K), regulate the tension and travel of the floating block. The knurled adjusting sleeve (L) is used to make final adjustment of the pointer to zero when the comparator is being set.

Any small movement of the pointer moves the target in the field of light so that its shadow moves upon the scale. The scale is graduated in divisions representing the vertical movement of the spindle. The shadow of the target is adjusted to zero at the center of the scale when the instrument is set to the master. When parts are being checked, any movement of the shadow to the right or left of zero on the scale represents the amount, plus or minus, that the part's dimension differs from the master or the basic dimension specified on the drawing.

The total magnification of the instrument equals the magnification of the reed mechanism times that of the light beam. For example: if the magnification of the reed mechanism is 50x and that of the light-beam lever is 40x, then the total magnification of the instrument is 50 times 40, or 2,000x. Reed comparators are available with magnifications from 500x to 20,000x.

SCALES. Figures 330A and 330B are typical of the scales found on reed comparators. Figure 330A is the scale of the 1,000-to-1 comparator which has a magnification of 1,000x. The scale is marked 1/10,000, meaning that

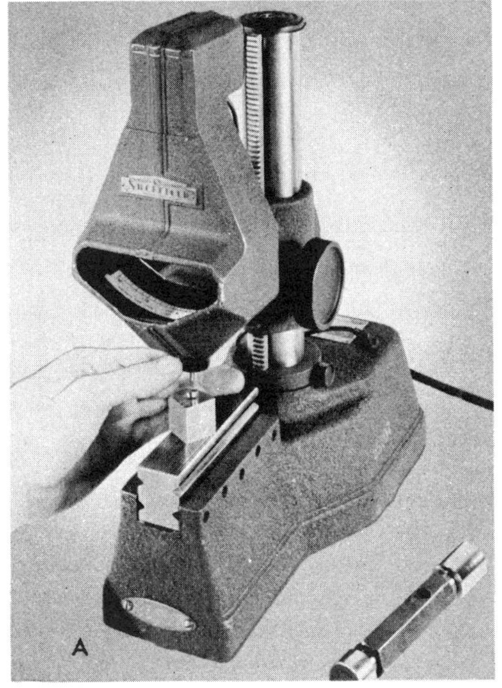

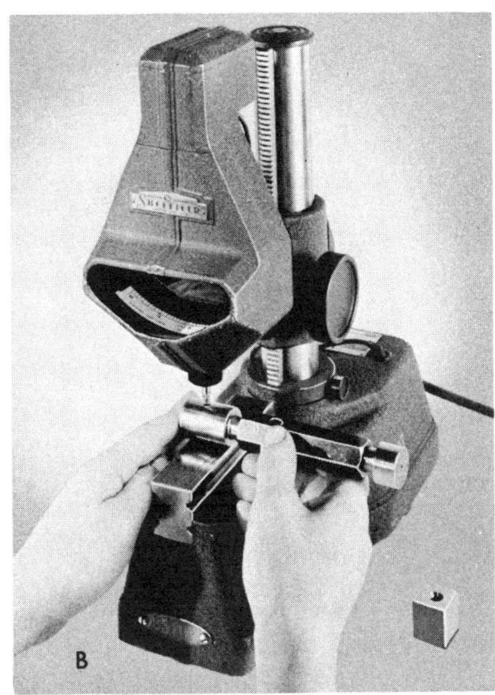

Figure 331. Checking a Plug Gage on Reed Comparator

the smallest division of the scale represents .0001". Each numbered division represents .001"; so the range of the scale is .002" plus and .002" minus, or a total of .004".

Figure 330B is the scale of the 2,000-to-1 reed comparator which has a magnification of 2,000x. The scale is marked 50/1,000,000, meaning that the smallest division on the scale represents .00005"; a double division represents .0001". Each numbered division represents .0005"; so the range of the scale is .001" plus and .001" minus, or a total of .002".

APPLICATION. Figures 331A and 331B show use of the reed comparator in checking the accuracy of a plug gage. After the gaging head is raised so that the spindle will clear the gage blocks, a stack of gage blocks corresponding to the size stamped on the plug end is wrung to the anvil (Figure 331A). The gaging head is then lowered until the spindle just touches the blocks, which is evidenced by the shadow of the target just entering the left end of the scale. The gaging head is then clamped to the column, and further adjustment is made with the adjusting sleeve until the shadow coincides with the zero of the scale

Remove the gage blocks and carefully slide the plug end between anvil and spindle as shown in Figure 331B. The shadow will again enter upon the scale and will indicate by its position whether the plug end is the correct size. The shadow will stop short of the zero position if the plug is undersize, and will override the zero position if oversize.

ELECTROLIMIT COMPARATOR

The Electrolimit comparator (Figure 332) employs an electrical device for magnifying very small movements of the spindle. The electric circuit consists of three units: the gaging head, the power unit, and the microammeter.

A cross-section of the gaging head is shown in Figure 333. The spindle (A) operates against a steel armature (B) which is hinged at one end by a flat strip of spring steel (C).

The other end of the armature is free to move between two induction coils (D) and (E), through which an electric current is flowing so that when the armature is exactly midway between the two coils no current will pass through

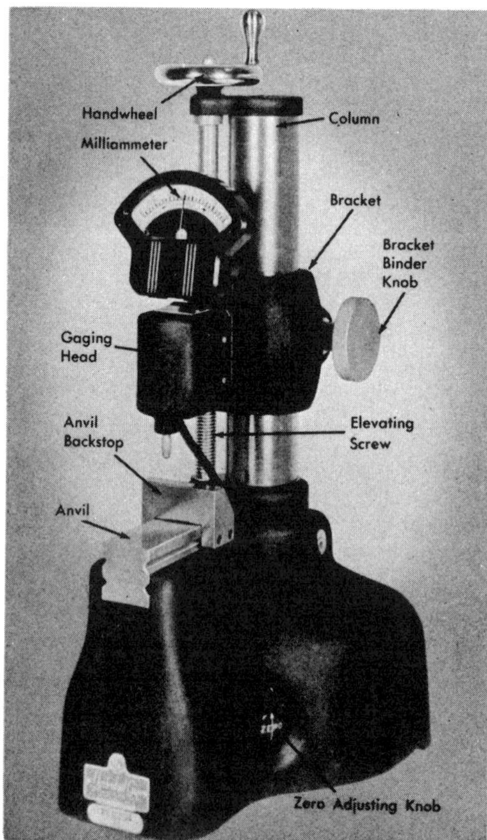

Figure 332. Universal Electrolimit Comparator

a microammeter connected in the circuit.

The slightest movement of the armature closer to one coil than to the other, however, will upset this balanced circuit, and current will pass through the ammeter and move the pointer on the scale in proportion to the amount the spindle has moved. The ratio between the spindle movement and the movement of the pointer is the magnification factor of the comparator.

The power unit, which is usually built in the base, but may be housed separately, contains:

1. A voltage regulator, which prevents line-voltage fluctuation from disturbing the accuracy of measurements.

2. The step-down transformer, which reduces voltage from 110 to 11.

3. A balancing transformer, part of the circuit.

4. A rectifier unit, which converts the alternating current to direct current for the

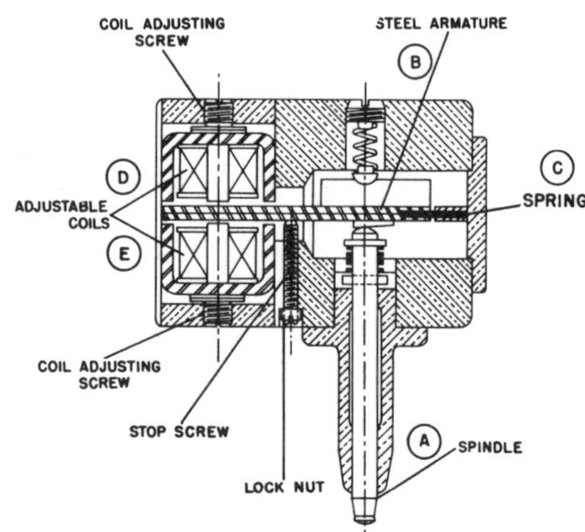

Figure 333. Electrolimit Gaging Head

ammeter.

Electrolimit comparators fall into two general classes according to the type of gaging head provided on the instrument:

1. Universal comparator
2. Master comparator

The principal difference between universal and master comparators is the latter's higher magnification and greater sensitivity.

The gaging head of the universal Electrolimit comparator (Figure 332) has a magnification factor of 1000 to 10,000x. A sensitive ammeter (called a milliammeter) is supplied with the gaging head.

The scales of the milliammeter are graduated to read to an accuracy of .00001" (10 millionths of an inch). Typical scales used on the universal comparator are shown in Figure 334. The full range of the standard scale is between .0005" and .004", which represents the maximum tolerance that can be measured on this comparator. This range is ample for any work requiring the measuring accuracy of the Electrolimit comparator.

Typical applications of the Universal Electrolimit gage are:

1. Checking gears, bushings, shafts, punches, and other production parts requiring the accuracy of the Electrolimit gage.

2. Checking working gages, including plug gages, flat gages, and masters for setting gaging fixtures.

The gaging head on the master comparator (Figure 335) has a magnification factor of from

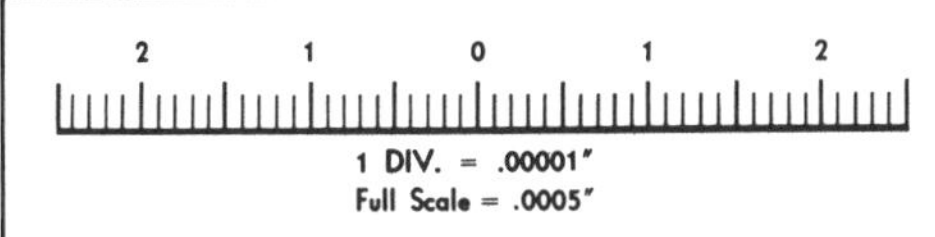

Magnification = 10,000 ×

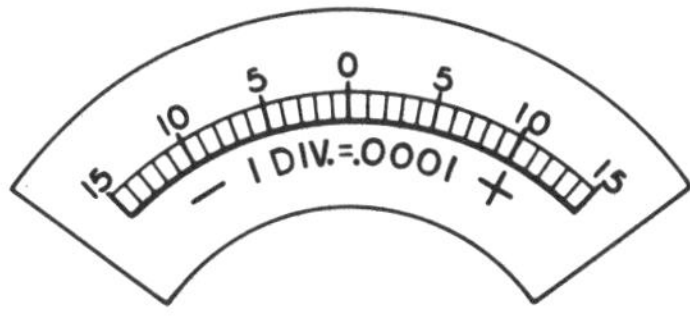

Magnification = 1,000 ×

Figure 334. Scales of Universal Electrolimit Comparator

10,000x to 110,000x. An extra-sensitive ammeter (called a microammeter) is supplied with this instrument.

Figure 336 shows a typical scale used on the microammeter of the master comparator. Each division is equal to .0000025" (2-1/2 millionths of an inch). The full range of this scale is .0002" which represents the maximum tolerance it will measure; so it is used in the more accurate field of measurement, involving tolerances of tenths of a thousandth or less.

Typical applications of the master Electrolimit comparator are:

1. Checking working sets of gage blocks against a master set to determine wear and classification.

2. Checking master plug gages.

The comparator should warm up for about five minutes before it is used or before any adjustments are made. Due to the short warm-up period, it is not necessary to have the comparator connected at all times, except for convenience.

Approximate adjustment to zero is made with the handwheel (Figure 332). Lower the gage head until the spindle barely touches the master (indicated by a fluctuation of the pointer near the left end of the scale) and lock the head in this position by means of the bracket binder knob. For fine adjustment of the pointer, use the zero-adjusting knob on the right side of the base.

The master Electrolimit is operated in the same manner as the universal, except that the zero-adjusting knob is in the meter cabinet.

Figure 335. Master Electrolimit Comparator

MAGNIFICATION CHECK. Variations in power supply affect the magnification factor of the electric circuit. To check magnification:

1. Select two stacks of gage blocks having a difference in height equal to the full value of the scale.

2. Place the minimum block on the anvil and adjust until the pointer indicates correctly at the left end of the scale.

3. Then insert the maximum block between the spindle and the anvil. If the pointer overrides the right end of the scale, the magnification is too great; if it falls short of the right of the scale, the magnification is too small. Adjust any difference with the magnification adjusting knob.

TWO-WIRE METHOD FOR CHECKING GEARS

The two-wire method for checking the pitch diameter of gears is simple and accurate, and lends itself particularly to comparison measurement. Two wires of predetermined diameter are placed in tooth spaces--diametrically opposite

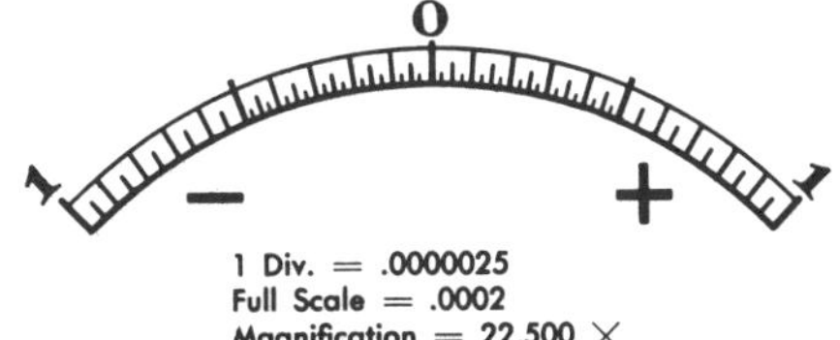

Figure 336. Scale of Master Electrolimit Comparator

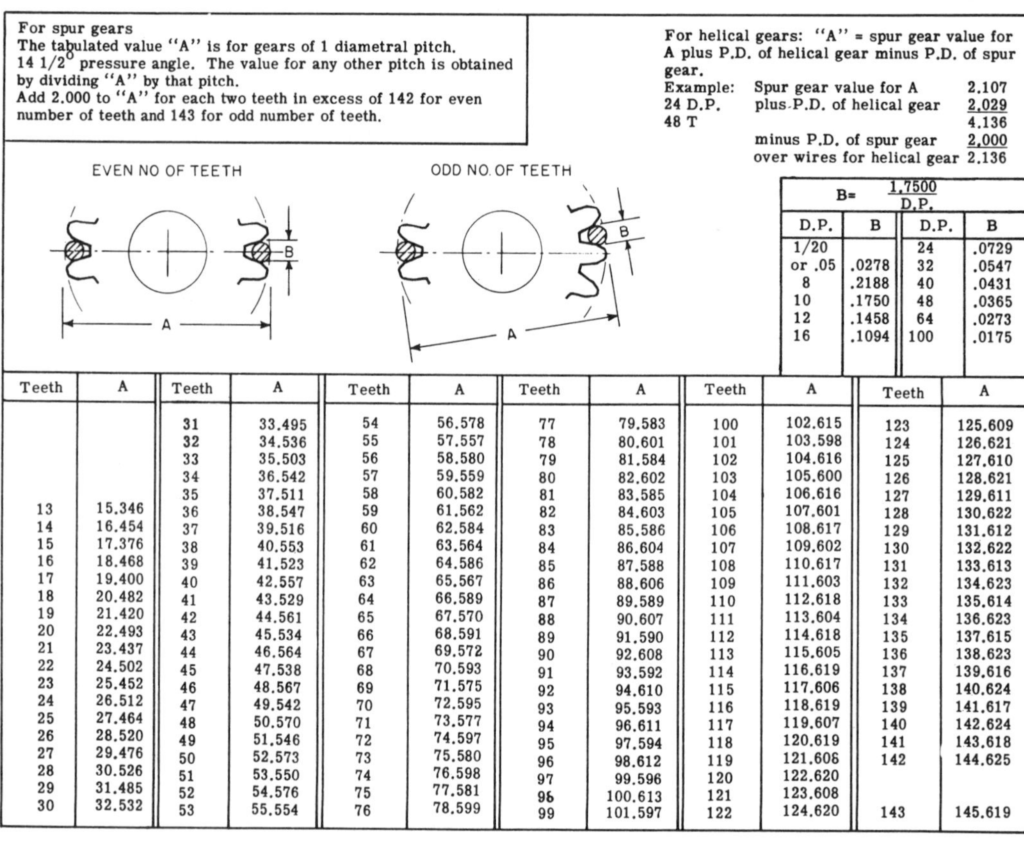

For spur gears
The tabulated value "A" is for gears of 1 diametral pitch.
14 1/2° pressure angle. The value for any other pitch is obtained by dividing "A" by that pitch.
Add 2.000 to "A" for each two teeth in excess of 142 for even number of teeth and 143 for odd number of teeth.

For helical gears: "A" = spur gear value for A plus P.D. of helical gear minus P.D. of spur gear.

Example:	Spur gear value for A	2.107
24 D.P.	plus P.D. of helical gear	2.029
48 T		4.136
	minus P.D. of spur gear	2.000
	over wires for helical gear	2.136

$B = \frac{1.7500}{D.P.}$

D.P.	B	D.P.	B
1/20 or .05	.0278	24	.0729
		32	.0547
8	.2188	40	.0431
10	.1750	48	.0365
12	.1458	64	.0273
16	.1094	100	.0175

Teeth	A	Teeth	A	Teeth	A	Teeth	A	Teeth	A	Teeth	A
		31	33.495	54	56.578	77	79.583	100	102.615	123	125.609
		32	34.536	55	57.557	78	80.601	101	103.598	124	126.621
		33	35.503	56	58.580	79	81.584	102	104.616	125	127.610
		34	36.542	57	59.559	80	82.602	103	105.600	126	128.621
		35	37.511	58	60.582	81	83.585	104	106.616	127	129.611
13	15.346	36	38.547	59	61.562	82	84.603	105	107.601	128	130.622
14	16.454	37	39.516	60	62.584	83	85.586	106	108.617	129	131.612
15	17.376	38	40.553	61	63.564	84	86.604	107	109.602	130	132.622
16	18.468	39	41.523	62	64.586	85	87.588	108	110.617	131	133.613
17	19.400	40	42.557	63	65.567	86	88.606	109	111.603	132	134.623
18	20.482	41	43.529	64	66.589	87	89.589	110	112.618	133	135.614
19	21.420	42	44.561	65	67.570	88	90.607	111	113.604	134	136.623
20	22.493	43	45.534	66	68.591	89	91.590	112	114.618	135	137.615
21	23.437	44	46.564	67	69.572	90	92.608	113	115.605	136	138.623
22	24.502	45	47.538	68	70.593	91	93.592	114	116.619	137	139.616
23	25.452	46	48.567	69	71.575	92	94.610	115	117.606	138	140.624
24	26.512	47	49.542	70	72.595	93	95.593	116	118.619	139	141.617
25	27.464	48	50.570	71	73.577	94	96.611	117	119.607	140	142.624
26	28.520	49	51.546	72	74.597	95	97.594	118	120.619	141	143.618
27	29.476	50	52.573	73	75.580	96	98.612	119	121.606	142	144.625
28	30.526	51	53.550	74	76.598	97	99.596	120	122.620		
29	31.485	52	54.576	75	77.581	96	100.613	121	123.608		
30	32.532	53	55.554	76	78.599	99	101.597	122	124.620	143	145.619

Figure 337. Two-wire Method for Checking Gears

each other when the gear contains an even number of teeth, and as nearly opposite as possible when the gear contains an odd number of teeth (Figure 337). The measurement over the wires (A) is compared with figures obtained from the table. The wire diameter is obtained by using the formula for (B).

The setup on the comparator (Figure 338) involves certain features that insure accuracy of the measurement:

1. The surface of the anvil must be flat and smooth, without serrations, because the lower wire might lodge in one of the serrations and nullify accuracy of the measurement.

2. The surface of the anvil must be in a plane forming a right angle with the axis of the spindle; this requirement is common to all comparison measurements.

3. A backstop must be provided to keep the face of the gear perpendicular to the face of the anvil when the upper wire is moved under the spindle for a measurement.

The backstop has a slot and clampscrew to hold the lower wire flat on the anvil, thus relieving the inspector of the awkward task of trying to hold two wires in place at the same time.

MEASURING EXAMPLE

Objective: To check the pitch diameter of a gear (Figure 339) using the two-wire method.

Reference: Figures 337, 338 and 339.

Information: Number of teeth 56.

Diametral pitch 24.

Refer to Figure 337 to find the wire diameter with the formula B = 1.7500/D.P.

B = Wire diameter D.P. = Diametral pitch

Substituting:

B = 1.7500/24

B = .07291 (wire diameter)

Figure 338. Checking a Gear (Two-wire Method)

The measurement (A) over the wires is obtained by dividing the figure in the table opposite the number of teeth, by the diametral pitch.

Refer to the table, under number of teeth, to find opposite 56 the figure 58.580.

Therefore: A = 58.580/D.P.

A = 58.580/24

A = 2.4408 (measurement over wires)

The measurement is made in the usual way by selecting a stack of blocks equal to 2.4408" and setting the comparator to zero. Then, with the wires in place and the gear pivoted on the lower wire, move the upper wire under the spindle (Figure 338). The difference between the basic pitch diameter and that of the gear will be indicated on the ammeter.

The foregoing calculation does not take into consideration backlash or allowance sometimes made for free running of mating gears.

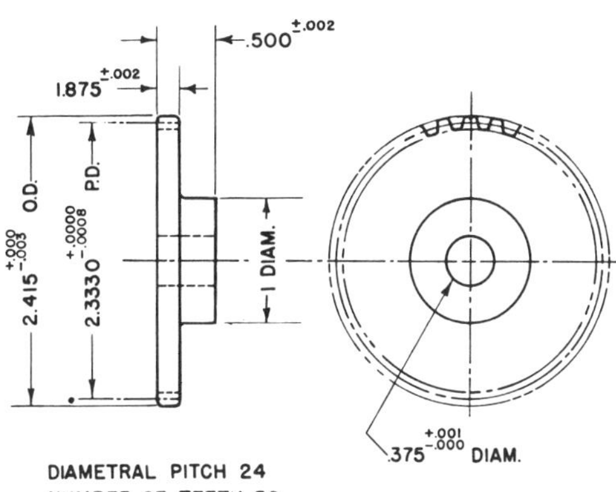

Figure 339. Spur Gear

ELECTROLIMIT INTERNAL COMPARATOR

Figure 340 shows a bench-type Electrolimit comparator designed to check internal diameters over a range of 1/4 to 3". The electrical units are the same as those used in the Electrolimit external comparators. This is a two-piece instrument, composed of the gage stand and the portable instrument cabinet which contains the indicating dial and all electrical units. The dial is calibrated .001" full scale, giving a magnification of approximately 4500x. Each small division represents .00002", and the divisions are spaced far enough apart so that estimated readings of .000010" can be obtained.

Three sets of gaging fingers to cover the indicated range are furnished as standard equipment and can be quickly changed for checking different diameters. The set number one covers a range from .240 to .510"; set number two, a range of .510 to 1.510"; set number three, a range of 1.510 to 3.000". One gaging finger is mounted rigidly with provision for adjustment over the indicated range. The other gaging finger is mounted flexibly; the movement of this finger is indicated on the meter. A gaging pressure of eight ounces is normally furnished.

In operation the gaging fingers are set to a stack of precision gage blocks assembled with outside caliper jaws and tie rods, or other master standards. The fingers are positioned approximately the correct distance apart by means of a left-hand screw operated by the finger-adjusting knob. The finger slide is locked in position by tightening the finger slide binder. Final setting is made with the zero-adjusting knob.

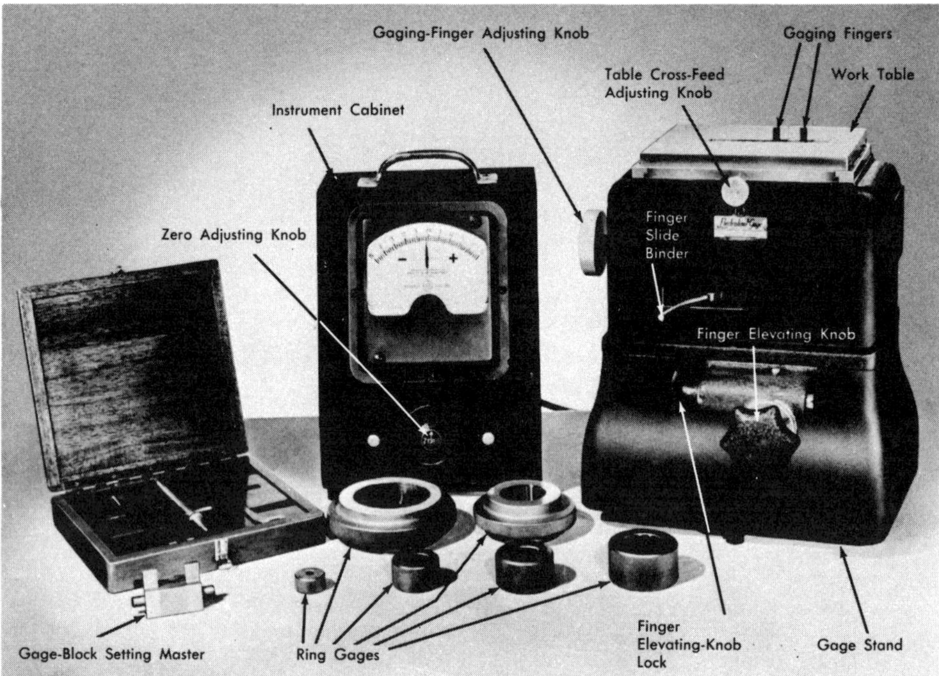

Figure 340. Electrolimit Internal Comparator

The work to be gaged is placed over the fingers on the floating work table. This table is mounted on balls to permit the work to position itself correctly in relation to the gage point. The table cross-feed adjusting knob positions the workpiece accurately to the greatest diameter. The combined float and cross-feed adjustment enables the operator to position the workpiece accurately without touching the part once it has been placed on the work table.

Errors due to taper are quickly and easily revealed. A tapered condition is detected by the receding feature of the gaging fingers. The work table and workpiece being checked remain at a constant elevation. The finger-elevating handwheel operates a rack and pinion for raising and lowering the gaging fingers in relation to the work table and workpiece.

The finger-elevating handwheel binder locks the elevating mechanism in position. Roundness can be checked by repositioning the workpiece and taking readings in various planes.

Because of the comparator's extreme sensitivity, its location is most important. The gage stand must be placed on a sturdy and level surface; a separate table used only for this equipment is strongly recommended. It should be installed where the temperature is fairly constant and should not be placed in the direct rays of the sun. Sensitive comparators of this type are mainly for use in gage rooms, and in tool inspection for checking internal diameters of master and working ring gages and parts having internal diameters that must be held to extremely close tolerances.

Figure 341 shows an Electrolimit internal comparator being used to check a ring gage.

ELECTRONIC COMPARATOR

The electronic comparator in Figure 342 provides reliable means for taking external and internal measurements electronically with a high degree of accuracy. This comparator employs the dependable Wheatstone bridge circuit in the form of a relatively new strain gage. As the gaging point of the comparator is deflected, minute changes in the electrical resistance of the bridge circuit create voltage signals that are amplified to operate the indicating meter. The complete instrument consists

Figure 341. Checking a Ring Gage on Internal Comparator

of a comparator or gaging stand and an amplifier to which a signal light attachment may be added.

The comparator or gaging stand consists of a base with a vertical column supporting the head. The head is adjusted on the vertical column through an elevating screw operated by the hand-wheel at the top of the column and then clamped in position by the knob on the lower right-hand side. The gaging spindle and gaging mechanism are in the head. For wear resistance under continued use, the measuring point is a diamond. An insulated grip is provided to lift the measuring point when gaging so requires. Pressure on the measuring point is adjustable from one to two pounds and is indicated by the calibrated adjustable plunger above the measuring point. The plunger is locked in place by the knurled thumbscrew at the front of the head.

The work surface, or anvil, is reversible, having a wide surface and a narrow one; the narrow surface is suited for measuring small parts or work of irregular shape; the wide surface is serrated.

This comparator has a self-checking device which enables it to be checked easily and quickly without gage blocks or masters. When turned, the knob on the right-hand side of the head provides the same effect as a .0002" movement of the gaging point. This movement is indicated on the amplifier when the knob is turned. If the reading does not correspond, adjustment can be made at the rear of the amplifier. The measuring point should rest on the work or the anvil when making this check. When the comparator is in normal use, the same mechanism can be used as a fine adjustment to locate the

Figure 342. Electronic Comparator

pointer setting at zero.

Through electronic means, the amplifier magnifies from 1800 to 18,000 times the movement of the gaging point of the comparator. The high degree of magnification is determined accurately by the setting of a graduation selector on the panel of the amplifier; the widely-spaced dial graduations represent values of from .0001 to .00001". This arrangement has a distinct advantage: it permits a setting to be made so that a predetermined number of graduations on the dial either side of zero represents the limits of tolerance on the work being measured; this means that tolerances as large as ± .001" and as small as ± .00001" may be used if desired. The amplifier, containing all of the important heat-generating equipment, is separate from the comparator to prevent temperature drift trouble; the amplifier is connected to the power supply and the comparator by means of electrical cord and plugs. Power is turned off or on by a switch at the front of the instrument.

A signal-light attachment (shown in Figure 342) may be used with the amplifier when it is desired to check work where it is not necessary to know the exact dimensions of the workpiece as long as its size falls within the limiting dimensions specified for it.

This attachment is fastened easily to the bottom of the amplifier; all electrical connections are made automatically when the two units are combined. The attachment has three lights

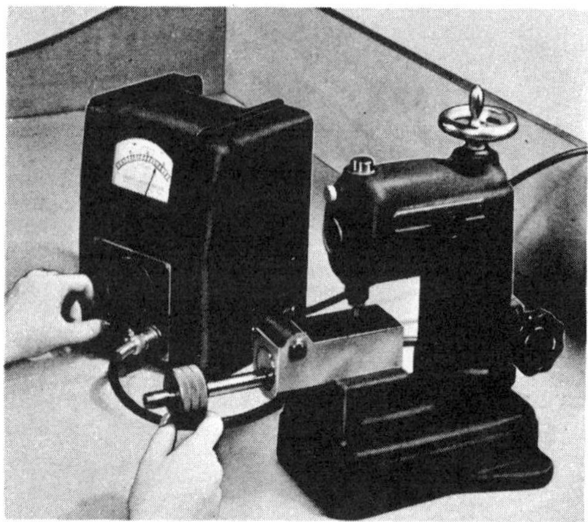

Figure 343. Electronic Comparator with Internal Measuring Attachment

placed in back of red, amber, and blue lenses, and tells at a glance whether or not a workpiece is within a prescribed tolerance. When set up for a specified tolerance, a workpiece placed under the measuring spindle of the comparator will cause the amber light to glow if it is within tolerance; the red light indicates the workpiece is too small; the blue light indicates that it is too large.

To operate the signal-light attachment:

1. Set the graduation selector on the amplifier to the desired sensitivity.

2. With a master or a workpiece of standard size under the comparator spindle, adjust the comparator and amplifier to bring the meter pointer to the low limit of the work tolerance.

3. Adjust the left-hand knob so that the red light just goes out and the amber light goes on as the meter pointer passes through the low limit when actuated by the zero-set knob on the amplifier.

4. Bring the meter pointer to the high limit.

5. Adjust the right-hand knob so that the amber light just goes out and the blue light goes on as the meter pointer passes through the high limit when actuated as in (3).

For measurement of internal diameters from 1/2 to 2", an attachment as shown in Figure 343 is added to the external comparator; the attachment makes possible the measuring or gaging of work in units from .0001 to .00001". The range of measurement of the internal attachment is made possible by the use of interchangeable plugs which fit into the attachment body; it is necessary to have the correct plug for a particular measurement. With the anvil removed, the attachment is placed in position in the external comparator so that its transmitting point is in alignment with and contacts the gaging point of the external comparator. The internal comparator attachment is set to a ring gage or master, and only one master is required in making the setting.

AIR GAGES

The air gage is a form of comparison measuring instrument originally designed to measure the inside characteristics of holes by indicating on a suitable scale the amount of air escaping between the sides of a hole and a gaging head which has been inserted in the hole. From this inside-diameter application have been developed gaging elements for checking outside diameters, concentricity, squareness, parallelism and many other conditions. Air gages are also combined with semi-automatic and fully automatic gaging machines involving the use of electrical and mechanical principles.

The fundamental principles of air gaging are the same for internal and external gaging and depend on the flow of air between the faces of jets in the gaging head and the part being checked. Both the velocity and the pressure of the air depend upon the clearance between the gaging head and the work being checked. The greater the clearance, the higher will be the velocity and the lower the back pressure; conversely, the smaller the clearance, the less the velocity and the greater the back pressure. Therefore, a comparison may be made by measuring either the velocity or the pressure of the air. Air gages which indicate the velocity of the air are called the flow type; those which indicate the pressure of the air are called the pressure type. Compressed air from the regular plant supply or that obtained from a portable compressor may be used to operate air gages.

The gaging heads may be of several different forms, depending upon the nature of the parts to be checked. They may be attached directly

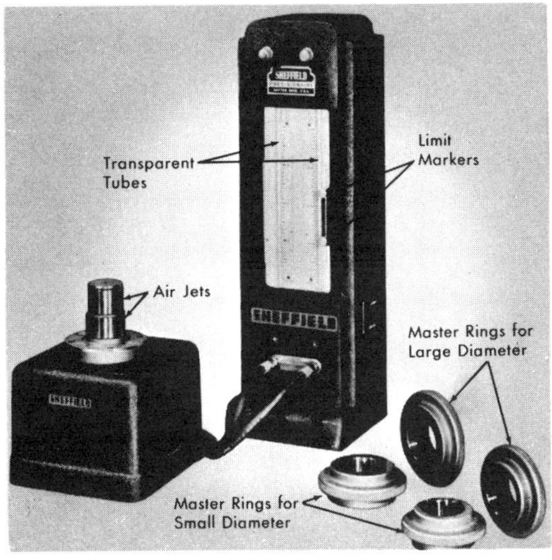

Figure 344. Precisionaire Gage

to the gage; attached to flexible tubing, which may be up to several feet in length; or built into special fixtures designed to check several dimensions simultaneously.

Some of the advantages of gaging by air are:

1. The air gage provides a direct and simple means for high amplification of minute dimensional variations on work held to close tolerances. Standard amplifications range from 1,000 to 1, to 40,000 to 1.

2. Less skill is required to do precise gaging than with fixed gages, such as plugs, rings and snaps.

3. Parts can be gaged without making any contact; this reduces possibility of marring soft or highly finished surfaces.

4. Air gaging heads will outlast many times the life of conventional fixed gages. Air spindles are made smaller in diameter than the minimum hole size. Airsnaps are set to clear the maximum limit of external dimension. In all gaging heads the jets are recessed below the surface to prevent them from coming in contact with the part being gaged.

5. Air gages can be used directly on the machine, or at the bench, with equal facility, as the gage can be presented to the work, or the work to the gage.

PRECISIONAIRE

The Precisionaire (Figure 344) is a two-column, flow-type air gage equipped with a gaging spindle for checking two inside diameters simultaneously. The gage is actuated by the velocity or flow of air escaping from the orifices of the gaging spindle. Gaging spindles may be attached to the gage as shown, or may be remotely attached by a flexible tube.

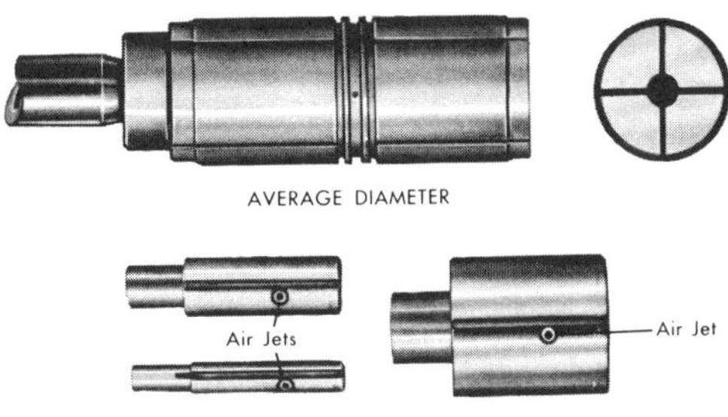

Figure 345. Air-gage Spindles

Compressed air enters the gage through compensating pressure regulators, passes through a vertical transparent tube in which a float registers the amount of flow, and passes out through the orifices in the gaging spindle.

The spindle consists of a cylindrical plug with a central air channel terminating in two or more jets in the side of the spindle. This spindle is connected to the indicating tube of the gage so that the air escaping from the jets or orifices creates a flow of air in the tube and causes the float to rise.

The gaging spindle takes on several forms, depending on the nature of the operation. When it is desired to check the average diameter of a hole, the air jets terminate in annular grooves which distribute the air evenly about the circumference. When diameter, out-of-round, bell mouth, or taper are involved, two diametrically opposite jets open directly in the spindle surface, permitting a point-to-point check. Figure 345 shows three standard spindles and one for checking average diameters of a hole.

SETTING UP THE PRECISIONAIRE

Master ring gages representing minimum and maximum tolerance limits are used to set the limit markers on the air gage when using the air spindle as the gaging element.

With the air pressure on, insert the gaging spindle into a master ring having the minimum

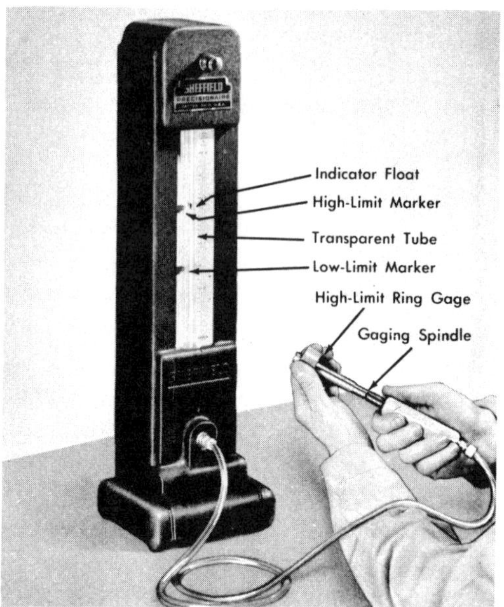

Figure 346. Setting up Precisionaire

Figure 347. Setting Adjustable Air-snap Gage

dimension of the hole to be checked. Adjust the air pressure so the float will come to rest near the central zone of the transparent tube. Set the lower limit marker opposite the position of the float. The marker will indicate the low limit of the dimension being checked.

Remove the minimum master and insert the spindle into a master ring having the maximum dimension of the hole. The float will rise to a new position above the minimum marker (Figure 346). Place the upper-limit marker opposite the new position of the float to indicate the high limit of the dimension being checked. The gage is now ready for making comparison measurements. Insert the spindle into the hole to be checked, and if the float comes to rest between the two markers the hole is the correct size.

ADJUSTABLE AIR SNAP

Figure 347 shows an adjustable air snap being set from a master setting disc. Adjustable air snaps used in conjunction with the air gage have a wide range of applications for checking the external dimensions of work in the machine or at the bench. They are also used to advantage on parts too large to be conveniently presented to the gage.

The adjustable air snap is first set up to a stack of gage blocks slightly larger than the maximum tolerance limit of the part to be checked; this prevents undue wear on the gaging surfaces. After the adjustable anvil is set to the oversize stack of gage blocks, the adjustable backstop is brought into contact with the anvil extension and locked in place. The actual setting of the maximum and minimum markers is then made with master setting discs or cylindrical plugs.

DIAL-TYPE AIR GAGES

Dial-type air gages, produced by several manufacturers of gaging equipment, are similar in size and operating principles. Compactness and small over-all size are outstanding features of dial-type air gages. They are easily portable for use at the inspection bench, or at the machine, and in many cases are attached directly to the machine so that work can be checked for size without removal from the machine. Dial-type air gages can be used to advantage in a variety of ways for checking internal and external dimensions on both single and multiple applications.

Figure 348 shows a typical dial air gage. This is a pressure-type air gage and differs from

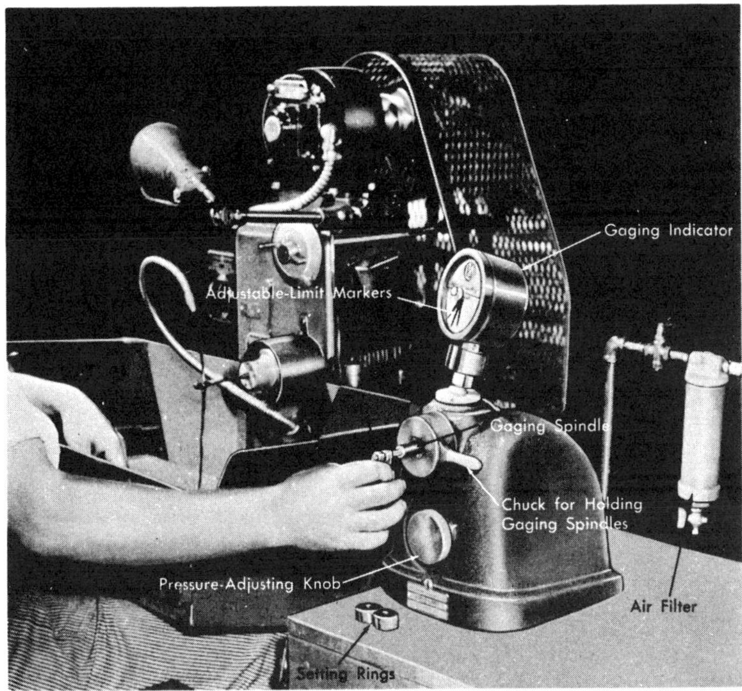

Figure 348. Dial-type Air Gage (Air-O-Limit)

the flow-type previously described because it registers the difference in pressure created by the air escaping between the gaging spindle and the surface being checked. In the illustration, the operator of a honing machine is checking the diameter of a hole in a part. The graduated scale on the gaging indicator is calibrated with minimum and maximum setting rings and limit markers set to the high and low limit. Then the size of the hole is read directly from the scale graduations.

Figure 349 is a schematic diagram of the air gage in Figure 348; the gage is a self-contained unit except for the air filter, which is

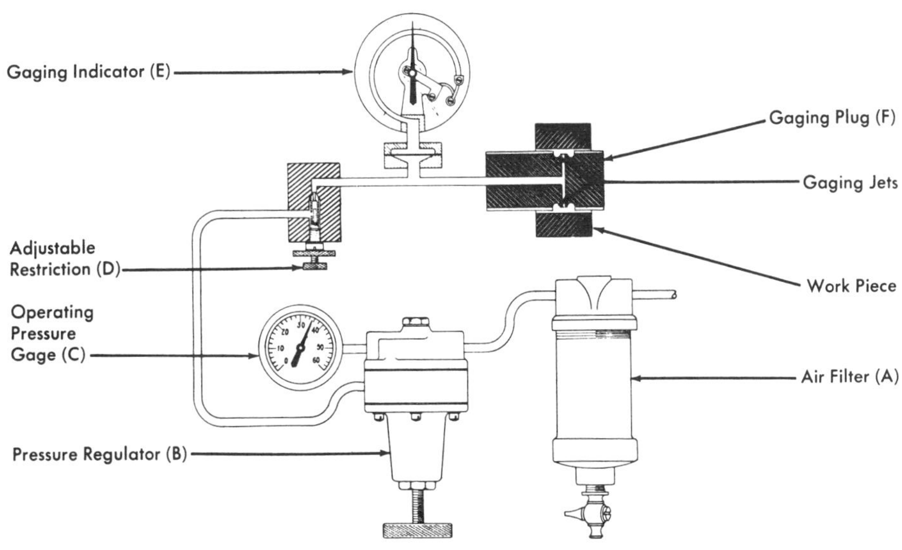

Figure 349. Schematic Diagram of Air-O-Limit Gage

Figure 350. Dial Air Gage Attached to a Machine

connected to the plant air supply. Air from the plant line passes through the filter (A) and a pressure regulator (B), which limits the pressure to 38 to 44 lbs. per square inch for operating the gage. The pressure regulator (B) contains a knob by which the gage indicator can be set to zero. An operating pressure gage (C) registers the air pressure being imposed upon the air gage. From the pressure regulator the air passes through an adjustable restriction tube (D), which serves to prevent pressure fluctuations, and thence past the gaging indicator (E) to the gaging spindle (F).

Figure 350 shows a dial-type air gage attached directly to a machine, where it is being used to check the internal diameter of the hole in a part. Here the operator can check and control size without removing the part from the machine. Applications of this kind are valuable where close tolerances must be held, reducing the need for 100 percent inspection and costly rework.

MULTIPLE-DIMENSION GAGING

Simultaneous inspection of multiple dimensions has made tremendous strides in the last decade. Two or twenty and even more dimensions may be checked simultaneously in about the time it formerly took to check one dimension. Advantages include greater accuracy, removal of the human element of error, large savings in time, and reduced inspection area.

The widespread use of multiple air and electric gages started a trend toward highly mechanical semi- and fully-automatic types of gaging machines. Automatic gages can be utilized to greatest advantage in measuring, classifying, and segregating high-production precision parts, such as pistons, piston pins, piston rings, shafts and bearing balls. The field for automatic gaging is almost limitless, each application being justified on a cost-per-piece basis.

AIR GAGING AND SURFACE ROUGHNESS

The quality of surface finish must be considered in air gaging. Fixed gages and mechanical-type indicating gages, will check the smallest diameter of a hole, or the largest diameter of a shaft; an air gage will give a reading somewhere between the smallest and the largest diameter. This tendency of air gages to give an average reading may be the

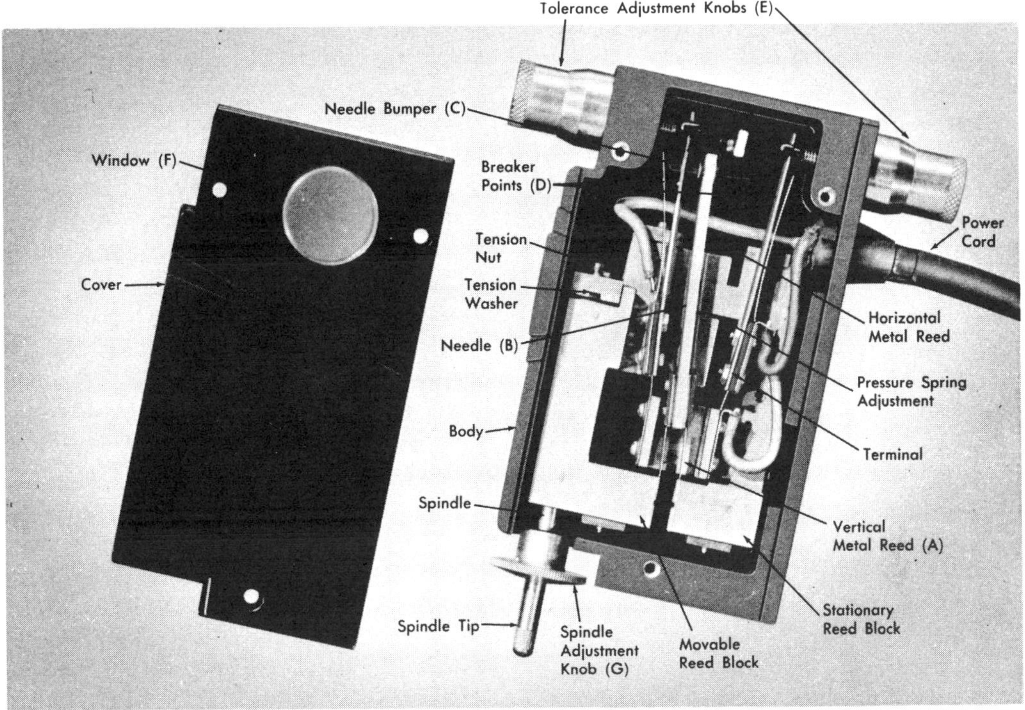

Figure 351. Electrichek Gage Head

cause of errors when surface finish is rough and tolerance small.

For air-gaging holes having a surface finish rougher than 50 to 100 microinches and tolerance of .005" or more, blade-contact-type air spindles are available. These spindles consist of a hardened steel body slightly smaller than minimum hole size; the spindle has a hardened-steel contact blade either pivoting or floating, whose action controls the flow of air through the jets, causing the blades to contact the sides of the hole, like an ordinary plug gage.

ELECTRICHEK GAGE HEAD

Figure 351 shows an Electrichek gage head which, because of its compactness, can be applied to a system of multiple-dimension gaging where a number of dimensions are gaged simultaneously.

The Electrichek gage head is a limit gage. It is similar in construction to the gage head previously described (see Figure 329)--because it utilizes the reed mechanism as a magnifying device--but it does not have the light beam and scale.

The vertical reeds (A) terminate in a pointer or needle (B) whose upper end contains a bumper (C) that breaks an electrical contact (D) at both extreme limits of its swing.

On one side the contact lights a red light; on the other a green light. When both lights are lit, their combination appears as an amber light. These lights are housed in a separate unit and are connected to the head by an electric cord. The contacts control the upper and lower limit of the tolerance zone. The positions of the contacts are controlled by the tolerance-adjusting knobs (E). The contacts can be viewed through a window (F) in the cover.

In setting the head for any given dimension, a master, equivalent to the minimum limit on the dimension, is placed in gaging position. By adjusting the spindle through knob (G), the needle (B) is moved toward the contact controlling the red light. When very close to the contact, further adjustment is made with the tolerance-adjusting knob (E) until the light just breaks from red to amber. A slight turn back will create a flickering state which marks the setting of the minimum dimension.

If the minimum master is removed and a master corresponding to the maximum limit is placed in gaging position, the needle will swing toward the contact which controls the green light. Adjust the contact on this side until the flickering occurs between amber and green; do not touch knob (G) which marks the setting of the maximum dimension.

The needle, swinging anywhere between these points, fails to break either contact, and the signal light will remain amber; but when the part being checked is larger than the maximum limit, the contact to the green light will be broken, and the red light will appear; when the part is smaller than the minimum limit, the contact of the red light will be broken, and the green light will appear.

Figure 352 shows a fixture designed to check eight different diameters of a shell casing at the same time. Eight individual gaging heads, one for each diameter, are being set with one of the masters. Each head actuates a signal light in the panel above. Lines from the signal lights to a diagram of the shell on the panel board indicate to the operator the dimension controlled by a particular light. A master light near the top of the panel will remain white if all dimensions check within the specified limits. If any one of the dimensions is in error, the

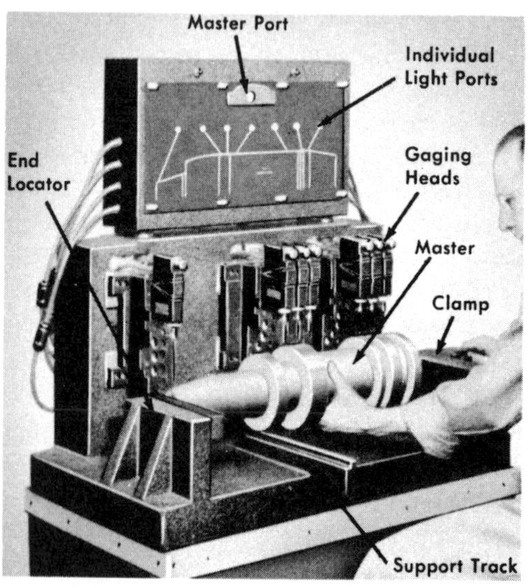

Figure 352. Multichek Gaging Fixture

master light will turn red. The inspector then refers to the individual lights to determine which dimension is in error.

Multiple-dimension gaging greatly reduces the number of inspection operations and reduces the possibility of error by eliminating the human factor. It is fast and accurate and conserves floor space. It requires a special fixture, however, and the number of pieces produced must warrant the investment.

Another example of inspection of multiple dimensions is shown in Figure 353. This is a multiple-column Precisionaire inspection machine, designed for checking several dimensions on a crankshaft.

The five snap-type air gaging heads, seen projecting from the top front edge of the machine, are elevated to their inspection positions by raising the counterbalanced-bar handle at the front of the machine. The five main bearings of the crankshaft are checked simultaneously by these built-in air gaging heads. Each head has three pairs of jets located diametrically opposite each other so that the diameters of each main bearing are gaged midway between the flanges and at 1/8" from each flange. While in the gaging position, the snap-type heads can float forward or backward to take care of any misalignment. The various diameters are indicated on the fifteen Precisionaire tubes at the top of the machine.

Simultaneously, the oil-seal diameter of the crankshaft is inspected at one location, and the width of the rear main bearing is checked on one of the Precisionaire units seen at the top of the machine.

In the illustration, the five gaging heads for checking the diameter of the main bearings have been lowered from their gaging positions. The diameter of the pin bearings is being checked manually by means of a portable snap-type air gaging head shown in the operator's right hand. This air snap also has three pairs of jets, and the diameters being checked are indicated on the three-column Precisionaire unit at the left, as can be seen from the position of the floats in the transparent tubes.

Figure 354 shows a specially designed air-

Figure 353. Multiple-dimension Checking with Air Gage

electric gaging machine for automatically checking all critical dimensions of automotive valve tappets and properly segregating the acceptable, rejected, and salvaged parts. The human element of error is entirely eliminated in the measuring of each dimension and in the proper segregation of parts. One inspector operates the machine, which has a cycle capacity of 2,000 parts per hour. All checks and segregations are made automatically at a rate exceeding the time required for each of several of the individual checks, thus saving time and cost.

This automatic gaging and segregating machine consists basically of a loading station, five gaging stations, five segregating stations, and two unloading stations. A walking beam lifts and carries the parts progressively through a series of tungsten carbide-faced vees which act as loading platforms for the gaging and segregating stations.

Parts are loaded into the machine with the open end to the right. A fixed-type snap gage is built into the top of the loading chute to prevent excessively-oversize parts from entering. If loaded incorrectly, the machine will stop automatically and a red light flashes on, indicating a jam at the first vee.

A plunger pushes each part from the bottom of the loading chute onto the first vee where the walking beam carries it to the first gaging station. Another plunger loads the part into an air ring where outside diameter at each end is checked and taper is gaged. The part is then pushed out of the air ring by a plunger onto the vee, from which the walking beam carries it to the segregating station. If the part is outside the tolerance limits, it will automatically be ejected from the machine through one of three chutes:

1. Reject on the O.D.
2. Salvage on O.D.
3. Reject on taper

If the part is acceptable, it remains on the vee, and the walking beam carries it to the second gaging station where a blade-type contact spindle checks the major internal diameter, At the next segregating station the part is either ejected through one of two chutes--"Reject" or "Salvage"--or remains to be carried to the

Figure 354. Air-electric Automatic Gaging Machine

next gaging station.

At this third gaging station, a two-step spindle checks the depth of the major internal diameter and also depth of the smallest I.D. At the next segregating station, the part may be ejected for:

Reject, depth 1
Salvage, depth 1
Reject, depth 2
Salvage, depth 2

If the part is good, it proceeds to station four, where squareness of the solid end face with the bore is checked, and over-all length is gaged. The segregating station ejects the following:

Rejects on length
Salvage on length
Reject on squareness

At the next station, alternate good parts are ejected from the machine. The other 50 percent of the parts proceed to a final gaging station for a check of concentricity of I.D. with O.D. and also for out-of-round. At the segregating station the unacceptable parts are ejected in separate chutes for:

Reject on concentricity
Reject on out-of-round

The acceptable parts are chuted on a conveyor and carried outside the machine.

Protective switches at each gaging and segregating station stop the machine immediately in case of a jam, and a red light on the panel shows exact location of the jam. The machine is built for continuous mass-production usage and is designed to require the minimum of maintenance.

QUESTIONS AND PROBLEMS

1. Explain what is meant by comparison measurement.
2. Why is magnification or amplification necessary in comparison measurement?
3. Why are locating surfaces important when making comparison measurements?
4. Name three advantages in using comparison fixtures for inspection of parts.
5. What is a light-beam lever? Explain how it works.
6. How is the magnification of the light-beam lever determined?
7. If the magnification of the light-beam lever were 20x and the reed mechanism were 50x, what would be the total magnification?

8. A scale has 10 divisions each side of zero, and the scale is labeled 50/1,000,000. What would be the range of the scale?
9. How does the Electrolimit head differ from the gaging head of the reed comparator?
10. Name the two general classes of Electrolimit comparators. Which is the more accurate?
11. Give two applications of the universal Electrolimit comparator; the master Electrolimit.
12. How may the magnification be checked on an Electrolimit comparator?
13. For what is the Electrolimit internal comparator used? What is its range?
14. Give two uses of the electronic comparator. How are these obtained?
15. What values can the graduations on the scale of the Electronic comparator be made to represent? How can this be done?
16. When is the signal light attachment used with the electronic comparator?
17. What is the principal difference between the flow-type and pressure-type air gage?
18. Give some advantages of air gaging.
19. How is the difference between the master and the part indicated on the Precisionaire gage? The Air-O-Limit gage?
20. How does the Electrichek head differ from the standard gaging head of the reed comparator?
21. In multiple-dimension checking with Electrichek gaging heads, what governs the number of gaging heads employed?
22. State three advantages of multiple-dimension gaging.

Chapter XI

OPTICAL INSTRUMENTS

IN PRECISION measurement, optical instruments are used to project and magnify an image of an object so that otherwise invisible surface characteristics can be inspected and measured with the eye. These instruments do this by controlling light waves.

LIGHT

Light is a form of radiant energy which, by action upon the eye, enables the eye to perform the function of sight. Light travels from its source in a straight line, and at a constant velocity, unless intercepted by a medium different from that in which it is traveling. The velocity of sunlight is approximately 186,000 miles per second in air.

LIGHT BEAMS AND RAYS

A light beam is a portion of light emanating in a single direction from a source. The beam is composed of innumerable lines of light called light rays. Each ray travels in its individual straight line or path; therefore, an unrestricted beam of light assumes the form of a cone, the apex of which is at the source, and the base of which increases in diameter in proportion to its distance from the source.

If a source of light were enclosed within a spherical hood, and if a hole were punctured in the hood, a small portion of the light would be projected through the hole, and in the darkness would appear as a beam extending off into the atmosphere in a single direction (Figure 355). If a card is placed in the beam, an image of the hole will be projected upon the card, but the image will be larger than the hole. Also, as the card is moved away from the hood, the projected image of the hole becomes larger as its distance from the hood is increased.

Light is invisible unless directed to the eye. A beam of light in the atmosphere is visible because the light is reflected to the eye from minute particles in the atmosphere.

REFLECTION

Light directed against a surface will be reflected in proportion to the smoothness of the surface and the nature of the substan. ,on which it is directed. The portion of light not reflected either passes through the substance or

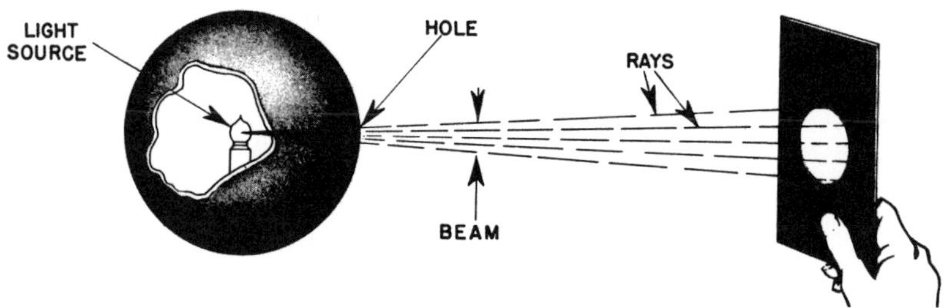

Figure 355. Light Beam and Rays

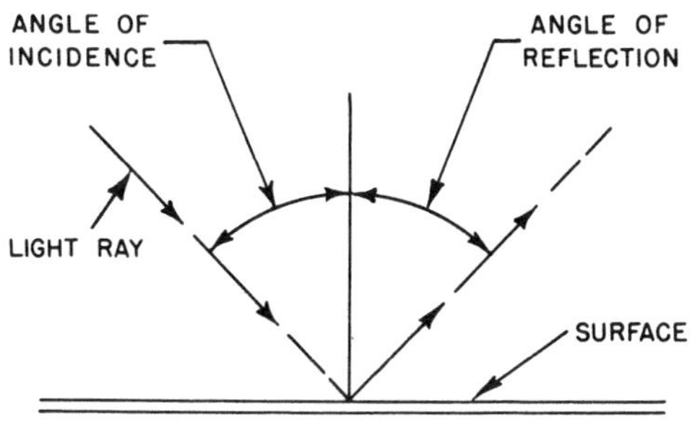

Figure 356. Reflection

is absorbed by the substance.

The angle at which the light approaches a surface is called the angle of incidence. The angle at which light is reflected from a surface is called the angle of reflection. The angle of reflection is always equal to the angle of incidence. Angles of incidence and reflection are measured to a line perpendicular to the surface.

A ray of light, striking a flat surface at a given angle to a line perpendicular to the surface at the point of incidence, will be reflected from the surface in the same direction and at an equal angle to the perpendicular, on the opposite side of the perpendicular (Figure 356).

REFRACTION

Refraction is the deflection, or bending from a straight line, of a ray of light in passing obliquely from one medium into another medium in which its velocity is different. The medium through which the light is passing affects the velocity of the light. When light passes from the air into glass, its velocity is reduced. This reduction of velocity is the direct cause of refraction, or bending of the ray of light. When a ray of light enters, at an angle, a block of glass having parallel surfaces, the ray is bent both when entering the block and when leaving it (Figure 357).

It is obvious that the portion of the ray at (A) reaches the block before the portion of the ray at (B). Therefore, the lower side of the ray at (A) is retarded before the upper side of (B) reaches the block, causing a bending of the ray in proportion to the angle of incidence. Refraction is the direct result of the difference in velocity between the edges of the ray of light.

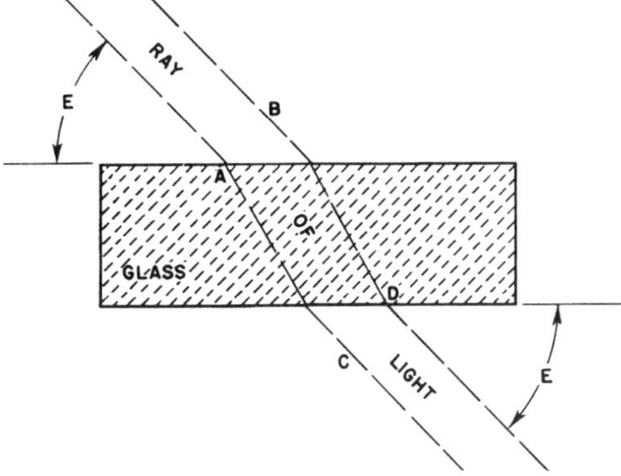

Figure 357. Refraction

When the ray of light leaves the block, the portion of the ray at (C) is free before that at (D), and so gains velocity before (D) reaches the surface. The ray of light is bent back again and assumes the direction in which it approached the glass on the opposite side.

LENSES

A lens is a transparent body having two curved surfaces, or one plane and one curved surface (Figure 358). The curved surface is usually a portion of a sphere, and the lenses are known as convex or concave according to whether the curvature is away from the medium or toward the medium.

When a beam of light passes through a double convex lens parallel to the axis of the lens (Figure 359), the rays will be brought together at a certain point (F), called the focal point of the lens. Since the center of the lens is much thicker than the edge, the beam of light is retarded more, near the center of the lens than

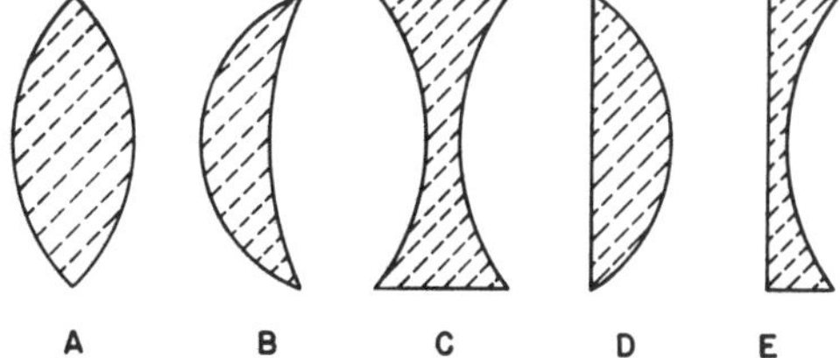

Figure 358. Lenses

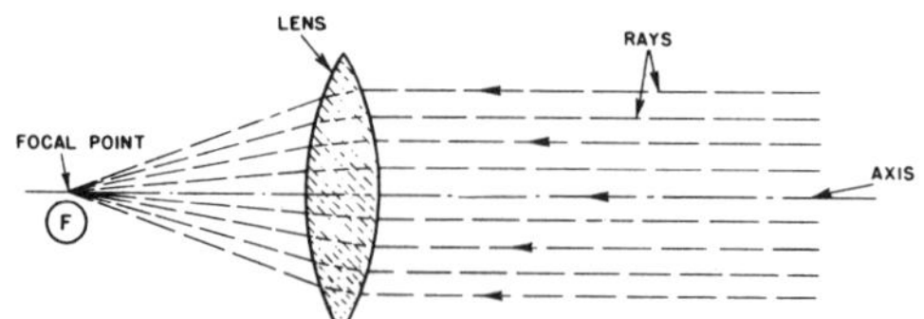

Figure 359. Focal Point

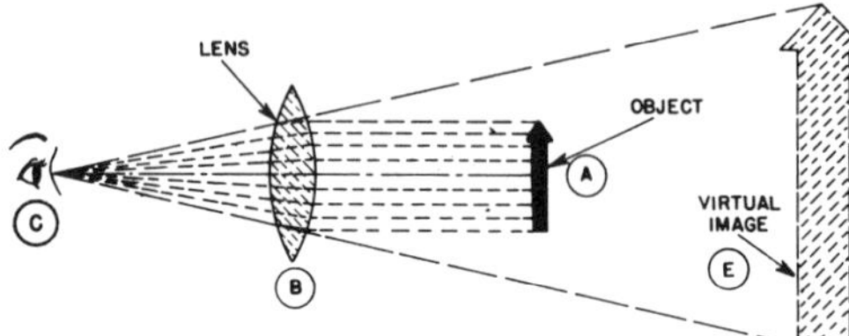

Figure 360. Magnification

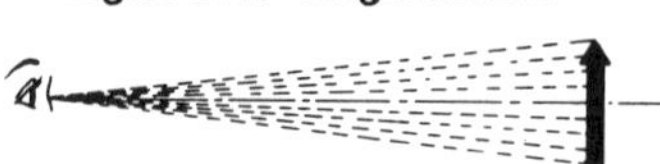

Figure 361. Normal Impression

near the edge; so both at incidence and departure, rays are bent toward axis of the lens, meeting at (F).

MAGNIFICATION

If the eye is placed at the focal point, it will have impressed upon it the image of the object from which the light rays are emanating. However, the image will appear much larger than the actual size of the object. This enlarged image is called the virtual image and is the result of the converging of the rays of light as shown in Figure 360.

The rays from object (A) are picked up by lens (B) and focused upon the eye at (C). Due to the angle at which the rays approach the eye, the impression is created that the rays are coming from an object at (E). The creation of the virtual image is a natural reaction of the eye. If the lens is removed, the eye receives rays directly from the object as shown in Figure 361, and the impression assumes its normal size.

THE MICROSCOPE

The term microscope is applied to a single lens or a combination of lenses used to magnify an object. The ordinary reading glass is a microscope in this sense. If two reading glasses are used together as shown in Figure 362, the object will be further magnified; and, because the focal point of the first glass is not passed before the image is picked up by the second glass, the image will not be inverted. Also,

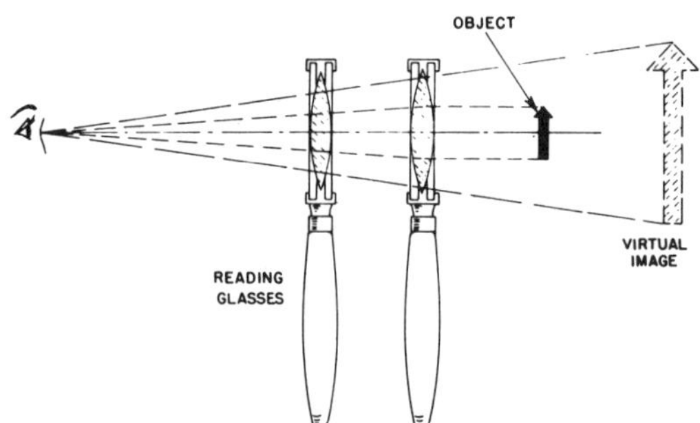

Figure 362. Image Not Inverted

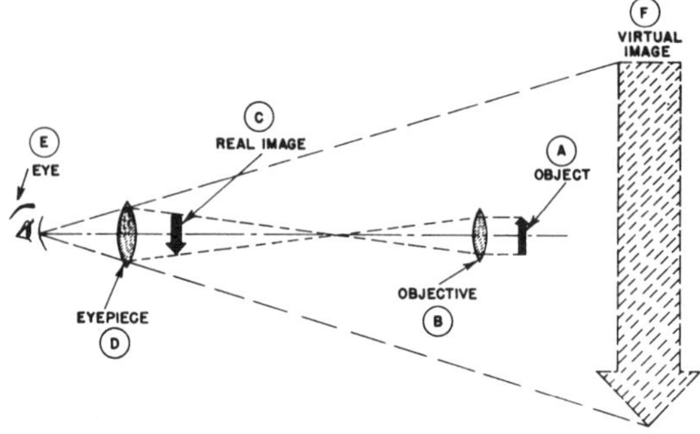

Figure 363. Principle of Compound Microscope

the magnification will not be great because of the long focal length of the glasses, which impresses the image upon the eye at a small angle.

If lenses with short focal length are substituted, the second lens will pick up the image after it has passed the focal point as shown in Figure 363. The resulting virtual image will be inverted; this is the principle of the compound microscope.

COMPOUND MICROSCOPE

Figure 363 shows how two lenses in proper relation with each other will magnify an object to many times its original size. The convex lens of short focal length (called the objective) is placed at (B). This lens forms a real image of the object at (C). Another convex lens (called the eyepiece) placed at (D) picks up the real image and focuses it so that the eye at (E) sees a virtual image at (F) magnified many times. The virtual image is inverted because the eyepiece picks up the image after it passes the focal point of the objective. The mechanical design of a simple compound microscope, and the arrangement of lenses, are shown in Figure 364.

A cylindrical tube provides a means of mounting the various elements in relation to one

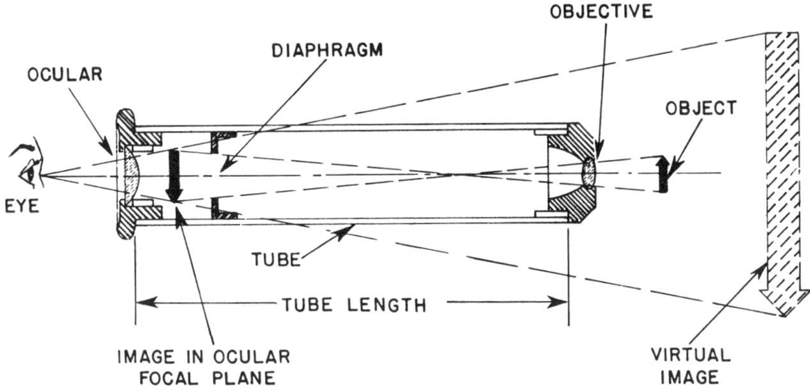

Figure 364. Compound Microscope

another; note, however, that the length of the tube affects the degree of magnification. If the tube were longer, the image projected in the ocular focal plane would be larger, and so the magnification would be greater. Originally, the length of tube was standardized at 10", the natural distance from the normal eye for best vision. This length proved cumbersome in modern designs; so the length was reduced and was standardized at 160 mm (6.3").

OBJECTIVE LENS

The objective mounted in the lower end of the tube varies in design according to the degree of magnification desired. Because the application of the modern microscope involves a variety of magnifications, several objectives of different power are mounted on a "nosepiece" so that they can be moved into position one at a time without disturbing the setup of the microscope. This feature will be observed on microscopes described later.

OCULAR OR EYEPIECE

The ocular or eyepiece is mounted in the opposite end of the tube. It picks up the image projected by the objective and focuses it upon the eye. The degree of magnification of the microscope is approximately the product of the magnification of the objective and the magnification of the ocular, and is usually indicated by the symbol X--i.e., 10X, 20X, 30X, etc.

DIAPHRAGM

The diaphragm mounted within the tube a short distance from the ocular limits the size of the field to that which can be picked up by the ocular. A transparent scale or chart placed at this position is superimposed upon the image so that the microscope becomes either a measuring instrument or a comparator.

ILLUMINATION

When an object is magnified, the light from the object is diffused and absorbed in proportion to the magnification; therefore, it is necessary in high-power magnification that the object be strongly lighted, otherwise the image will appear dim. Light is either directed upon the object from an external source, or a lamp is incorporated in the design of the microscope so that its light can be directed upon the object from below, above, or at an oblique angle.

SHOP MICROSCOPE

The shop microscope shown in Figure 365, although simple in design and application, contains all of the essential features of the compound microscope. As outlined before, these consist of objective, ocular, diaphragm, scale, and source of illumination.

A combination column and stand (A) provides means of raising and lowering tube (B) by means of clampscrew (C). The ocular (D) can be adjusted to the proper focus by simply turning to right or left. A source of light in the column is directed upon the object through the opening at (E), and is turned on or off by means of a switch controlled by a knob (F).

A micrometer scale (Figure 366) mounted

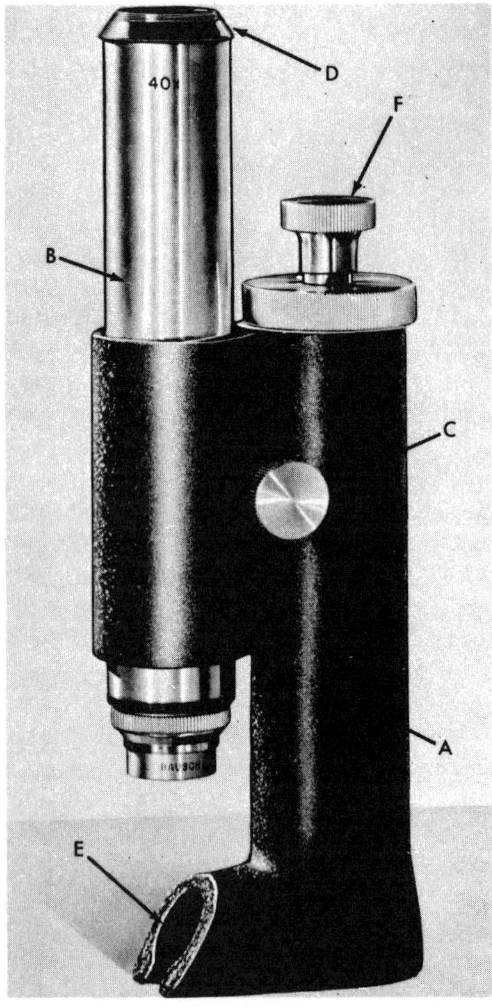

Figure 365. Shop Microscope

with the ocular remains in focus with the object after adjustment to the eye, and is used in obtaining direct measurements of any visible part of the object. The scale represents a total of 1/10" and is divided into 100 divisions; so measurements can be made to an accuracy of .001", and subdivisions of .0001" can be estimated.

The shop microscope is comparatively small, measuring 7-3/4" in height. It can be conveniently carried from place to place; and as the base is small, it is particularly adapted to inspecting partly enclosed areas. It has a field up to 7/32" diameter and magnifies 40X. The field of the microscope is the diameter of the area that can be observed through it in a single position. The greater the magnification, the smaller the field; also, the greater the working distance (distance of objective from the object), the greater the field.

High-powered laboratory microscopes have very small working distances; therefore, their field is very small. Those used in industry for inspection work are usually low power and have a large field so that areas under observation can be readily oriented by noting their relation to areas surrounding them.

APPLICATION

The shop microscope is used principally for the inspection of surfaces for quality and defects, and for measuring surface characteristics. It is used for the examination of the edges of cutting tools, for detecting flaws, cracks, and blowholes in surfaces; for measuring the diameter of small holes, width of scribed lines, and other small dimensions which must be held to an accuracy of .001" and require inspection during machining operation. Two particularly important applications of the shop microscope are:

1. Checking the thickness of the case on case-hardened parts. A cross-section of the part is obtained by cutting the part in half, after which the thickness of the case can be readily measured with the scale of the shop microscope.

2. Measuring the diameter of the impression made by the penetration of the ball when making a hardness test with the Brinell hardness tester. This instrument is described later in this text under the subject of "Hardness Testing."

CARE AND OPERATION

1. Handle with care. Understand the instrument and be familiar with its operation. See that lenses and working parts are clean. Remove dust and lint with a camel-hair brush. Wipe lenses with clean, soft cloth or lens tissue saturated with an approved cleansing solution. Dirt contains abrasive and will scratch a lens if rubbed across the surface.

2. Place the instrument on a piece of white paper, illuminate the field by turning switch (F) (Figure 365), apply the eye to the ocular (D),

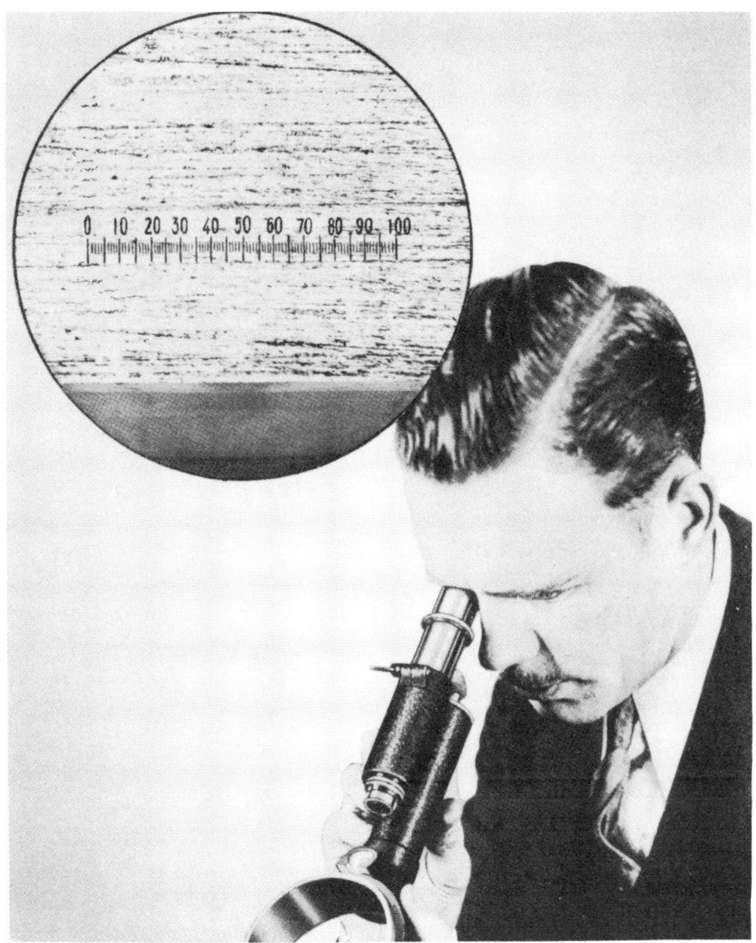

Figure 366. Inspecting Surface of a Ring with Shop Microscope

and bring the scale into focus by adjusting the ocular. Leave the ocular in this correctly-adjusted position.

3. Place the instrument on a flat surface which contains uniform markings, such as a piece of fabric or printed paper. Loosen clamp-screw (C) and adjust tube (B) until the figures on the surface are sharply in focus. When the scale and figures are seen simultaneously in focus, tighten the clampscrew, and the instrument is in adjustment and ready for use. When examining surfaces for flaws and cracks, rotate the instrument slowly about its vertical axis. In this way the illumination will cast shadows of the imperfections, and make them easier to detect. Size of imperfections, holes, or small sections can be read directly on the scale.

The design of the microscope is governed by the particular kind of work for which it will be used. Standard microscopes, however, are somewhat similar in appearance and design, differing only in special features that adapt them to a particular field of application.

The wide-field microscope shown in Figure 367 is particularly suited to industrial requirements for inspection of surfaces and examination of raw materials. It is a general-purpose instrument with a magnification sufficient to meet the requirements of production and inspection. Some of its distinctive features are:

1. WIDE FIELD. When examining surfaces for cracks, flaws, and finish, it is important that the field be large enough to orient the area being inspected so that flaws can be studied in

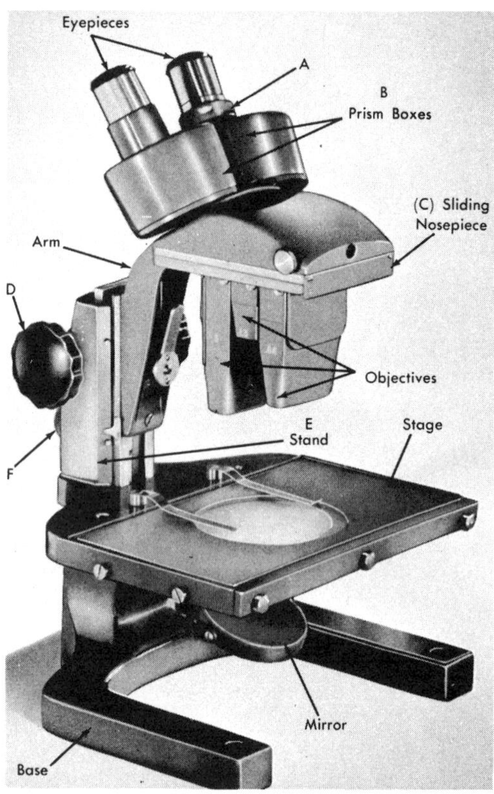

Figure 367. Wide-field Microscope

relation to the surface surrounding them, and uniformity of surface finish can be examined.

2. LONG WORKING DISTANCE. The long working distance permits a large range of adjustment and allows ample room for manipulation of large specimens, when examination involves moving the specimen about or turning it over for general inspection.

3. BINOCULAR VISION (stereoscopic effect). Each eye is provided with its own complete microscope. This feature lends itself to better vision by eliminating the strain and discomfort experienced in extended examination of a specimen using but one eye. The paired objectives are mounted together at the approximate angle of binocular vision (15 degrees) for a 10" object distance. One eyepiece is provided with a focusing collar (A) for the individual eye. When bifocular instruments are being adjusted to the eye, it is usually necessary to adjust the focus to one eye at a time because of the individual characteristics of the eyes.

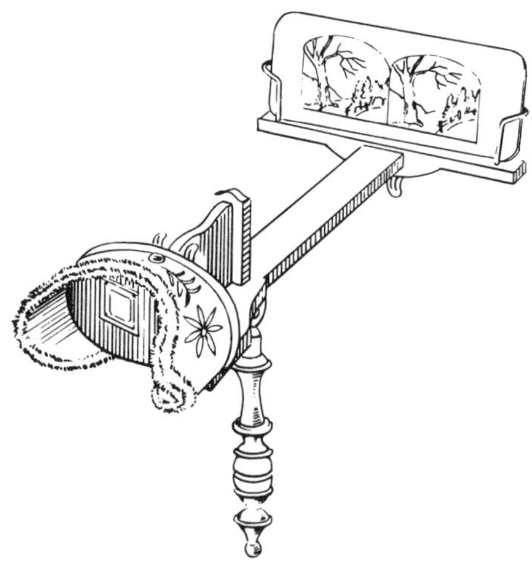

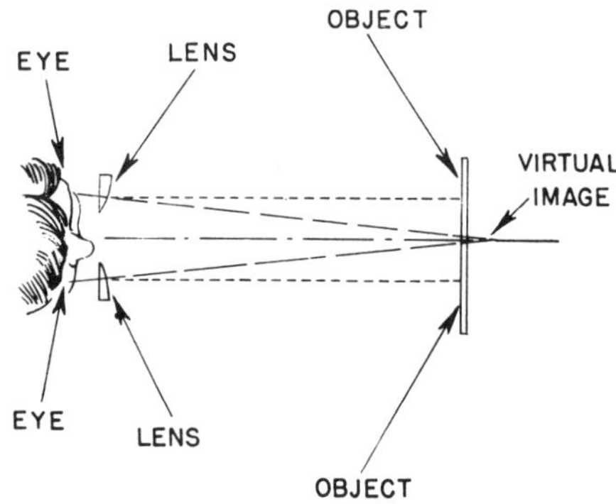

Figure 368. Stereoscope

A decided stereoscopic effect is obtained as a result of the double point of view. The stereoscope (Figure 368), popular years ago, uses two photographs at slightly different angles; these photographs are viewed through lenses so arranged that the images are combined into one virtual image; this gives the impression of depth or distance away from the observer. Stereoscopic effect is particularly desirable in some applications of the microscope such as examining fabrics, surfaces, and small objects.

4. PORRO-ERECTING PRISM. The rays from the objective pass through porro-erecting prisms similar to those employed in binocular field glasses (Figure 369). These prisms correct the image so that it is not inverted. As in binocular field glasses, the prisms reduce the length of the microscope by redirecting the rays of light in three directions in a horizontal plane. The distance, or the length of travel,

in this horizontal plane can be deducted from the straight-line distance between the objective and the eyepiece required in a tubular microscope of the same power. The prism boxes (Figure 367) are individually rotatable about the optical axis so that they can be adjusted to the interpupillary distance between the eyes.

5. SLIDING NOSEPIECE. The sliding nosepiece (C in Figure 367) holds three paired objectives to permit instant change of magnification. Positioning in the optical axis is precisely effected by check-stops. The convergence of the optical axes and accuracy of alignment required make this a distinctive feature on binocular microscopes. The optical unit is mounted on an arm adjustable vertically by means of a rack and pinion controlled by knob (D). The stand (E) is so designed that the microscope including the stage can be separated from the base by turning the knob (F) counterclockwise. The base contains a plano-concave mirror adjustable in two planes, for lighting the object from below.

MAGNIFICATION

The equipment includes two sets of paired eyepieces (10X and 15X) and three sets of paired objectives (0.66X, 1.33X and 3.0X), giving a range of magnification of 6.6 to 45 diameters. For example, if 15X eyepieces were used in combination with 1.33X objectives, total magnification would be 15 x 1.33 = 19.95X (diameters).

Most applications in the shop require illumination from above the object; therefore, the upper part of the microscope, including arm and stage, can be used separately. The stage is so designed that glass and opaque background can be removed for the inspection of large surfaces, It is similar in this respect to the shop microscope.

OPERATION

1. In using the binocular microscope, first set the eyepiece to the proper interpupillary distance to suit the eyes of the person using the instrument. This adjustment is made by sighting into the instrument and turning the prism boxes on their axes until the two fields coincide.

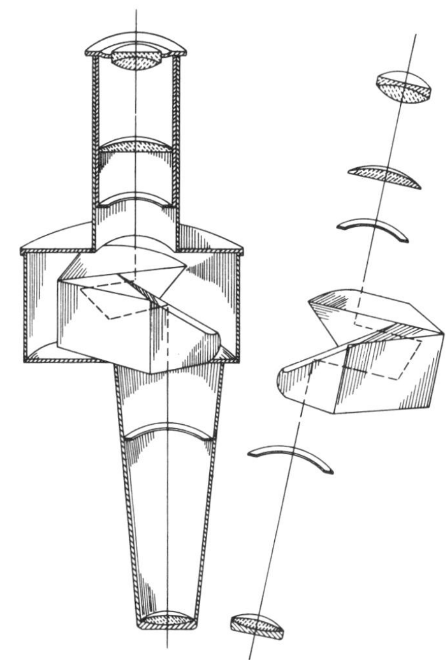

Figure 369. Poro-erecting Prisms

2. Next, select the proper magnification. As stated, there is a choice of either of two sets of eyepieces in combination with any one of three objectives. For a magnification of 30X, select the 10X eyepiece and 3.0 objective.

3. Third, focus the instrument on the object. This is done by raising or lowering the optical unit by means of the knobs (D) (Figure 367) until the object is in focus. (Care should be taken in making this adjustment that the objective does not strike the specimen; many objectives have been scarred or broken by careless operators adjusting the objective into the specimen or the stage.)

When this adjustment brings the specimen into focus, adjust the left ocular by turning it by means of the knurled section, to compensate for difference in the eyes. This adjustment may involve a slight change in the vertical adjustment of the optical unit.

TOOLMAKER'S MICROSCOPE

The toolmaker's microscope (Figure 370) was designed for use in the shop, toolroom, and mechanical laboratory. It is rugged in design, has high-power magnification, and is adaptable to measuring linear and angular dimensions,

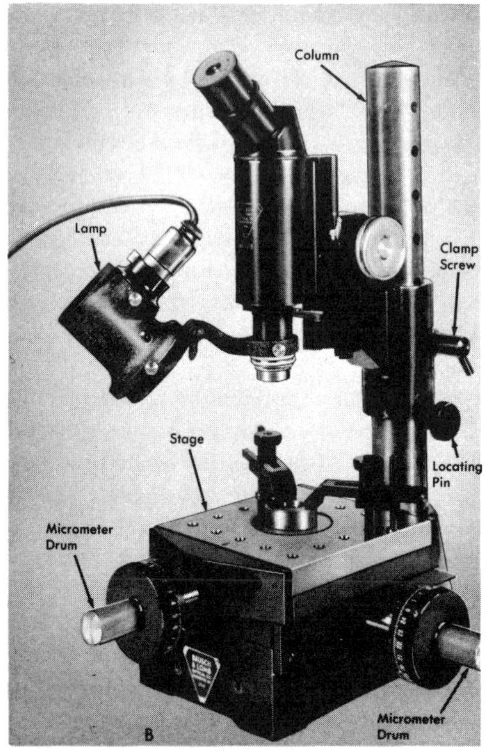

Figure 370. Toolmaker's Microscope (Bausch & Lomb)

and a variety of other applications that make it a valuable aid in inspecting small tools and parts. One of its principal applications is that of a comparator, where a thread form or small tool form is compared with a chart mounted in the eyepiece of the instrument. The outline on the chart is superimposed directly upon the virtual image of the object, thus making it possible to check the contour of the object to .0001".

OPTICAL UNIT

The optical unit (C) is the monocular type having but one eyepiece. A prism mounted in the tube redirects the projected image so that it can be viewed from an angle of 45 degrees, which is more convenient than the vertical position required by the straight-tube microscope.

The optical unit is supported by an arm (D) which can be positioned on a column (E) by means of a locating pin inserted in equally-spaced holes in the column. Further adjustment is obtained through the knob (F) that turns a pinion engaged in a rack on the optical unit.

The stage (G) is adjustable in two directions horizontally by means of micrometer drums (H). The drums are graduated so that the stage can be moved to an accuracy of .0001" in either direction. More accurate adjustment of the stage can be obtained by inserting gage blocks against stop (J). The stage is mounted on ball bearings to obtain a frictionless movement in response to the turning of the micrometer drums.

ILLUMINATION

A reflector mounting (K) picks up light from an external source and by means of mirrors and prisms directs it against the bottom of the object through an opening in the center of the stage at (L). A glass cover for this opening provides a transparent stage for small objects.

Figure 371 is a view of the bottom of the instrument showing how the light is directed from the reflector (K) to mirrors (M) and (N) which project the light through the glass stage in the center of the table. Objects are held in position on the stage by means of clamps (P) (Figure 370A) mounted in tapped holes in the stage.

Figure 371. Toolmaker's Microscope (Bottom View)

Illumination from above the object is provided by an auxiliary lamp (R) mounted on the optical unit and which can be directed at an angle upon the object as in Figure 370B. For vertical illumination, an illuminator attachment (S) (Figure 370A) is combined with the objective so that light for illumination is directed through the lenses of the objective and focused upon the object from above.

The principle of the vertical illuminator is shown in Figure 372. Light from source (A) is projected through diaphragm (B) and is reflected by glass (C) into objective (D). The lenses of the objective condense the light and focus it

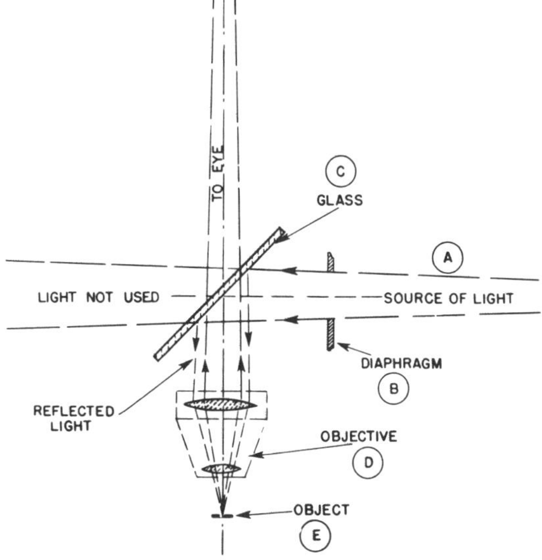

Figure 372. Principle of Vertical Illumination

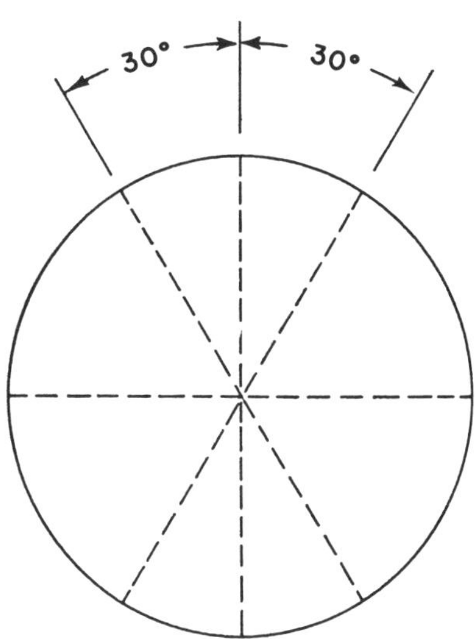

Figure 373. Crosslines in Eyepiece

upon the object at (E). These same lenses project the image of the object back through the microscope to the eyepiece.

OPERATION

The working distance of the toolmaker's microscope is approximately 6" maximum, for examination of small objects. However, the usual precaution must be taken not to adjust the objective into the specimen. After mounting the specimen in the center of the stage, rough adjustment of the optical unit is made by means of a locating pin in the column. Further adjustment is made with knob (F) (Figure 370A) until the specimen is in focus.

The eyepiece contains four crosslines as shown in Figure 373. Angles of 30, 60, 90, 120 and 150 degrees can be measured directly with the regular eyepiece.

For linear dimensions, the stage is adjusted until one side of the part to be measured is in line with a crossline. The reading of the micrometer is taken and recorded; then the stage is moved by means of the micrometer until the other side of the part is in line with the same crossline. Again, a reading is taken, and the difference between the two readings is the distance the stage has moved between readings, or the length of the part being measured.

Figure 374 shows how the crosslines appear

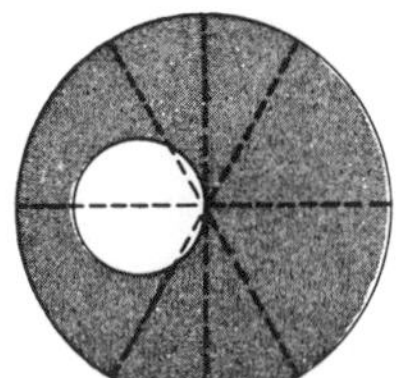

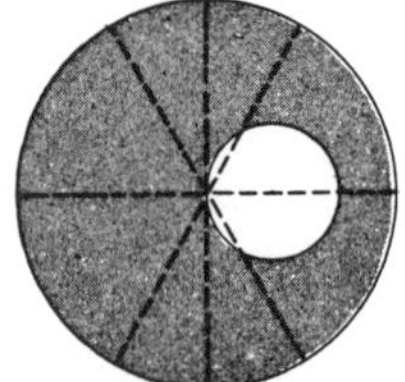

FIRST POSITION
MICROMETER READING: .4375

SECOND POSITION
MICROMETER READING: .4862

DIAMETER OF HOLE = .4862 – .4375 = .0487

Figure 374. Measuring Diameter of a Small Hole

when the diameter of a small hole is being measured.

To measure with gage blocks, release the micrometer screw to its fullest extent so that the reading is more than one inch. In this position, the stage rests against the gage-block stop. Place the object on the stage and align the image of one edge of the section to be measured with the crossline in the eyepiece. Select gage blocks to correspond to the dimension; and after moving the stage by means of the micrometer drum a distance of slightly more than the length of the gage blocks selected, insert the gage blocks between the stage and the stop. Now release the micrometer so that the stage will rest against the blocks and, if the dimension of the object is correct, the crossline will coincide with the other edge of the section being measured.

PROTRACTOR EYEPIECE

The protractor eyepiece (Figure 375) is an accessory of the toolmaker's microscope, by means of which angles can be measured to an accuracy of one minute.

Figure 375. Protractor Eyepiece

It contains two eyepieces: one views the image of the object in relation to two interrupted crosslines and a scale in the eyepiece; the other is used to read a scale engraved on the outside of the accessory.

Figure 376 is a sectional drawing of the construction of the protractor eyepiece. The eyepiece at (A) contains two crosslines--one stationary at (B) and the other at (C) movable by means of a knurled ring at (D). The angular relation of these two lines can be varied between 40 and 70 degrees on an inside scale visible in the field of the microscope (see Figure 378). This scale registers the value of the angle between the crosslines to an accuracy of 20 minutes.

On the outside of the accessory at (E) (Figure 376) is another scale upon which the angular relation between the stationary crossline and the object on the stage of the microscope can be read by means of a vernier to an accuracy of one minute.

The knurled ring (F) moves eyepiece (A), including crosslines and inside scale, in relation to the object. By means of a clamp at (G) the eyepiece can be held in fixed relation to the object so that successive objects, or parts, can be checked to an initial setting without further adjustment.

A section of the outside scale including the vernier is shown in Figure 377. One main-scale division represents 20 minutes. The vernier divides each main-scale division into 20 parts; therefore, the scale can be read to an accuracy of one minute. The reading in the figure is 66 degrees, 28 minutes.

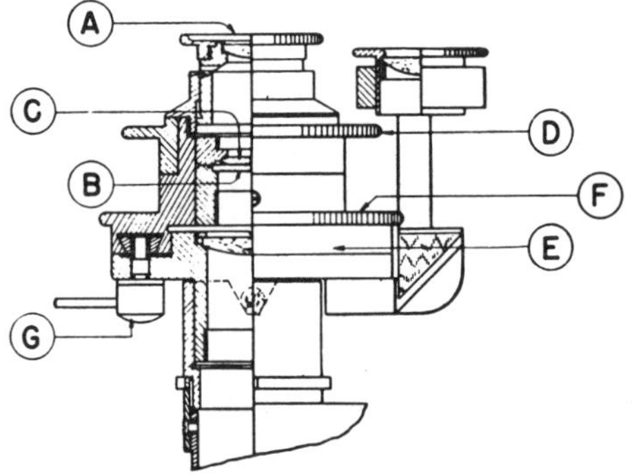

Figure 376. Section of Protractor Eyepiece

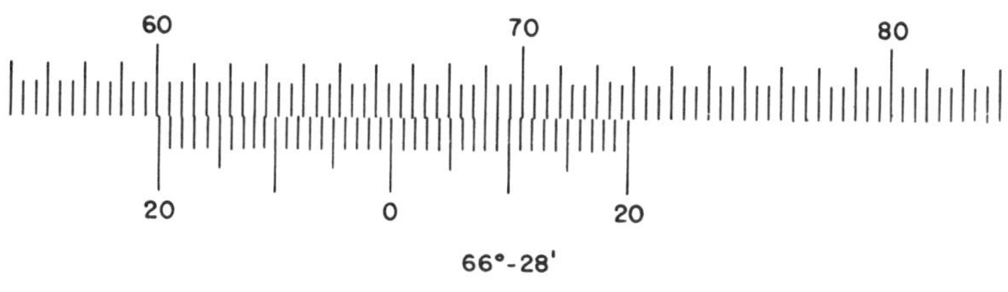

Figure 377. Outside Scale of Protractor Eyepiece

Either of the scales can be used independently of the other. They can also be used in combination with each other. For example, in checking the thread angle of a 60-degree thread, the angle between the two crosslines is set at 60 degrees by means of the knurled ring at (D). They appear in the eyepiece as shown in Figure 378A. To bring the crosslines in proper relation to the thread--the axis of which is parallel to the movement of the table--the eyepiece is set to half the angle of the thread (30 degrees) on the outside scale. Since the angle between the sides of the thread is 60 degrees, each side makes an angle of 30 degrees with a line perpendicular to the axis of the screw.

The image of the thread is then adjusted, by means of the micrometer adjustment on the stage of the microscope, until the thread form coincides as nearly as possible with the crosslines as shown in Figure 378B.

Figure 379A shows a condition in which the thread angle is in error. The actual amount of error can be measured by adjusting both crosslines until they coincide with the sides of the thread image. The angle of the thread can then be read to an accuracy of 20 minutes on the scale in the eyepiece (Figure 379B), or half the error can be calculated to an accuracy of one minute by subtracting the reading on the outside scale from 30 degrees. The outside scale always indicates half the angle between the crosslines when a line bisecting the angle is perpendicular to the movement of the stage. The correct setting of the outside scale would be 30 degrees for a 60-degree thread angle; therefore, any variation in this reading from 30 degrees will indicate an error in half the thread angle.

Figure 380A shows an error of "lean." In

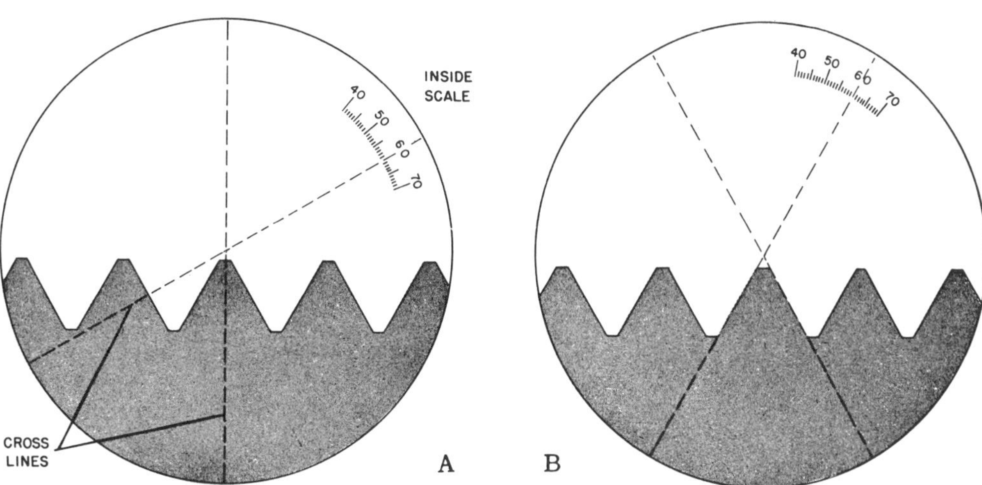

Figure 378. Measuring Thread Angle

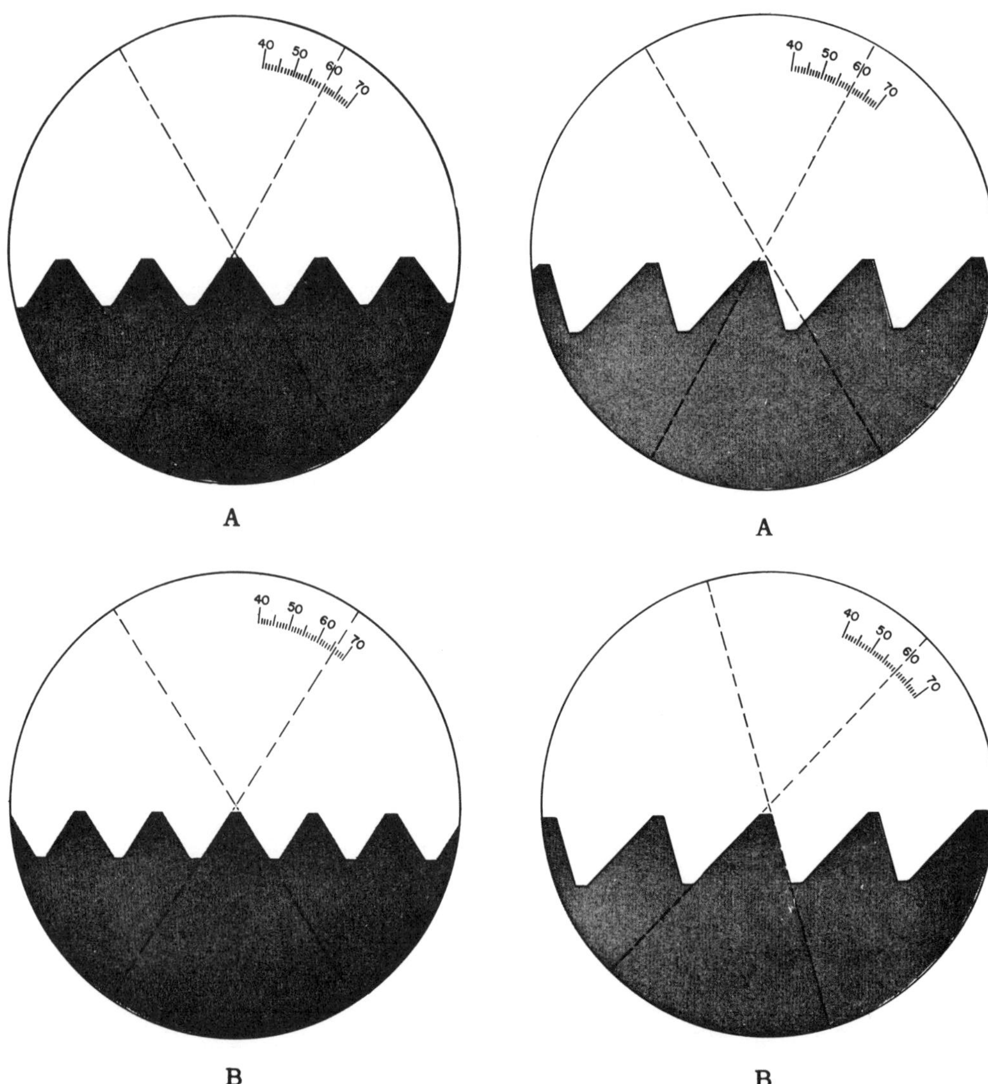

Figure 379. Measuring Error in Thread Angle

Figure 380. Measuring Error of Thread "Lean"

this case the thread angle is correct, but the thread is leaning toward the right. The amount of lean can be measured by setting the crossline at 60 degrees and adjusting the eyepiece with knurled ring (F) until the crosslines coincide with the image of the thread as in Figure 380B. Then calculate the amount of lean by subtracting the reading on the outside scale from 30 degrees.

Pitch diameter, major and minor diameters, pitch, and lead can be measured by means of the micrometer attachment on the stage of the microscope. The pitch of the thread is measured by lining up the crosslines with the outline of one thread, noting the reading on the micrometer, then adjusting the stage until the crosslines line up with the outline of the next adjacent thread. The difference between successive readings will be the pitch of the thread.

In all thread measurements it is essential that the axis of the thread be parallel to the movement of the stage. Special attachments are available for mounting parts in the correct position for making measurements. To obtain a clear image of the outline of a thread, the axis of the thread must be set at an angle to the surface of the stage; this is equal to the

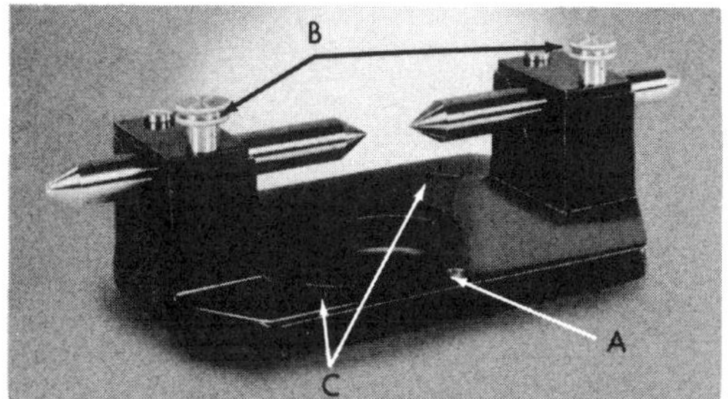

Figure 381. Lead Measuring Attachment

helix angle of the thread.

Setting the thread at an angle slightly distorts the outline as seen through the microscope. The error involved in this distortion is quite small and is sometimes neglected in the measurement of standard threads. In all accurate measurements of threads, however, a correction factor must be used to compensate for the error introduced by setting the part to the helix angle.

LEAD-MEASURING ATTACHMENT

The lead-measuring attachment (Figure 381) provides a means of mounting, on the stage of the toolmaker's microscope, parts which require male or female centers. It is used principally for the measurement of the lead of threads on screws, taps, and male thread gages.

A short flange on the bottom of the attachment fits into the circular opening in the center of the stage provided for the glass disc, and in conjunction with a stud at (A) (Figure 381) locates the attachment so that the centers are in line with the movement of the table. Two thumbscrews at (B) clamp the centers in position after they have been adjusted to the work.

The glass disc is transferred from the stage to a similar opening in the base of the attachment. Two slots at (C) provide a means of clamping the attachment to the stage of the microscope. Two studs mounted in tapped holes in the stage extend through these slots, and the attachment is held down with thumb nuts.

To measure the lead of a thread, mount the part between centers as shown in Figure 382. Focus the work by adjusting the optical unit vertically. Adjustment is made by locating the arm to the nearest hole in the column with the locating pin, clamping the arm to

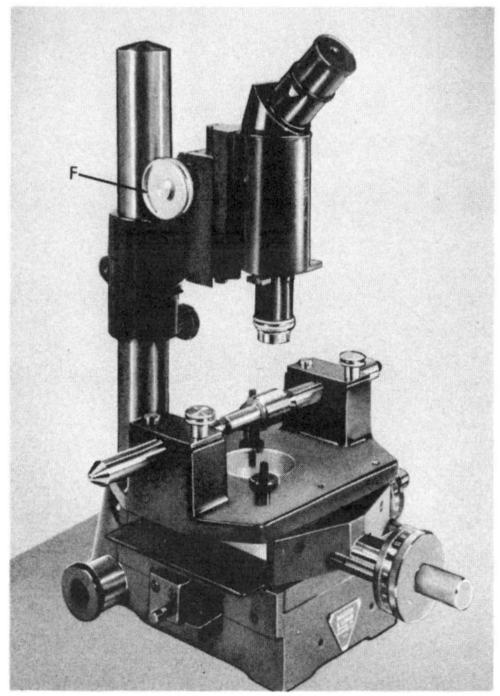

Figure 382. Measuring Lead of Thread

the column, and making further fine adjustment with knob (F).

After focusing the crosslines upon the part, adjust the stage until the crosslines coincide with the outline of the thread; then take a reading of the micrometer. Move the stage with the micrometer screw until the adjacent thread form coincides with the crosslines; take a second reading of the micrometer. The difference between the two readings will be the pitch or lead of a single thread.

When measuring a double thread, the lead will be twice the pitch (or the distance from one thread to the second thread), and the helix angle will be approximately twice that of a single thread of the same pitch. If the thread has a large helix angle, one side of the thread will be sharper than the other; therefore, when measuring the lead of a thread with a large helix angle, the microscope should be focused upon the sharp side of the thread form.

THREAD-MEASURING ATTACHMENT

To obtain a clear image of the whole outline of a thread, it is necessary to mount the

Figure 383. Thread Measuring Attachment

threaded part at an angle equal to the helix angle of the thread. The thread-measuring attachment (Figure 383) provides a convenient and accurate means of mounting threaded parts so that various elements of the threads can be inspected and measured on the toolmaker's microscope. Some of the screwthread elements measured with the aid of the thread-measuring attachment are pitch diameter, major diameter, minor diameter, thread angle, helix angle, lean, pitch, and lead. The attachment is mounted directly upon the stage of the microscope and held in alignment by two flanges on the base at (A) (Figure 383). The slight clearance allowed for convenience in mounting the attachment is taken up by two setscrews in one of the flanges.

The V-block is adjustable laterally on a dovetail slide by adjusting screw (B). The slide is pivoted at (C) so that the V-block can be set to the helix angle of the thread being measured by the angle-setting device at (D). This device consists of a block (E) which can be turned a quarter turn in either direction by knob (F). Adjustably mounted on the block (E) is a yoke (G) containing an index line which indicates the angle on a scale on block (E). The angle is set with thumbscrew (H) which moves the yoke in relation to the block.

The yoke contains a pin which rides in a slot in the V-block slide so that when the knob (F) is turned a quarter turn, it tilts the V-block to an angle corresponding to the amount the pin is off the center of the knob (F).

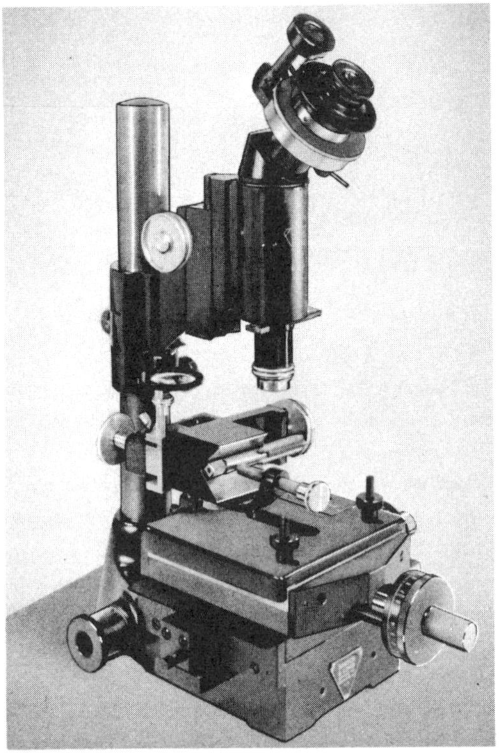

Figure 384. Measuring a Right-hand Thread

When the index line on the yoke is opposite zero of the scale, the knob and pin are concentric, and the V-block does not move when the knob is turned. When the index line is in any other position, the V-block will move to the angle indicated on the scale.

Knob (F) can be turned a quarter turn in either direction from the neutral position. When it is turned to the right, the V-block is automatically set to the helix angle of a left-hand thread, and the surface of the yoke marked (A) is under the magnifying glass. When turned to the left, the V-block is set to the same helix angle for a right-hand thread, and the surface of the yoke marked (B) is under the magnifying glass. Figure 384 shows a threaded part in position for making a measurement of a right-hand thread.

The toolmaker's microscope shown in Figure 385 consists of a microscope with a protractor head, a support column with adjustable tilt for measuring the flank angle of screw threads, and a mechanical stage permitting lengthwise, cross motion, and angular measurement. It is equipped

Figure 385. Toolmaker's Microscope (Gaertner)

with a removable cradle and other accessories for holding various types of work on the stage. Figure 386 shows this microscope with a projection attachment for projecting an image of objects in the field of view onto a ground-glass screen, similar to a projection comparator. The toolmaker's microscopes described here are only two of several excellent instruments of this type commercially available.

OPTICAL FLATS

As stated before, light is a form of radiant energy which proceeds in a straight line unless interrupted or restricted by a change of medium. Light, like a radio impulse, travels in waves. These waves have properties similar to waves in water. For example, if a stone is tossed into the still surface of a body of water, the energy of the stone is imparted to the water, resulting in a series of concentric waves moving outward from the point of impact. The force of the impact will be represented by the height of the waves. The more energy put into throwing the stone, the higher will be the waves.

If a cork is floating in the path of the waves,

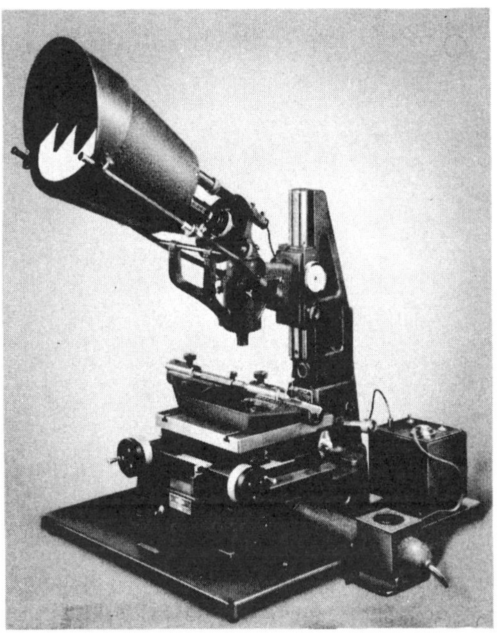

Figure 386. Toolmaker's Microscope with Projection Attachment

part of the energy of the waves will be used in raising and lowering the cork (Figure 387). If another wave-train one-half wave length behind the former wave-train is started, the waves of the second train will attempt to lower the cork at the same time the first wave-train is raising it. Thus, the waves of the first wave-train will be neutralized by those of the second, and the cork will not move up and down. This neutralizing action is known as interference.

Therefore, if two light waves are arranged so that one will interfere with the other, they will neutralize each other, and no light will be seen.

The optical flat (Figure 388) is a glass disc whose surfaces are so perfectly flat that, when placed in contact with another flat surface, the disc has the property of dividing a ray of light into two components. The component waves--one of which is reflected from the bottom surface

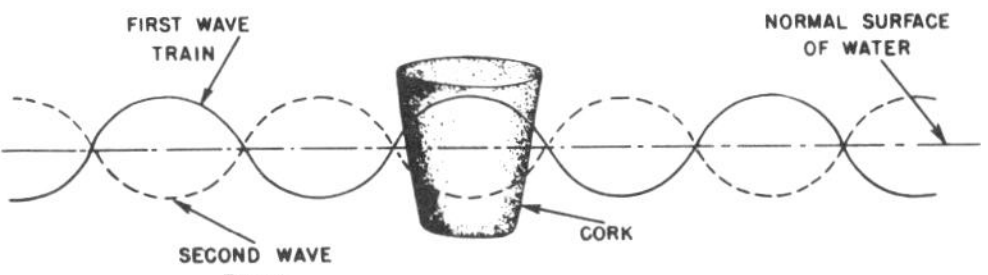

Figure 387. Interference of Waves

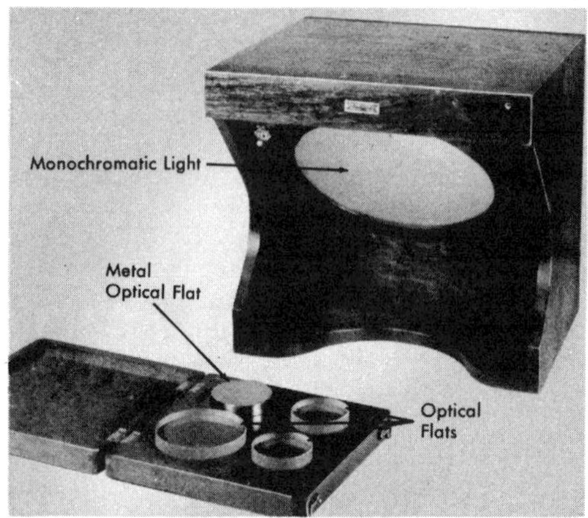

Figure 388. Optical-flat Measuring Equipment

of the optical flat, the other from the flat surface of the part--will either join each other or neutralize each other, depending upon the minute space between the optical flat and the flat surface.

The function of the optical flat is shown in Figure 389. A ray of light penetrates the optical flat. When it reaches the lower surface of the flat, it is divided into two components. One component is reflected back to the eye, and the other continues downward to the surface of the work, from which it, too, is reflected back to the eye.

As both components have the same wave length, the second component will lag behind the first an amount equal to twice the space between the optical flat and the surface of the work. If this space is equal to half of a wave length, the second component will be in step with the first when it joins it again as in (A) Figure 389. If the space is equal to three-fourths of a wave length, the second component will be half a wave length out of step with the first and will neutralize the first component when it joins it, as at (B). Again, if the space between the flat and the part is equal to a full wave length, the first and second components will be in step when they join, as at (C).

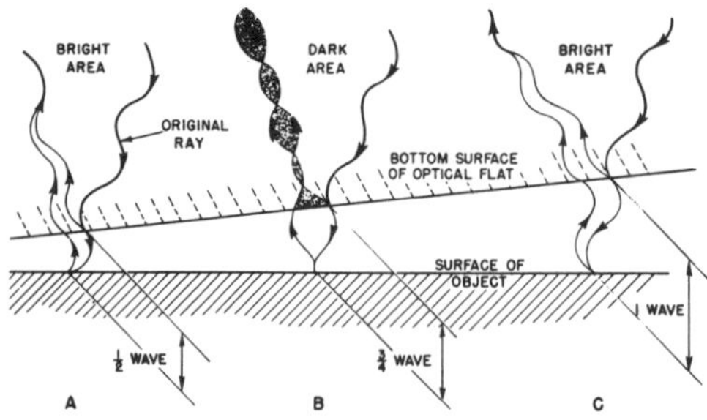

Figure 389. Theory of Optical Flats

So if the space between the optical flat and the surface of the work is wedge-shaped, there will appear to the eye a series of alternate bright and dark bands, whose number will indicate the distance in half wave lengths between the optical flat and the surface of the work at the widest part of the wedge.

THE SPECTRUM

When a rainbow appears in the sky, few stop to question the source of the various colors appearing in this presentation of the solar spectrum. Sunlight is composed of a great variety of different colors of light, each color having its individual wave length. Some of these colors are not visible to the human eye, but those visible range between violet with a wave length of .0000167", to red with a wave length of .0000275". Sunlight itself has an average wave length of .000020"; and when it passes through the optical flat, it is broken up into the various colors of the spectrum, and the bands appear in the colors of the rainbow.

Figure 390 is a diagram of the solar spectrum with the wave length of each color of light. The colors of the spectrum blend gradually one into

WAVE LENGTHS OF COLORS

	INCHES	MILLIONTHS OF AN INCH (MICROINCHES)	MILLIMETERS
ULTRA VIOLET	.0000157	15.7	.0004000
VIOLET	.0000167	16.7	.0004240
BLUE			
	.0000193	19.3	.0004912
GREEN	LIGHT EMITTED FROM MONOCHROMATIC LAMP .00002313	23.1	.0005876
	.0000226	22.6	.0005750
YELLOW	.0000236	23.6	.0006000
ORANGE			
	.0000254	25.4	.0006470
RED			
	.0000275	27.5	.0007000
INFRA RED			

Figure 390. Spectrum Chart

the other so that there is no definite limit to each band; for convenience, however, they are here separated into eight separate bands. The two extreme bands (ultra-violet and infra-red) are invisible to the eye but are very useful in scientific research.

MONOCHROMATIC LIGHT

Monochromatic light consists of rays of one definite wave length. Each of the colors of the spectrum might be considered in itself monochromatic light and as such is much more desirable for use with optical flats than ordinary sunlight.

The scale of the optical flat is the number of bands appearing upon the surface of the master or the work. One band includes both the light and dark components. Usually the dark components are counted when counting bands; if the pattern starts and ends with a dark component, however, the first dark component is ignored.

When sunlight is used for measuring with optical flats, the light is broken up into the colors of the spectrum, and the counting of the bands becomes involved with the separation of the various colors; consequently, the counting is much more difficult, and there is a greater possibility of error. When using monochromatic light, the bands appear light and dark, and are easily counted.

The separating of the monochromatic bands in sunlight for use with optical flats might prove difficult; but fortunately, there are several gases with the property of emitting light of a few wave lengths when electricity is passed through them. These gases include neon, helium, mercury vapor, and sodium vapor.

The gas best suited for optical flats is helium. Although the light emitted by helium gas contains several wave lengths of light, that portion of the light with a wave length of 23.2 millionths of an inch, is so much stronger than the others that they can hardly be observed. In Figure 390 this wave length corresponds to the yellow band in the spectrum, which accounts for the golden-yellow appearance of the helium lamp.

Measurements with optical flats may be either simple comparison measurements or measurements of surface flatness. When used for comparison measurements, the number of bands indicates the difference in height between the work and a master containing the basic dimension of the work. When used for measuring flatness, the shape and number of bands indicate the variation of the surface from a true plane, or perfectly flat surface. This variation may be in the form of high spots, low spots, ridges or grooves. The interpretation of the patterns appearing upon the surface enables the operator to determine the exact nature of the surface being measured.

COMPARISON MEASUREMENT

In making comparison measurements, the number of bands appearing in the length of the master is an indication of the thickness, at the wide end, of a wedge of air between the optical flat and the master. The difference in the height of master and part can then be calculated if the distance between the contact points on master and part is known.

In Figure 391 a comparison measurement is being made to determine just how much the diameter of a steel ball differs from the height of a gage block having the basic dimension specified for the ball. The equipment necessary for making this measurement includes a toolmaker's flat (A) (or similar flat surface), an optical flat (B), and a master (C).

To calculate the difference between the height of the ball and the height of the master, the distance between the points of contact of the optical flat with the ball and master must be known. In Figure 391 the distance between the point of contact on the ball to the farther edge of the master, which is also a contact point, is 1-7/16". The width of the master is 15/16", and the number of bands appearing in the width of the master is six. Because a complete band includes the area from one dark band to the next, the first dark band in Figure 391 is not counted because it is not a complete band.

The difference between the height of the ball and the height of the master is calculated by the formula which follows. If the ball is

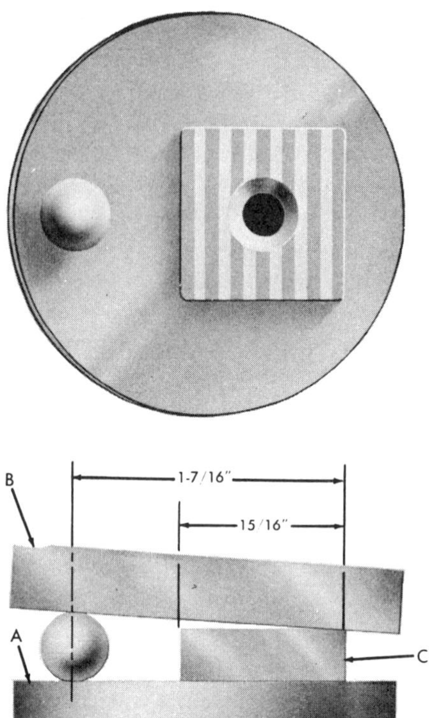

Figure 391. Measuring a Ball with Optical Flats

smaller than the height of the master, conditions will be as shown in Figure 392A; if the ball is larger than the master, conditions will

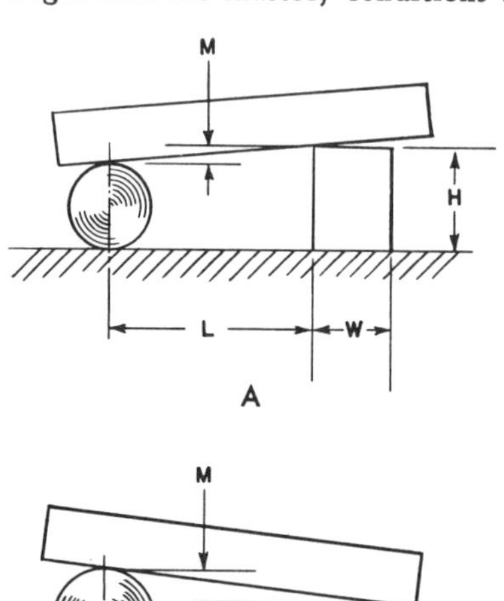

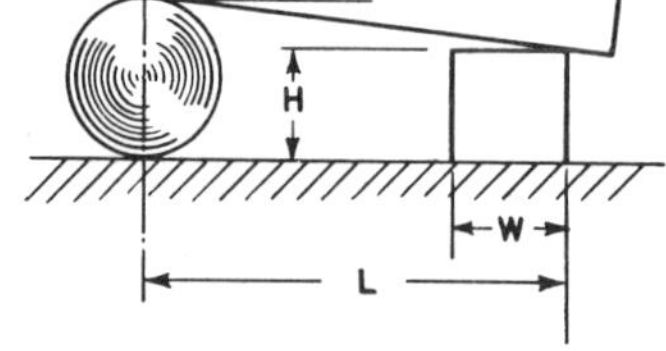

Figure 392. Using Optical Flats to Measure Diameter

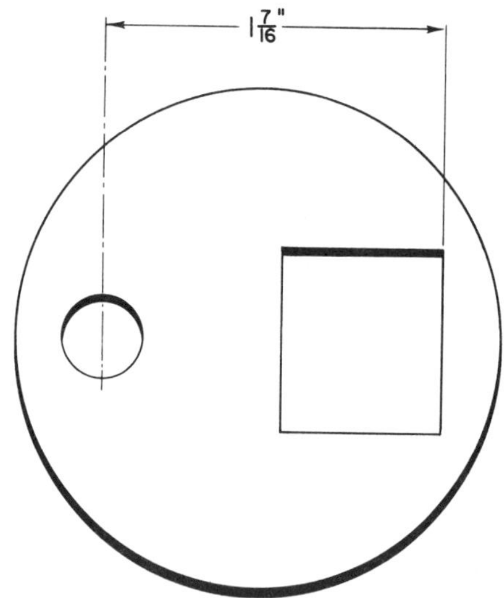

Figure 393. Fixture for Holding Master and Part

be shown in Figure 329B.

To illustrate use of the formula, the difference in the heights of ball and master shown in Figure 391 are calculated.

Refer to Figure 329B and the formula:

$$M = \frac{N \times L \times 11.6}{W}$$

M = difference in heights of ball and master

N = number of bands appearing in width of master

L = distance between contact points

W = width of master

11.6 = half wave length of monochromatic light.

Substituting:

$$M = \frac{6 \times 1\text{-}7/16 \times 11.6}{15/16}$$

$$= 6 \times \frac{23}{16} \times \frac{16}{15} \times 11.6$$

$$= \frac{6 \times 23 \times 16 \times 11.6}{16 \times 15}$$

= 106.7 millionths of an inch

The fixture shown in Figure 393 suggests a convenient method of establishing the distance between master and part.

POINT OF CONTACT

Figure 394 is a view of the bands resulting from a check on an ordinary flat surface. There are ten bands, indicating an air wedge of a

Figure 394. Interference Bands on a Flat Surface

thickness of 10 x 11.6 = 116 microinches at one edge of the optical flat. The question: which edge of the flat is making contact with the work?

A simple solution is to press the finger alternately at opposite edges of the optical flat. If the bands remain the same or increase in number, this condition shows that the contact side is being pressed. If the bands seem to spread and become fewer in number, the pressure is on the edge not touching the surface of the work.

In Figure 395 half wave lengths are represented by lines parallel to the under-surface of the optical flat. In Figure 395A, the lines intersect the surface of the work at (C, D, E, F, G). If the finger is pressed as shown in Figure 395B, the wedge will tend to become wider, and the parallel lines will intersect the surface of the work at (H, J, K, L, M, N, O). Note that the bands have increased in number and are closer together.

Conversely, if the optical flat is allowed to return to normal and the opposite edge is pressed (Figure 395C), the wedge will become narrower, and the bands will intersect the surface of the work at (P, R, S). Note that the bands have spread and decreased in number. So, the open end of the air wedge is indicated by the finger in Figure 395C.

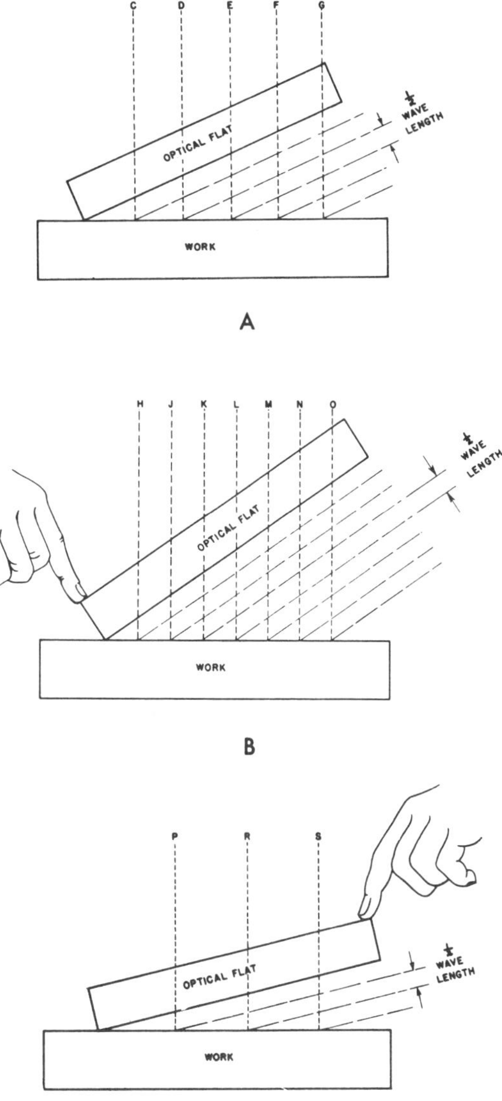

Figure 395. Determining Point of Contact

SURFACE FLATNESS

The interference bands are like contour lines on a contour map of the work surface. Straight bands indicate a flat surface; curved bands indicate a concave or convex surface. Spots with concentric bands are either high points or low points.

In Figure 396A the bands are straight, indicating that the surface is flat; however, there is a wedge of air between the optical flat and the surface of the work. The open end can be determined by pressure of the finger, as previously explained.

In Figure 396B the surface is flat, but there

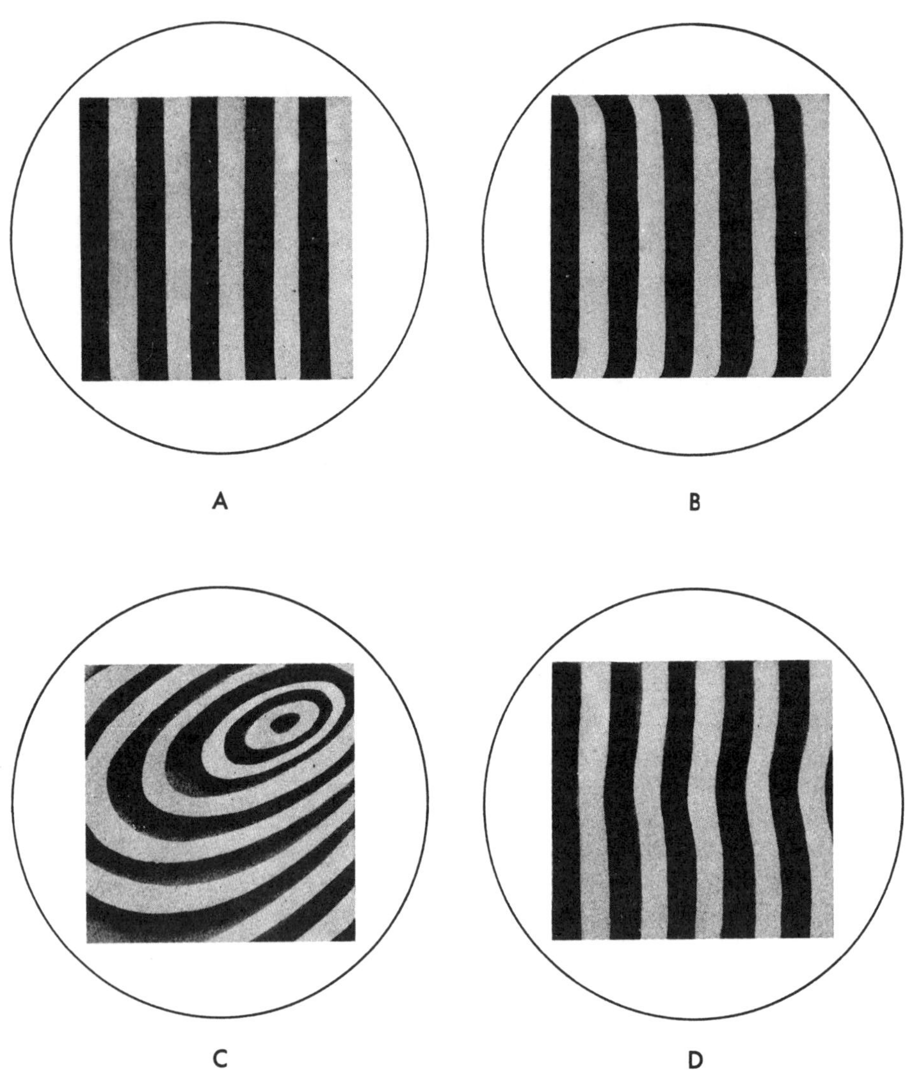

Figure 396. Patterns of Interference Bands

is a slight taper or radius along the edges.

In Figure 396C the surface contains either a high or low spot. If pressure upon the optical flat over the spot causes broadening of the bands, the surface is concave.

The pattern in Figure 396D represents either a ridge or a valley. Here again, pressure of the finger will aid in interpreting the pattern. Curvature due to a hill always points toward the open part of the air wedge; that due to a valley always points toward the closed part of the wedge. Pressure of the finger will indicate the thick side of the wedge.

CYLINDRICAL SURFACES

Figure 397 shows the wave pattern of a cylindrical surface. That the surface is convex is determined from the fact that the curvature of the pattern points toward the open part of the wedge. The spacing (A, B, C, D, etc.) indicates that these points are the same height, and the increase in number of bands toward the sides (G and H) indicates that the surface is gradually falling away toward the upper and lower edges of the optical flat.

Figures 398A and 398B show how spherical convex and concave surfaces are interpreted

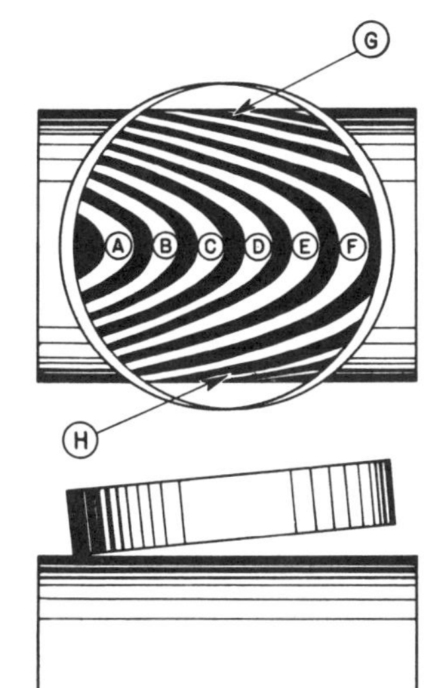

Figure 397. Interpretation of Cylindrical Surface

by finger pressure. Figure 398A represents a spherical convex surface. Note how the center of the pattern follows the point of contact between the optical flat and the spherical surface. Figure 398B represents a spherical concave surface. Here the optical flat does not rock, and pressure near the center causes it to deflect slightly, and the bands increase in width and spacing.

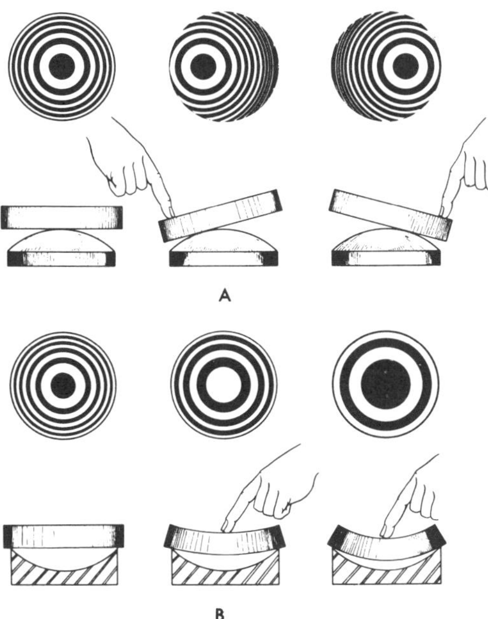

Figure 398. Interpreting Spherical Surfaces

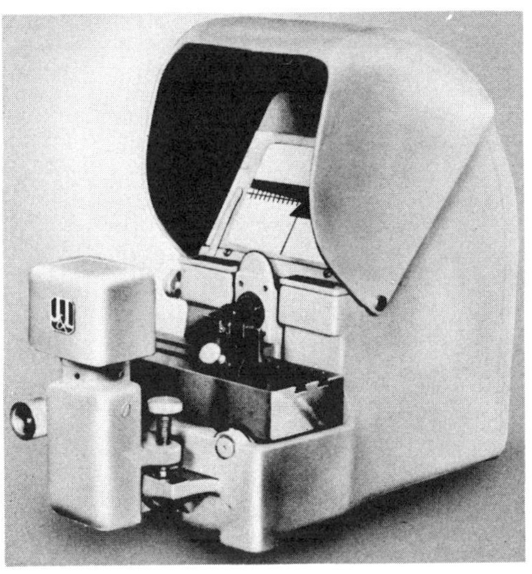

Figure 399. Bench-type Comparator

OPTICAL COMPARATORS

The optical comparator is a measuring instrument which projects an enlarged shadow of the part being measured upon a screen, there the shadow is compared with lines, on a chart; these lines correspond to the limits of the dimensions or the contour of the part being checked.

Optical comparators may be divided into two classes--bench-type and floor-type. Both classes are identical in principle and similar in construction, differing principally in that one is designed to be placed upon a bench and the other to be placed upon the floor. The floor-type is usually more elaborate, having additional features and accessories that give it a much wider application.

BENCH-TYPE COMPARATOR

Figure 399 shows a bench-type comparator, and Figure 400 is a line drawing of the comparator with main parts labeled. The principle of operation can best be understood by following through, alphabetically, the functions of the various parts.

Light from a lamp (A) (Figure 400) is condensed by a condenser (B) and projected against an object at (C). The shadow of the object is picked up by a lens system (D), and an enlarged image is projected to mirror (E) where it is reflected to the chart at (F). A yoke (G), holding

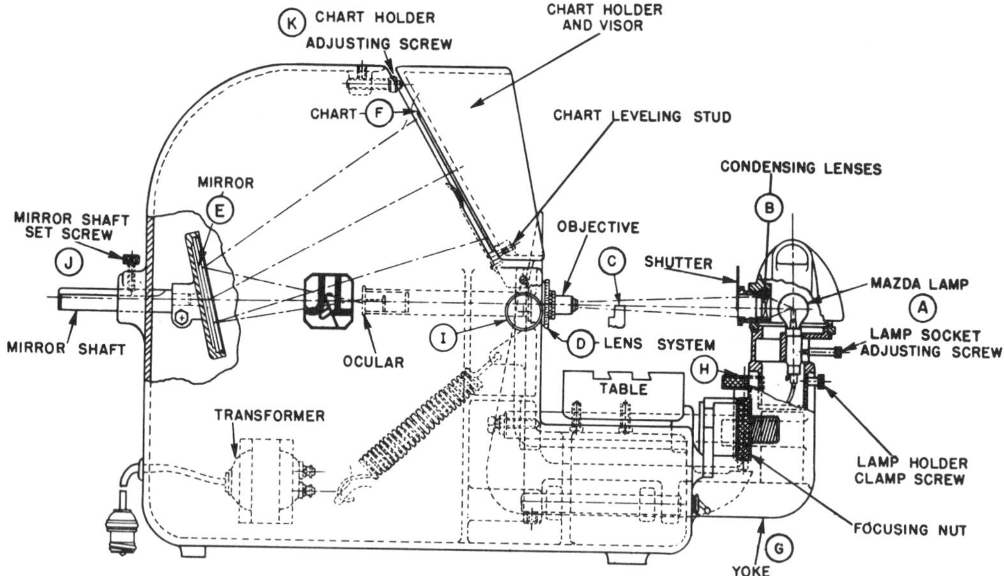

Figure 400. Schematic Diagram of Bench-type Comparator

the lamp housing and lens system in alignment, is adjustable vertically by adjusting screw (H) and laterally by adjusting screw (I). The mirror is adjustable at (J), and the chart holder is adjustable at (K).

ADJUSTING THE LIGHT BEAM

To secure proper screen illumination, the light beam should be central with the projection lens. To center the light beam, place a card or paper so that it covers half the face of the objective lens (Figure 401). Set the eccentric lamp holder (A) (Figure 400) so that the lamp-socket adjusting screw points toward the operator, and clamp in this position with the lamp-holder clampscrew. Swivel the lamp housing until the light beam is central with the objective lens laterally, and lock the lamp housing in this position with the lamp-holder clampscrew.

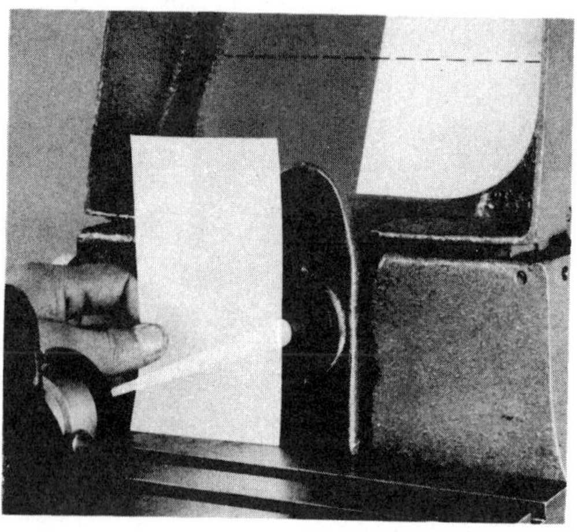

Figure 401. Adjusting Light Beam

The light beam now can be adjusted vertically with the lamp-socket adjusting screw until the light beam is central vertically with the projection lens. The light beam should be approximately the same size as the projection lens opening. To increase or decrease the diameter of the beam, adjust the condenser (B) toward or away from the objective lens.

MAGNIFICATION

The magnification of the projection comparator is the ratio of the size of the shadow on the screen to the size of the object on the table. The magnification depends primarily upon the projection lens system and secondarily upon the position of the mirror. Interchangeable lens systems are available, with magnifications of 5, 10, 20, 31.25, 50, 62.5, 90, 100 and 125X. The magnification required for a particular setup is decided in advance, and a chart containing an outline of the object or portion of the

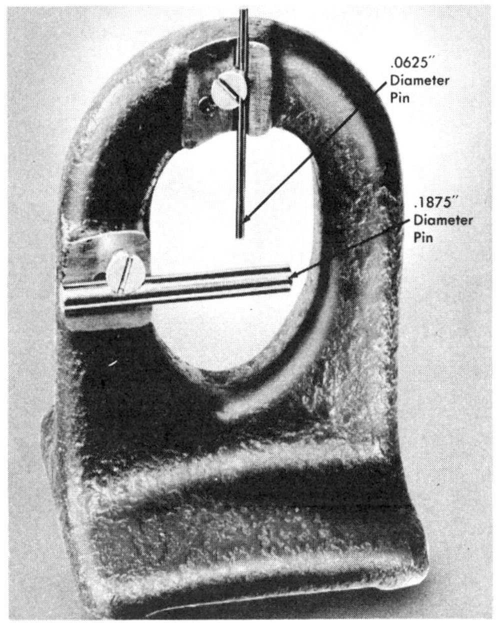

Figure 402. Test-pin Holder

object to be checked is drawn to the same scale as the selected magnification.

CHECKING MAGNIFICATION

Magnification of the comparator must be checked for every setup and every time the lens system is changed. For convenience in making this check, a small fixture called a test-pin holder (Figure 402) is supplied with each instrument. The test-pin holder has a .0625" diameter pin in the vertical position and a .1875" diameter pin in the horizontal position. It is placed on the table so that the shadow of the pin is projected upon the screen (Figure 403). The shadow is brought into sharp focus by moving the holder in line with the beam of light until the edges of the shadow appear clear and sharp.

It is general practice to measure the shadow of the test pin with an ordinary scale; however, because the usual scale is limited to 1/64" graduations, this method is not accurate enough for applications requiring smaller magnifications. For example, using a lens system and chart of 10X magnification, 1/64" error in the magnification would represent an error of .01625/10", or .0016", in the measurement.

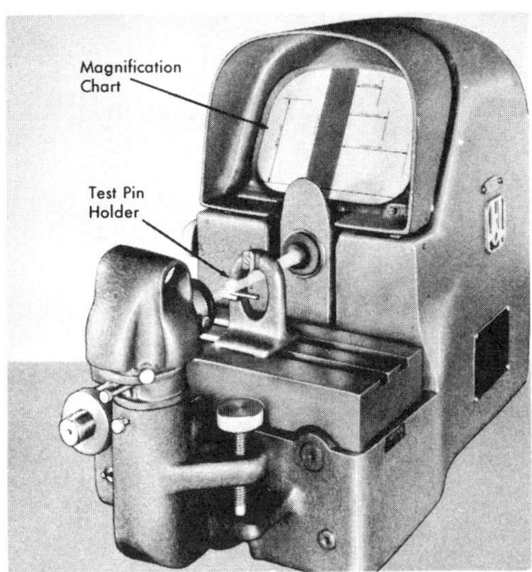

Figure 403. Checking Magnification

MAGNIFICATION CHART

The magnification chart in Figure 404 was designed to eliminate errors in setup and provide a convenient method of checking magnification. Line (A) is the reference line for the .0625" diameter test pin, and line (B) is the reference line for the .1875" diameter test pin.

When the .0625" diameter pin is being used, the left edge of the shadow is lined up with the reference line (A). For 20X magnification the shadow should extend exactly from (A) to (C). For 31.25X it should extend from (A) to (D); and for 50X from (A) to (E). In Figure 403 the comparator is set up for 20X magnification, and the shadow of the .0625" test pin reaches exactly from line (A) to line (C). Should the shadow be greater or smaller than the space between the two lines corresponding to the magnification, it can be reduced or enlarged by adjustment of the mirror at (J) (Figure 400). Should the shadow on the chart be wider at the top than at the bottom, adjust the chart holder away from the operator with adjusting screw (K). If the shadow is wider at the bottom, adjust the chart holder toward the operator.

MEASURING EXAMPLE

A typical setup on the bench comparator is shown in Figures 405A and 405B. The part (Figure 406) is a detent pawl. The latching

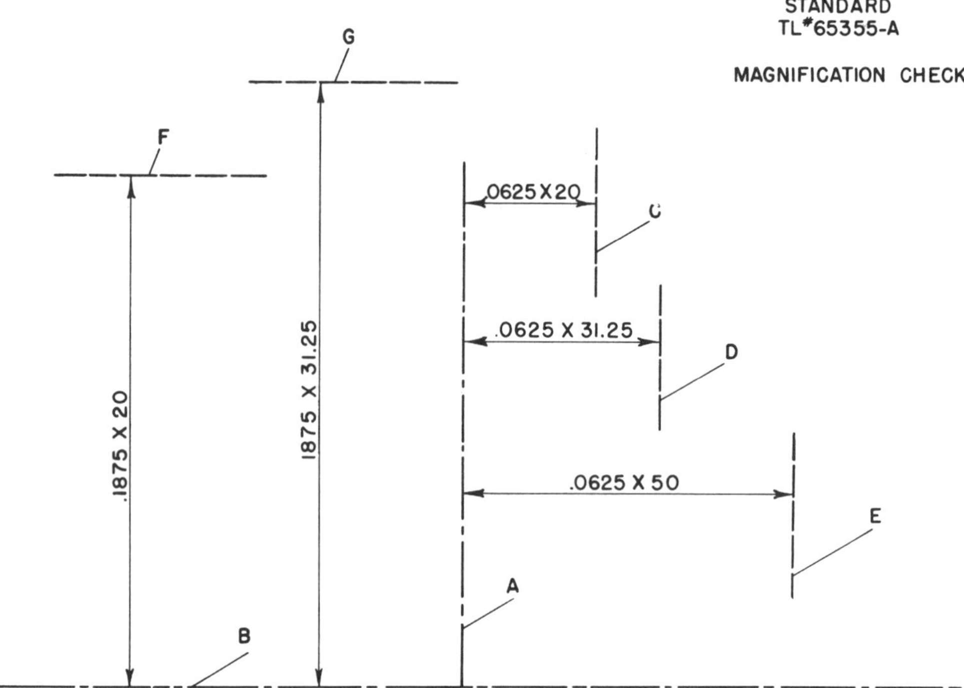

Figure 404. Magnification Chart

point at (A) contains three finished surfaces which must be held in close relation to one another and to the .377" diameter hole at (B). The chart (Figure 407) contains two setup lines, one horizontal and the other vertical, and two sets of tolerance lines that mark the limits of the surfaces being checked.

The complete fixture (Figure 408) consists of three main parts: the base, which is standard and will accommodate a variety of fixtures; the fixture itself, which provides location for the part; and the master, which has the same locating points as the part to be checked and is used for locating the fixture in correct relation

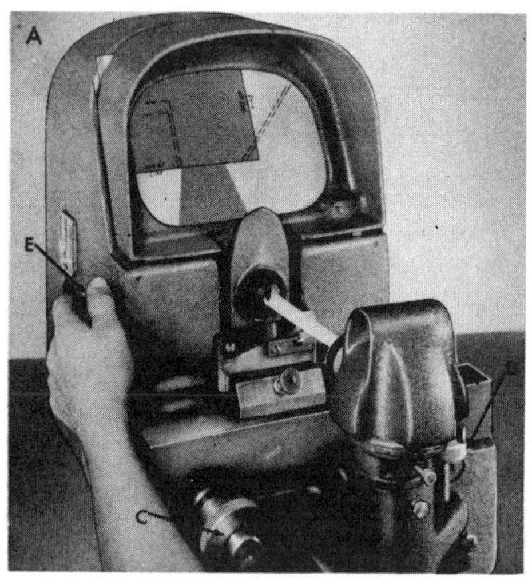

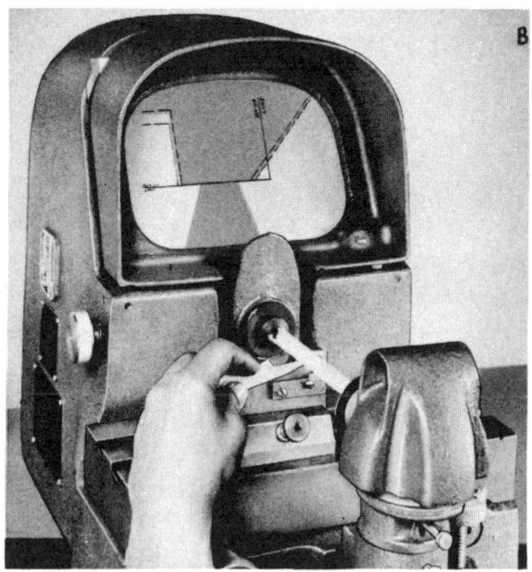

Figure 405. Checking a Part on a Bench-type Comparator

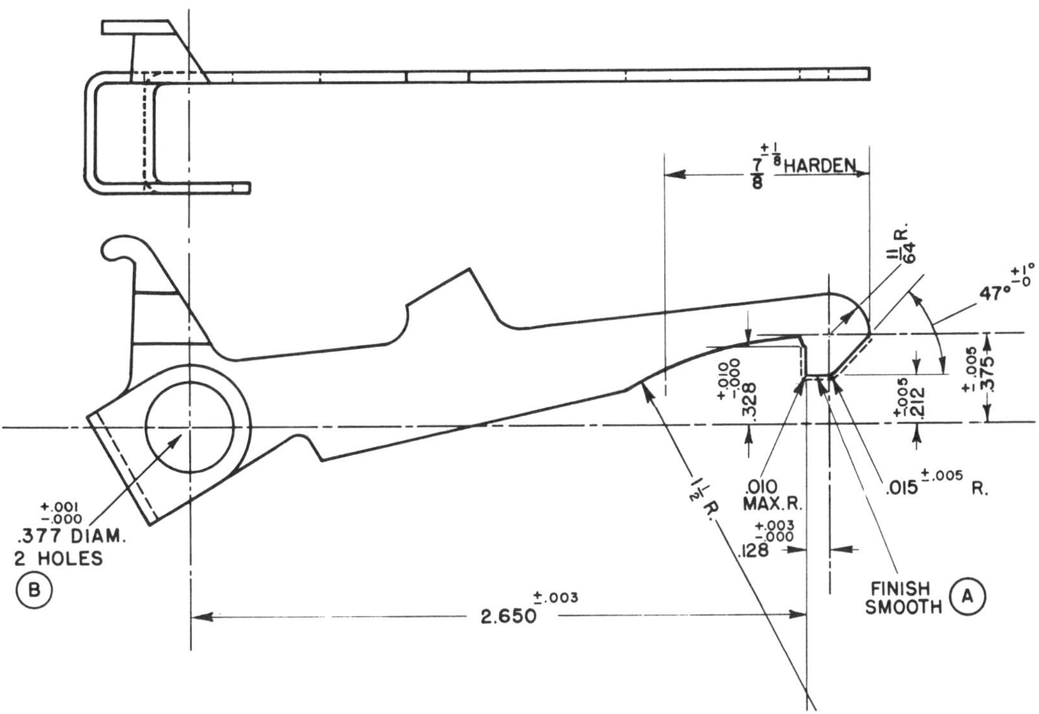

Figure 406. Print of Detent Pawl

to the chart. The base is designed to locate in a slot of the table and is clamped in place with thumbscrew (B) (Figure 408).

When magnification has been checked, the complete fixture is mounted on the table in approximately the same location as the test-pin holder. The fixture has an aperture at (A) through which the shadow of the master is projected into the lens system. This aperture must be adjusted into the light beam so that the shadow of the locating surfaces of the master coincide exactly with the two setup lines on the chart.

It is important that the projected shadow be sharp and clear before final adjustment is made. As the fixture is clamped to the table, the shadow is focused by adjusting the table toward or away from the lens system by adjusting screw (C) (Figure 405A). After the shadow has been focused, it is adjusted so that it lines up with the two setup lines as shown. This is done by adjusting the yoke with the vertical and horizontal adjusting screws (D and E).

The master is removed, and successive parts--locating on the same locators--are po-

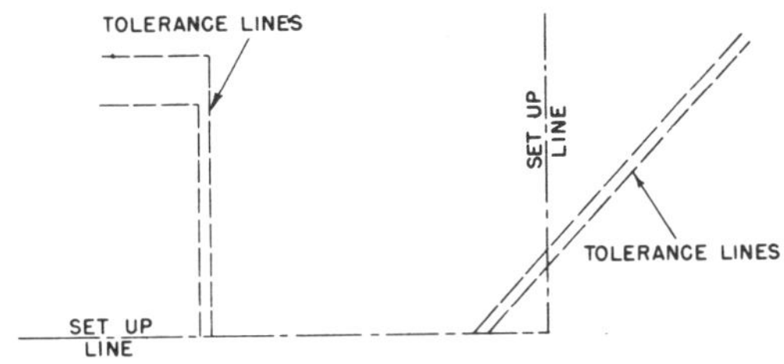

Figure 407. Comparator Chart for Detent Pawl

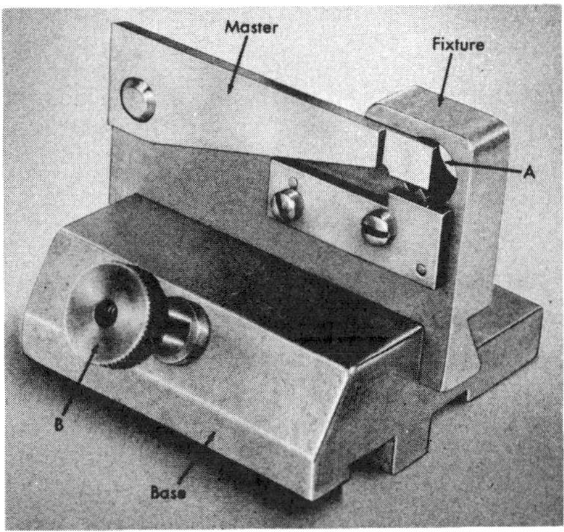

Figure 408. Fixture for Detent Pawl

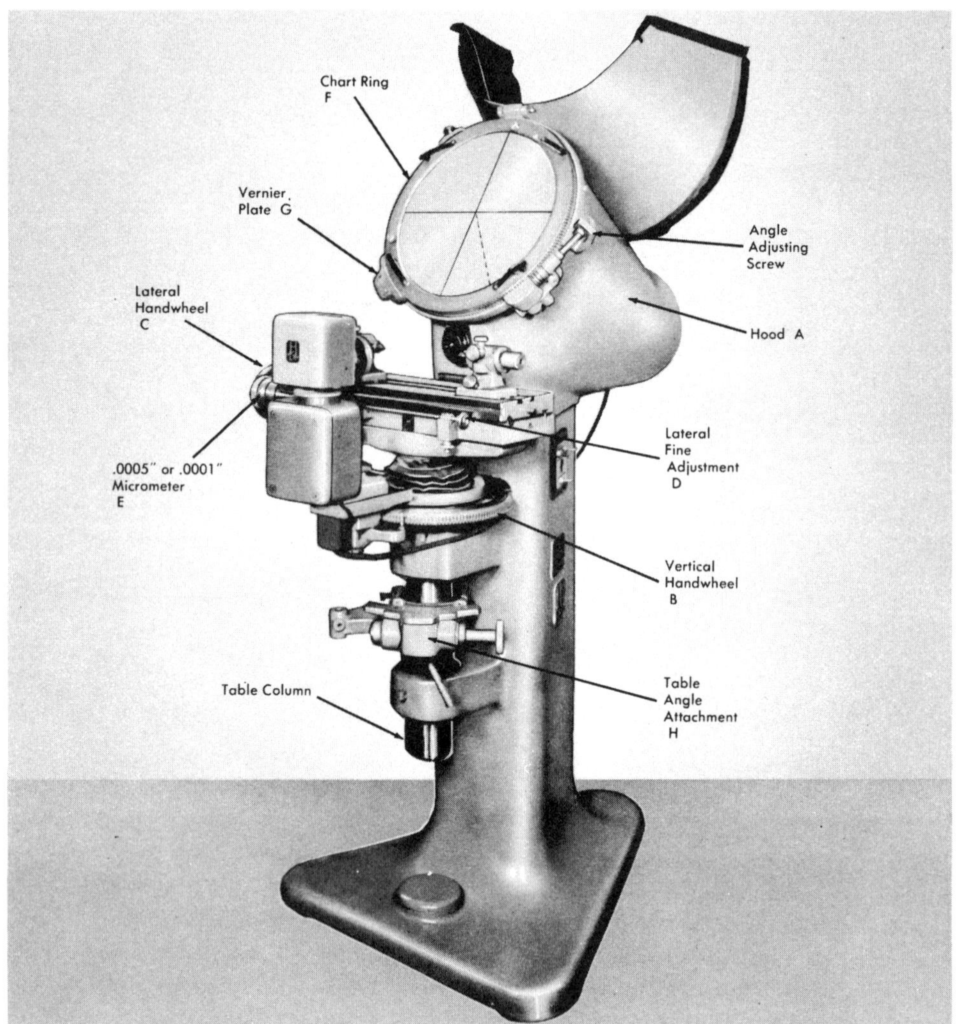

Figure 409. Pedestal-type Comparator

sitioned on the fixture (Figure 405B). Edges of the projected shadow of the latching point must fall between broken lines indicating the limits of the dimensions which locate the finished surfaces; otherwise the part must be rejected. In Figure 405B the projected shadow of each of these surfaces falls within the limits indicated; so the part is passed by the inspector.

PEDESTAL COMPARATOR

The pedestal comparator (Figure 409), designed to rest on the floor, consists of a pedestal upon which provision has been made to support the table and the lamp house, and a hood (A) which contains the lens system, mirror, and chart.

The table is mounted on a round-column that slides vertically in two bearings supported by lugs on the pedestal. It is adjustable vertically by means of handwheel (B), right and left by means of handwheel (C), and toward or away from the lens system by a focusing nut, as described for the bench-type comparator. The table is supported and guided on balls; this permits rapid and easy traverse. A counterweight holds the table against the micrometer anvil, assuring uniform pressure between measuring points. Fine vertical adjustment of the table is made with an adjusting screw which can be disengaged when rough vertical adjustment is being made with the handwheel. Fine lateral adjustment is made with the lateral anvil-adjusting screw (D).

Although the pedestal comparator is a projection comparator upon which the shadow of a part can be compared with an outline on the chart, it is primarily a measuring machine. The movement of the table vertically and laterally can be measured to an accuracy of .0001". By rotating the chart ring (F), angular measurements can be made to an accuracy of one minute.

Vertical movement of the table is measured by means of a scale on the handwheel (B). This scale is graduated in .0001" and is read to this accuracy without the aid of a vernier. Lateral movement of the table is measured with a micrometer head at (E). The micrometer head is graduated in either .0005" or .0001" and is read directly to this accuracy.

Angular measurements are made by means of an angular scale on the chart ring and a vernier (G) graduated to read to an accuracy of one minute. The table can be set at an angle, horizontally, by means of the table-angle attachment (H). This feature is used when projecting the outline of a part containing surfaces at an angle to the normal axis of the part. When projecting the outline of screw threads upon a chart to obtain a clear outline of the thread, for example, the table must be set at an angle equal to the helix angle of the thread.

MAGNIFICATION

As with the bench comparator, interchangeable lens systems are available for different magnifications. Whenever the lens system is changed, the mirror must be adjusted to the change in magnification.

The adjustment of the mirror is made with an attachment at (C) (Figure 410), containing an individual adjustable stop for each magnification. An extension of the shaft supporting the mirror is located against the stop corresponding to the magnification of the lens system in place.

GAGE-BLOCK TROUGH

A self-aligning V-trough is provided (Figure 411A) for inserting gage blocks or measuring bars between the micrometer measuring points. An adjustable anvil block attached to the table supports a hardened anvil directly in line with the micrometer spindle.

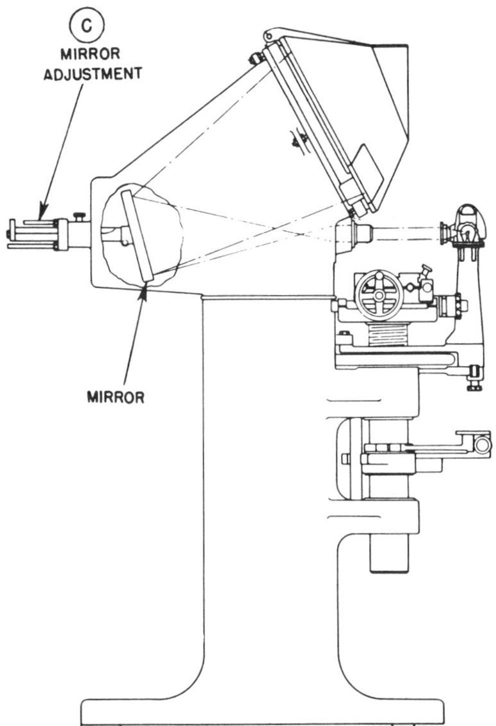

Figure 410. Schematic Diagram of Pedestal-type Comparator

When gage blocks or measuring bars are being used, the micrometer is set to zero. The object is then staged on the table so that the shadow is in approximate position on the chart. The anvil block is adjusted until the anvil makes contact with the micrometer spindle and is clamped in this position. Final adjustment of the shadow is made with adjusting screw (D) (Figure 409). From this initial position dimensions to other portions of the part can be measured by inserting measuring bars or gage blocks corresponding to the various dimensions.

MEASURING EXAMPLE

An example of length measurement using gage blocks is shown in Figures 411A and B. The problem is to measure the spacing, width, and depth of slots in a roll (Figure 412). These

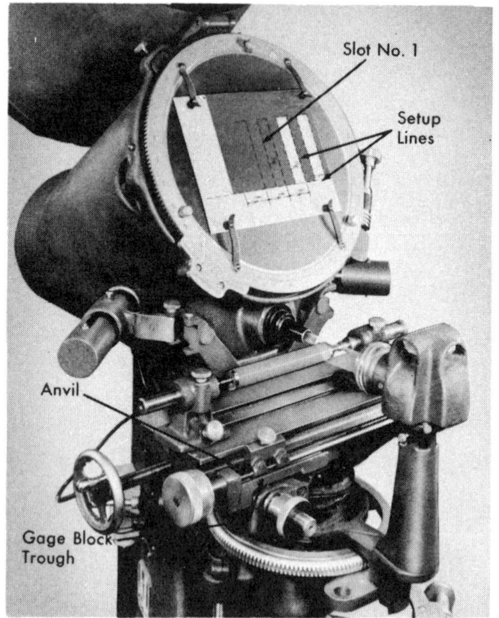

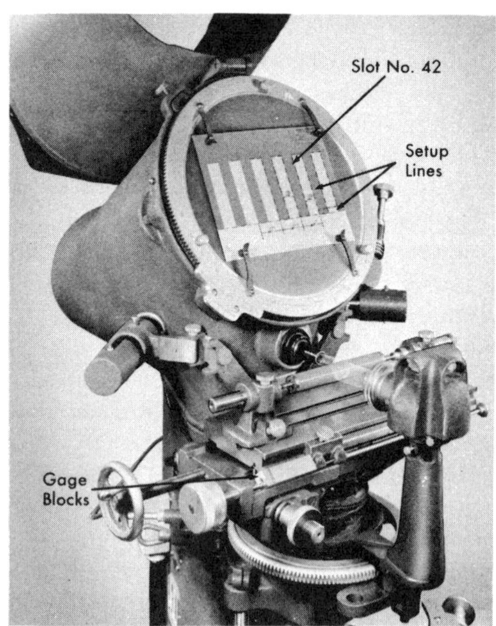

Figure 411. Checking Slots in Guide Roll

slots are milled in two operations: there are 80 slots, 40 of which are milled in the first operation, and 40 in the last operation. The spacing between adjacent slots is checked by comparing the shadow with the outline on the chart; but to check the relation of the slots of the first operation to those of the second, it is necessary to measure the distance between two slots 41 spaces apart.

Figure 411A shows the shadow of the first slot lined up with the setup lines on the chart. In this position the micrometer is set to zero, and the anvil is resting against the spindle. The distance between slot 1 and slot 42 (41 spaces) is 3.567". The table is moved to the right far enough to permit insertion of 3.567" in gage blocks between the spindle and anvil of the micrometer. With the blocks in place, the shadow of slot 42 should line up with setup lines (Figure 41 1B). Tolerance lines indicate whether this dimension checks within limits specified on the drawing.

REFLECTION ATTACHMENT

The reflection attachment (Figure 413) makes it possible to measure and inspect surfaces such as type faces, punches and stamps, which cannot be projected by the shadow method, Light from two lamps (A and B) is directed upon the face of the object (C) located at the focal point in front of the objective lens. The face of the object is thus brightly illuminated and

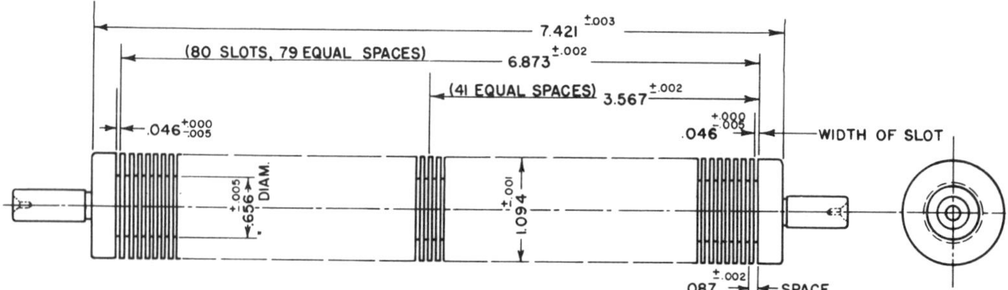

Figure 412. Print of Guide Roll

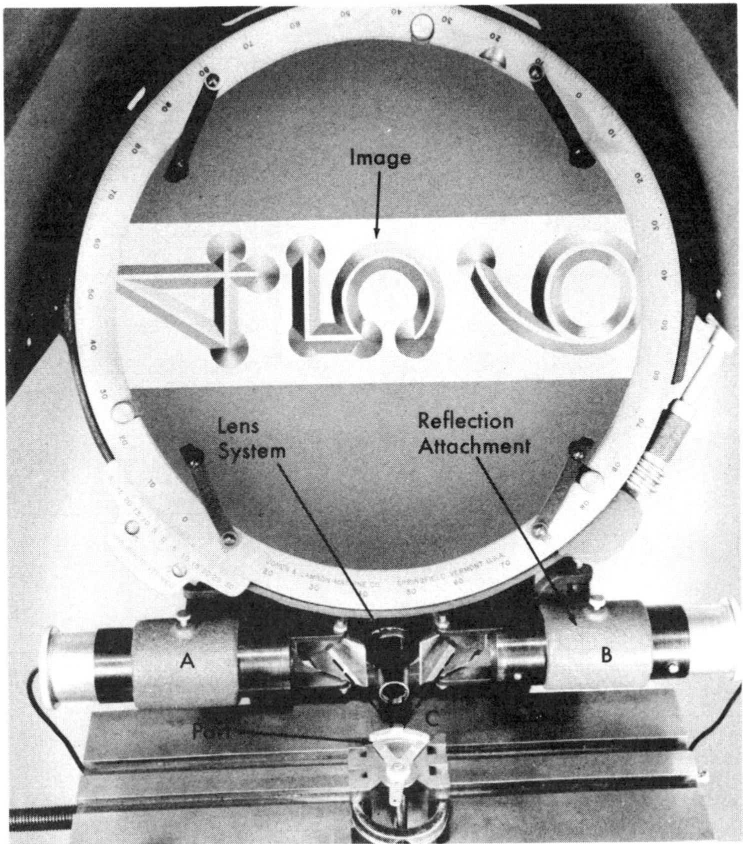

Figure 413. Reflection Attachment

the reflection is projected upon the chart by the lens system and mirror.

CONTOUR-MEASURING PROJECTOR

The contour-measuring projector (Figure 414) is a floor-type projection comparator similar in principle and application to the projection comparators previously described. It is 72" high, and occupies a floor space 28" by 62".

OPTICAL SYSTEM

The optical system (Figure 415) consists of a light source (1), which passes through a condenser at (2), through two water cells at (3), and then is projected upward by a mirror at (4). The water cells, which are removable, absorb the heat of the light beam and are used when measuring delicate parts which might be affected by increase in temperature.

The light beam passes through a glass plate in the table at (5) and projects the shadow of the object (6) into the lens system (7). The image is then reflected horizontally by a roof prism (8) to the two reflectors at (9) which reflect the image back upon the screen at (10). The image appearing on the screen is not reversed. It appears in the same relation to the eye as the object when viewed from the position of the operator.

CONSTRUCTION

The contour-measuring projector (Figure 414) is a single unit. The cabinet consists of three sections: the hood (A), the lower front compartment (B), and the lower rear compartment (C).

The hood encloses part of the projection system and contains at the front end a frame in which a variety of interchangeable screens can be mounted. The lower front compartment

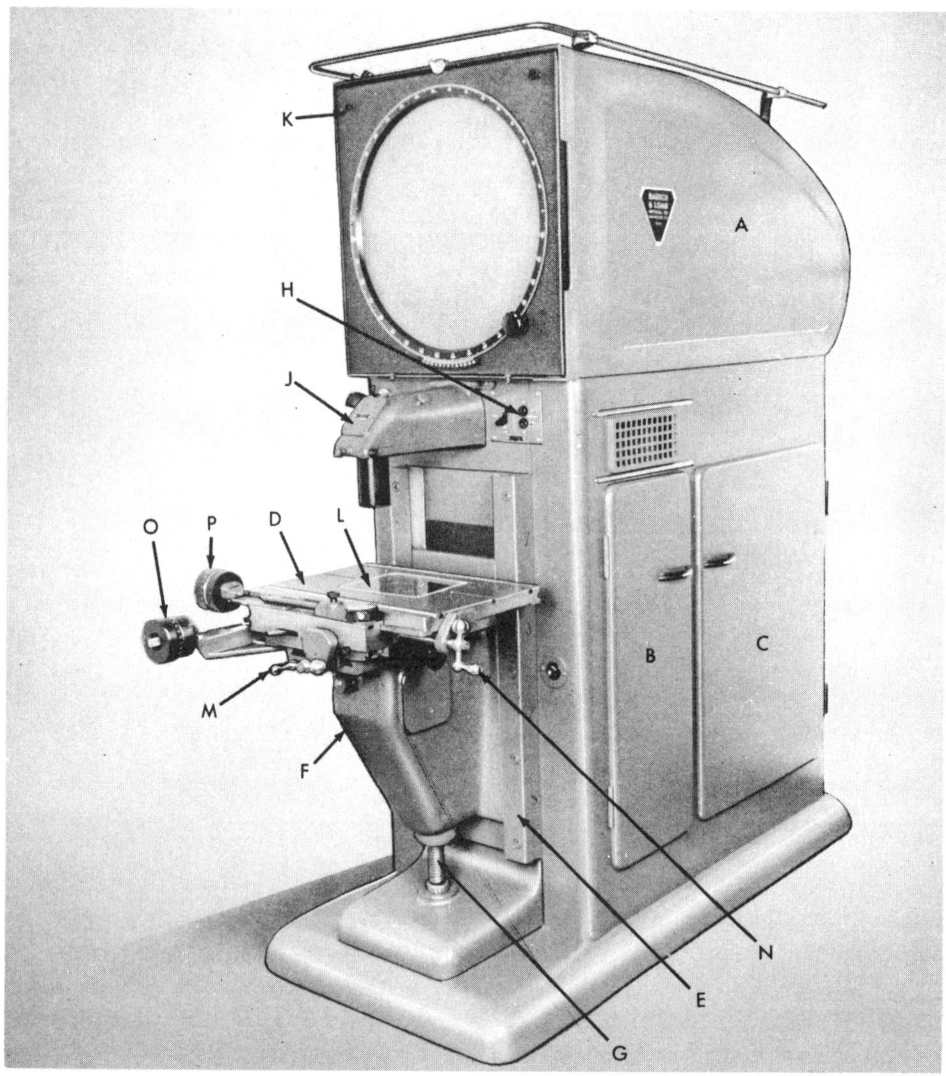

Figure 414. Contour Measuring Projector

carries the table (D) with the focusing mechanism which raises and lowers the table automatically.

The front wall of this compartment consists of a heavy casting having dovetail guides (E). Accurately fitted to these guides is a hollow knee-formed casting (F) which supports the table and is actuated vertically by an elevating screw (G). Power is supplied by a motor inside the compartment, and the movement of the table is controlled by two push buttons at (H).

Directly above the table at (J) is the projection lens bracket in which the lens system is mounted. To change the lens system, raise the cover, lift out one lens system and replace it with the other (Figure 416). Lens systems are available from 10 to 100X. The lower rear compartment is partitioned off to provide convenient storage space for interchangeable screens, lens systems, and other accessories.

WORK TABLE

The work table (D) (Figure 414) contains a glass plate (L) through which the vertical light beam is projected. This arrangement has the following advantages:

1. Many objects can be placed directly

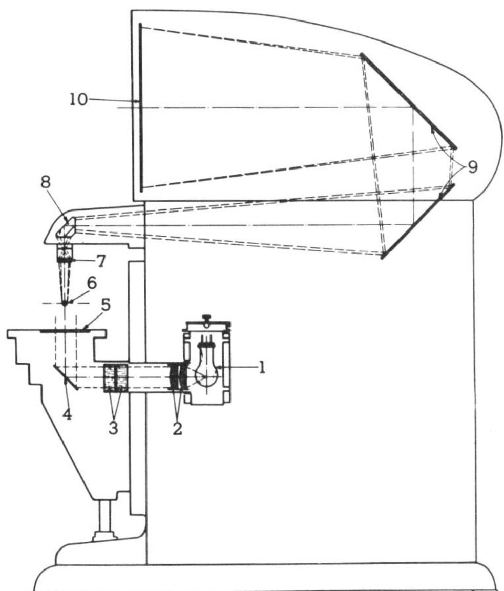

Figure 415. Schematic Diagram of Contour Measuring Projector

upon the glass without fixtures or clamps, thus eliminating special tools for holding the object.

2. The time required for setup is reduced.

3. The object is automatically placed at a right angle to the beam of light.

4. While in focus, the object can be moved about by hand to line up the image with setup lines or the outline on the chart.

MICROMETER MEASUREMENT

The table is adjustable horizontally by feed handles (M) and (N) (Figure 414) and can be clamped in position in either direction. Two measuring units--one for front and back measurements at (O) and the other for right and left measurements at (P)--permit the measurement of the table movement with the micrometers or gage blocks.

The measuring unit is composed of a gage-block tray (A) (Figure 417) at one end of which is mounted a micrometer drum (B); the other end contains a dial-gage assembly (C) for determining the zero setting. The range of the micrometer is one inch. Any linear measurement within this range can be made with the micrometer without the use of gage blocks. To make a right and left measurement, proceed as follows:

1. Line up the left edge of the image with a vertical line on the screen.

2. Set the micrometer drum at zero.

3. Bring the dial-gage assembly (C) up to the micrometer spindle, and adjust it so that the pointer makes approximately one-half turn; clamp it in this position with thumbscrew (D).

4. Rotate the face of the dial gage until it reads zero. Recheck the setting by rotating the micrometer drum, to assure that the micrometer and dial gage read zero at the same time.

Figure 416. Changing Lens

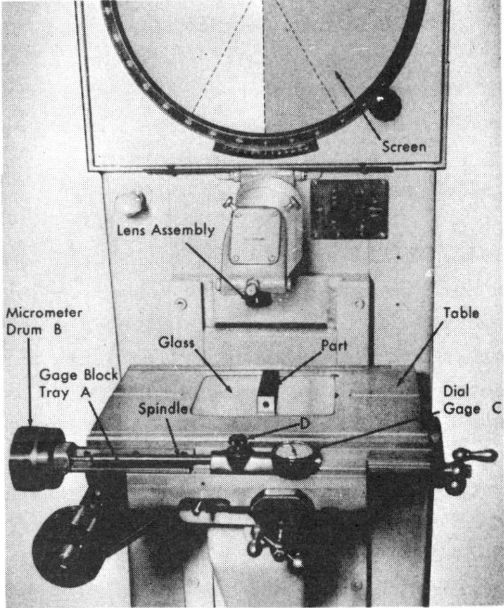

Figure 417. Micrometer Measuring Attachment

Figure 418. Center Supports

5. Feed the table to the left until the right edge of the image lines up with the same vertical line on the screen. Because the table carries the micrometer, and the dial-gage assembly is stationary, the micrometer spindle will move away from the dial-gage spindle.

6. Rotate the micrometer drum until the spindle again makes contact with the spindle of the dial gage and actuates the pointer to zero. Care must be taken that the pointer does not make a complete revolution, thus reaching zero twice. The reading of the micrometer is the dimension of the object. If the dimension being measured is greater than the range of the micrometer, gage blocks equivalent to the difference are placed between the micrometer spindle and the dial-gage spindle. Dimensions up to six inches can be measured with the right and left measuring unit.

ACCESSORIES

1. CENTER SUPPORTS. These are two identical fixtures (Figure 418) mounted on the table to hold cylindrical parts. The fixture consists of a base (A) and a sliding V-block (B). The base has a V that fits in a V-slot of the table for alignment. The V-block (B) is adjustable on the base and is clamped in position by means of a T-slot. A clamp (C) can be attached to the V-block for holding cylindrical parts or center bearings (D) when centers are required. The centers (E) are adjustably mounted on the center bearings. They are lined up by projecting an image of each center on the screen and lining up the side of the center with a line on the screen.

Figure 419. Measuring a Screw Thread

2. SCREW-THREAD ACCESSORY. This fixture (Figure 419) for measuring the various properties of screw threads consists of a base (A) containing a pivot upon which a frame (B) is so mounted that it can be swung in a vertical plane. By means of an angular scale and a vernier the frame can be accurately set at the helix angle of the thread and clamped in position.

Mounted on a dovetail-way on the frame is a cross-slide (C) containing two V-blocks in which center bearings can be mounted in the same way they are mounted in the center supports. Movement of the cross-slide can be measured with a measuring unit mounted on the fixture. The measuring unit is similar to that mounted on the table, and micrometer measurements are made in the same manner as previously described.

Figure 419 also shows a threaded part in position for making a measurement. The part is first mounted between the centers, and the frame is set to the helix angle of the thread to bring the threads parallel to the light beam. In this position the image of the thread is projected on an outline on the screen, and the various features of the thread are accurately measured.

SURFACE ILLUMINATOR

To make possible the projection of surfaces, a surface illuminator (Figure 420) is available. This self-contained apparatus, mounted on the body of the projector by adjustable clamp brackets, is easily manipulated to direct light upon an object. In use, a powerful beam of light is directed downward on the object so that the surface is brightly illuminated. This surface is reflected on the screen, and is measured or compared in the same manner as a shadow image.

CONTOUR PROJECTOR

The contour projector shown in Figure 421 is an optical comparator and measuring instrument similar in operation and application to those previously described. When fully equipped with the available measuring attachments, this comparator combines the simplicity of a bench-type comparator with the versatility of a contour-measuring projector. Figure 422 shows a cutaway view of the arrangement of the optical system. For acquaintance with the main components, a few simple methods of operation are given.

Figure 420. Surface Illuminator

HORIZONTAL BEAM

The light beam for silhouette projection originates in the horizontal-beam lamp house and is directed through the horizontal-beam collimator, across the work table, and into the relay lens assembly. From there the light is reflected twice by a penta-mirror assembly so that it is directed upward through the projection or magnifying lens. The light is then reflected horizontally toward the screen by a 45-degree mirror. These reflections provide a screen image oriented correctly up and down, right and left, at all magnifications. Just before coming to focus on the screen, the light passes through a special Ektalite field lens. This lens, built in behind the screen, tends to converge the light rays toward the operator's eyes, greatly increasing the brightness of the screen as viewed from the operator's position. To meet the need of the operator or the work, the intensity of the light can be adjusted with a brightness-control switch.

When an object is placed in the correct focal plane on the work table so that the object or a portion of the object intercepts part of the light beam from the lamp house, a magnified shadow of the object appears on the screen or a comparator chart placed on the screen (Figure 423A). The object is located eight inches from the front relay lens. This working distance is maintained at all magnifications.

Instead of a number of interchangeable lenses for different magnifications, the comparator can be equipped with a six-piece lens turret indexed by a magnification-selector handwheel to bring a lens of the desired magnification into position for use. No adjustment of mirrors is necessary when changing from one magnification to another. The optical system and the magnification lenses are so corrected that there is no measurable distortion anywhere on the screen. Thus, comparative measurements

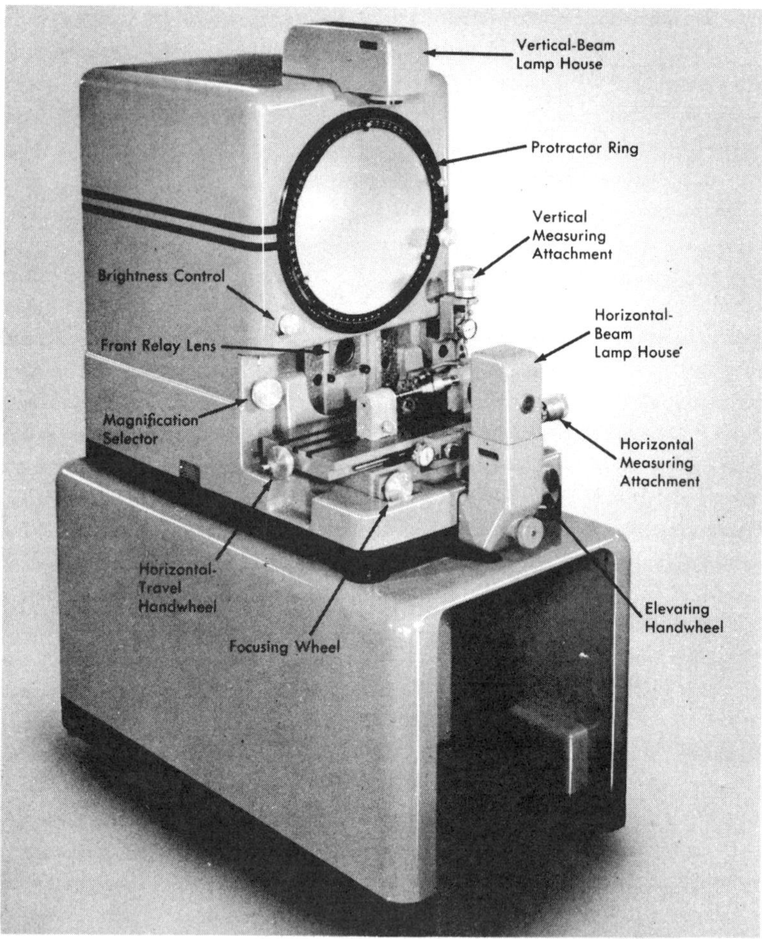

Figure 421. Contour Projector

can be taken anywhere on the screen area. Projection lenses are available in magnifications of 10, 20, 31.25, 50, 62.5 and 100X.

SURFACE ILLUMINATION OR REFLECTION

To measure and inspect surfaces which cannot be projected by the shadow method, the surface-illuminator lamp inside the cabinet is switched on. This lamp directs a beam of light through a condensing lens to a 45-degree telecentric mirror in the relay-lens tube. From this mirror the light is reflected out through the front of the relay lens assembly to the face of the object on the work table. The light is then reflected back from the object into a relay lens assembly; it passes through the hole in the telecentric mirror and eventually reaches the screen to form a magnified reflection in full color (Figure 422). Figure 423B shows a surface reflection of a portion of a die-cast housing.

When taking measurements without the use of specially prepared charts, a screen having perpendicular crosslines is used, and the amount of vertical and horizontal table travel is measured with the micrometer and dial-indicator attachments. By means of the protractor ring, angles can be measured to an accuracy of one minute. The above procedure is similar to that described for the pedestal comparator, Figure 409.

VERTICAL BEAM

A vertical-beam unit consisting of a vertical-

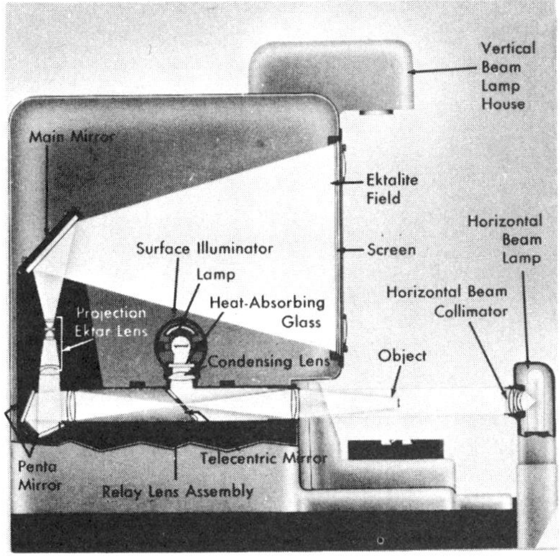

Figure 422. Arrangement of Optical System (Contour Projector)

beam lamp house and lamp, a vertical-beam stage, and a vertical-beam mirror is available for use on the comparator. This unit is used for staging and projecting contours and surfaces of small flat pieces that can be inspected best when laid on a horizontal surface. The vertical beam, as well as the horizontal beam, can be used with surface illumination. Shadow silhouettes from the horizontal or vertical beam can be used in simultaneous combination with full-color images from the surface illuminator.

CARE OF PROJECTION COMPARATORS

Too much emphasis cannot be placed upon the importance of keeping measuring instruments clean. Dust and dirt will jeopardize accuracy of the instrument by collecting between contact surfaces and causing wear and corrosion. A thin film of oil should cover all steel parts not protected by paint or plating. Lubricate moving parts as recommended by the manufacturer.

When connecting the comparator or any other measuring instrument to a source of electrical current, be sure the voltage and frequency correspond to that specified for the instrument.

Caution should be used in handling the mirror; its face is coated with aluminum oxide and should not be touched with the fingers. Use a clean camel-hair brush for removing dust and lint. Should the surface become coated with a film, clean with a piece of cotton saturated with

A

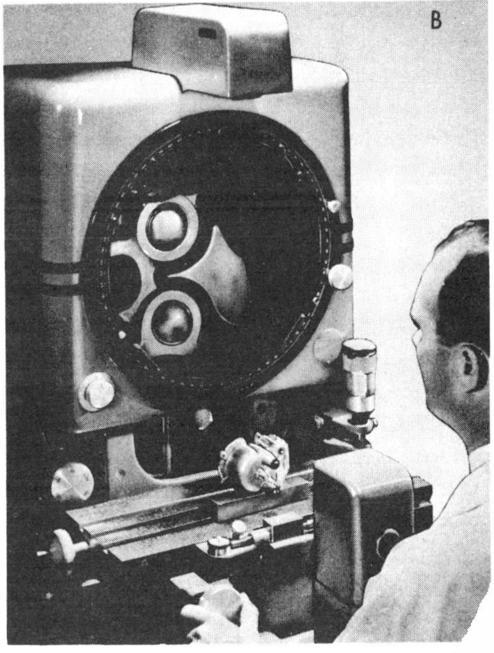

Figure 423. Silhouette and Reflection on Contour Projector

grain alcohol, ether, or other approved cleansing solution. Wipe the mirror surface gently. Do not rub.

The lens system of the comparator is an optical instrument. It can easily be damaged by carelessness and abuse. When not in use, interchangeable lenses should be wrapped in a clean cloth or tissue and put in a safe place. Should the lens become coated with dust or grease, clean with a camel-hair brush or an approved cleaning solution--and with the same care as outlined for the mirror. Dust and dirt contain abrasive. Optical glass is comparatively soft; therefore, rubbing the lens will scratch the surface and reduce its efficiency. Finger marks are particularly difficult to remove from lenses. Lenses should never be touched with the fingers.

When adjusting the table of the comparator to bring the work into focus, be careful that the fixture or the work does not touch the lens system. The operator usually has his eyes and attention focused upon the image on the screen and neglects to observe whether the fixture or part is approaching the lens system.

COMPARATOR CHARTS

The comparator chart is the master with which the enlarged image, projected on the screen, is compared. The contour measuring chart (Figure 424) consists of a sheet of stiff translucent material upon which an outline of the ideal image is drawn or etched in black lines, and to a scale corresponding to the magnification of the lens system with which the chart will be used.

MAGNIFICATION

If a chart is drawn to a 20X scale, it can be used only with a 20X lens system; if drawn to a 50X scale, it must be used with a 50X lens system. So, when setting up the comparator, important that the lens system corresponds gnification marked on the chart.

every chart is a fixture or of the machine, pro- ter (the projected

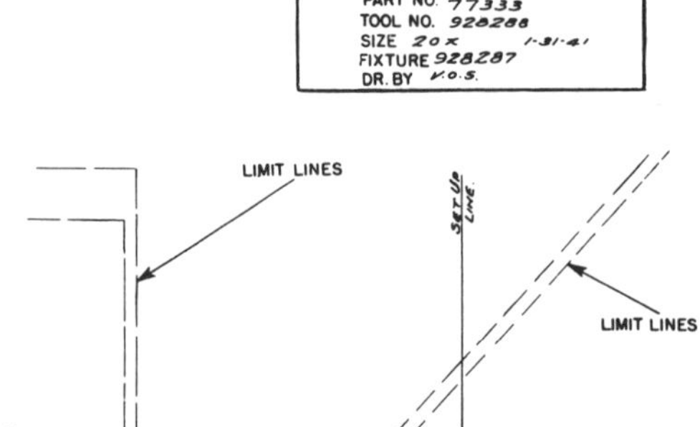

Figure 424. Contour Measuring Chart

image of which is used to locate the chart) and a location for successive parts to be compared with the master through the medium of the chart. The contour-measuring chart contains two lines at right angles to each other, marked setup lines (Figure 424). These setup lines--one vertical, the other horizontal--are superimposed upon the outline drawn on the chart.

The master contains two surfaces which, when projected upon the chart, are lined up with the setup lines, thus locating the outline on the chart in correct relation to the fixture on the table. The outlines of important surfaces of the part are drawn on the chart as double broken lines; these mark the limit of the dimensions governing the location of the surfaces (Figure 424). Therefore, when the master is replaced by a part to be checked, the projected image of these important surfaces must fall between the two broken lines, representing the limits of these surfaces; otherwise the part must be rejected.

LINEAR MEASUREMENT CHART

The chart in Figure 425 was designed for the purpose of making linear measurements of small portions of a part, without recourse to table movement and micrometer reading. The chart is drawn to 20X scale and must be used with a lens system of 20X magnification. The length of the diagram, horizontally, represents a distance of .250", divided into five major divisions of .050" each; these major divisions are subdivided into ten divisions of .005" each. Vertically, the diagram is .200" wide, containing four major divisions of .050" each; these, in turn, are subdivided into ten divisions of .005" each.

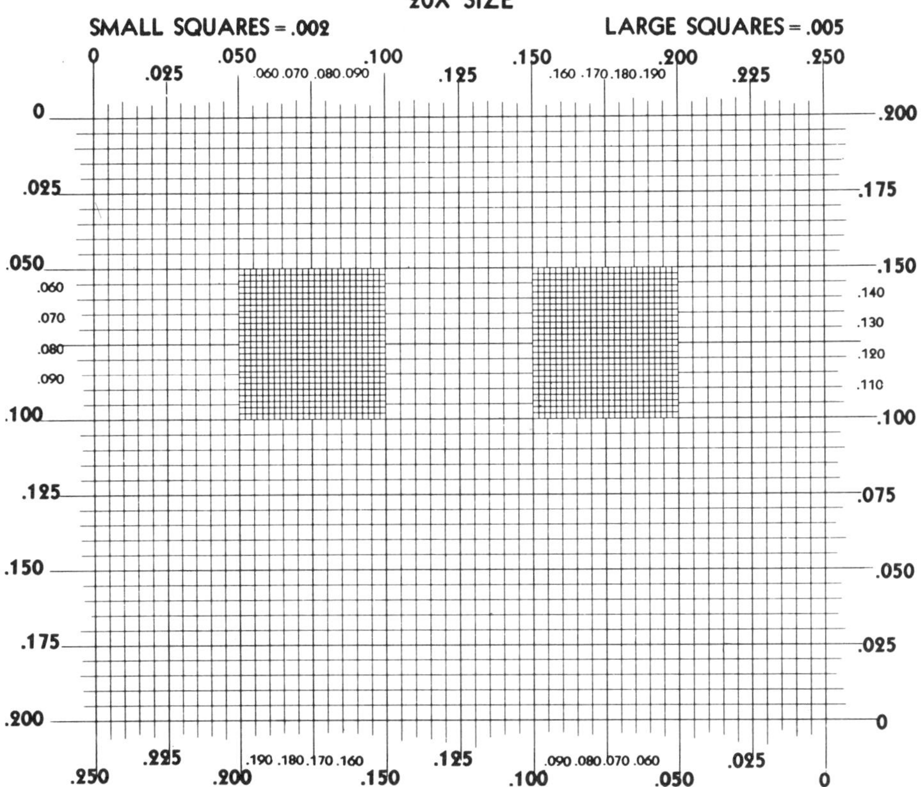

Figure 425. Linear Measuring Chart

The smallest division either way on the larger scale is .005"; therefore, it can be read to an accuracy of .005". Near the top of the diagram, between the .050" and .100" division, a portion of the diagram has been divided into divisions of .002" each. This portion of the diagram is used when it is desirable to read the scale to an accuracy of .002".

APPLICATION

Figure 426 shows imaginary images superimposed upon the linear measurement chart for the purpose of illustrating the application of the chart. The image at (A) is of the corner of a part, whose drawing specifies a chamfer of 1/32" on the corner shown. To measure the chamfer, the image of the corner is projected upon the chart, and the amount of chamfer is read as .035" vertically and .030" horizontally.

Figure 426B shows the image of a narrow flange projected upon the same chart. Width of the flange is specified as .0625" plus or minus .005". As measured on the chart, the width of the flange is read as slightly more than .060".

TAPERS. Any angle can be specified in terms of taper per inch. For any given angle the taper per inch is equal to the tangent of the angle: therefore, an angle of 30 degrees is equivalent to a rise or taper per inch equal to the tangent of 30 degrees or .577". Figure 426C shows the image of a 30-degree angle projected upon the chart. Note that in a horizontal distance of .100", the rise is .057", which is equivalent to a rise of .570" in one inch.

GLASS CHARTS

The material commonly used for comparator charts is a form of translucent cellulose acetate similar to celluloid. It is non-inflammable, tough, resilient, and presents a surface upon which drawings in pencil or ink can be readily made. However, it has a tendency to warp, and is affected by temperature changes.

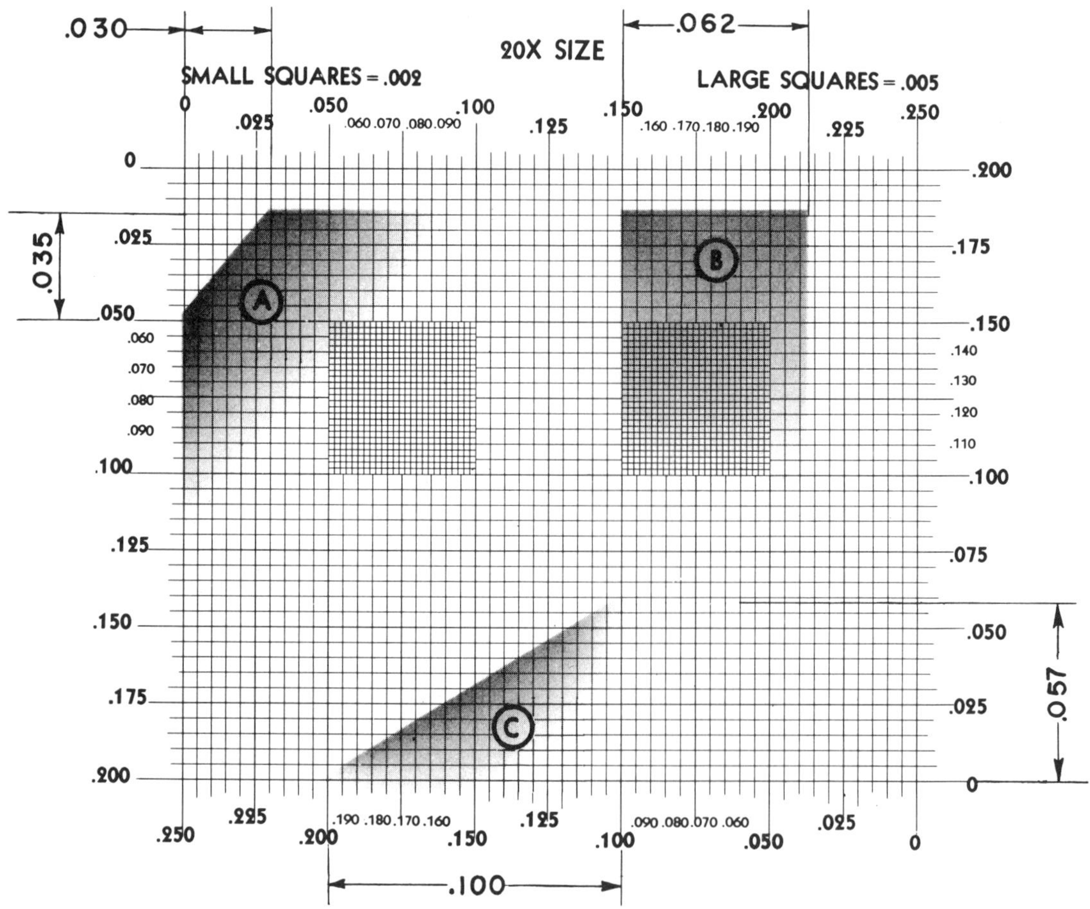

Figure 426. Application of Linear Measuring Chart

A process has been developed by which outlines are etched upon the surface of a ground-glass plate. The glass chart eliminates all of the disadvantages of the flexible chart and adds advantages of permanency, clarity, and ease in cleaning. Its only disadvantage is that it is fragile and must be handled with care. However, the fact that it is hard, permanent, and easy to read, makes it the ideal chart for comparator measurements.

RADIUS CHART

An example of the glass chart is the radius chart shown in Figure 427. These charts are available in magnifications corresponding to the various magnifications of the lens systems used on the projection comparator. The particular chart shown, however, has no magnification. The circles are spaced .100" apart so that in measuring an image which has been magnified 10X, the actual size of the object can be obtained merely by changing the decimal point. All magnifications that are a multiple of 10 can be mentally calculated. To check the radius of a part, the image of the radius is projected upon the chart and adjusted to coincide as nearly as possible with the circle corresponding to the radius specified on the drawing, multiplied by the magnification of the lens system used (Figure 427A).

Linear measurements can be made, at any angle, on a line passing through the center of the chart. In Figure 427 the width of the image at (B) is found by lining up one edge of the image with the center of the chart and reading the dimension on the circle nearest to the other

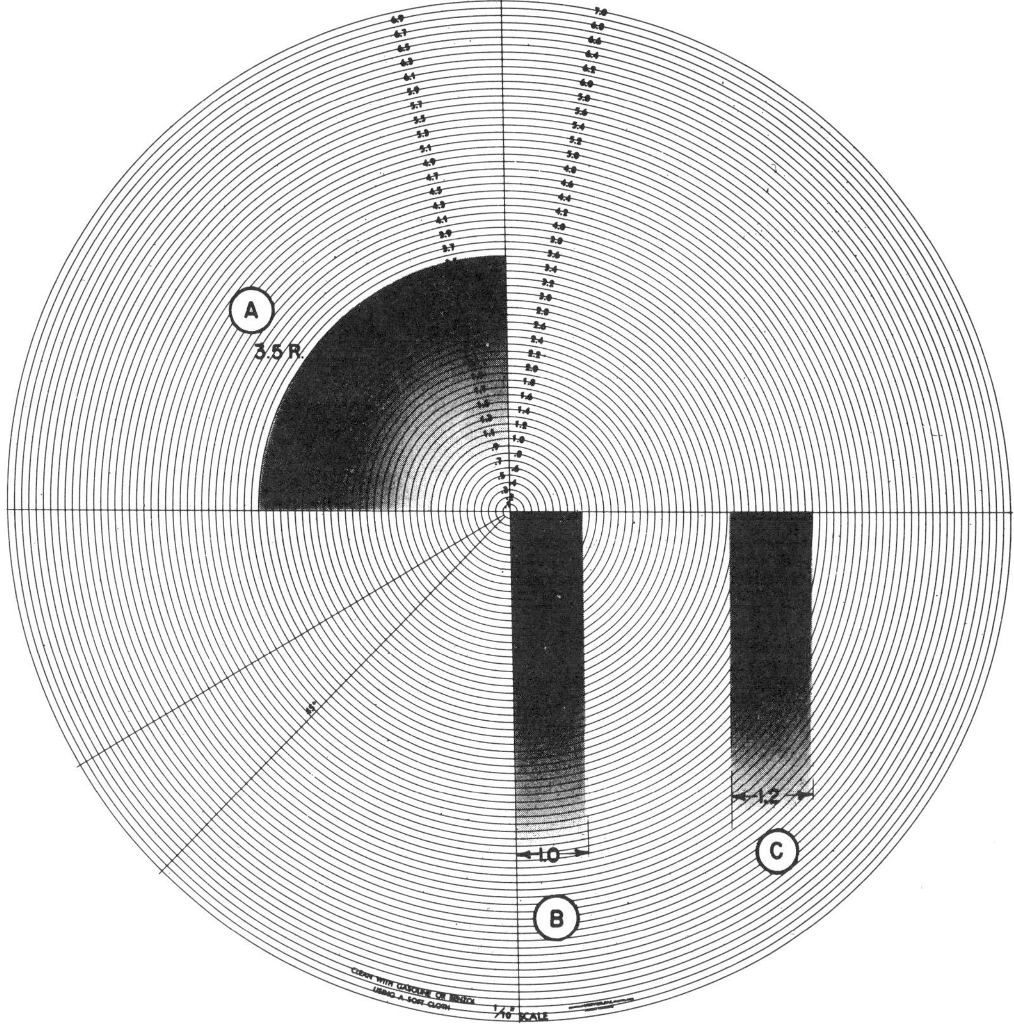

Figure 427. Radius Chart

edge of the image. Similar measurements can be made by lining up the image between any two circles and finding the difference between the dimensions on the respective circles (Figure 427C).

MICRO-GAGE SCREENS

Figure 428 shows a glass comparator screen, having what is called a Micro-Gage line. Accurate comparator measurements to split-thousandths are possible with screens and charts of this type. Instead of the usual thin, continuous or broken line, a double row of alternately-spaced, heavy (about .032" wide), parallel rectangles are used. The inside edges of the rectangles comprising one line coincide exactly with the inside edges of the rectangles comprising the opposite line. Between them, in effect, this creates a line without width and with which the edge of the projected shadow can be brought into precise coincidence.

The Micro-Gage line or bridge can be used to advantage for production parts when dimensions are to be checked to closer tolerances than can easily be seen on the "go" and "not-go" lines on a conventional comparator chart.

Figure 429 shows how the Micro-Gage "bridge" works when used on a chart for checking a specific dimension. The contrast of light and shadow is amplified by means of the bridge;

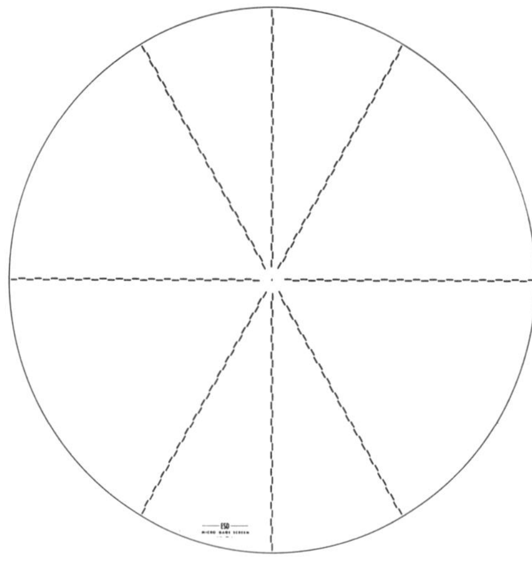

Figure 428. Micro-Gage Screen

light must be seen within the indentations of the bridge, but must not be seen below the "footing" of the bridge. In (A) the part is within tolerance, as light is seen within the bridge. The depth of the Micro-Gage bridge is equivalent to "go" and "not-go" limits. In (B) the part is oversize and a reject, as it excludes light in the bridge. (C) is also rejected, because it is undersize and permits light to appear below the Micro-Gage limit lines.

Standard Micro-Gage screens and specially-made comparator charts incorporating the Micro-Gage bridge principle, are commercially available to fit all optical comparators and contour projectors.

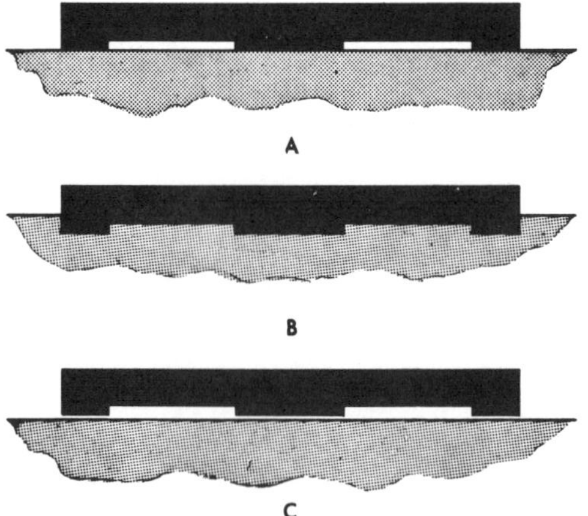

Figure 429. Principle of Micro-Gage Bridge

Figure 430. Universal Projector Rule

PROJECTOR RULES

If the magnification of the lens system used in making a projection upon the comparator screen is known, any portion of the image can be measured with an ordinary scale. This measurement, if divided by the magnification, will be the actual size of the portion of the part represented by the image. For example: if, when using the 20X lens system, the image measures 1-3/32" (1.093"), the actual size of the part is:

$$\frac{1.093''}{20} = .0546''$$

The Universal projector rule shown in Figure 430 was designed to meet the need for a

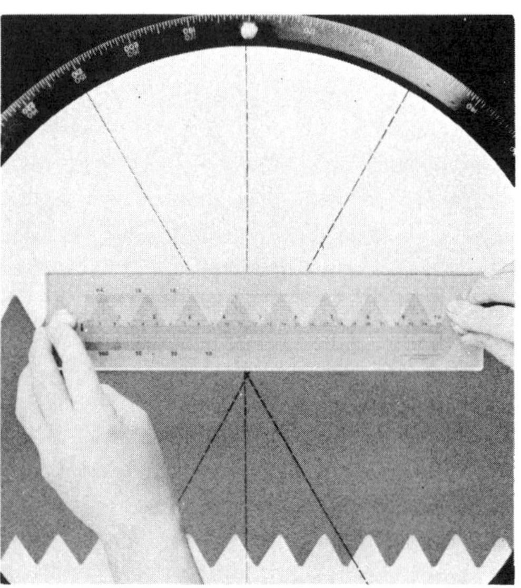

Figure 431. Application of Projector Rule

convenient rule for making direct measurements on the screen and checking the accuracy of the outlines on the charts. It consists of a transparent glass plate fitted with handles and containing, on the lower side, a scale graduated in 10, 16, 20, 32, 50, 64, and 100 divisions to the inch, with an additional scale graduated in millimeters.

Figure 431 shows an application of the projector rule in measuring the elements of a screw thread. The rule has no magnification of the lens system being used. Projector rules graduated to correspond to the magnification of various lens systems can be supplied by the manufacturer of this rule. These magnifications include 10X, 20X, 31.25, 50X and 62.5X.

QUESTIONS AND PROBLEMS

1. Define the following: light, light beam, light rays.
2. What happens to a light beam that is directed against a surface?
3. Define reflection; refraction.
4. What is a lens?
5. Describe the following lenses: double convex, concavo-convex, plano-concave.
6. What is the focal point of a lens; the focal length?
7. What is meant by the term "virtual image"?
8. How does the length of the tube of a microscope affect the magnification?
9. Define the terms "objective" and "ocular" as applied to a microscope.
10. Why is strong illumination necessary when observing an object with the microscope?
11. What is the field of a microscope? Is there any relation between the magnification and the field?
12. How should the lenses of a microscope be cleaned?
13. Name the principal applications of the shop microscope.
14. What is the function of the porro-erecting prism in the wide-field microscope?
15. Name three uses of the toolmaker's microscope.
16. How can measurements greater than one inch be made on a toolmaker's microscope?
17. Name three accessories on the toolmaker's microscope and their uses.
18. What is an optical flat? How is it used?
19. What are interference bands?
20. Define monochromatic light.
21. How is the point of contact determined when using optical flats?
22. Explain how you can determine whether the pattern in an optical flat indicates a ridge or a valley.
23. In an optical flat pattern of a flat surface, seven dark bands and six light bands appear. Which figure would you use in determining the height of the open end of the air wedge? Why?
24. Define projection comparator.
25. How is magnification obtained in projection comparators?
26. When and how is magnification checked on the bench-type comparator?
27. What is the purpose of the setup lines on the chart? The limit lines?
28. What is the purpose of the micrometers on a projection comparator?
29. What is the purpose of the reflection attachment on the pedestal-type comparator?
30. Give the advantages of the glass plate in the table of the contour measuring projection.
31. Give some of the advantages of the comparator shown in Figure 421.
32. What precautions should be observed in using and caring for projection comparators?
33. Differentiate between the contour measuring chart and the linear measuring chart.
34. State the advantages of glass charts.
35. Describe the universal projector rule and give one example of its application.

Chapter XII

MEASURING MACHINES

THE TERM "measuring machines" is here applied to a class of machines built for making very accurate linear measurements. These machines usually incorporate the principle of the micrometer screw, scale, and vernier; they have means of (1) controlling measuring pressure so that measurements can be duplicated to a high degree of accuracy, and (2) varying measuring capacity to permit measurements greater than the range of the micrometer head, which is ordinarily limited to one inch.

In making any accurate linear measurement, the main factors beyond accuracy of the machine itself are measuring pressure and uniformity of temperature. Accuracy of a measuring machine depends upon the accuracy with which a measurement can be duplicated; so, a definite measuring pressure must be established and adhered to throughout the measurement.

All solid materials expand or contract when subjected to temperature change. The amount a unit-length of the material expands for a rise in temperature of one degree is called the coefficient of linear expansion of that material. Because different materials have different coefficients of expansion, it has been necessary, for accurate linear measurement, to establish a standard temperature at which all measurements should be made.

The National Bureau of Standards has established standards for both measuring pressure and measuring temperature. Measuring pressures vary according to the material and nature of the work, but the measuring temperature has been established at 20 degrees centigrade or 68 degrees Fahrenheit.

Control of measuring pressure is usually incorporated in the machine itself. Temperatures of machine and work are usually controlled by room temperature. If the room is held at a constant temperature, the machine, the master, and the work, will eventually assume the same temperature, changed only by contact with the human body or some other source of heat, such as an illuminating lamp, direct rays of the sun, or electrical apparatus.

An important feature of a measuring machine is its working distance or capacity for measuring long pieces of work. The working distance is usually adjustable; so the machine must have accurate means of adjustments to take larger or smaller work. These adjustments are made either in reference to a standard measuring bar or with the aid of standard gage blocks.

THE SUPERMICROMETER

The Supermicrometer (Figure 432) bridges the wide gap between the hand micrometer and the more elaborate and costly measuring machine. It is light, sturdy, and adaptable to everyday shop use. It consists of a cylindrical bed (A), one end of which is supported by a micrometer headstock (B), and the other by an adjustable foot (C), by which the micrometer may be leveled. A pressure tailstock (D), mounted on the bed, is adjustable to varying work distances by a rack-and-pinion arrangement and is held in place by a clamp at (E). A slot in the bed serves as a guide to keep the micrometer headstock and tailstock in line.

The range of the micrometer head is one

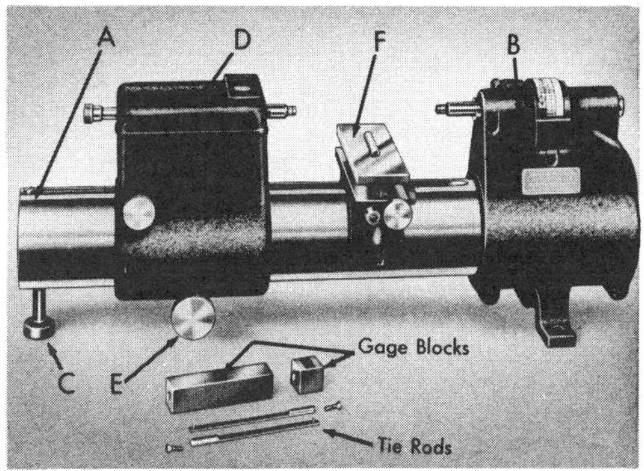

Figure 432. Supermicrometer

inch. Greater working distances, up to ten inches, are set with the aid of precision gage blocks.

An elevating table (F) is mounted on the bed to support the work during the measuring operation.

The measuring pressure control, which controls pressure on the anvil, is housed in the tailstock. This control provides a constant pressure of either 1 or 2-1/2 lbs. These pressures are recommended by the National Bureau of Standards for the measurement of threads by the three-wire system.

Threads finer than 20 per inch are measured with a pressure of 1 lb.; threads 20 per inch or coarser are measured with a pressure of 2-1/2 lbs. The 2-1/2-lb. pressure is necessary to centralize these coarser pitches in the measuring machine. If greater pressure were used either for fine or coarse threads, there would be danger of deforming the measuring wires, thus introducing error into the measurement.

PRESSURE TAILSTOCK

Construction of the pressure tailstock is shown in Figure 433. The anvil (A) is rigidly mounted in two blocks (B); these, in turn, are mounted on two flat springs (C). The flat springs permit the anvil to move laterally but hold it in strict alignment horizontally. The anvil bears against a plunger, in the barrel (D), which imposes a continuous pressure upon the anvil.

The pressure is set to either 1 or 2-1/2 lbs. by means of the knurled knob (E). The knob is mounted on a barrel containing circumferential

Figure 433. Pressure Tailstock

lines marked 1 and 2-1/2, respectively. When the edge of the knob coincides with the line marked 1, one-pound pressure is imposed on the anvil; when the edge of the knob coincides with the line marked 2-1/2, two and one-half pounds pressure is imposed upon the anvil.

The pointer (F) is suspended between two pivots; one pivot is mounted on the arm (G); the arm, in turn, is mounted on block (B) and actuates the pointer when the anvil is moved against the pressure of the plunger. Exact measuring pressure is obtained by lining up the tip of the pointer with an index line at (H).

MICROMETER HEADSTOCK

The micrometer headstock (B, Figure 432) contains the micrometer unit and is rigidly attached to one end of the barrel. Accuracy of the instrument is in the micrometer unit. The micrometer spindle, which does not revolve, is graduated in .050" divisions read in relation to an index line at (A) in Figure 434. The micrometer drum (B) contains 50 divisions, each representing a spindle travel of .001", and is turned by means of a knurled band on the drum itself. The scale on the drum is read in relation to a vernier at (C); by this means travel of the

Figure 434. Supermicrometer Headstock

spindle can be read to an accuracy of .0001". The relation of the vernier and the drum is adjustable by turning knob (D) and can be secured by two screws at (E).

The final adjustment when setting the micrometer for a measurement is made with the vernier. The range of the micrometer is one inch; therefore, when dimensions over one inch are to be measured, the whole inches are first set with gage blocks (Figure 435). If more than one block is used, the blocks should be fastened together with tie rods and screws. Tighten the screws lightly. Clean both ends of the block to remove all foreign matter. To avoid raising block temperature above normal room temperature, gage blocks should not be handled any more than necessary.

Assuming that the dimension to be measured is 4.587", set the desired measuring pressure on the tailstock; set the micrometer by turning the headstock dial until the spindle is at the outer end of its travel and reads **zero (nearest** headstock). Zero on the headstock dial must also line up with the zero on the vernier. Release clamp (E) and adjust the tailstock with knob (J) so that a four-inch gage block can be

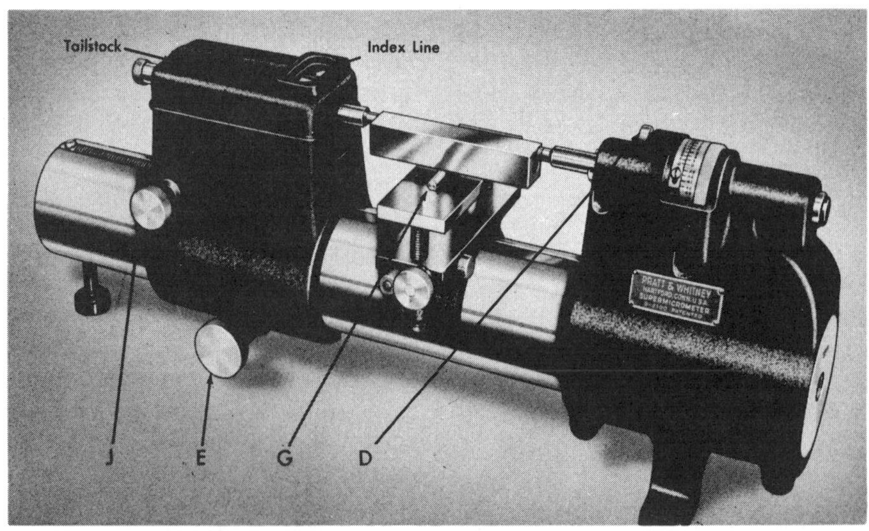

Figure 435. Setting Supermicrometer with Gage Blocks

placed between the spindle and the anvil. Place the gage block on the flatted cylinder (G) and raise or lower the elevating table until the spindle and anvil make contact with the end surfaces of the gage block. Continue adjusting the tailstock until the pointer on the tailstock coincides with the index line; then clamp the tailstock in place. Gently move the gage block between the anvil and the spindle to make sure it is squarely placed. Any change in pressure caused by clamping the tailstock can be compensated for by turning the micrometer slightly and adjusting the vernier to zero with knob (D). Remove the gage block and the flatted cylinder; the Supermicrometer is ready for any measurement between four and five inches. The one-inch travel of the headstock spindle provides the final measurement.

THREE-WIRE MEASUREMENT OF SCREW THREADS

Measurement of screw-thread pitch diameter with the three-wire method is outlined in Chapter IV and is reviewed here to demonstrate a three-wire measurement on the Supermicrometer.

In the three-wire method of measuring pitch diameter, small hardened-steel cylinders or wires of correct size are placed in the thread groove, two on one side of the screw and one on the opposite side (Figure 436). The size of

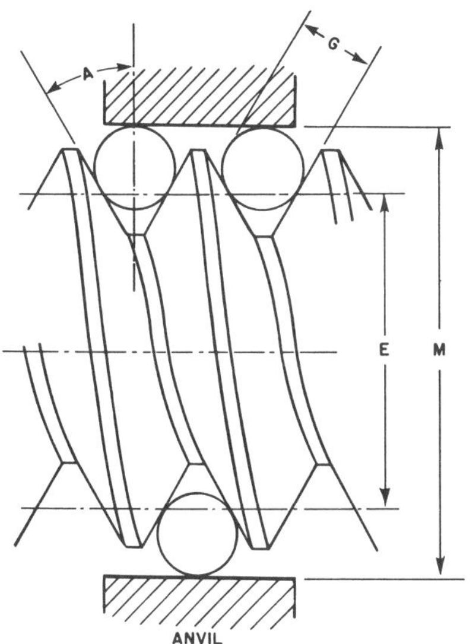

Figure 436. Three-wire Method of Thread Measurement

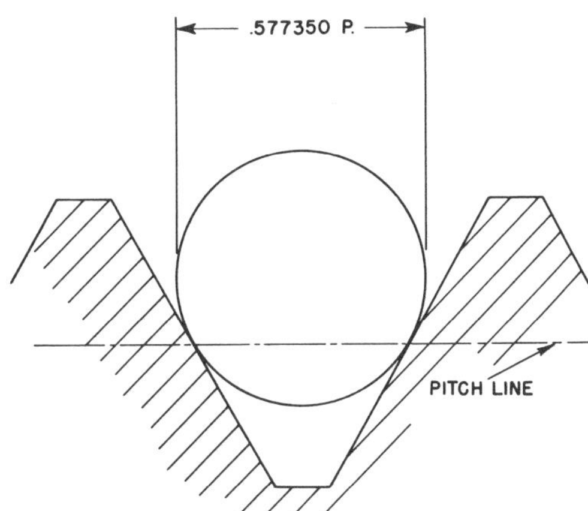

Figure 437. Best Wire Size

wire touching the side of the thread exactly on the pitch line is known as the best-size wire. Other sizes approximating the best-size wire may be used when measuring the pitch diameter of the screw; but the best-size wire is preferred, because it minimizes the possibility of measurement error that may be due to error in thread angle. Stated in terms of the pitch of the thread (P), the best wire-size for American National thread form is .577350 P (Figure 437).

Frequently, when the best wire-size is not available, it is desirable to measure the pitch diameter with wires of a size other than the best wire-size. The minimum-size wire is limited to that which extends above the crest of the thread; the maximum-size wire to that which will rest on the side of the thread, just below the crest of the thread. Stated in terms of the pitch of the thread, for American National thread form, the maximum diameter of the wire is 1.01036 P (Figure 438).

This does not mean that the maximum wire-size should be selected. The maximum wire-size is based on the basic thread form; so allowance must be made for variation in thread form within limits specified for a particular class of thread.

In terms of thread pitch, the minimum-size wire for the American National thread form is .505182 P (Figure 439). This minimum-size wire is also based on the basic thread form, and allowance must be made for variations within the limits specified. The margin between maximum and minimum wire-size is so great

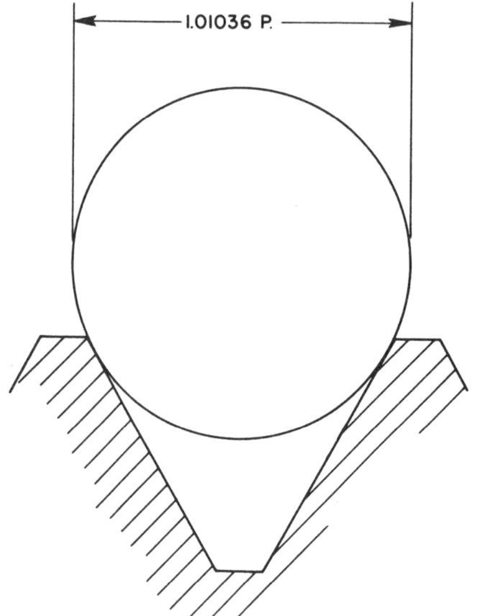

Figure 438. Maximum Wire Size

that there should be no need for even approaching the maximum or minimum when selecting the wires. Tables are available in standard handbooks giving best, maximum, and minimum wire-size for all standard threads.

For the American National Acme Thread form, the best wire-size is .516450 P, the maximum wire-size is .650013 P, and the minimum wire-size is .487263 P.

FORMULA FOR AMERICAN NATIONAL THREADS

The general formula for determining pitch diameter of any thread, whose sides are symmetrical about a line perpendicular to the axis of the screw, is:

$$E = M + \frac{\cot A}{2n} - G\left(1 + \operatorname{cosec} A + \frac{S^2}{2}\cos A \cot A\right)$$

E = Pitch diameter

M = Measurement over wires

A = 1/2 thread angle

$$\cot A - \frac{1}{\tan A}$$

n = Number threads per inch

G = Wire size

S = Tangent of helix angle

$$\operatorname{cosec} A = \frac{1}{\sin A}$$

This formula is basic and is presented merely

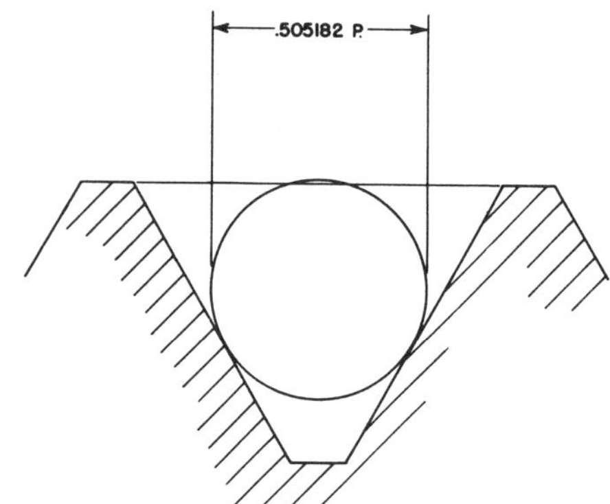

Figure 439. Minimum Wire Size

to show the derivation of the formulas which follow.

For the standard American National thread form (60°), the foregoing formula reduces to:

$$E = M + \frac{1.73206}{2n} - G\,(1 + 2) - G\left(\frac{S^2}{2} \times .86603 \times 1.73206\right)$$

$$= M + \frac{.86603}{n} - 3G - .75000\,G\,S^2$$

Since the helix angle of the American National thread series is small, the latter part of the formula ($-.75000\,G\,S^2$) is usually neglected, and the formula becomes:

$$E = M + \frac{.86603}{n} - 3G$$

However, when the helix angle of the thread is large (as in coarse-pitch screws of small diameter or as in double, triple or quadruple threads), the complete formula must be used to obtain accurate measurement.

For example: The helix angle of a 3/8-16 American National screw thread is $3^\circ\ 24'$. Tan $3^\circ\ 24' = .05941$. The best wire-size is .03608 diam. So, the error involved by dropping the latter part of the formula is:

$.7500 \times .03608 \times .05931^2 = .0000955$ which is negligible.

When error is greater than .00015", however, it is advisable to include the complete formula for accurate measurements.

An example of a condition requiring a complete formula is:

Required, the pitch diameter of a screw,

.250" diam., 20 threads per inch, double thread

Helix angle $= 8^\circ 19'$

Tan $8^\circ 19' = .14618$

Best wire-size = .02887" diam.

Actual error involved by not using latter part of formula would be:

$.7500 \times .02887 \times .14618^2 = .000462"$

FORMULA FOR AMERICAN NATIONAL ACME THREADS

For American National Acme threads, the basic formula resolves to:

$$E = M \frac{1.93336}{n} - 4.99393\,G + 1.87178\,G\,S^2$$

As in the case of the American National form of thread, when the latter part of the formula ($+ 1.87178\,G\,S^2$) is less than .00015", it is usually neglected; when it amounts to more than .00015", it should be included for accurate measurement of AmericanNational Acme threads.

HOLDING WIRES

When making three-wire measurements of threads, holding the wires--until the micrometer spindle can be adjusted to clamp them in place--is usually awkward. A small rubber band, as shown in Figure 440, is a convenient means of holding wires when measuring with a hand micrometer. But, when using the Supermicrometer with the part supported on the table, the spindle can be advanced so that the space be-

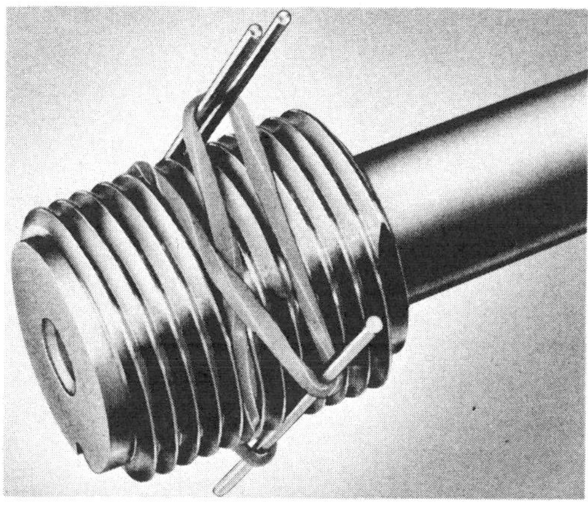

Figure 440. Holding Three Wires with a Rubber Band

Figure 441. Inserting Wires

tween the spindle and anvil is just sufficient to allow the wires to be inserted in the thread grooves (Figure 441), and the wires will remain in the grooves until they are clamped by further advance of the spindle. When using any auxiliary method of holding wires in grooves, care should be taken that wires are not sprung (Figure 442), as this condition will affect the accuracy of the measurement.

Special wire holders (Figure 443) are available for use on the Supermicrometer. These holders can be mounted on the spindle and anvil and held with a setscrew. Their main function is to prevent wires dropping when released by the spindle, but they can be used to hold wires against the face of the spindle (Figure 444). Here again, care must be taken that wires are not sprung. It is good practice to release the pressure on the wires after they are held in place by the spindle, so that the wires will locate properly in the thread grooves.

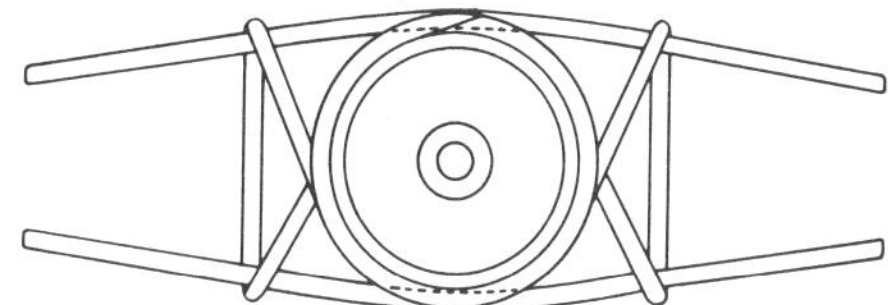

Figure 442. Springing Wires

Figure 443. Wire Holders

MEASURING EXAMPLE

Figure 445 shows a setup for measuring the pitch diameter of a 3/4-16 NF - 3 "go" thread gage.

For class 3 fit, it is recommended that the accuracy of the gage be class X, which requires a tolerance limit of + .0003, - .0000 on the pitch diameter for 16 threads per inch (reference Screw-Thread Standards for Federal Services). The limits on the pitch diameter, therefore,

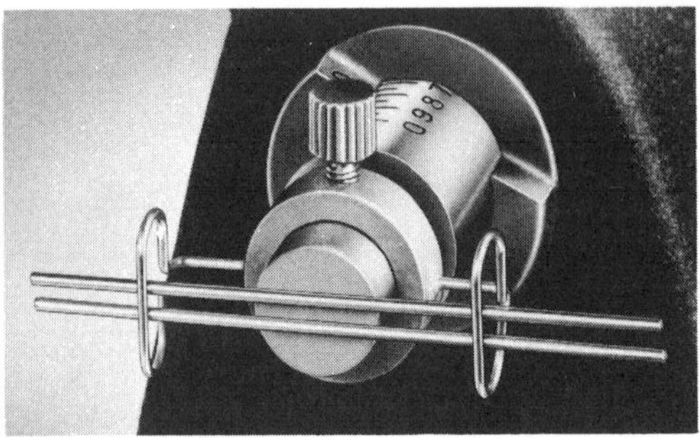

Figure 444. Holding Wires on Face of Spindle

are .7097 and .7094. Measurement over the wires is .7636. Calculating the P.D. by formula:

$$E = M + \frac{.86603}{n} - 3G$$

$$n = 16$$

$$G = .03608$$

gives

$$E = .7636 + .05413 - .10824$$

$$E = .70949$$

which is well within the limits specified for a class X thread gage.

The tangent of the helix angle is .0281, and the error involved by not using the latter part of the formula is:

$$.7500\, G\, S^2$$

$$= .7500 \times .03608 \times .0281^2$$

$$= .0000214$$

which is negligible.

Figure 445. Three-wire Measurement with Supermicrometer

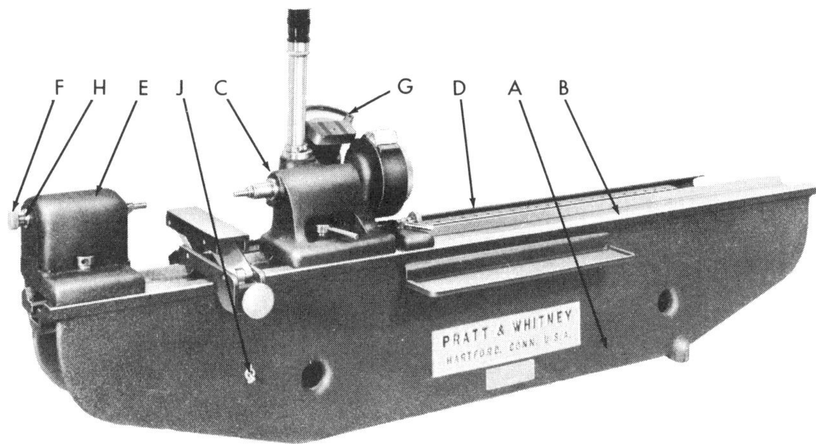

Figure 446. Standard Measuring Machine

STANDARD MEASURING MACHINE

The standard measuring machine (Figure 446) consists of a heavy base (A) containing accurate ways (B) upon which a micrometer headstock with microscope (C) can be positioned in relation to a measuring bar (D) attached to the side of the base. A pressure tailstock (E), also adjustable on the ways, provides a means of adjusting the measuring pressure to any value between 8 oz. and 2-1/2 lbs.

This machine is similar to the Supermicrometer in appearance and operation, yet its extreme accuracy (.000010) involves special features and adjustments, whose operation requires knowledge of the machine and meticulous care. For accurate measurements, the manufacturer recommends that the machine be operated only by one or two dependable operators who thoroughly understand its principle and operation.

PRESSURE TAILSTOCK

The pressure tailstock is similar in appearance and operation to the Supermicrometer, but the principle of the mechanism differs.

While the Supermicrometer tailstock has a purely mechanical means of pressure adjustment, this tailstock registers the adjustment through an Electrolimit unit similar to the Electrolimit gaging head described under Electrolimit comparators. A steel armature, actuated by the movement of the anvil, upsets the inductive balance between two magnetic coils so that when the pressure on the anvil corresponds to that set by the adjusting knob (F), the pointer of an ammeter (G) will register on zero. Any slight departure from the set pressure will immediately be indicated by the movement of the pointer either plus or minus on the meter.

Adjustment of the pressure is made with knob (F) by lining up the circumferential lines on the shaft with the edge of the bushing at (H). Turning the knob advances or retracts the shaft, thus increasing or diminishing the pressure on the anvil. Figure 447 is a close-up of the pressure-adjusting knob, showing the scale on the shaft.

Figure 447. Pressure-adjusting Knob

Figure 448. Master Bar (Standard Measuring Machine)

MASTER BAR

The basic feature of the standard measuring machine is the master bar (Figure 448). All measurements are made in relation to it, and the accuracy of measurements is dependent upon it. The master bar is graduated at each one-inch interval. These graduations are microscopic reference lines on highly polished plugs (A in Figure 448). The graduations are transferred from a master--calibrated at the National Bureau of Standards--whose errors are known and compensated for when transferring the graduation to the master bar on the measuring machine.

MEASURING HEAD

The measuring head (Figure 449) contains the micrometer unit, microscope, and ammeter. It is adjustable on the ways of the bed and clamped in position by a binder at (A). Fine adjustment is made by means of adjusting block (B). The measuring head is always located in relation to one of the index lines on the measuring bar. Approximate location is made with both binder clamps (A) and (C) open, after which the adjusting block is clamped to the ways with clamp (C); further adjustment of the measuring head is made with adjusting screw (D) until the index line on the master bar is in correct relation to two hairlines in the microscope.

Figure 449. Measuring Head (Standard Measuring Machine)

Range of the micrometer is one inch. Measurements greater than one inch are made by setting the measuring head in relation to an index line on the measuring bar corresponding to the whole number of inches in the dimension.

The measurement is read directly from three dial scales at (E) (Figure 449). Reading from left to right, the first scale indicates tenths and hundredths of an inch (Figure 450); the second scale indicates thousandths, as well as ten-thousandths of an inch; the third or vernier scale subdivides the ten-thousandths divisions into ten parts, providing a vernier reading in 10 millionths of an inch.

Figures 450A and B show two readings on the dial scales. The reading on the first dial (Figure 450A) is made in reference to an index line to the left at (C) and is .980. The reading on the second dial is made in reference to the zero line of the vernier and is .0018, making a total thus far of .9818. The reading of the vernier is .00003 which, added to .9818, makes a total of .98183. The reading on the first dial (Figure 450B) is .130, that on the second dial is .0048, and the vernier reading is .00007, making a total of .13487.

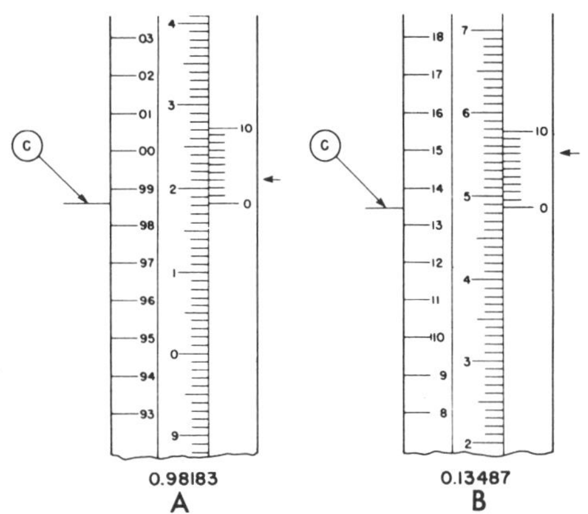

Figure 450. Scales of the Standard Measuring Machine

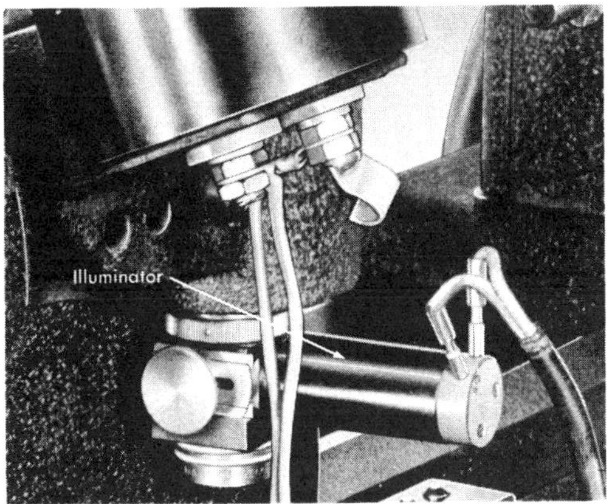

Figure 451. Illuminator on Standard Measuring Machine

ADJUSTING TO ZERO

Before the standard measuring machine can be used for measurement, it must be accurately adjusted to zero. This adjustment is made as follows:

1. Locate the measuring head so as to bring the microscope approximately over the plug, in the master bar, marked zero (0), but do not clamp it to the base.

2. Turn on the illuminator (Figure 451) with switch at (J) (Figure 446), and adjust the illuminator so that the zero knob on the master bar is illuminated. (In this connection it will be well to refer to Figure 372 in the previous chapter for instructions in adjusting the illuminator, as the vertical illuminator of the measuring machine is similar to that of the toolmaker's microscope.)

3. Adjust the microscope by turning the eyepiece at the extreme top of the microscope until the two parallel cross-lines appear clear and sharp. Place a heavy screwdriver in the slot of the screw which holds the microscope to the measuring head on the rear of the machine. Hold the microscope with the left hand so that it will not slide down when released. Loosen the screw and raise or lower the microscope until the index line on the master bar appears in the field of the microscope. Tighten the binding screw when the index line is clear, and adjust collar (F)(Figure 449) against eccentric bushing (G) so that the microscope can be relocated when removed at any time. If the lines in the microscope are not parallel to the index line, turn the microscope head until they are parallel.

4. Adjust the measuring head on the ways until the index line is midway between the two parallel lines in the microscope. To make this adjustment, clamp adjusting block (B) (Figure 449) and adjust the head with the fine adjusting screw (D). When the index line is exactly centered, clamp the head in position.

5. Set the micrometer to zero by turning the micrometer drum until zero of the first dial is opposite the index line (C, Figure 450). Then move the micrometer drum carefully until the anvils touch (do not jam), and continue turning until the pointer of the ammeter registers zero. Now adjust the vernier to bring the zero of the vernier and the zero of the second dial in line.

Check again to see that the index line of the master bar is properly centered in the microscope and the micrometer zeros are opposite their respective index lines. Also see that the ammeter registers zero. If all scales are at zero, the measuring head is in the true zero position and should seldom require readjustment.

The measuring head is now set for measurements within one inch. If measurements greater than one inch are required, loosen the clamps on the adjusting block and measuring head. Clean the plug on the master bar, corresponding to whole inches in the dimension being measured, with a camel-hair brush and approved

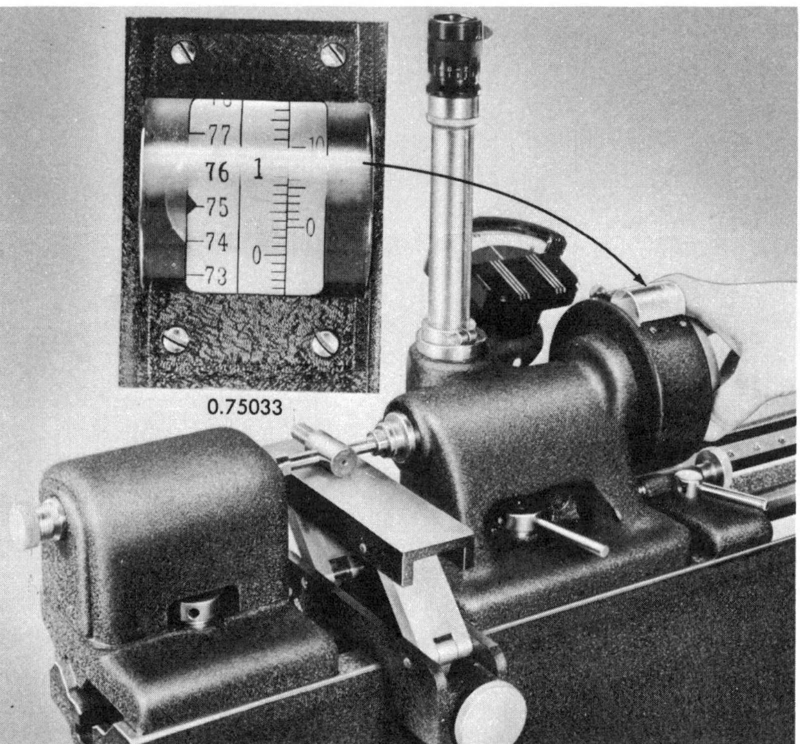

Figure 452. Measuring a Plug Gage on Standard Measuring Machine

cleaning fluid. Cover the zero plug with white petrolatum, and locate the measuring head over the plug in new position by clamping the adjusting block and adjusting the measuring head until the new index line is midway between parallel lines in the microscope. Clamp the measuring head in place, and the machine is set for new measurement.

After it is once set, do not move the tailstock. Do not adjust the microscope. Any change in the tailstock location or any adjustment of the microscope will involve resetting the machine to zero.

Figure 452 shows a setup on the standard measuring machine to check the diameter of a plug gage. The plug gage is specified as a "go" gage, .7503 diameter, class X accuracy, which permits a gage maker's tolerance of .00004". As the plug gage is a "go" gage, the tolerance will be applied plus; therefore, the gage must check between .75030 and .75034. The reading on the micrometer indicates the gage is .75033 diameter, thus checking within the tolerance permitted for X accuracy.

ELECTRO-MECHANICAL LEAD TESTER

This instrument was designed to measure the lead of external threads, both straight and tapered. By definition, the lead of a screw is the advance, parallel to the axis of the screw, a single thread makes in one complete turn of the screw. For a single thread, this distance is equal to the pitch of the thread; for a double thread, the lead is equal to twice the pitch; for a triple thread, three times the pitch, etc.

The Electro-Mechanical lead tester (Figure 453) consists essentially of a precision, direct-reading, measuring head (A) similar to that of the standard measuring machine, which operates in relation to an axially movable carriage (B) supporting a 15-inch sine bar (C). The sine bar contains two center heads (D) for mounting the work.

Exact locations are obtained by means of an Electrolimit thread-locating head (E), mounted

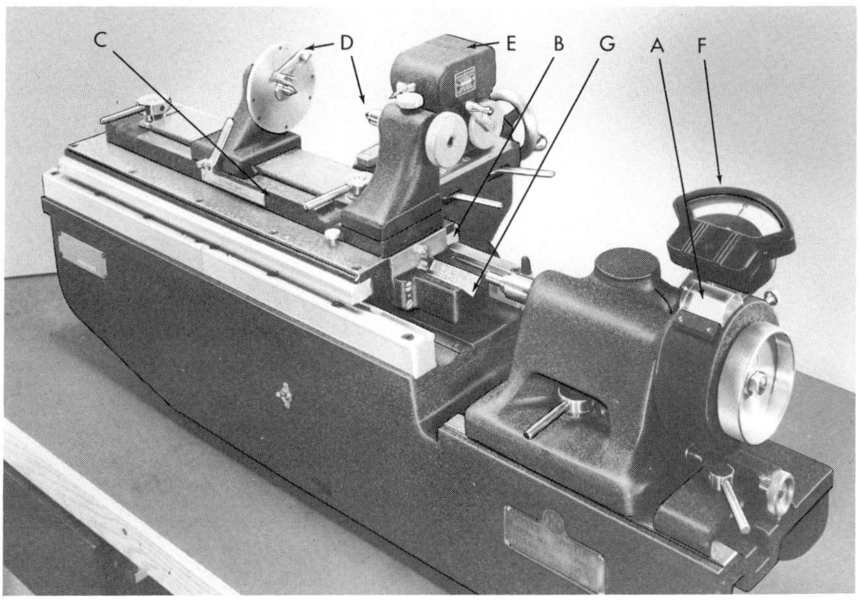

Figure 453. Electro-mechanical Lead Tester

directly on the base of the machine so that any movement of the carriage is made in relation to the spindle of the locating head. A milliammeter (F) mounted on the measuring head indicates, by the zero position of the pointer, when the thread-locating head is properly located in the thread groove.

MEASURING HEAD

The measuring head is identical to that of the standard measuring machine except that it does not have the microscope. Measurements less than one inch are made as described under "Standard Measuring Machine." Measurements greater than one inch are obtained by setting the whole inches with gage blocks. A gage block trough (G) (Figure 453) is provided for this purpose. Measurements can be made to an accuracy of .00001".

SINE BAR CARRIAGE

The sine bar carriage (Figure 454) carries a 15-inch sine bar pivoted at (A) and clamped in position by a clamp at (A) and (B). The angle of the sine bar, which is usually one-half the included angle of a tapered thread, is set by means of gage blocks at (C). A .100" gage block between the sine bar and the stop provides a means of accurately setting the sine bar to zero position. Therefore, when calculating the height of gage blocks for a given angle, .100" must be added into the calculation to replace this original .100" gage block. The formula for calculating the required stack of gage blocks is:

Height of gage blocks = 15 x (sine 1/2 included angle of taper) plus .100".

For example, the standard taper of the American National pipe thread is 3/4" per foot. This taper represents an included angle of 3° 35'.

Half the angle = 1° 47' approx.

Sine 1° 47' = .03112

Substituting in the above-mentioned formula:

H = 15 x .03112 + .100

= .5668 (height of stack of gage blocks)

Two center heads (D and F, Figure 454) are mounted on the sine bar. The center head (D) is rigidly mounted on the sine bar, but the center itself is adjustable by knob (E). The center head (F) is mounted on a dovetail slide to accommodate various work lengths. It is clamped in position by a clamp at (G). An index plate at (H) provides means of positioning the work for angular location and holding it in place with a dog. To insure accuracy of readings, constant pressure is maintained against

Figure 454. Sine-bar Carriage on Electro-mechanical Lead Tester

the anvil of the measuring head by means of a counterweight attached to the carriage.

AUTOMATIC STOP

The anvil, which makes contact with gage blocks or the spindle of the measuring head, is in the form of an adjustable contact stud (A, Figure 455). To safeguard the measuring-head spindle so that the carriage will not strike the spindle, a ratchet arrangement (called the automatic stop) has been installed (B, Figure 455). When the carriage is pushed to the left, the pawl (C) rides over the teeth of the rack (D). When the pressure is released, the pawl engages the rack immediately and holds the carriage in position. To release the carriage, press the carriage-release button (E), being careful to hold the carriage with one hand whenever automatic stop is released.

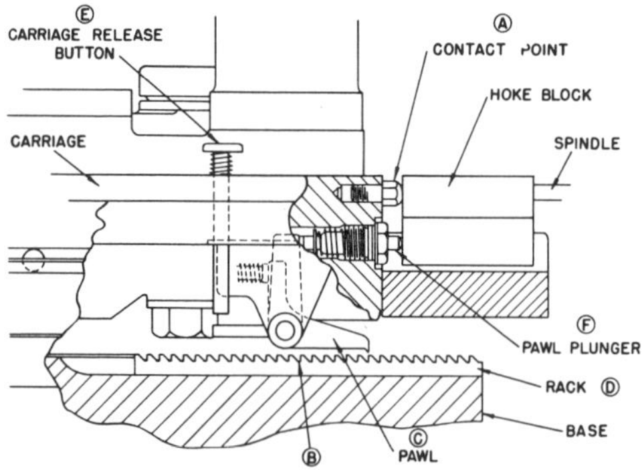

Figure 455. Automatic Stop

When gage blocks are in position, a pawl plunger (F) releases the automatic stop so that the anvil will make free contact with the gage blocks.

ELECTROLIMIT THREAD-LOCATING HEAD

The Electrolimit thread-locating head is mounted by means of two vertical reeds on a small cross-slide upon which it is moved in and out of contact with the thread by means of a handwheel (A, Figure 456). It is clamped in position with two clamps at (B).

A small crank at (C) provides means of moving the spindle of the locating head in and out of the thread groove without using the cross-slide; thus, the spindle can be moved clear of the thread while the carriage is adjusted to bring successive threads in line with the spindle.

The Electrolimit locating head operates on the same principle as other Electrolimit units discussed previously. In this particular unit, movement of the spindle is sidewise instead of in and out; however, the electrical principle involved is exactly the same.

A milliammeter located on the measuring head indicates when the exact measuring point has been reached. When making a measurement, the carriage is moved by means of the microm-

Figure 456. Thread-locating Head

eter drum on the measuring head until the spindle of the locating head is correctly positioned in the thread groove; at this position the pointer of the milliammeter will register zero. The coils of the Electrolimit measuring head are adjusted so that the movement of the spindle is magnified 3,000 times when read on the milliammeter.

The spindle of the locating head is equipped with a removable ball point which locates in the thread groove. A set of seven different ball points is available for different thread sizes. The table (Figure 457) shows the proper ball point to select for a particular thread size.

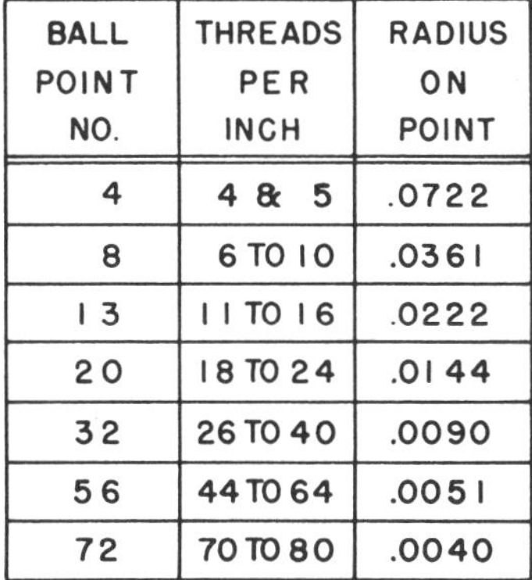

BALL POINT NO.	THREADS PER INCH	RADIUS ON POINT
4	4 & 5	.0722
8	6 TO 10	.0361
13	11 TO 16	.0222
20	18 TO 24	.0144
32	26 TO 40	.0090
56	44 TO 64	.0051
72	70 TO 80	.0040

Figure 457. Sizes of Ball Points for Electro-mechanical Lead Tester

Figure 458 shows how the ball point is mounted in the spindle of the locating head. The holder is first removed from the spindle; then the ball point is removed. Since the ball point has a tapered shank, it can be removed with a drift (Figure 459). The ball point corresponding to the size of the thread to be measured can then

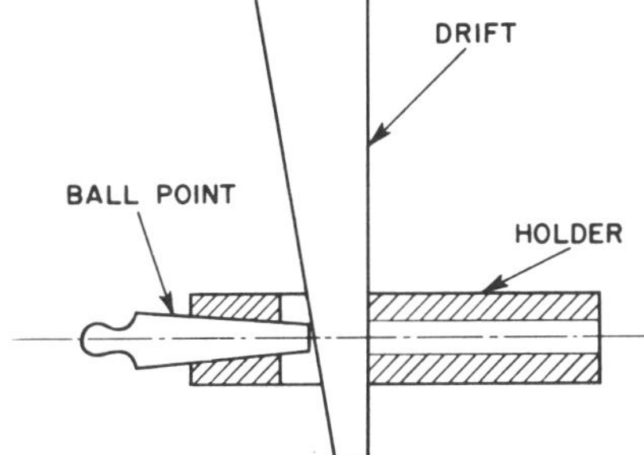

Figure 459. Removing Ball Point

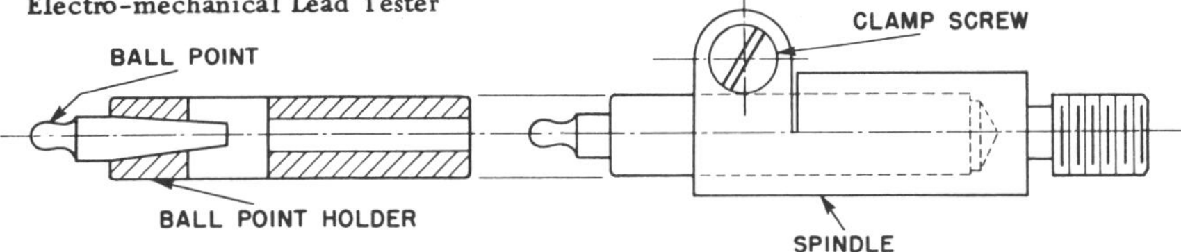

Figure 458. Ball-point Holder

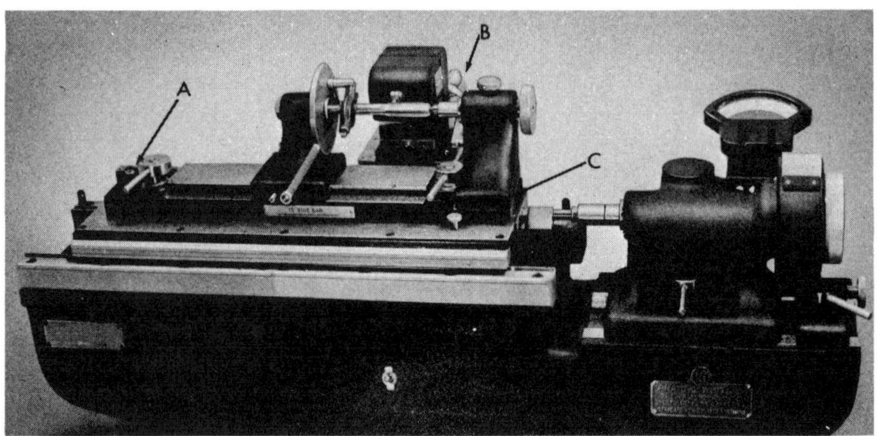

Figure 460. Measuring Lead of a Tap

be inserted into the holder and secured by lightly tapping it on wood or other soft material.

MEASURING STRAIGHT THREADS

Figure 460 shows a straight-thread tap in position for making a lead measurement.

Specified size of tap: 5/8 - 11 NC-3

Lead: .09091

Max. tolerance on lead: .0012 (consuming half the pitch diameter tolerance over the length of engagement)

Procedure:

1. Place a .100" gage block between the roller on the sine bar and the stop (at A, Figure 460). Move the sine bar carefully up to the gage block so as not to damage block or roller, and until proper contact is made between roller and gage block. Contact can be checked by sliding the gage block between the roller and stop, and adjusting the sine bar until a good sliding fit is obtained. Clamp the sine bar in this position, and the centers are in line with the axis of the machine.

2. Select a ball point corresponding to the pitch of the thread, which according to the table is no. 13. Insert the ball point in the spindle of the locating head as previously described. Bring the ball point to within 1/8" of the work by turning handwheel (B) of cross-slide. See that the crank is released so that the locating head is floating freely.

3. Release the automatic stop by pressing release plunger (C) and position the carriage so that the first full thread on the work is opposite the ball point. Insert the one-inch gage block between the anvil of the carriage and the spindle of measuring head. Adjust the measuring head so that the spindle makes contact with the block. Lock the adjusting block of the measuring head, move the measuring head with the fine adjusting screw, and at the same time adjust the ball point into the thread groove, making sure that the ball point contacts both flanks of the thread. Since the locating head is floating on two reeds, the spindle is safeguarded against impact, and end pressure is controlled by the flexibility of the reeds; however, care should be taken that the locating head is advanced far enough for good contact and not farther than necessary.

If, after the ball point is located in the thread groove, the pointer of the milliammeter is off-scale to the left, turn the adjusting screw on the measuring head adjusting block clockwise; if to the right, turn the adjusting screw counter-clockwise until the pointer is on zero.

Lock the measuring head in position; then make final adjustment of the pointer to zero with micrometer drum. Take the reading of the measuring head. The first reading is .70621.

4. With the crank on the side of the locating head, retract the ball point from the thread and move the carriage with the micrometer drum until the ball point is opposite the next thread to be checked. Assume it is the next adjacent thread, which should measure .09091" from the first setting. Release the

crank, allowing the ball point to enter the thread groove at the new location. Adjust the location with the micrometer drum until the pointer of the millammeter is on zero, and take the reading on the measuring head scales. This reading is .61528". Subtracting the readings (.70621 - .61528), gives .09093", which is well within the tolerance specified.

MEASURING TAPER THREADS

When measuring tapered threads, it is necessary to set the sine bar to one-half the included angle of taper. As previously stated, the formula for determining the stack of gage blocks for a particular setting of the sine bar is:

H = 15 (sine 1/2 included angle) + .100

which, for the standard pipe thread taper is .5057.

Inserting a stack of gage blocks of this dimension between the roller of the sine bar and the stop will bring one side of a tapered-pipe thread parallel to the movement of the carriage. Procedure for measurement of the lead is then identical to that of measuring straight threads; but because the axis of the work is now at an angle to the movement of the carriage, the reading must be corrected for this angle by using the following formula: Actual lead of thread = (measured lead) x (cosine of 1/2 included angle of taper). The angle of taper on the standard pipe thread is 3° 35'; half the angle is 1° 47' (approximately).

$$\cos 1^{\circ}\,47' = .99951$$

Therefore, for standard pipe threads, the foregoing formula resolves to:

Actual lead = .99951 x measured lead.

CARE OF MEASURING MACHINES

Accuracy of a measuring machine depends upon the care it receives. It should never be operated by anyone unfamiliar with its construction and operation. For best results, it should be operated by a limited number of persons who have been trained in its use.

A measuring machine should be kept clean at all times, and all exposed surfaces should be protected with a film of oil. When the machine is not being used, all clamps should be released, and it should be protected with the cover that is provided.

The ideal location for a measuring machine is an air-conditioned, enclosed room where temperature and humidity can be controlled. Relative humidity of the atmosphere should preferably be kept under fifty percent to lessen possibility of corrosion. If such a room is not available, care in locating the machine is essential. Location should be free from dust, grit, drafts, and direct sunlight; constant temperature should be maintained if possible.

SPRING TESTING

Springs are important components of many machines and other devices, and industry is becoming increasingly conscious of new methods and instruments for spring testing. To make springs to exact specifications that assure long and satisfactory service, it is necessary to have an accurate means for checking them; this is usually done with specially designed machines, called spring testers.

COIL SPRINGS

Coil springs are formed by winding round or rectangular wire or bars into cylindrical coils. Springs of small capacity are usually made of tempered-steel wire called music wire. Heavier springs are formed while the steel is hot, and tempered after forming.

Coil springs have the property of exerting axial pressure or tension in a direction opposite to a force acting upon them. Those which react against compression are called compression springs, and those which react against tension are called tension springs.

Tension springs (Figure 461A), sometimes called closed springs, are wound so that the turns lie one against the other, usually with an initial pressure between the turns. End turns are usually formed into eyes or hooks for convenience in attaching to machine parts. When a tension spring is subjected to tension or pull, the turns are pulled apart; if the elastic limit of the material is not exceeded, the spring will return to its original form when the tension

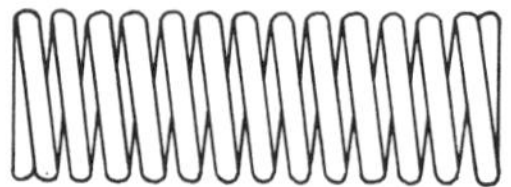

Figure 461. Coil Springs

is removed.

Compression springs (Figure 461B), sometimes called open springs, are wound so that there is a space between the turns. The two end turns are usually closed in, and the end is ground flat to provide a seat for the spring. When a compression spring is subject to compression, the turns are pressed closer together; if the elastic limit of the material is not exceeded, the spring will return to its original form when the load is removed.

In the design of machinery requiring the use of springs, it is essential that a spring have certain characteristics constant and within the limits specified. A typical specification for a coil compression spring is as follows:

Over-all length 1-7/8"
Outside diameter 15/64"
Diameter of wire .031 ± .0005"
Number of turns 31
Spring must compress to 1-1/4" ± 1/32" under a load of four pounds.

So it is necessary to be able to measure these characteristics accurately and quickly.

The spring tester in Figure 462 is designed to measure the characteristics of coil springs, both compression and tension. It measures the extension or compression of a spring, in relation to the load applied, to an accuracy of 1/32" on the main scale, and to an accuracy of .001" with the aid of a dial indicator attachment. The frame consists of a base, two columns, and a head. The base contains the mechanism for applying the load. The head contains the balancing mechanism, and the columns merely tie the base and head together.

The load is applied through a lever (A) which moves the lower compression plate (B) up and down by means of a rack and pinion. Compression springs are placed between the lower compression plate (B) and the upper compression plate (C). Tension springs are attached between two hooks (D) and (E), and the specified loads are placed on the tray at (F). The lower compression plate can be clamped in position by the star wheel at (G).

Figure 463 is a schematic drawing of the balancing unit. A knife-edge fulcrum (A) is attached rigidly to the head of the machine. A lever (B) is so balanced on the fulcrum at (C) that a downward pull at (D) is equal and opposite to an upward pull at (E). So, it is obvious that a downward pull at (D), the point where a tension spring is attached, will create the same amount of movement in a pointer at (F) as an equal

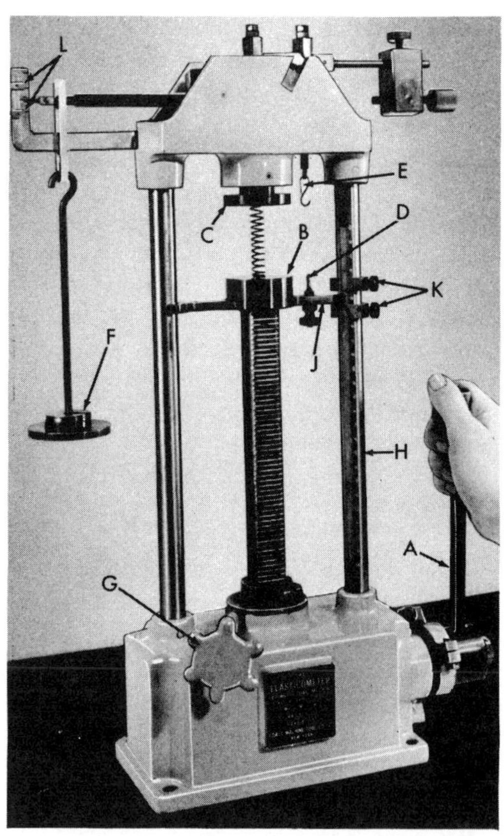

Figure 462. Spring Tester (Elasticometer)

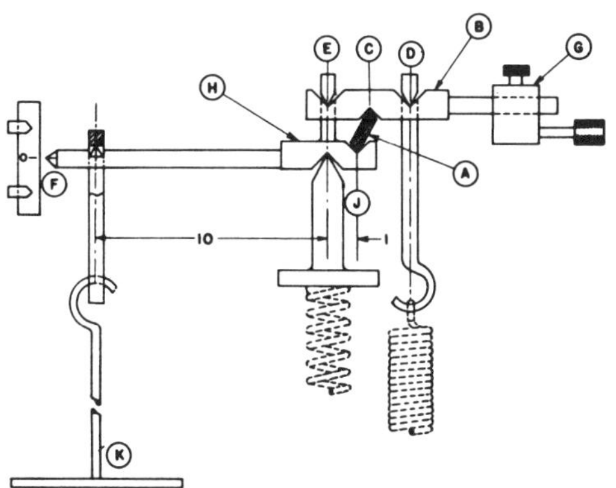

Figure 463. Schematic Drawing of Balancing Unit of Elasticometer

upward pressure of a compression spring at (E). Therefore, the same scale can be used for either tension or compression springs. An adjustable counterweight at (G) counterbalances the weight of the empty tray and provides a means of adjusting the pointer to zero before making a measurement.

A second lever at (H), having a fulcrum at (J), provides a ratio of 10 to 1 between the tension or compression of the spring and the weights placed on the tray at (K); thus, loads of 1 to 75 lbs. can be applied to the springs by placing on the tray weights equal to one-tenth of the applied load.

The scale is engraved on the right-hand column at (H) (Figure 462) and is read in relation to an index line on a yoke (J) attached to the lower compression plate. When the two pressure plates are touching each other, the index line is opposite the zero of the scale; therefore, the space between the pressure plates (or hooks) is indicated directly in inches on the scale.

Two adjustable markers at (K) are provided for setting the length of the spring on the scale. If a tolerance on the length of the spring is specified, the markers can be placed to indicate the limits on the length of the spring. If the tolerance is on the load, only one marker is used, and it is placed to indicate the specified length of the spring. The two adjustable markers at (L) can be placed to indicate the limits on the load specified for the spring.

Measuring Example

Objective: To check a tension spring on the Elasticometer.

Specifications for spring:

Free length 3"

Outside diameter 7/16"

Diameter of wire .050"

Number of turns 42

Must stretch to 4-1/16" under 6 lbs. 4 oz. load. Tolerance ± 2 oz.

PROCEDURE (Figure 464)

1. Hang the spring on the hook (E) before setting the machine to zero, so that the weight of the spring will not counterbalance part of the load, thus introducing an error in the measurement. Set the pointer to zero by adjusting the counterweight at (M); then place on the tray weights corresponding to the specified load 6 lbs. 4 oz.

2. With lever (A) raise the lower compression plate until the lower hook of the spring

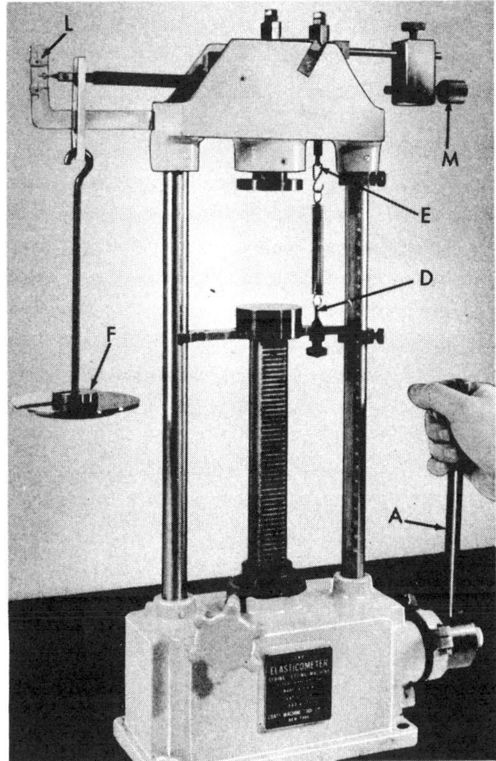

Figure 464. Measuring a Tension Spring

can be attached to extension (D); then apply load to the spring by lowering the compression plate until the index line registers 4-1/6" on the scale. In this position, the pointer should register near zero at (L).

3. To set tolerance markers for load. Select a spring that registers zero at (L) under the specified load 6 lbs. 4 oz. at the specified spring length of 4-1/16". With the spring under load and the pointer at zero, select a 2-oz. weight and add it to the weight on the tray. The pointer will now move down to a new position. Mark this new position of the pointer with the lower marker and place the upper marker an equal distance above the zero position. The markers now mark the limits of the load specified for the spring.

4. Remove the 2-oz. weight from the tray, leaving 6 lbs. 4 oz., and the machine is ready for checking the remainder of the springs made to the foregoing specifications.

If the springs are too strong, the pointer will move up and beyond the upper marker; if too weak, the pointer will fall below the lower marker. All springs registering between the two markers will be within the tolerance specified.

BUCKLING

When measuring compression springs, whose lengths are four times the diameter or greater, the springs have a tendency to buckle under load (Figure 465). Under this condition, an error will be introduced in the measurement, because the actual length of the spring will be longer than that indicated on the scale of the machine.

To prevent long compression springs from buckling, provision has been made in the compression plate for mounting a centering pin, whose diameter approximates the inside diameter of the spring (Figure 465C).

The diameter at (D) fits the hole in the lower compression plate; the flange at (E) keeps the pin upright, and the length and diameter at (F) vary according to the length and diameter of the spring.

Caution: when using pins of the type shown in Figure 465, the spring rests upon the flange;

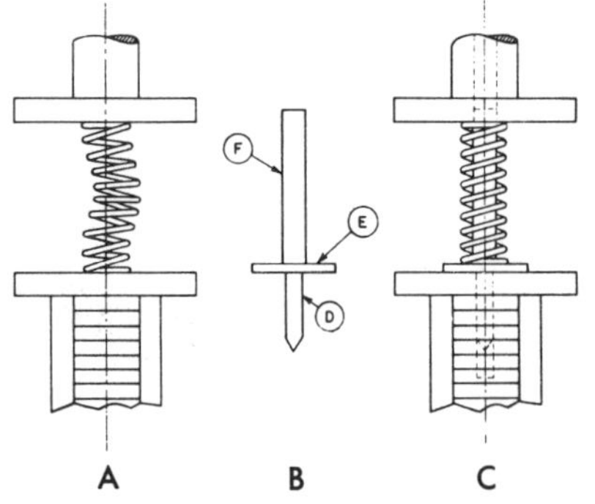

Figure 465. Buckling of Compression Springs

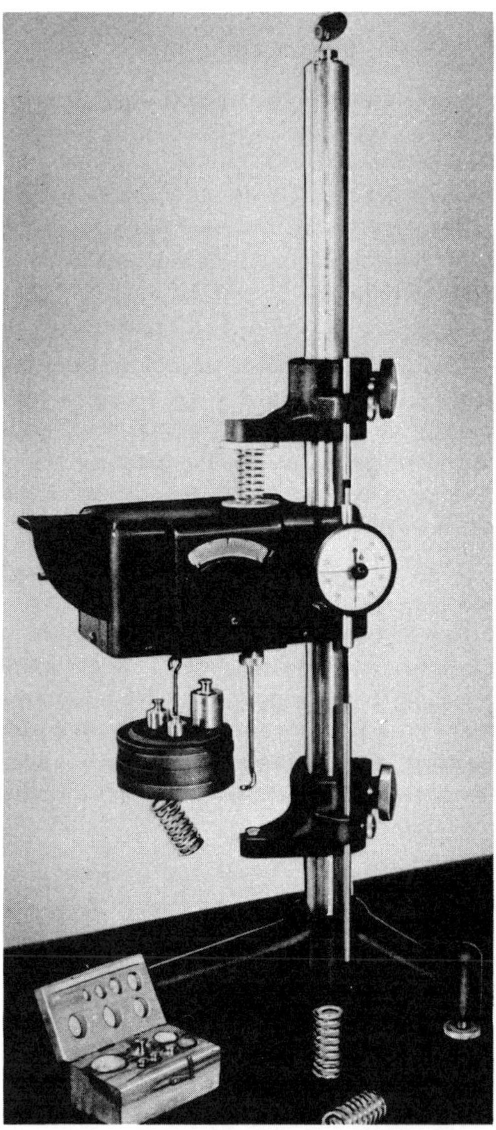

Figure 466. Spring Tester (Baldwin-Hunter)

therefore, the thickness of the flange at (E) must be added to the dimension specified for the length of the loaded spring and the marker on the main scale set accordingly.

When checking a large number of springs to the same specifications, an adjustable collar at (M) (Figure 462) can be set in relation to a stop on the base so that the stop is reached when the specified loaded length of the spring is indicated on the scale. This feature enables the operator to check the springs without referring to the scale on the column. The lever is adjustable at (N) so that it will not be in an awkward position at the end of its travel.

The spring tester shown in Figure 466 is similar in operation to the one shown in Figure 462 and is designed for the accurate testing of small compression and extension springs up to 5 lbs. load capacity and 12" in length. This instrument is sensitive to load changes of less than 50 milligrams, and the load applied to the weighing head is accurate to $\pm$ 0.1 percent. The dial indicator reads the spring length to $\pm$ .001".

The tester may be used in three different ways, depending on which one best suits the particular need: (1) to determine the exact loading of each spring required to compress or extend it to a specified length; (2) to find the exact compressed or extended length of a spring at a specified load; (3) to use for "go" "not-go" testing by reading for each spring being tested, the position of the beam pointer with both the beam loading and the position of the compressed or extension head fixed and invariable. The position of the pointer is read between limits, indicated by red tolerance markers.

QUESTIONS AND PROBLEMS

1. What effect has temperature change on all metal parts? How is this controlled when making an accurate linear measurement?
2. Why is the control of measuring pressure necessary when making accurate linear measurement?
3. How does the Supermicrometer differ from the ordinary micrometer?
4. What is the range of the Supermicrometer head and how accurately can it be read?
5. When making three-wire measurement of threads, what might happen if the measuring pressure were too great?
6. When are gage blocks used with the Supermicrometer?
7. Why is the vernier adjustable on the Supermicrometer?
8. When must the complete formula for measuring threads with the three-wire system be used?
9. How does the standard measuring machine differ from the Supermicrometer? What particular feature have they in common?
10. Describe the measuring bar on the standard measuring machine and explain how it is used.
11. How accurately can measurement be made on the standard measuring machine?
12. Name five steps in setting the standard measuring machine.
13. Define the "lead" of a thread.
14. How does the measuring head of the Electro-mechanical lead tester differ from that of the standard measuring machine?
15. What is the purpose of the sine bar on the Electro-mechanical lead tester?
16. What is the purpose of the .100" gage block? How does it enter into the calculation of the height of blocks required to set the sine bar to a given angle?
17. How are measurements greater than one inch obtained with the Electro-mechanical lead tester?
18. The included angle of a tapered thread is 3 degrees, 36 minutes. At what angle would you set the sine bar when measuring the lead of the thread?
19. Why is a variety of ball points provided for use with the locating head? What governs the size ball point that should be used?
20. What two important characteristics of coil springs are measured on the spring tester?
21. What is the purpose of the marker on the scale? Is the same scale used for tension and compression springs?
22. Give three important rules that should be observed in the care and operation of measuring machines.

Chapter XIII

SURFACE ROUGHNESS

ALL MECHANICAL devices involve the movement of one surface upon another; this movement causes friction which creates wear and heat. The amount of friction governs to a great extent the speed of the machine and the load that may be imposed upon it.

The smoother the surfaces, the less friction--and the greater the permissible speed and load. A loaded canalboat weighing many tons can be drawn through the smooth water of a canal by a single horse; yet to draw the same boat over rough ground would require so many horses and would entail so many difficulties that it would be entirely impractical.

Lubrication of two metal surfaces in moving contact reduces friction by interposing a film of oil between the surfaces; here again, however, the smoothness of the surfaces governs the speed and the load that can be applied to the surfaces without breaking down the oil film between them.

Surface roughness is a condition; therefore, it cannot be specified as a linear dimension. So that the quality of surface roughness can be measured and specified in definite terms, special instruments have been designed, and a unit of measurement has been adopted.

NOMINAL SURFACE

To measure the characteristics of a surface, it is usually necessary to have a reference plane to which these measurements can be referred. This reference plane, called the nominal surface (Figure 467), is the two-dimensional boundary of separation between a theoretically perfect surface and the surrounding medium; this boundary corresponds to the plane represented by an object line on a drawing. Any departures from the nominal or theoretically perfect surface are called surface deviations, and take the forms of surface irregularities such as, flaws, waviness, roughness, lay, and profile.

SURFACE FLAWS are irregularities that occur at one place or at relatively infrequent intervals. They may take the form of a scratch, ridge, hollow, crack or similar surface flaw.

WAVINESS is a kind of surface deviation consisting of recurrent or random irregularities on a surface in the form of waves, the peak-to-peak distance of which may be from 1/32" to several inches (Figure 468).

ROUGHNESS is the relatively finely-spaced irregularities superimposed on a flat or wavy surface (Figure 469). On surfaces produced by machining and abrasive operations, roughness includes the irregularities produced by the cutting action of tool edges and abrasive grains, and by the feed of the machine tool. Roughness

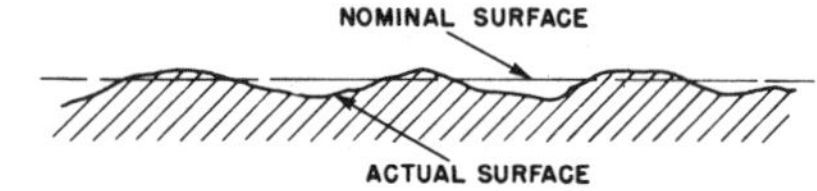

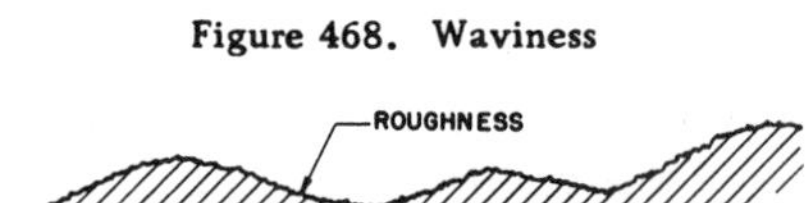

Figure 467. Nominal Surface

Figure 468. Waviness

Figure 469. Roughness

measurement deals principally with the measurement of these finer irregularities. They are rarely more than a few thousandths of an inch in width and much less in depth, generally between 0.000001" and 0.0005".

Height of surface irregularities is the most important factor in classifying and rating the surface in terms of roughness; for all practical purposes this value may be expressed directly in inches. The height and depth of the individual peaks and valleys, however, are not as important as a running average in a distance of 1/8" or more of the surface. A comparison of such averages measured at different locations on the surface is far more useful than the observation of the individual peaks and valleys.

LAY is a term used to designate the direction of the predominant surface pattern. For example, milling, turning, and grinding produce definite direction of lay, while a lapped finish does not and so is designated as multi-directional.

PROFILE is the contour of the roughness and the waviness of a surface.

MICROINCH is a unit of measurement equal to one millionth of an inch (0.000001").

The symbol recommended by the American Standards Association to designate surface irregularities is the check mark and a horizontal extension. Figures 470A and B illustrate the meaning of the symbols and the definitions of waviness, roughness, and lay.

Roughness measurements are taken across the lay of a surface, or in the direction that indicates the greatest roughness, unless otherwise specified. Surface flaws are not considered a part of the roughness measurement. The table (Figure 471) shows the approximate range of roughness produced by various machining operations. For example, a milling machine may be expected to produce a roughness between 16 and 250 depending on the speed, feed, type of cutter and other variables encountered in the operation of the machine.

ROOT-MEAN-SQUARE AVERAGE

Surface roughness is usually specified as a number representing the root-mean-square (r.m.s.) average of surface roughness expressed in microinches. If a cross-section were made of a typical surface, the irregularities would appear somewhat as shown in Figure 472. As an average of the extreme heights and depths of the roughness would not represent a true average of the irregularities, a root-mean-square average of all the individual irregularities is used as a basis for roughness measurement.

Figure 472 shows an example of a horizontal center line drawn through the outline half way between the peaks and valleys at equally-spaced intervals; vertical lines a, b, c, d, e, f, g, etc., represent the amount the actual surface varies from the nominal surface at these points. A root-mean-square of these values is obtained as follows:

$$\text{Root-mean-square} = \sqrt{\frac{a^2 + b^2 + c^2 + d^2 + e^2 \ldots}{n}}$$

n = number of values used

For practical purposes the root-mean-square average equals about one-third of the maximum value.

ROUGHNESS MEASUREMENT

There are numerous methods of determining the quality of surfaces, including microscopic comparison, chemical analysis, electron diffraction, and profile measurement. Instruments that employ a point or stylus to investigate the irregularities and record them greatly magnified are the most practical for shop applications. Two such instruments are the Profilometer and the Brush surface analyzer, both of which will be described as examples of roughness measuring instruments.

THE PROFILOMETER

The Profilometer is a mechanical-electronic instrument for measuring surface roughness in microinch units (millionths of an inch); this instrument employs the stylus or tracer-point method of measurement.

The Type Q Amplimeter described here has no calibration circuit; in older models this circuit was required for adjusting the instrument to the sensitivity of individual tracers. All tracers can be used with the older model amplimeters, P and PAC, which have calibration

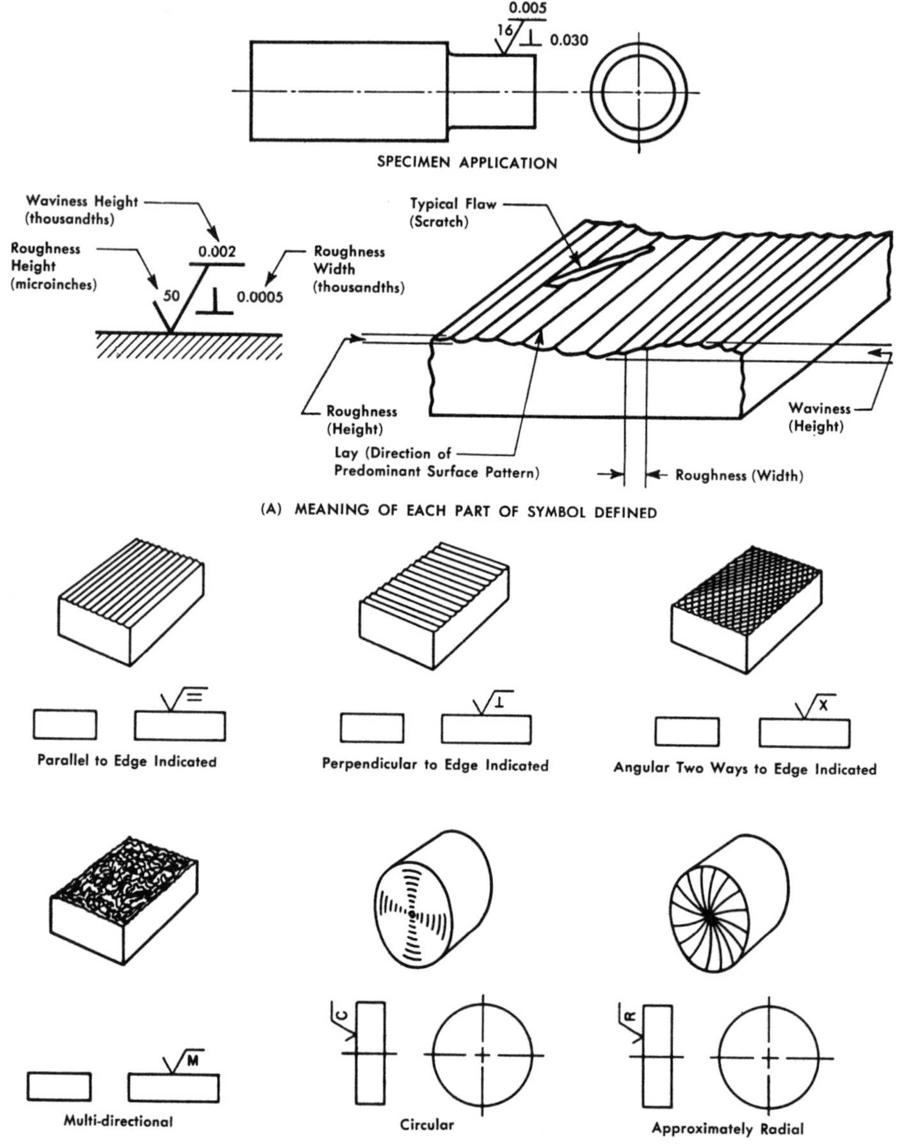

Figure 470. Definitions of Surface Roughness

circuits. Tracers with a calibration of 40 are used with the newer Q amplimeters.

Figure 473 shows the Type Q Profilometer. It consists of two main units--a tracer and an amplimeter.

The tracer is moved over the surface being measured, either by hand or by an accessory motor-driven tracing mechanism. The tracer has a tracing point which rides lightly on the work surface and converts the vertical movement of the tracing point into a small fluctuating voltage.

The amplimeter consists of an electronic amplifier, inside the case, which amplifies the voltage produced by the tracer so that it actuates the meter on the panel. The meter shows the r.m.s. average height of the surface irregularities directly in microinches. A scale selector switchon the panel provides six full-scale meter readings of from 3 to 1,000 microinches.

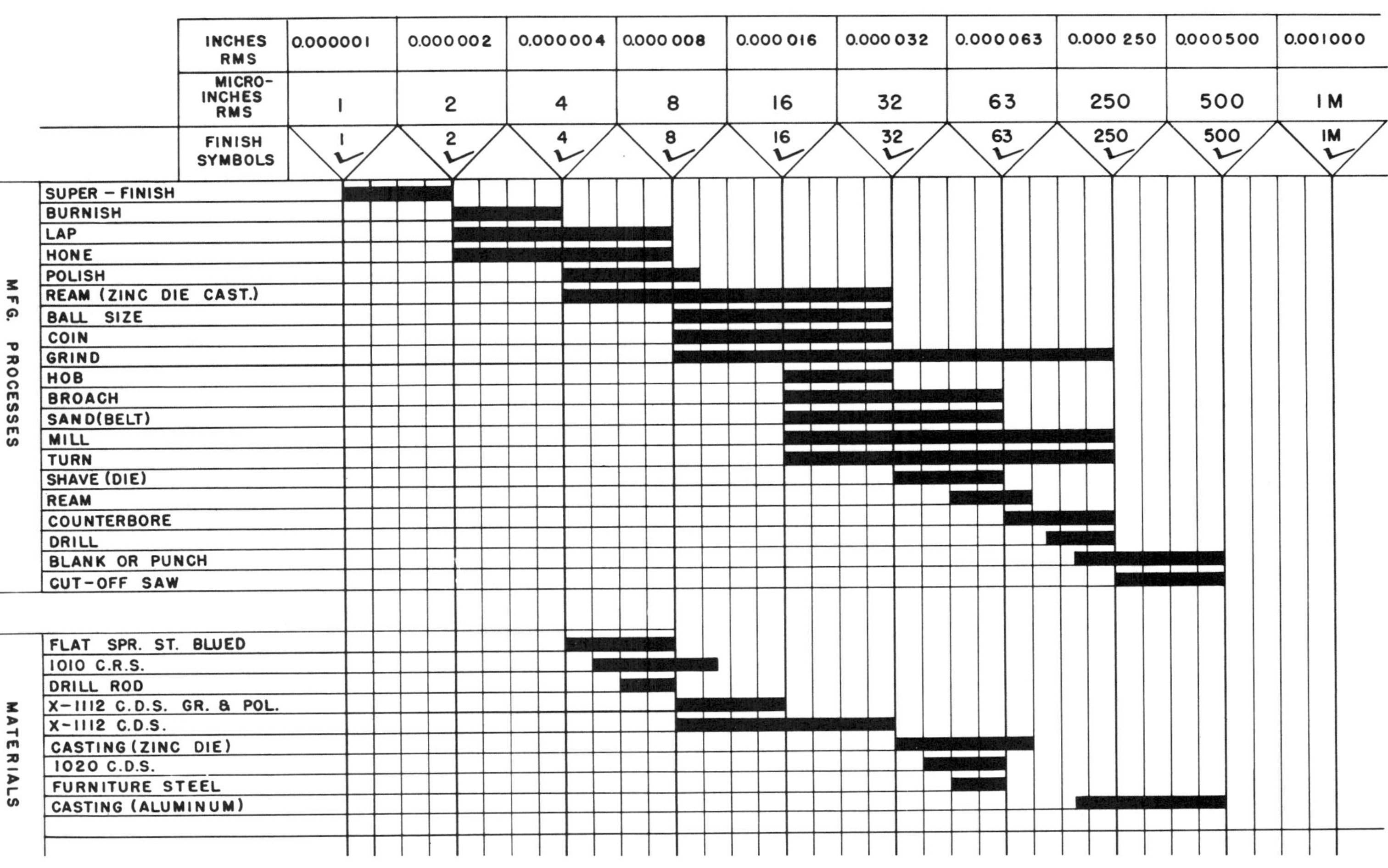

Figure 471. Approximate Range of Surface Roughness

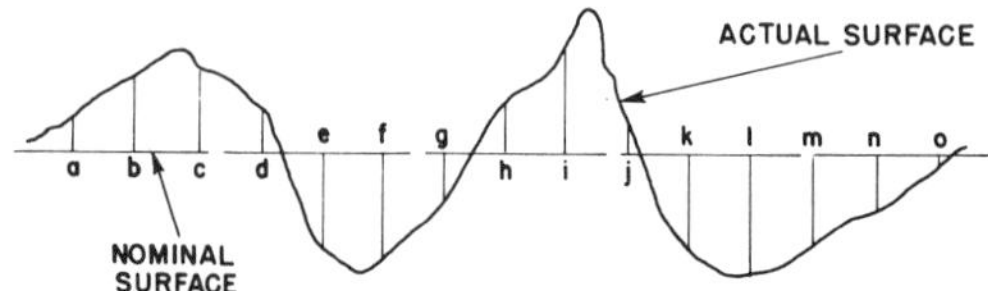

Figure 472. Root-Mean-Square Average

TRACERS

The M.A. tracer unit (Figure 473) is sufficiently small to hold in one hand for movement over a surface, and can be operated mechanically if desired.

Figure 474 shows a cross-section of the tracer unit. It consists of skids or feet which establish the reference plane, and a stylus which, as it moves up and down in relation to the skids, explores the roughness of the surface. A coil mounted on the stylus is energized by moving through a magnetic field created by the magnet and pole piece; the amount of this induced current is in proportion to the vertical movement of the coil, and is a direct measurement of the roughness.

Figure 475 shows how the skids establish the reference plane by riding over the roughness, while the stylus, with its fine diamond point, explores the hills and valleys and moves up and down while the tracer is being moved over the surface.

The diamonds used for the tracer points are small pieces of about the same grade as that used for jewelry. These small pieces are formed into circular cones having an included angle at the tip of 90 degrees. This cone is terminated in a spherical tip having a radius of .0005". To obtain a rugged point, it is desirable to have as large a tip radius as possible. All factors considered, the tip radius of .0005" is sufficiently small to bottom the principal irregularities of the surface, although it would not be small enough to give a true condition of the surface character if profile records were required.

A variety of interchangeable skids is available for adapting the tracer to various-shaped surfaces. Cemented-carbide skids support the tracer on the work. In Figure 476, A is used for flat surfaces or surfaces of large radii; B is used for O.D.'s from 3/8" to flat; C is identical to A, except that it is cut away to permit measuring close to shoulders of 7/32" height or less and can also be used on concave surfaces 1-13/16" in diameter and larger. Other skids are available for measuring smaller diameters and bearing

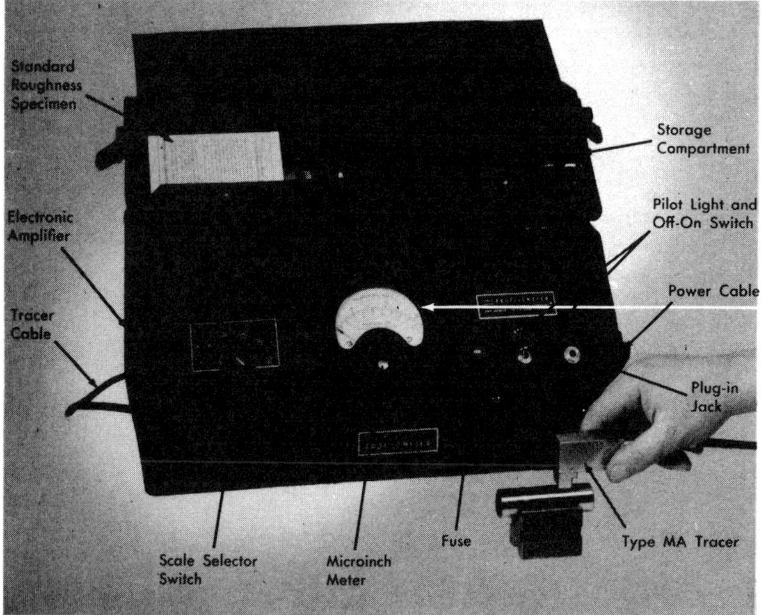

Figure 473. Type Q Profilometer and MA Tracer

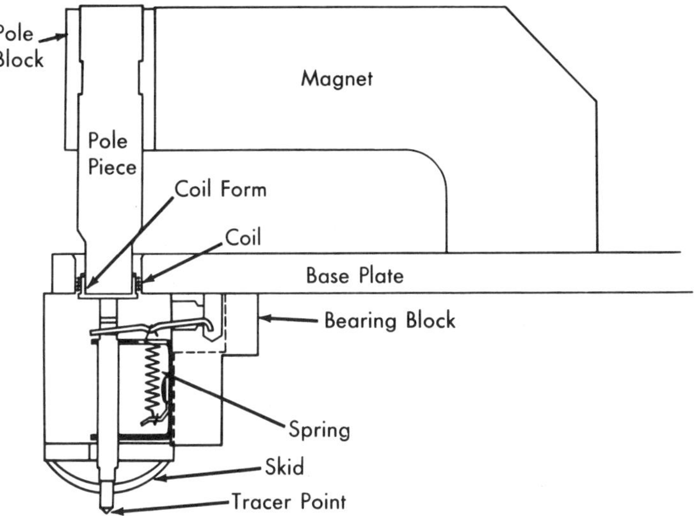

Figure 474. Cross-section of Profilometer Tracer Unit

balls up to 1" diameter. The standard M.A. tracer can be used on both straight and tapered surfaces, internal and external, and is completely self-adjusting to all surfaces within its range.

Several types of special tracers are available for measuring small internal and external surfaces, such as holes as small as 5/64" diameter, and slots as small as 1/16".

Because of its speed and all-around convenience, manual operation of the tracer is desirable in many cases, and makes possible the measuring of many surfaces that cannot otherwise be measured, such as parts being machined, out-of-the-way bearing surfaces on large parts, and surfaces up to several feet in length. When measuring manually, the tracer can be moved in any desired direction at a rate of about one inch in three seconds.

Mechanical operation of the tracer by means of a Mototrace or Pilotor is the most practical method in other cases--for example, with short tracing stroke (below 1/4") and on extra smooth surfaces, where a steady vibration-free tracing motion is required for greatest accuracy. Mechanical tracing permits the use of tracers without supporting skids, for entering small holes and other hard-to-reach surfaces. All Profilometer tracers with skids can be operated mechanically.

Figure 477 shows the Type A Linear Pilotor in use with a Type KA tracer. This fixture is used for operating tracers which do not have skids. It includes a vertical column and a base plate.

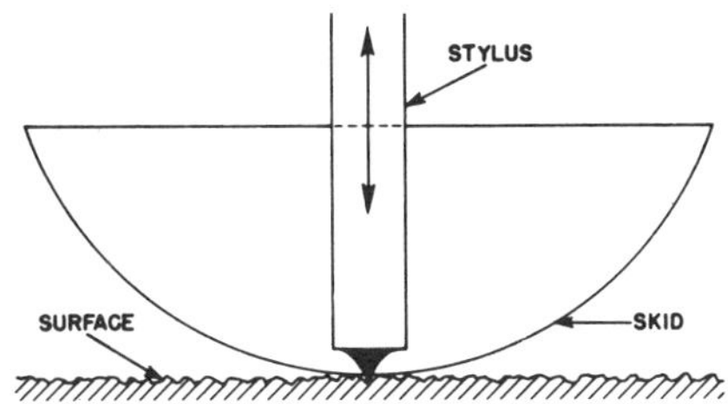

Figure 475. Skid Establishes Reference Plane

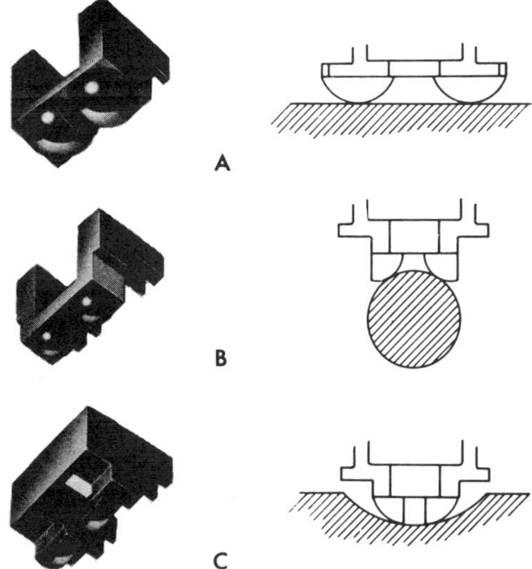

Figure 476. Interchangeable Skids

Figure 477. Linear Pilotor with KA Tracer

The column carries a saddle, adjustable vertically by the handwheel; the saddle, in turn, carries a motor-driven reciprocating slide which supports and moves the tracer over the work. Tracers used with this fixture frequently are adjusted to a trace length of less than 1/4". For this reason, the speed of the slide is 0.1" per second instead of the normal tracing speed of about 0.3" per second. At this slower speed the tracer will not read irregularities spaced more than 0.010" apart; consequently, this Pilotor is not recommended for surfaces much rougher than 100 microinches. Any desired length of trace is available from 1/32" to 1-1/8". The Type KA tracer measures in small holes or slots and on all flats and O.D.'s straight or tapered where the parts can be mounted with the surface horizontal.

Figure 478 shows the Type V Mototrace and the Type J tracer being used to check the surface roughness of the hole in a gear. The Mototrace is similar in operation to the Pilotor. It is a compact and portable motor-driven mechanism for mechanical tracing with several types of tracers and permits Profilometer readings to be taken right on the job as parts are produced. It is equipped with a link arm to which the tracers and amplimeter cable are connected. A coupling attaches the link arm to the Mototrace post, and has a horizontal swivel, clamped bv an angular adjustment clamp, and swivels freely in the vertical plane to permit tracing on tapered or non-horizontal surfaces. The tracing speed is about 0.3" per second over any length from 1/16 to 2-3/4".

The Type J tracer shown in Figure 478 is used with a variety of interchangeable "snouts" which protect the tracer beam and carry the skids, and permit changing the working range of the tracer for different applications, such as holes, O.D.'s and flat surfaces. This tracer is usually operated mechanically by the Mototrace and link arm. With suitable skill, however, manual tracing can be done, with the tracer attached to a standard stiff-arm or handle, which facilitates measuring into cylinder bores or other hard-to-reach surfaces, by adding about 9" to the reach of the tracer.

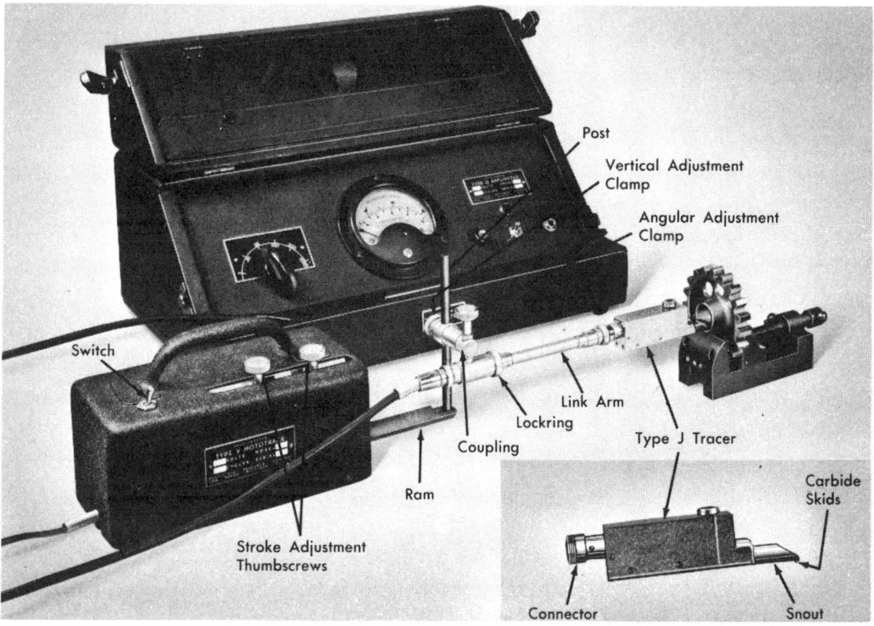

Figure 478. Type V Mototrace and Type J Tracer

SETUP AND OPERATION

Accurate readings with the Profilometer are readily obtained by this sequence of instructions.

1. Place the amplimeter on a sturdy work bench or similar support, located as far as possible from vibrating machinery and at least 10 feet away from strong magnetic fields (heavy power lines, transformers, generators, etc.).

2. Plug the power cable into an AC outlet providing the voltage and frequency specified on the plate on the meter panel. The amplimeter is normally furnished for 115-volt, 50/60-cycle power supply; but check the data plate to be sure. Variations in line voltage between 90 and 130 volts will not affect the operation.

NOTE: The amplimeter should not be connected to power lines supplying heavy induction equipment such as induction furnaces, welders and induction motors which are frequently started and stopped. Such equipment may cause erratic jumping of the meter pointer.

3. Turn the off-on switch on the meter panel to "on" position and let the amplimeter warm up for about five minutes before taking roughness readings. It is best to leave the instrument turned on throughout the working day.

4. Plug the tracer cable into the required tracer and tighten the knurled locking ring on the connector.

5. Prepare the work to be measured, keeping the following points in mind:

(a) The surface over which the tracer will move should be thoroughly cleaned of foreign particles such as dust, metal chips or shavings, and abrasive grit. Otherwise, the meter pointer will jump erratically during the tracing stroke, and the tracer will be subjected to excessive wear. Excessive oil or grease should also be wiped off so that they cannot get on the tracer point and gum up the moving parts. However, a thin film of clean light oil is not objectionable and often permits smoother tracing.

(b) Provide a steady support for the work, with the surface to be measured held approximately horizontal. Many parts are large and heavy enough to stay in position by themselves while the tracer is being moved across the surface. Light parts, or parts which have a tendency to roll, may be held on a surface plate with a dab of Plasticene or modeling clay; cylindrical parts can be placed in a V-block. Parts of irregular shape can be held in a vise.

(c) For particularly accurate measurement in all cases, and especially for very smooth surfaces (five microinches or less), the parts should be placed on a surface plate supported on sponge rubber pads about 1" thick to reduce the effects of extraneous vibration. The amplimeter can be conveniently located behind or beside the surface plate.

Mechanical tracing with a Type V Mototrace or a linear pilotor is recommended for surfaces smoother than five microinches. The Mototrace and the part to be measured must be placed on the same surface plate; and with a linear pilotor, the work must be placed on the pilotor base plate. Do not place resilient material between the Mototrace and the surface plate or between the work and surface plate or pilotor base plate. Otherwise, excessive vibration will be set up between the Mototrace or pilotor and the work.

6. Check the setup for vibration. To do this, place the tracer on the standard roughness specimen, set the scale selector switch at 3, and observe the microinch meter with the tracer stationary. If the meter reading is less than one-fourth of the average roughness value of the part to be measured, the error in roughness readings caused by vibration will be negligble and can be ignored. (This fact can be confirmed mathematically or by a comparison test.) If the reading is greater than that amount, vibration should be reduced, either by relocating the setup or by using a rubber-mounted surface plate as described above.

7. With the scale selector switch at 3, move the tracer slowly over the work and watch the meter. If the meter reads full scale, move the scale selector switch to the next higher setting. Then trace the work again and watch the meter. Move the scale selector switch to another setting if necessary, until the meter gives as high a reading as possible without the pointer going off the scale.

Where the approximate roughness of the part is known, this procedure can be shortened by turning the scale selector switch to the setting required by the roughness of the work. In any case, it is a quick and simple matter to arrive at the proper setting.

8. Trace the work and observe the roughness reading.

TRACING THE WORK

Either manual or mechanical tracing can be used with the general-purpose Type MA tracer, and the tracer can be moved in any desired direction--sideways as well as back and forth. Other tracers are normally used under conditions which require mechanical tracing.

Detailed instructions are given in separate manuals for each tracer. Before tracing the work, the operator should refer to the proper tracer manual and carefully follow the instructions given.

READING THE METER SCALES

As shown in Figure 473, the microinch meter has two scales: (1) the upper scale, graduated from 0 to 30, and (2) the lower scale, graduated from 0 to 10.

The scale selector switch has two corresponding scales: (1) the upper scale, with settings of 3, 30, and 300, and (2) the lower scale, with settings of 10, 100, and 1,000.

Note that the upper scale on the meter is used with the upper settings of the scale switch (3, 30, and 300); the lower scale on the meter is used with the lower settings of the scale selector switch (10, 100, and 1,000); and the figure at which the scale selector switch is set shows the FULL-SCALE or MAXIMUM READING of the meter in microinches.

To insure correct reading of the meter, first note the setting of the scale selector switch.

When the scale selector switch is set at 3, 30, or 300, read the upper scale on the meter as follows:

3-scale: read 10 as 1, 20 as 2, 30 as 3, etc.

30-scale: read 10 as 10, 20 as 20, 30 as 30, etc.

300-scale: read 10 as 100, 20 as 200, 30 as 300, etc.

When scale selector switch is set at 10, 100 or 1,000, read the lower scale on the meter as follows:

10-scale: read 2 as 2, 4 as 4, 6 as 6, etc.

100-scale: read 2 as 20, 4 as 40, 6 as 60, etc.

1,000-scale: read 2 as 200, 4 as 400, 6 as 600, etc.

The operator should make a few practice readings with the scale selector switch at the various settings before proceeding with his work. Otherwise serious errors may be made in measuring surface roughness.

TAKING THE ROUGHNESS READING

As the tracer is moved along the work, the meter pointer does not stay steadily at one position. Instead, the pointer moves back and forth within a certain range of roughness values, as indicated in Figure 476. This is because the work itself varies in roughness from point to point along its surface. So, the operator must watch the meter and estimate the average position of the pointer. This is readily done as follows:

When tracing a typical surface, the pointer might "wobble"--for example, between 25 and 35 microinches. If it fluctuates evenly throughout this range, the correct roughness reading will be half way between 25 and 35 microinches--that is, 30 microinches. If the pointer stays most of the time in the neighborhood of 25 and swings to 35 only occasionally, the correct reading will be correspondingly lower--say 27 or 28 microinches. After a little practice it is easy to arrive at the proper average reading.

In many cases, the range of pointer movement will vary considerably from point to point along the surface traced. For example, the pointer may fluctuate between 25 and 35 on one part of the work, and between 30 and 40 on another part. Such variations are generally caused by variations in the machining, grinding, or finishing process. In such cases, when the average roughness over about one-quarter of the surface is greater than the maximum roughness specified, the part is considered too rough and is rejected. Conversely, when a minimum roughness is specified, if about one-quarter of the surface shows an average roughness less than that specified, the part is considered too smooth.

ROUGHNESS TOLERANCE

In general, good machined or finished surfaces will give a range of pointer movement of about 10% above and below the specified roughness--that is, a variation of plus or minus 10%. Variations smaller than this figure are exceedingly rare. Variations of plus or minus 15% are commonly obtained under average conditions; while a variation of plus or minus 20% indicates a surface of below-average uniformity.

These same figures commonly apply to variations in the average roughness along the length of trace. Thus, good machined or ground surfaces may vary about 10% above and below the specified roughness from one end to the other. Variations of plus or minus 15% are commonly obtained and are generally considered satisfactory. Surfaces whose average roughness varies more than plus or minus 20% from end to end are of large non-uniformity. Note that in any case, where more than about one-quarter of the surface is beyond specifications, the part is usually rejected.

These figures are offered only as an approximate guide to accepted practice. In any given case the allowable variations both as to range of pointer movement and average roughness along the work must be determined by the requirements of the workpiece; the allowable variations should be specified on the drawings for the guidance of the Profilometer operator.

CHECKING PROFILOMETER ACCURACY

The condition of both the amplimeter and tracers is most readily checked by tracing the standard roughness specimen furnished with the amplimeter. The range of roughness of this specimen is marked on the label on the back, and the reading should be within those values. If the reading is outside these limits, take another reading using a different tracer. If the same "wrong" reading is then obtained, the amplimeter is not operating correctly; and if a correct reading is obtained with the second tracer, the first tracer is at fault.

NOTE: The roughness specimen must be thoroughly clean before it is used, for a dirty specimen will give a wrong reading. Also, make sure that the amplimeter has been turned on for at least five minutes before making this check; otherwise the reading may be incorrect.

Glass is used for the roughness specimen because this provides the most uniform surface obtainable. Furthermore, it is hard and has good wearing qualities; it does not corrode; and, being brittle, its surface has sharper valleys than any metal surface and thus more readily detects a dull tracer point.

OPERATING PRECAUTIONS

Type Q amplimeters and tracers are sturdily constructed and require only the usual care and handling given any precision tool or instrument. The most delicate and easily damaged part of the equipment is the tracing point and associated mechanism at the nose of the tracer; precautions to be observed are given in separate instruction manuals furnished with each tracer.

THE BRUSH SURFACE ANALYZER

The Brush surface analyzer (Figure 479) is an instrument that measures and records the irregularities of surface roughness. It is similar in principle to the Profilometer because it employs a stylus which explores the surface and amplifies the minute impulses electrically so that they can be recorded or indicated on an electrical meter. It differs from the Profilometer in the design of the analyzer unit, which corresponds to the tracer unit of the Profilometer, and in the method of recording the roughness of the surface.

The surface analyzer makes instantaneous, permanent chart records of irregularities in surface roughness. It can be used on metal, glass, plastics, paper, and plated or painted surfaces. The surface under test is traced with a fine stylus, and the motion of the stylus is electrically magnified and automatically recorded on a paper chart in such a manner that the chart contains a permanent record of:

1. The number of irregularities traversed by the stylus.
2. The depth of each irregularity.
3. Whether irregularities are above or below the normal surface.

The surface analyzer (Figure 479) consists of four principal units: a surface plate (A), an analyzer head (B), an amplifier (C), and an oscillograph (D).

SURFACE PLATE

The function of the surface plate (A) (Figure 479) is to establish a solid reference surface for work and analyzer head. In all measurements with the surface analyzer, it is important that the work and the analyzer head be on the same solid support, so that the record will not be affected by vibration or variations between the

Figure 479. Brush Surface-Analyzer

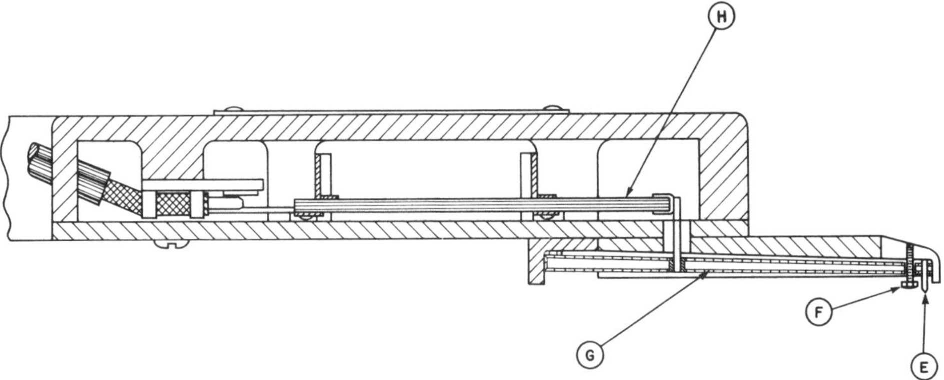

Figure 480. Surface-Analyzer Tracer Arm

support of the work and that of the head.

ANALYZER HEAD

The analyzer head (B) (Figure 479) consists of a tracer arm and a drive unit, mounted on an adjustable stand. The stand can be moved about on the surface plate to accommodate various sizes of work.

The tracer arm (Figure 480) contains a stylus at (E) which explores the roughness of the surface and a positioning stud at (F) which establishes the reference plane or normal surface with which the movement of the stylus is compared. The positioning stud performs the same function as that of the skid of the Profilometer tracing unit.

The movement of the stylus is transmitted by means of a lever (G) to one end of a piezoelectric crystal at (H) which has the property of generating a small electrical current when it is distorted. This electrical current is proportional to the amount of distortion; hence, the electrical impulses are in proportion to the roughness of the surface.

The tracer arm is pivoted on a carriage at (J) (Figure 481) so that the arm can be lifted from the work, and will not be damaged when adjusting the head on the column. A bracket (K) prevents the arm from dropping more than a few degrees below the horizontal position and prevents accidental contact with the surface plate when work is not in position. To avoid damage to the stylus, the arm should be handled carefully so as not to drop it on the surface of the work or hit it against projections on the work.

The carriage, in the form of a shaft projecting below the analyzer head, is actuated by a synchronous motor so that it moves the tracer arm parallel to the surface of the surface plate at a constant speed of approximately 1/8" in ten seconds--1/16" forward and 1/16" return--thus completing one cycle in ten seconds. A dial on the front of the analyzer head at (L) (Figure 481) indicates the travel position of the arm in its movement back and forth.

When measuring flat surfaces, the tracer arm should be parallel to the surface plate so that the stylus will make proper contact with the work when the positioning stud is resting on the surface. Adjustment of the tracer arm is made with the two knobs (M) and (N) (Figure 481). Release the analyzer head by turning

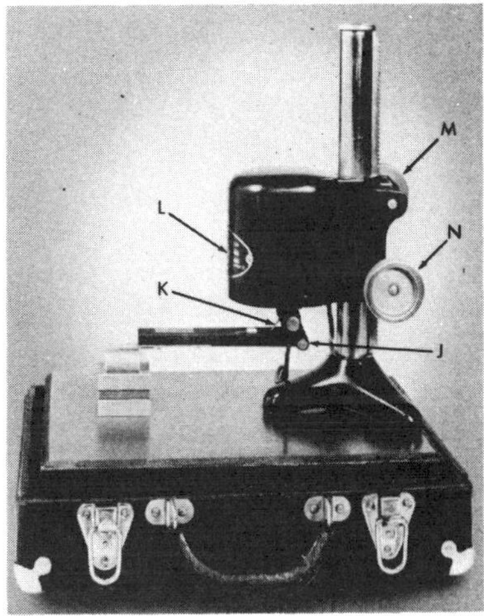

Figure 481. Surface-Analyzer Analyzing Head

Figure 482. Surface-Analyzer Amplifier

knob (M) on left of the head one turn; then adjust vertical position of the tracer arm with knob (N) until it is horizontal. Clamp the head in this position with knob (M).

THE AMPLIFIER

The small electrical impulses generated by the piezoelectric crystal must be amplified many times before they can actuate the oscillograph; therefore, they are directed into the amplifying unit (Figure 482) where they are amplified from 40 to 40,000 times, depending upon the setting of the attenuator, or variable rheostat (K), mounted in the amplifying unit.

The amplifier consists of three stages of amplification similar to the stages of amplification in a radio set; in addition, the amplifier contains a calibrating circuit for introducing a known input voltage for the calibration of the amplifier and the oscillograph, and for checking power-line voltage.

THE OSCILLOGRAPH

The oscillograph (Figure 483) provides a means of recording, on a continuously moving paper chart, the instantaneous fluctuations of the amplified impulses impressed upon it.

A piezoelectric crystal picks up the amplified electrical impulses and actuates a pen, which, riding on the paper chart, records the roughness of the surface being explored by the stylus of the tracer arm.

A synchronous motor moves the chart in

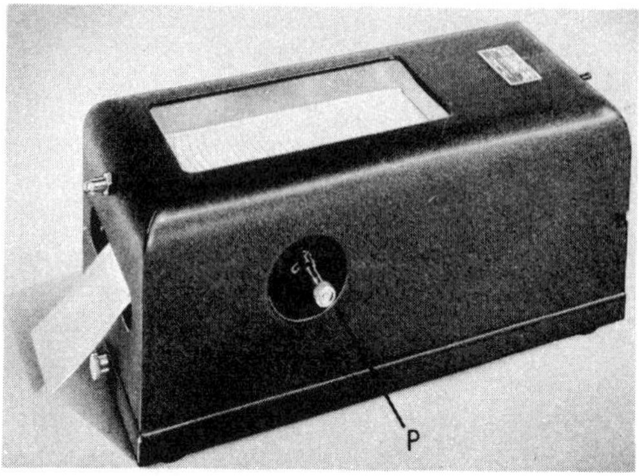

Figure 483. Surface-Analyzer Oscillograph

direct relation to the speed of the tracer arm; therefore, the width of the record on the chart represents the depth of the roughness, and the spaces between the recorded impulses represent the spacing of the individual irregularities on the surface being measured.

Three chart speeds are available and are controlled by the selector knob (P) (Figure 483). They are: 5 millimeters (approximately 1/5") per second, 25 millimeters (approximately 1") per second, and 125 millimeters (approximately 5") per second. The intermediate speed (25 millimeters per second) is usually preferred, as it provides a clear record and conserves paper. (Caution: before attempting to shift the speed of the chart, the oscillograph should be turned off.) Both the amplifier and the oscillograph are turned on and off with a toggle switch to the rear of the respective units. As the analyzer head and the oscillograph are synchronized, the switch on the oscillograph controls the motor of the analyzer head.

Figure 484 is a view of the panel on the amplifier. The voltmeter at (B) performs the double duty of indicating the voltage imposed upon the tracer arm and checking the line voltage. When the knob (S) marked "cal. volt" is turned all the way to the left, the voltmeter needle should register in the red area marked "line" on the scale. When the instrument is being calibrated, the meter registers the voltage being applied to the tracer arm and should be adjusted with knob (T) until the voltage corresponds to that specified for the instrument. (See instructions with the instrument.)

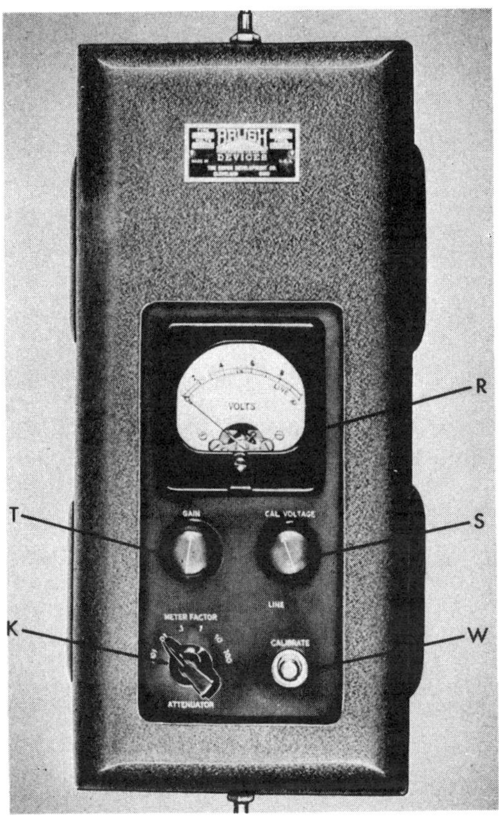

Figure 484. Surface-Analyzer Amplifier Panel

The button at (W) (Figure 484) actuates the pen of the oscillograph and is adjusted to the work, and the operator is ready to make a record on the chart. It is also used when calibrating the instrument.

As previously mentioned, the attenuator (K), controls the amplification, and its setting is governed by the roughness of the surface being measured.

The RMS meter (Figure 479) is for use with the surface analyzer for giving a quick, visual indication of surface roughness. The meter may be quickly inserted into the standard surface analyzer setup by means of connecting cables; the meter can be used either along with the chart recording, or separately, where a chart profile is not required.

For instructions on the operation and calibration of the surface analyzer, refer to operating instructions with each instrument.

SURFACE ROUGHNESS COMPARATOR BLOCKS

While surface roughness may be measured directly and accurately by various instruments designed expressly for that purpose, another method widely used is that of visual and tactual comparison. For production applications, where sight and feel methods of comparison are sufficient, standard surface roughness comparator blocks may be used (Figure 485). These blocks are furnished in sets of varying numbers; each block, marked with its degree of roughness in microinches (r. m. s.), represents a particular type of surface finish, such as: ground, milled, turned, shaped, broached, and lapped.

When making comparisons with the standard roughness specimens, it is important that the comparison be made with the surface whose lay most nearly resembles the work surface. Generally comparisons are made by sight and feel. By running the fingernail or tip of the finger over the work surface and over a standard specimen, the degree of roughness can be estimated.

FAXFILM

One of the newer methods developed for the evaluation of surfaces is called the Faxfilm Surface Comparator. The instrument shown in Figure 486 is a portable unit for the comparison of Faxfilms at 30 diameter magnification. Images of the surfaces being compared appear side by side in the screen on the bottom of the cabinet.

Faxfilm is a method of surface examination which employs a plastic replica to produce a projected image of the surface being inspected. The information obtained is primarily qualitative, such as the lay, chatter, porosity and uniformity of a machined finish, rather than a direct measurement, as might be obtained from the Surface Analyzer and other instruments for measuring surface roughness.

The process offers the advantages of low cost and relative simplicity of use. A clear plastic film is moistened with a solvent, which may be applied either to the film or to the surface being examined. The film is then

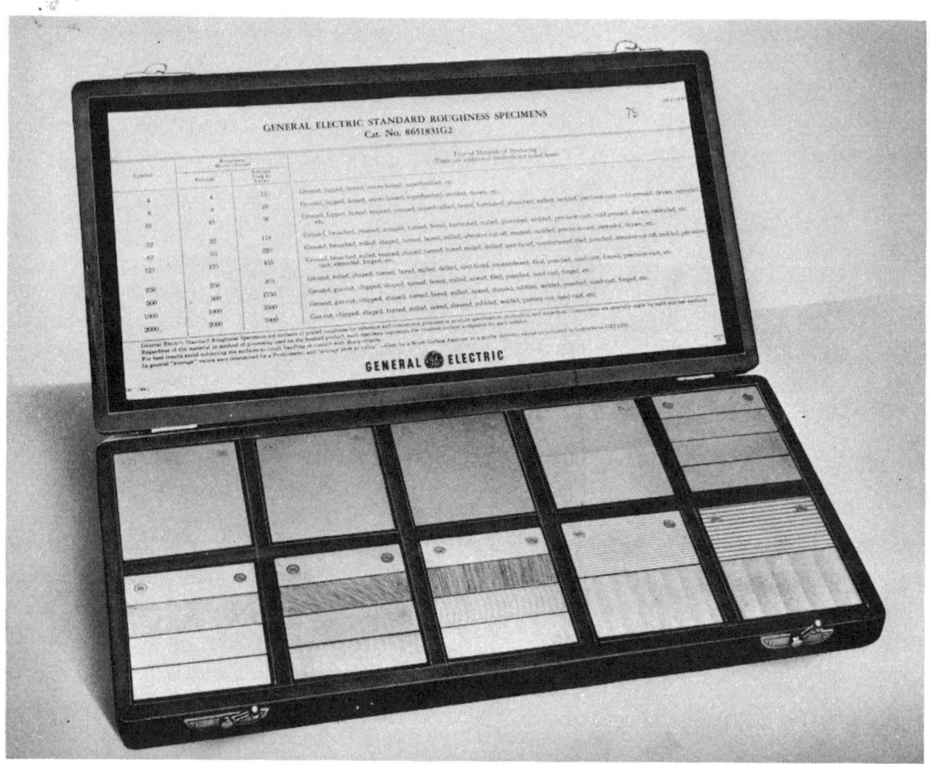

Figure 485. Surface Roughness Comparator Blocks

pressed against the surface, where it remains for less than a minute, forming an exact reverse replica of the original surface (see Figure 487). Each little irregularity serves as a tiny prism in the replica, which is then projected in a microprojector; each prism bends the rays of transmitted light from the projector and, by this refraction, creates a magnified image of the surface on the screen. The Faxfilm replica may be made of curved or irregular shapes and, in projection, will give a flat image of these surfaces.

The applications for Faxfilm can be compared roughly with those for a comparison microscope but with a number of important differences. Faxfilms can be made of a surface in one location and taken to another location for viewing, thus eliminating the problems of handling bulky pieces or cutting them for setup under a microscope. The images appear side by side on a screen, as shown in Figure 486, where a number of people can view and discuss them together, rather than appear to only one person at a time in an eye piece. The microscopist's problems of lighting and focus on curved or irregular shapes, particularly on highly reflective surfaces, are eliminated with Faxfilm. An outstanding advantage of Faxfilm is that it provides a permanent, easy-to-file record of the surface without photography, although photographs may be made from Faxfilms, using them much as negatives.

The variety of applications for Faxfilm is too great to be more than suggested here. The machining characteristics of a material and the best machining method to produce a desired finish can be established by taking the Faxfilm unit to the machine and experimenting with various feeds, speeds, cutting tools and cutting oils. As soon as a small area of the material has been machined, a Faxfilm may be made of the finish and compared with finishes produced under slightly different conditions.

Wear tests provide another valuable application for Faxfilm. The Faxfilms may be made of the surface at the beginning of the test and at

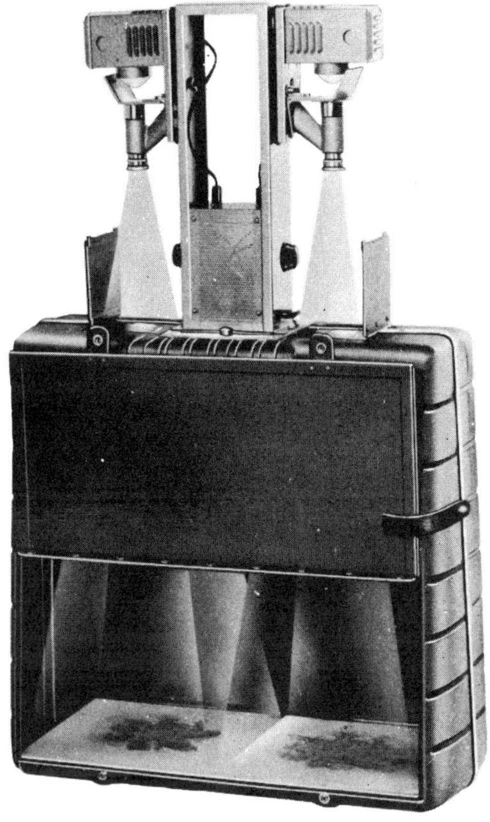

Figure 486. Faxfilm Comparator

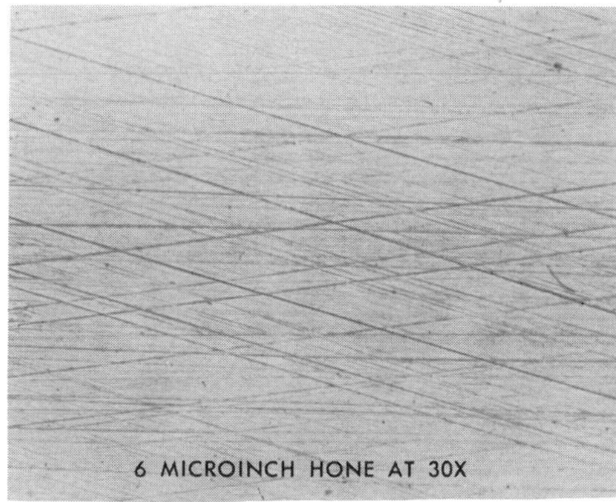

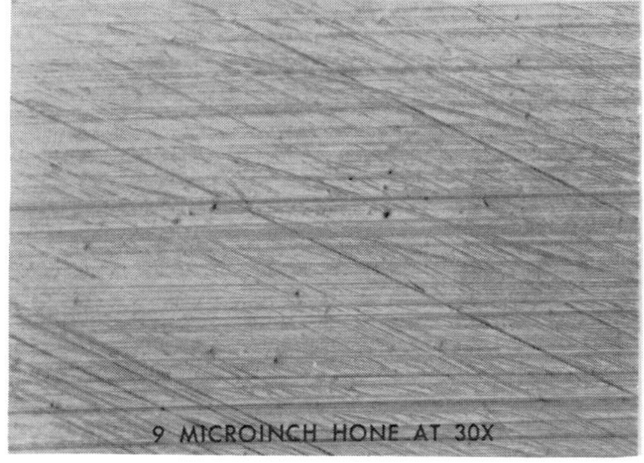

Figure 488. Photographs of Faxfilms

intervals to its conclusion. When the test is completed, these Faxfilms may be compared and studied and even saved for comparison with later tests. If necessary for a report, photographs can be made from these Faxfilms (Figure 488). Other applications include studies of grinding belt life and wear, flexing tests in plating, and similar life or corrosion tests.

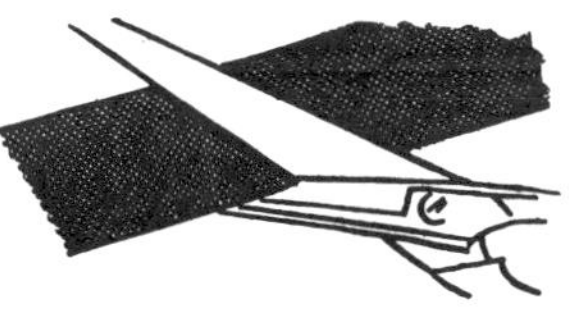

Cut one inch of film from roll

Wet surface of subject with solvent

Press film on to surface

Peel film from specimen

Moisten inside surfaces of a frame coated with glue

Insert and seal in moistened gummed Faxfilm frame

Figure 487. Application of Faxfilm

The uses of Faxfilm in inspection are fairly obvious. While the Faxfilm does not give a direct quantitative measure of a surface, comparison projection of the work specimen with a known standard makes possible a reasonably accurate estimate of the roughness of the work specimen, especially at the thirty and one hundred diameter magnifications obtainable in standard Faxfilm comparison projection units. In addition, the qualitative information shown by Faxfilm can be more important than the possible error in estimates of roughness. For example, a scratch or ridge in a metal seal could cause it to leak, or a bearing with a uniform finish could be a better bearing than one having a lower average roughness but high peaks. This is the type of information which Faxfilm will give.

QUESTIONS AND PROBLEMS

1. Define the following: nominal surface; waviness; roughness.
2. What is the approximate difference between the root-mean-square average depth of roughness and the maximum depth of roughness?
3. Define the term profile as used in reference to surface roughness.
4. Explain the principle of the Profilometer.
5. At what reading would you set the selector switch of the Profilometer for measuring surface roughness between $\sqrt{10}$ and $\sqrt{20}$; $\sqrt{30}$ and $\sqrt{40}$?
6. What is the size of the point of the stylus used on the Profilometer?
7. When is the upper scale on the Profilometer used? The lower scale?
8. What is the average tolerance on surface roughness?
9. How may the accuracy of a Profilometer be checked?
10. In what particular feature does the Brush surface analyzer differ from the Profilometer?
11. On what materials can the Brush analyzer be used to check surface finish?
12. What are the four principal units of the Brush surface analyzer and their use?
13. What are the speeds of the oscillograph?
14. What is Faxfilm?
15. How does the Faxfilm surface comparator method of surface roughness inspection differ from that of the Profilometer or Brush analyzer method?
16. What are some of the advantages of the Faxfilm method of inspection?
17. What are standard surface roughness comparator blocks, and how are they used?

Chapter XIV

HARDNESS TESTING

HARDNESS IS that property of solid materials which endures and resists any physical force tending to wear it away or permanently deform it. There is a relation between the hardness of a material and its strength, toughness, and wearing quality. Fortunately the hardness of steel, the most-used metal in industry, can be controlled.

STEEL

Steel is an alloy (combination) of iron, carbon, and other elements that are present as impurities or are added to produce a particular quality such as hardness, toughness, or machinability. Steels may be roughly classified as straight-carbon steels and alloy steels. A straight-carbon steel is one that owes its special properties to the amount of carbon it contains and does not contain substantial amounts of other alloying elements. Carbon is the principal hardening agent in steel.

Straight-carbon steels may be further classified as: (1) low-carbon steels containing .25 percent or less carbon; (2) medium-carbon steels containing .25 to .60 percent carbon; (3) high-carbon steels containing .60 to 1.30 percent carbon.

Alloy steels are those that contain alloying elements such as chromium, vanadium, nickel, molybdenum, and tungsten, and that owe their special properties to these elements. Some of these properties are: greater strength, toughness, machinability, and ability to retain hardness at high temperatures.

HEAT-TREATMENT

Heat-treatment is a process by which the physical properties of a metal are changed by heating it to a specified temperature and cooling it, either slowly or rapidly, according to the qualities desired.

CRITICAL TEMPERATURE RANGE

This is a range of temperature in which definite physical changes take place in steel. If a piece of steel is heated to a white heat and allowed to cool, certain changes will be observed during the heating and cooling process. While the steel is being heated, definite temperatures will be reached at which the steel will momentarily cease to rise in temperature--and may even drop slightly before continuing to rise.

In the cooling process, a reversal of these changes will be observed. Definite temperatures will be reached at which the steel will momentarily cease to drop in temperature, and may even rise slightly before continuing to cool. The range of temperatures in which these changes take place is called the critical temperature range. The critical temperature range of steel is important because all heat-treating operations, except drawing and tempering, involve the use of temperatures in the critical range, or slightly above or below it.

HARDENING

To harden steel, it must be heated to above its critical temperature range and cooled in some medium such as water, brine, oil, or--

in the case of some alloy steels--in air. In general, the more rapidly the steel is cooled from a temperature above the critical range, the greater hardness it will have.

TEMPERING

Because fully-hardened steel is much too brittle for most applications, the hardness must be modified to make the steel generally useful. If hardened steel is heated to a temperature below the critical range and allowed to cool, the resulting hardness and toughness will be in proportion to the temperature to which the steel was heated. This second heating, used to relieve internal stresses and make the steel softer, more ductile, and tougher, is called drawing or tempering.

NORMALIZING

Normalizing consists in heating steel above the critical temperature range and cooling it in still air. This treatment is to prepare the steel for further heat-treatment or to relieve stresses set up in the steel by forging, rolling, or drawing.

FULL ANNEALING

In full annealing, the steel is heated slowly to a point above its critical temperature range and then cooled slowly, usually in a furnace where the cooling rate can be controlled. This treatment softens the steel and removes stresses which may be present.

CASEHARDENING

Casehardening is a process by which carbon is introduced into the surface of low-carbon steel by heating it to a high temperature while it is in contact with carbonaceous material. The surface of the steel absorbs part of the carbon and, when cooled rapidly, becomes hard while the core remains soft and ductile.

Advantages of casehardening are: (1) low-carbon steels can be substituted for high-carbon steels in applications requiring surface hardening only, thus reducing material and machining costs; (2) surface wear resistance can be obtained while leaving the core tough and ductile to resist shock.

FILE TEST FOR HARDNESS

One of the simplest methods of testing hardness is to draw a file across the surface of the material. File testing is an art acquired by experience. It must be fitted to the needs of the particular product manufactured and is rarely used as a standard for specifications. Standard test files are available for making the file test, which, when made by the skilled workman, is a fast and useful means of determining approximate hardness of material.

HARDNESS TESTING MACHINES

Hardness testing machines discussed in this book fall into two general classes: (1) those by which the hardness of the material is indicated by the amount of deformation resulting from pressing a standard penetrator into the surface of the material, under specified load; (2) those by which the hardness is measured by the rebound of a small hammer or "tup" dropped from a standard height.

ROCKWELL STANDARD HARDNESS TESTER

The hardness value of the Rockwell hardness tester (Figure 489) is based upon the depth of penetration made by a standard penetrator, under a standard load. When testing soft materials, a standard 1/16" diameter ball penetrator is used. When testing hard materials, a sphero-conical diamond penetrator (called a "brale") is used.

The test is made in two stages. First, a minor load is applied to overcome the usual surface imperfections in the material. The dial gage is then set back to zero and a major load is added to the minor load. The additional depth of penetration caused by the application of the major load is indicated on the dial gage and is a measure of the hardness of the material. A minor load of 10 kilograms and a major load of 60, 100, or 150 kilograms is used with the Rockwell standard hardness tester.

OPERATION

The specimen is placed upon the anvil at (A) and is raised with the handwheel at (B) until

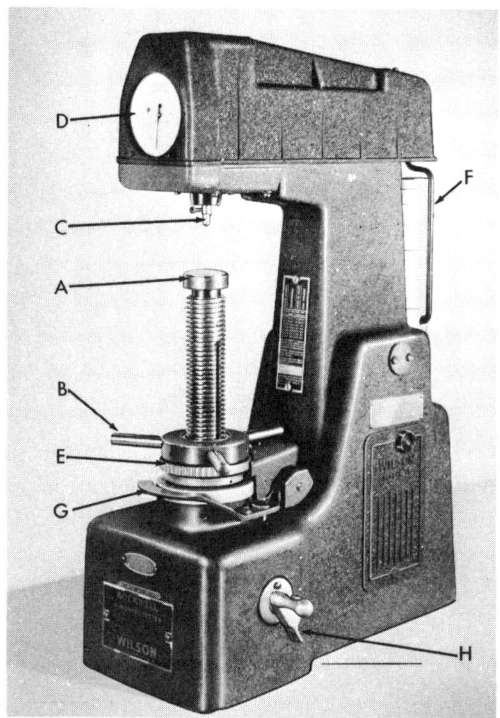

Figure 489. Standard Rockwell Tester

it makes contact with the penetrator at (C). The minor load is then applied by continuing to raise the specimen until a small hand on the dial gage at (D) coincides with a small black dot on the face of the dial.

Meanwhile, the depth of penetration of the minor load has been registered on the dial gage. The hand of the dial gage is returned to zero by turning the knurled ring at (E); then the major load (F) is applied by pressing the yoke at (G).

The major load is imposed upon the minor load, causing an additional penetration into the specimen. This additional penetration is indicated on the dial gage; however, part of the depth of penetration is due to the elasticity of the material. So that the reading of the dial gage will not be influenced by the elasticity of the material, the major load is now removed with the crank (H), resulting in a new reading on the dial gage, which is the Rockwell hardness number of the material. The specimen is now lowered until it is clear of the penetrator, and the machine is ready for another test.

Figure 490 is a schematic drawing of the sequence of operations in making a Rockwell hardness test.

At E, the work is placed on the anvil preparatory to applying the minor load. Note the position of the lever which actuates the dial gage and has a pivot at (K).

At F, the minor load has been applied by raising the anvil, and the dial gage has been set to zero. The depth of penetration of the load is represented by the space A-B.

At G, the major load has been applied and the additional depth of penetration registered on the dial gage. The additional penetration is represented by the space B-C.

At H, the major load has been removed, and the penetrator reacts slightly to the elasticity of the material. This reaction is represented by the space D-C. The dial gage now shows the actual static penetration and is read as the Rockwell hardness number, represented by B-D.

At J, both minor and major loads have been removed, and the machine is ready for another test.

Figure 491 shows the principle of the Rockwell hardness test--the positions of the penetrator under minor and major load, the depth of the impression, and the increment used as the basis of Rockwell hardness numbers.

ROCKWELL SCALES

Hardness values on a Rockwell hardness tester are designated by a letter and a number. The letters A, B, C, D, etc., indicate the type of penetrator and the load applied in the test; the number is the hardness value as read on the dial gage.

Figure 492 shows the dial gage of the Rockwell Standard hardness tester. It contains two scales: the outside scale marked C printed in black; the inside scale B is printed in red. When the diamond penetrator, or "brale," is used with 150-kg load, the outside or C scale is used. When the 1/16" diameter ball penetrator is used with the 100-kg load, the inside or B scale is used. The remaining scales (A, D, E, F, etc.) are considered special and are read from either the red or black scales on the dial as

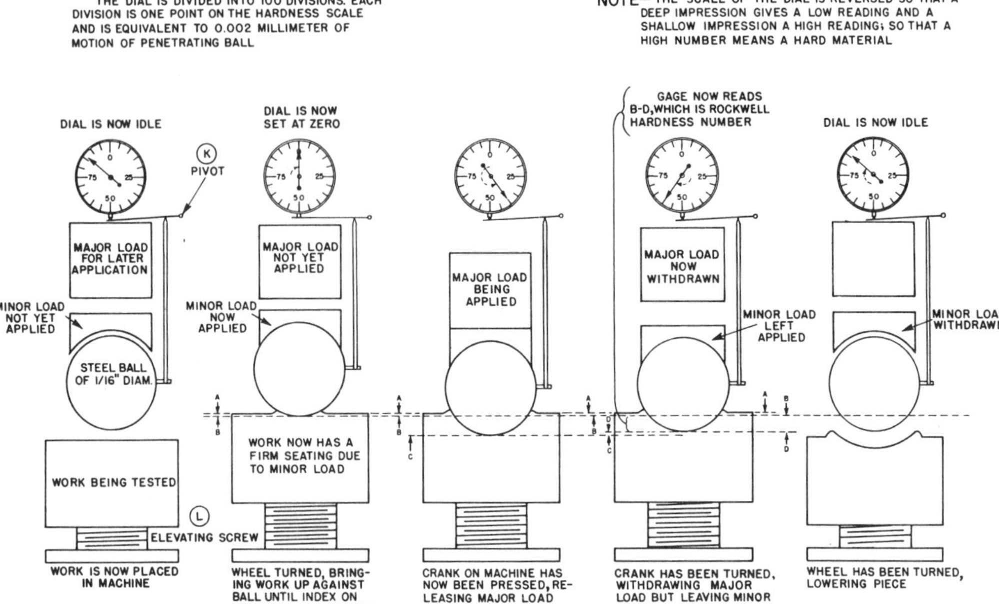
THE DIAL IS DIVIDED INTO 100 DIVISIONS. EACH DIVISION IS ONE POINT ON THE HARDNESS SCALE AND IS EQUIVALENT TO 0.002 MILLIMETER OF MOTION OF PENETRATING BALL
NOTE—THE SCALE OF THE DIAL IS REVERSED SO THAT A DEEP IMPRESSION GIVES A LOW READING AND A SHALLOW IMPRESSION A HIGH READING; SO THAT A HIGH NUMBER MEANS A HARD MATERIAL
DIAL IS NOW IDLE
K
PIVOT
MAJOR LOAD FOR LATER APPLICATION
MINOR LOAD NOT YET APPLIED
STEEL BALL OF 1/16" DIAM.
WORK BEING TESTED
L
ELEVATING SCREW
WORK IS NOW PLACED IN MACHINE
E
DIAL IS NOW SET AT ZERO
MAJOR LOAD NOT YET APPLIED
MINOR LOAD NOW APPLIED
WORK NOW HAS A FIRM SEATING DUE TO MINOR LOAD
WHEEL TURNED, BRINGING WORK UP AGAINST BALL UNTIL INDEX ON DIAL READS ZERO. THIS APPLIES MINOR LOAD
F
MAJOR LOAD BEING APPLIED
CRANK ON MACHINE HAS NOW BEEN PRESSED, RELEASING MAJOR LOAD
G
GAGE NOW READS B-D, WHICH IS ROCKWELL HARDNESS NUMBER
MAJOR LOAD NOW WITHDRAWN
MINOR LOAD LEFT APPLIED
CRANK HAS BEEN TURNED, WITHDRAWING MAJOR LOAD BUT LEAVING MINOR LOAD
H
DIAL IS NOW IDLE
MINOR LOAD WITHDRAWN
WHEEL HAS BEEN TURNED, LOWERING PIECE
J

Figure 490. Steps in Rockwell Testing

shown in the table (Figure 493), which specifies the penetrator, load, and color of scale for both the standard and special alphabetical scales.

Tables XIV-A and XIV-B give approximate relationship between hardness values determined on "Rockwell" and "Rockwell superficial" hardness testers and values determined on other testers.

DEPTH OF PENETRATION

When using the Rockwell standard hardness tester, the depth of penetration varies between .0025" and .010", depending upon the type of penetrator used and the load applied. Because the hardness test is influenced by the hardness of the material to a depth of ten times the depth of penetration, it is obvious that the minimum thickness of the specimen is limited to approximately ten times the depth of penetration.

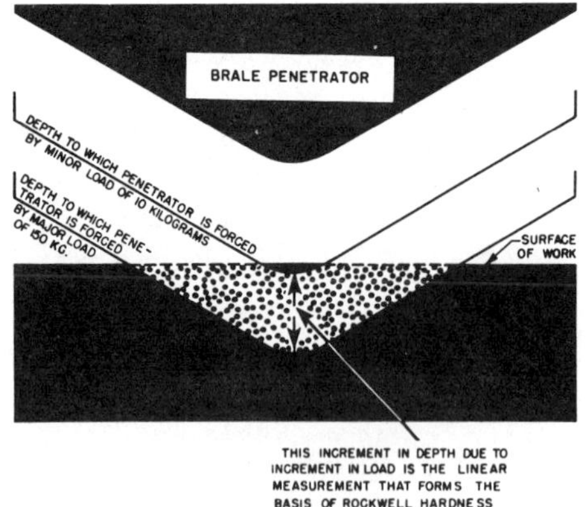
BRALE PENETRATOR
DEPTH TO WHICH PENETRATOR IS FORCED BY MINOR LOAD OF 10 KILOGRAMS
DEPTH TO WHICH PENETRATOR IS FORCED BY MAJOR LOAD OF 150 KG.
SURFACE OF WORK
THIS INCREMENT IN DEPTH DUE TO INCREMENT IN LOAD IS THE LINEAR MEASUREMENT THAT FORMS THE BASIS OF ROCKWELL HARDNESS READINGS

Figure 491. Rockwell Impression

Figure 492. Dial of Rockwell Standard Hardness Tester

	SYMBOL	PENETRATOR	LOAD IN KG.	SCALE
STANDARD	B	1/16" BALL	100	RED
	C	BRALE	150	BLACK
SPECIAL	A	BRALE	60	BLACK
	D	BRALE	100	BLACK
	E	1/8" BALL	100	RED
	F	1/16" BALL	60	RED
	G	1/16" BALL	150	RED
	H	1/8" BALL	60	RED
	K	1/8" BALL	150	RED

Figure 493. Rockwell Scales

ROCKWELL SUPERFICIAL HARDNESS TESTER

The Rockwell superficial hardness tester (Figure 494) was designed especially for making hardness tests on thin materials or casehardened parts. The maximum depth of penetration of the superficial hardness tester is less than .005". The principle of operation is exactly the same as that of the standard hardness tester, except that a minor load of only 3 kg and a major load of 15, 30, or 45 kg is used for the test.

The ball penetrator (1/16" diameter) used on the superficial hardness tester is the same as that on the standard hardness tester. The "brale" penetrator used on the superficial tester is the same as on the standard tester, except it is ground with greater precision. It is marked "N-brale" to distinguish it from the standard "brale."

SUPERFICIAL HARDNESS DESIGNATION

The dial gage of the superficial hardness tester differs from that of the standard hardness tester in that it contains only one scale whose graduations are numbered 1 to 100. The superficial hardness number is made up of three parts: (1) the load applied; (2) the letter N when the "brale" is used; the letter T when the standard 1/16" ball penetrator is used; (3) the reading of the dial gage.

For example:

15-N-50 is interpreted:

15-kg applied load, "brale" penetrator, and dial reading 50.

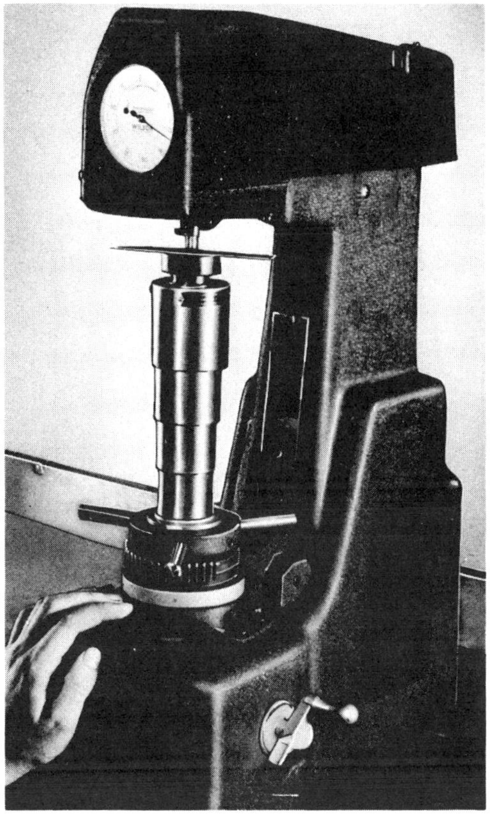

Figure 494. Rockwell Superficial Hardness Tester

30-T-60 is interpreted:

30-kg applied load, 1/16" ball penetrator, and dial reading 60.

In all tests on the superficial hardness tester, care should be taken that the hardness specification contains both the type of penetrator and the load applied.

The approximate relation of the hardness values of the superficial hardness tester and the standard hardness tester is shown in Figures 495 and 496. Similar conversion charts for all scales of the standard hardness tester are furnished with each superficial hardness tester.

ROCKWELL MOTORIZED HARDNESS TESTER

The Rockwell motorized hardness tester in Figure 497 is identical to the standard hardness tester, except that the major load is applied automatically. The minor load is applied with

TABLE XIV-A. ROCKWELL CONVERSION CHART FOR HARDENED STEEL AND HARD ALLOYS

C.	A.	D.	15-N **	30-N **	45-N **	Diamond Pyramid	BRI-NELL	G.	Tensile Strength
ROCKWELL HARDNESS TESTER C scale BRALE penetrator—150 kg load	ROCKWELL HARDNESS TESTER A scale BRALE penetrator—60 kg load	ROCKWELL HARDNESS TESTER D scale BRALE penetrator—100 kg load	ROCKWELL SUPERFICIAL 15-N scale N BRALE penetrator—15 kg load	ROCKWELL SUPERFICIAL 30-N scale N BRALE penetrator—30 kg load	ROCKWELL SUPERFICIAL 45-N scale N BRALE penetrator—45 kg load	4 sided 136° diamond pyramid 10 kg load—Measurement of 2 diagonals by microscope. Hardness numbers computed by formula.	Brinell Standard Type, 3000 kg load Hultgren 10 mm ball penetrator	ROCKWELL HARDNESS TESTER G scale 1/16″ ball penetrator—150 kg load	Thousand lbs. per sq. in.
80	92.0	86.5	96.5	92.0	87.0	1865			
79	91.5	85.5	—	91.5	86.5	1787			
78	91.0	84.5	96.0	91.0	85.5	1710			
77	90.5	84.0	—	90.5	84.5	1633			
76	90.0	83.0	95.5	90.0	83.5	1556			
75	89.5	82.5	—	89.0	82.5	1478			
74	89.0	81.5	95.0	88.5	81.5	1400			
73	88.5	81.0	—	88.0	80.5	1323			
72	88.0	80.0	94.5	87.0	79.5	1245			
71	87.0	79.5	—	86.5	78.5	1160			
70	86.5	78.5	94.0	86.0	77.5	1076			
69	86.0	78.0	93.5	85.0	76.5	1004			
68	85.5	77.0	—	84.5	75.5	942			
67	85.0	76.0	93.0	83.5	74.5	894			
66	84.5	75.5	92.5	83.0	73.0	854			
65	84.0	74.5	92.0	82.0	72.0	820			
64	83.5	74.0	—	81.0	71.0	789			
63	83.0	73.0	91.5	80.0	70.0	763			
62	82.5	72.5	91.0	79.0	69.0	739	—	—	—
61	81.5	71.5	90.5	78.5	67.5	716	—	—	—
60	81.0	71.0	90.0	77.5	66.5	695	614	—	314
59	80.5	70.0	89.5	76.5	65.5	675	600	—	306
58	80.0	69.0	—	75.5	64.0	655	587	—	299
57	79.5	68.5	89.0	75.0	63.0	636	573	—	291
56	79.0	67.5	88.5	74.0	62.0	617	560	—	284
55	78.5	67.0	88.0	73.0	61.0	598	547	—	277
54	78.0	66.0	87.5	72.0	59.5	580	534	—	270
53	77.5	65.5	87.0	71.0	58.5	562	522	—	263
52	77.0	64.5	86.5	70.5	57.5	545	509	—	256
51	76.5	64.0	86.0	69.5	56.0	528	496	—	250

Note:—General Conversion Values on material harder than about C-70 on the ROCKWELL Hardness Tester will probably be long in dispute because as yet there are few materials hard enough to be brought into that study and because some of those few do not flow nicely under the testing load but have a tendency to crumble and others do not approach homogeneity.

INEXACT AND ONLY FOR STEEL

Courtesy of Wilson Mechanical Instrument Co., Inc.

TABLE XIV-A (continued)

C.	A.	D.	15-N **	30-N **	45-N **	Diamond Pyramid	BRI-NELL	G.	Tensile Strength
ROCKWELL HARDNESS TESTER C scale BRALE penetrator—150 kg load	ROCKWELL HARDNESS TESTER A scale BRALE penetrator—60 kg load	ROCKWELL HARDNESS TESTER D scale BRALE penetrator—100 kg load	ROCKWELL SUPERFICIAL 15-N scale N BRALE penetrator—15 kg load	ROCKWELL SUPERFICIAL 30-N scale N BRALE penetrator—30 kg load	ROCKWELL SUPERFICIAL 45-N scale N BRALE penetrator—45 kg load	4 sided 136° diamond pyramid 10 kg load—Measurement of 2 diagonals by microscope. Hardness numbers computed by formula.	Brinell Standard Type, 3000 kg load Hultgren 10 mm ball penetrator	ROCKWELL HARDNESS TESTER G scale 1/16" ball penetrator—150 kg load	Thousand lbs. per sq. in.
50	76.0	63.0	85.5	68.5	55.0	513	484	—	243
49	75.5	62.0	85.0	67.5	54.0	498	472	—	236
48	74.5	61.5	84.5	66.5	52.5	485	460	—	230
47	74.0	60.5	84.0	66.0	51.5	471	448	—	223
46	73.5	60.0	83.5	65.0	50.0	458	437	—	217
45	73.0	59.0	83.0	64.0	49.0	446	426	—	211
44	72.5	58.5	82.5	63.0	48.0	435	415	—	205
43	72.0	57.5	82.0	62.0	46.5	424	404	—	199
42	71.5	57.0	81.5	61.5	45.5	413	393	—	194
41	71.0	56.0	81.0	60.5	44.5	403	382	—	188
40	70.5	55.5	80.5	59.5	43.0	393	372	—	182
39	70.0	54.5	80.0	58.5	42.0	383	362	—	177
38	69.5	54.0	79.5	57.5	41.0	373	352	—	171
37	69.0	53.0	79.0	56.5	39.5	363	342	—	166
36	68.5	52.5	78.5	56.0	38.5	353	332	—	162
35	68.0	51.5	78.0	55.0	37.0	343	322	—	157
34	67.5	50.5	77.0	54.0	36.0	334	313	—	153
33	67.0	50.0	76.5	53.0	35.0	325	305	—	148
32	66.5	49.0	76.0	52.0	33.5	317	297	—	144
31	66.0	48.5	75.5	51.5	32.5	309	290	—	140
30	65.5	47.5	75.0	50.5	31.5	301	283	92.0	136
29	65.0	47.0	74.5	49.5	30.0	293	276	91.0	132
28	64.5	46.0	74.0	48.5	29.0	285	270	90.0	129
27	64.0	45.5	73.5	47.5	28.0	278	265	89.0	126
26	63.5	44.5	72.5	47.0	26.5	271	260	88.0	123
25	63.0	44.0	72.0	46.0	25.5	264	255	87.0	120
24	62.5	43.0	71.5	45.0	24.0	257	250	86.0	117
23	62.0	42.5	71.0	44.0	23.0	251	245	84.5	115
22	61.5	41.5	70.5	43.0	22.0	246	240	83.5	112
21	61.0	41.0	70.0	42.5	20.5	241	235	82.5	110
20	60.5	40.0	69.5	41.5	19.5	236	230	81.0	108

**The 15-T, 30-T, 45-T, 15-N, 30-N and 45-N values are in scales of the ROCKWELL Superficial Hardness Tester, a specialized form of ROCKWELL Tester, having lighter loads and more sensitive depth reading system, used where for one or another reason the indentation must be exceptionally shallow.

TABLE XIV-B. ROCKWELL CONVERSION CHART FOR UNHARDENED STEEL, STEEL OF SOFT TEMPER, GREY AND MALLEABLE CAST IRON, AND MOST NON-FERROUS METALS

B.	F.	G.	15-T **	30-T **	45-T **	E.	H.	K.	A.	BRI-NELL	BRI-NELL	Tensile Strength
ROCKWELL HARDNESS TESTER B scale 1/16" ball penetrator—100 kg load	ROCKWELL HARDNESS TESTER F scale 1/16" ball penetrator—60 kg load	ROCKWELL HARDNESS TESTER G scale 1/16" ball penetrator—150 kg load	ROCKWELL SUPERFICIAL 15-T scale 1/16" ball penetrator—15 kg load	ROCKWELL SUPERFICIAL 30-T scale 1/16" ball penetrator—30 kg load	ROCKWELL SUPERFICIAL 45-T scale 1/16" ball penetrator—45 kg load	ROCKWELL HARDNESS TESTER E scale 1/8" ball penetrator—100 kg load	ROCKWELL HARDNESS TESTER H scale 1/8" ball penetrator—60 kg load	ROCKWELL HARDNESS TESTER K scale 1/8" ball penetrator—150 kg load	ROCKWELL HARDNESS TESTER A scale BRALE penetrator—60 kg load	Brinell, Standard Type, 500 kg load 10 mm ball penetrator	Brinell, Standard Type, 3000 kg load 10 mm ball penetrator	Thousand lbs. per sq. in.
100	—	82.5	93.0	82.0	72.0	—	—	—	61.5	201	240	116
99	—	81.0	92.5	81.5	71.0	—	—	—	61.0	195	234	112
98	—	79.0	—	81.0	70.0	—	—	—	60.0	189	228	109
97	—	77.5	92.0	80.5	69.0	—	—	—	59.5	184	222	106
96	—	76.0	—	80.0	68.0	—	—	—	59.0	179	216	103
95	—	74.0	91.5	79.0	67.0	—	—	—	58.0	175	210	101
94	—	72.5	—	78.5	66.0	—	—	—	57.5	171	205	98
93	—	71.0	91.0	78.0	65.5	—	—	—	57.0	167	200	96
92	—	69.0	90.5	77.5	64.5	—	—	100	56.5	163	195	93
91	—	67.5	—	77.0	63.5	—	—	99.5	56.0	160	190	91
90	—	66.0	90.0	76.0	62.5	—	—	98.5	55.5	157	185	89
89	—	64.0	89.5	75.5	61.5	—	—	98.0	55.0	154	180	87
88	—	62.5	—	75.0	60.5	—	—	97.0	54.0	151	176	85
87	—	61.0	89.0	74.5	59.5	—	—	96.5	53.5	148	172	83
86	—	59.0	88.5	74.0	58.5	—	—	95.5	53.0	145	169	81
85	—	57.5	—	73.5	58.0	—	—	94.5	52.5	142	165	80
84	—	56.0	88.0	73.0	57.0	—	—	94.0	52.0	140	162	78
83	—	54.0	87.5	72.0	56.0	—	—	93.0	51.0	137	159	77
82	—	52.5	—	71.5	55.0	—	—	92.0	50.5	135	156	75
81	—	51.0	87.0	71.0	54.0	—	—	91.0	50.0	133	153	74
80	—	49.0	86.5	70.0	53.0	—	—	90.5	49.5	130	150	72
79	—	47.5	—	69.5	52.0	—	—	89.5	49.0	128	147	
78	—	46.0	86.0	69.0	51.0	—	—	88.5	48.5	126	144	
77	—	44.0	85.5	68.0	50.0	—	—	88.0	48.0	124	141	
76	—	42.5	—	67.5	49.0	—	—	87.0	47.0	122	139	
75	99.5	41.0	85.0	67.0	48.5	—	—	86.0	46.5	120	137	
74	99.0	39.0	—	66.0	47.5	—	—	85.0	46.0	118	135	
73	98.5	37.5	84.5	65.5	46.5	—	—	84.5	45.5	116	132	
72	98.0	36.0	84.0	65.0	45.5	—	—	83.5	45.0	114	130	
71	97.5	34.5	—	64.0	44.5	100	—	82.5	44.5	112	127	
70	97.0	32.5	83.5	63.5	43.5	99.5	—	81.5	44.0	110	125	
69	96.0	31.0	83.0	62.5	42.5	99.0	—	81.0	43.5	109	123	
68	95.5	29.5	—	62.0	41.5	98.0	—	80.0	43.0	107	121	
67	95.0	28.0	82.5	61.5	40.5	97.5	—	79.0	42.5	106	119	
66	94.5	26.5	82.0	60.5	39.5	97.0	—	78.0	42.0	104	117	
65	94.0	25.0	—	60.0	38.5	96.0	—	77.5	—	102	116	
64	93.5	23.5	81.5	59.5	37.5	95.5	—	76.5	41.5	101	114	
63	93.0	22.0	81.0	58.5	36.5	95.0	—	75.5	41.0	99	112	
62	92.0	20.5	—	58.0	35.5	94.5	—	74.5	40.5	98	110	
61	91.5	19.0	80.5	57.0	34.5	93.5	—	74.0	40.0	96	108	
60	91.0	17.5	—	56.5	33.5	93.0	—	73.0	39.5	95	107	
59	90.5	16.0	80.0	56.0	32.0	92.5	—	72.0	39.0	94	106	
58	90.0	14.5	79.5	55.0	31.0	92.0	—	71.0	38.5	92	104	
57	89.5	13.0	—	54.5	30.0	91.0	—	70.5	38.0	91	103	
56	89.0	11.5	79.0	54.0	29.0	90.5	—	69.5	—	90	101	
55	88.0	10.0	78.5	53.0	28.0	90.0	—	68.5	37.5	89	100	
54	87.5	8.5	—	52.5	27.0	89.5	—	68.0	37.0	87	—	
53	87.0	7.0	78.0	51.5	26.0	89.0	—	67.0	36.5	86	—	
52	86.5	5.5	77.5	51.0	25.0	88.0	—	66.0	36.0	85	—	
51	86.0	4.0	—	50.5	24.0	87.5	—	65.0	35.5	84	—	
50	85.5	2.5	77.0	49.5	23.0	87.0	—	64.5	35.0	83	—	

Even for steel, Tensile Strength relation to hardness is inexact, unless determined for specific material.

TABLE XIV-B (continued)

B.	F.	G.	15-T **	30-T **	45-T **	E.	H.	K.	A.	BRI-NELL
ROCKWELL HARDNESS TESTER B scale 1/16″ ball penetrator—100 kg load	ROCKWELL HARDNESS TESTER F scale 1/16″ ball penetrator—60 kg load	ROCKWELL HARDNESS TESTER G scale 1/16″ ball penetrator—150 kg load	ROCKWELL SUPERFICIAL 15-T scale 1/16″ ball penetrator—15 kg load	ROCKWELL SUPERFICIAL 30-T scale 1/16″ ball penetrator—30 kg load	ROCKWELL SUPERFICIAL 45-T scale 1/16″ ball penetrator—45 kg load	ROCKWELL HARDNESS TESTER E scale 1/8″ ball penetrator—100 kg load	ROCKWELL HARDNESS TESTER H scale 1/8″ ball penetrator—60 kg load	ROCKWELL HARDNESS TESTER K scale 1/8″ ball penetrator—150 kg load	ROCKWELL HARDNESS TESTER A scale BRALE penetrator—60 kg load	Brinell, Standard Type, 500 kg load 10 mm ball penetrator
50	85.5	2.5	77.0	49.5	23.0	87.0	—	64.5	35.0	83
49	85.0	1.0	76.5	49.0	22.0	86.5	—	63.5	—	82
48	84.5	—	—	48.5	20.5	85.5	—	62.5	34.5	81
47	84.0	—	76.0	47.5	19.5	85.0	—	61.5	34.0	80
46	83.0	—	75.5	47.0	18.5	84.5	—	61.0	33.5	—
45	82.5	—	—	46.0	17.5	84.0	—	60.0	33.0	79
44	82.0	—	75.0	45.5	16.5	83.5	—	59.0	32.5	78
43	81.5	—	74.5	45.0	15.5	82.5	—	58.0	32.0	77
42	81.0	—	—	44.0	14.5	82.0	—	57.5	31.5	76
41	80.5	—	74.0	43.5	13.5	81.5	—	56.5	31.0	75
40	79.5	—	73.5	43.0	12.5	81.0	—	55.5	—	—
39	79.0	—	—	42.0	11.0	80.0	—	54.5	30.5	74
38	78.5	—	73.0	41.5	10.0	79.5	—	54.0	30.0	73
37	78.0	—	72.5	40.5	9.0	79.0	—	53.0	29.5	72
36	77.5	—	—	40.0	8.0	78.5	100	52.0	29.0	—
35	77.0	—	72.0	39.5	7.0	78.0	99.5	51.5	28.5	71
34	76.5	—	71.5	38.5	6.0	77.0	99.0	50.5	28.0	70
33	75.5	—	—	38.0	5.0	76.5	—	49.5	—	69
32	75.0	—	71.0	37.5	4.0	76.0	98.5	48.5	27.5	—
31	74.5	—	—	36.5	3.0	75.5	98.0	48.0	27.0	68
30	74.0	—	70.5	36.0	2.0	75.0	—	47.0	26.5	67
29	73.5	—	70.0	35.0	1.0	74.0	97.5	46.0	26.0	—
28	73.0	—	—	34.5	—	73.5	97.0	45.0	25.5	66
27	72.5	—	69.5	34.0	—	73.0	96.5	44.5	25.0	—
26	72.0	—	69.0	33.0	—	72.5	—	43.5	24.5	65
25	71.0	—	—	32.5	—	72.0	96.0	42.5	—	64
24	70.5	—	68.5	32.0	—	71.0	95.5	41.5	24.0	—
23	70.0	—	68.0	31.0	—	70.5	—	41.0	23.5	63
22	69.5	—	—	30.5	—	70.0	95.0	40.0	23.0	—
21	69.0	—	67.5	29.5	—	69.5	94.5	39.0	22.5	62
20	68.5	—	—	29.0	—	68.5	—	38.0	22.0	—
19	68.0	—	67.0	28.5	—	68.0	94.0	37.5	21.5	61
18	67.0	—	66.5	27.5	—	67.5	93.5	36.5	—	—
17	66.5	—	—	27.0	—	67.0	93.0	35.5	21.0	60
16	66.0	—	66.0	26.0	—	66.5	—	35.0	20.5	—
15	65.5	—	65.5	25.5	—	65.5	92.5	34.0	20.0	59
14	65.0	—	—	25.0	—	65.0	92.0	33.0	—	—
13	64.5	—	65.0	24.0	—	64.5	—	32.0	—	58
12	64.0	—	64.5	23.5	—	64.0	91.5	31.5	—	—
11	63.5	—	—	23.0	—	63.5	91.0	30.5	—	—
10	63.0	—	64.0	22.0	—	62.5	90.5	29.5	—	57
9	62.0	—	—	21.5	—	62.0	—	29.0	—	—
8	61.5	—	63.5	20.5	—	61.5	90.0	28.0	—	—
7	61.0	—	63.0	20.0	—	61.0	89.5	27.0	—	56
6	60.5	—	—	19.5	—	60.5	—	26.0	—	—
5	60.0	—	62.5	18.5	—	60.0	89.0	25.5	—	55
4	59.5	—	62.0	18.0	—	59.0	88.5	24.5	—	—
3	59.0	—	—	17.0	—	58.5	88.0	23.5	—	—
2	58.0	—	61.5	16.5	—	58.0	—	23.0	—	54
1	57.5	—	61.0	16.0	—	57.5	87.5	22.0	—	—
0	57.0	—	—	15.0	—	57.0	87.0	21.0	—	53

Courtesy of Wilson Mechanical Instrument Co., Inc.

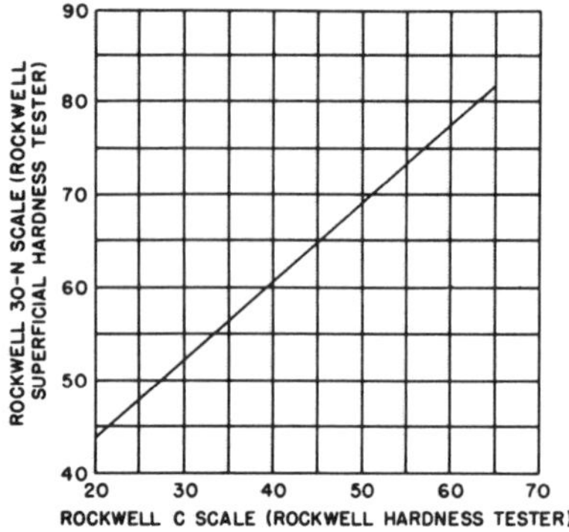

Figure 495

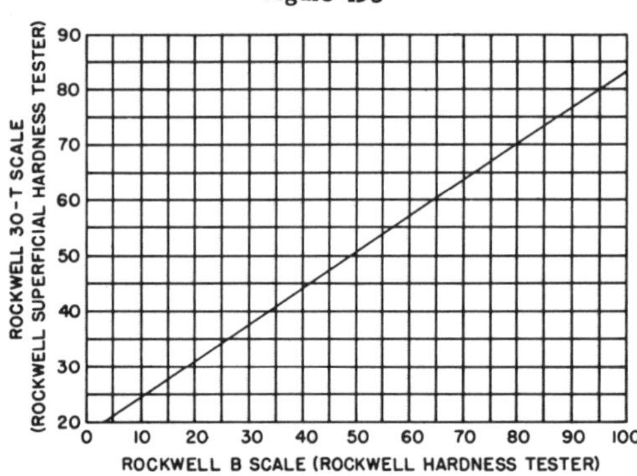

Figure 496

a handwheel (A), and the dial gage is set to zero with the knurled ring at (B). The major load is automatically applied long enough to form the impression and then removed by pressing the switch at (C).

The current for the motor is turned on or off with a switch at (D), and a warning light at (E) reminds the operator to turn off the current when the machine is not in use.

A small lever at (F) controls the depth of penetration. It can be placed in either of two positions--one for standard depth of penetration, the other for shallow penetration when thin specimens are being tested.

ANVILS. Figure 498 shows a variety of interchangeable anvils used on Rockwell hardness testers. The choice of anvil depends upon the size and shape of the work. In addition to those shown, various accessories are available, such as work supports for long rods, a larger Cylindron for supporting large cylindrical work, and large diameter anvils for use where a work support of large area is needed. Cylindrical

Figure 497. Rockwell Motorized Hardness Tester

parts should be supported by an anvil with a "V"; flat work on the "plane" anvil. The pedestal anvil is used when a particular spot on the work is to be supported; the Cylindron is used for cylindrical work.

PENETRATOR ADAPTER. The adapter shown in Figure 499 is available for checking the inside surface of rings, tubes, etc.

TEST BLOCK. A test block (Figure 500) is provided for checking the accuracy of the Rockwell hardness tester. Periodic check with this standard test block will assure that the testing machine is in proper working condition. The hardness, in Rockwell units, is stamped on each test block.

In the operation of the Rockwell hardness tester, certain precautions must be observed to assure accurate and consistent results:

1. When changing anvils, the first reading should be disregarded, as it requires one test to seat the anvil properly.

2. The surface tested must be comparatively smooth and flat.

3. The speed of application of the load is

Figure 498. Anvils for Rockwell Hardness Tester

controlled by a dashpot. When applying a major load of 100 kg, the elapsed time should be approximately five seconds; when applying the 150-kg load, the elapsed time should be approximately four seconds.

4. The under-surface of the work must not show the effects of the indentation.

5. Curvature of the surface may affect the reading. When necessary to test on curved surfaces, curvature of surface should be specified with reading.

6. Never make a repeat test on the same spot.

7. Metals that flow should be kept under major load for a definite time; this time should be specified with the reading.

8. On materials which contain graphite crystals, such as cast iron, the penetrator must be large enough to measure the average hardness of the material.

9. Accuracy of the machine should be checked every day.

10. Locate the machine in a place free from vibration.

11. Do not allow the penetrator to touch the anvil. Any slight indentation of the anvil will affect the accuracy of readings of thin sections.

Hardness measurement is a comparison measurement whose units apply only to the particular type of machine used for the measurement; therefore, the units used on various machines will not have identical values. They will, however, have comparative values, and comparisons of results from different machines can be compared on the basis of conversion tables prepared for most standard hardness testing machines.

BRINELL HARDNESS TESTER

The principle of the Brinell hardness tester (Figure 501) consists in pressing a hard steel ball into the surface of the material to be tested, then measuring the diameter of the impression with a microscope.

The spherical area of the impression is then

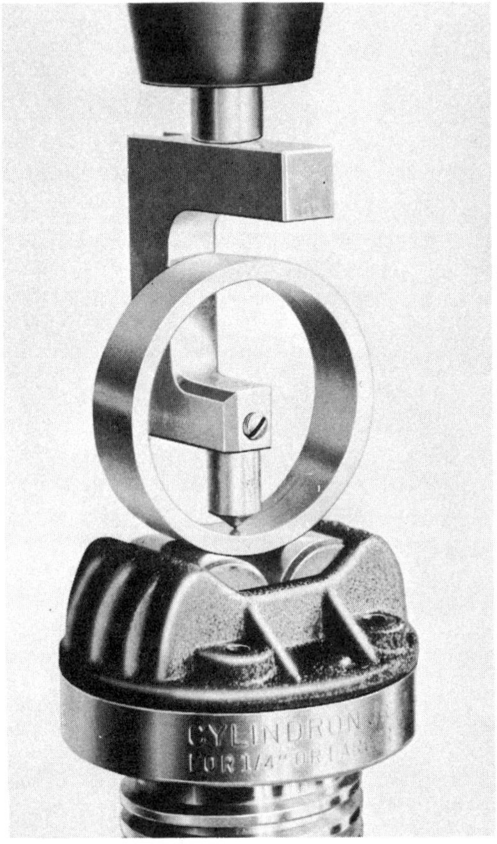

Figure 499. Rockwell Penetrator Adapter

calculated, and the Brinell hardness value is obtained by dividing the area of the impression by the load which was applied to the penetrator in making the impression. The Brinell hardness number, therefore, is the load in kilograms per square millimeter required to crush the material under test.

The Brinell microscope (Figure 502) has a magnification of 20X and a glass scale graduated in tenths of a millimeter. Estimations can be made to 1/20 mm. The Brinell impression is illuminated by a small electric bulb, operating from a small self-contained battery, or 110-volt current. A porcelain checking scale for checking the markings on the microscope scale

Figure 500. Rockwell Test Block

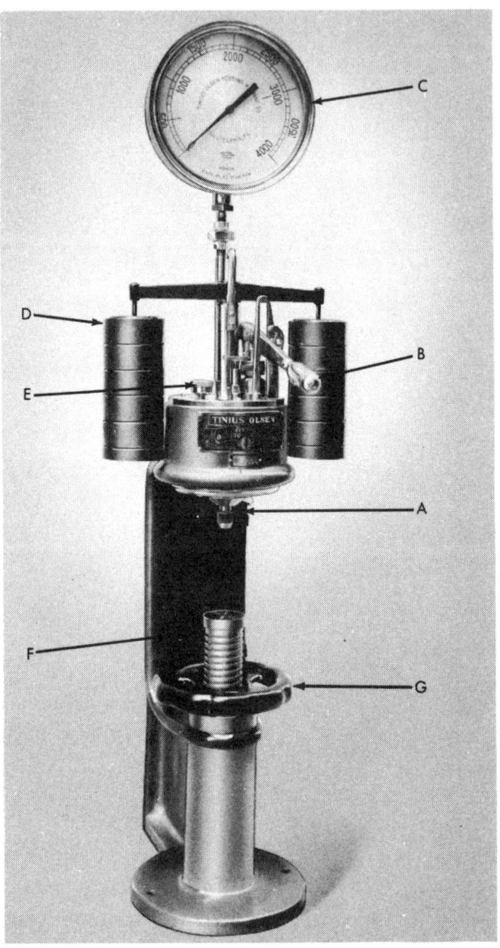

Figure 501. Brinell Hand-operated Hardness Tester

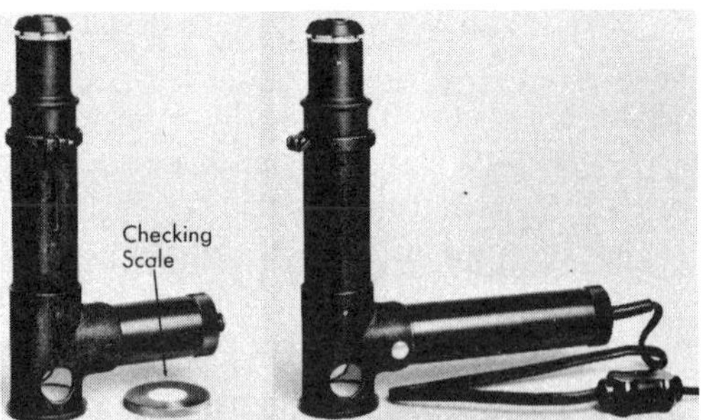

Figure 502. Brinell Microscopes

Diameter of Steel Ball--10 mm.

Diameter of Ball Impression	Hardness Number for a Load of KG.		Diameter of Ball Impression	Hardness Number for a Load of KG.		Diameter of Ball Impression	Hardness Number for a Load of KG.		Diameter of Ball Impression	Hardness Number for a Load of KG.		Diameter of Ball Impression	Hardness Number for a Load of KG.	
MM.	500	3000	MM.	500	3000	MM.	500	3000	MM.	500	3000	MM.	500	3000
2.00	158	945	3.00	69.1	415	4.00	38.1	229	5.00	23.8	143	6.00	15.9	95.5
2.05	150	899	3.05	66.8	401	4.05	37.1	223	5.05	23.3	140	6.05	15.6	93.7
2.10	143	856	3.10	64.6	388	4.10	36.2	217	5.10	22.8	137	6.10	15.3	92.0
2.15	136	817	3.15	62.5	375	4.15	35.3	212	5.15	22.3	134	6.15	15.1	90.3
2.20	130	780	3.20	60.5	363	4.20	34.4	207	5.20	21.8	131	6.20	14.8	88.7
2.25	124	745	3.25	58.6	352	4.25	33.6	201	5.25	21.4	128	6.25	14.5	87.1
2.30	119	712	3.30	56.8	341	4.30	32.8	197	5.30	20.9	126	6.30	14.2	85.5
2.35	114	682	3.35	55.1	331	4.35	32.0	192	5.35	20.5	123	6.35	14.0	84.0
2.40	109	653	3.40	53.4	321	4.40	31.2	187	5.40	20.1	121	6.40	13.7	82.5
2.45	104	627	3.45	51.8	311	4.45	30.5	183	5.45	19.7	118	6.45	13.5	81.0
2.50	100	601	3.50	50.3	302	4.50	29.8	179	5.50	19.3	116	6.50	13.3	79.6
2.55	96.3	578	3.55	48.9	293	4.55	29.1	174	5.55	18.9	114	6.55	13.0	78.2
2.60	92.6	555	3.60	47.5	285	4.60	28.4	170	5.60	18.6	111	6.60	12.8	76.8
2.65	89.0	534	3.65	46.1	277	4.65	27.8	167	5.65	18.2	109	6.65	12.6	75.4
2.70	85.7	514	3.70	44.9	269	4.70	27.1	163	5.70	17.8	107	6.70	12.4	74.1
2.75	82.6	495	3.75	43.6	262	4.75	26.5	159	5.75	17.5	105	6.75	12.1	72.8
2.80	79.6	477	3.80	42.4	255	4.80	25.9	156	5.80	17.2	103	6.80	11.9	71.6
2.85	76.8	461	3.85	41.3	248	4.85	25.4	152	5.85	16.8	101	6.85	11.7	70.4
2.90	74.1	444	3.90	40.2	241	4.90	24.8	149	5.90	16.5	99.2	6.90	11.5	69.2
2.95	71.5	429	3.95	39.1	235	4.95	24.3	146	5.95	16.2	97.3	6.95	11.3	68.0

Figure 503. Brinell Hardness Numbers

is supplied with the instrument.

The size of the ball penetrator is standardized at 10 mm diameter, and the load applied varies between 500 and 3,000 kg according to the nature of the material under test. A load of 3,000 kg is used for iron, steel, and similar hard materials; a load of 500 kg is used for soft metals, alloys, bronzes, etc.

Because calculation of individual tests would be laborious and time-consuming, tables have been prepared from which the Brinell hardness number, corresponding to the diameter of the impression made under a standard load, can be obtained directly. Figure 503 shows the standard Brinell hardness table for loads of 500 or 3,000 kg, using the standard 10-mm diameter penetrator. The corresponding Brinell hardness numbers for impression diameters from 2.00 mm to 6.95 mm can be read directly from this table. Should it be necessary to use loads other than the standard 500 or 3,000-kg loads, the corresponding hardness numbers can be calculated; if warranted, a table can be prepared on the basis of the load applied.

The table (Figure 503) was prepared on the basis of the standard Brinell formula given below. Tables for other pressures can be prepared by using the same formula.

Formula for determining the Brinell hardness number corresponding to the diameter of the impression:

$$H = K\left(r + \sqrt{r^2 - R^2}\right) \div 2\pi r R^2$$

H = Brinell hardness number

K = Load in kilograms

r = Radius of ball penetrator

R = Radius of impression

π = 3.1416

The load is applied hydraulically by means of a hand-operated pump or a motor-driven pump, and the pressure applied to the penetrator is indicated on a pressure gage. The load is applied long enough to fix the deformation of the material. For hard materials, the load is usually applied for about 30 seconds, but for soft materials it may be held as long as two minutes.

BRINELL HAND-OPERATED HARDNESS TESTER

Figure 501 shows the Brinell standard hand-operated hardness tester. The piston at (A) carries a standard 10-mm ball penetrator, and the pressure is applied by means of a hand pump at (B). The pressure is indicated in kilograms on a pressure gage at (C).

A dead-weight mechanism at (D) limits the applied load to that required by the test and safeguards the mechanism against excessive pressure. Weights corresponding to pressures of 500, 1500, 2000, 2500, and 3000 kg are furnished with each machine. The dead weight also checks the accuracy of the pressure gage. A relief valve (E) relieves pressure on the penetrator after the impression has been made. The valve is operated by turning the knurled knob clockwise to close the valve, and counterclockwise to open it. The anvil (F) is raised and lowered by turning handwheel (G).

OPERATION

The test piece should be flat and smooth at the spot where the impression is to be made.

1. On the dead-weight yoke, place weights corresponding to the desired load. Open the relief valve and place the test piece on the anvil so that the spot is under the penetrator. Raise with the handwheel until the test piece is in solid contact with the penetrator.
2. Close the relief valve and apply the load with the hand pump in gentle, even strokes until the desired load is indicated on the pressure gage. Apply the load for a period long enough to fix the impression (usually ten to thirty seconds).
3. Remove the load by opening the relief valve, allowing the liquid to return to the pump reservoir. The relief valve is left open until the next test piece is ready for the application of the load.
4. Remove the test piece and measure the diameter of impression with the microscope. The diameter of the impression should be measured in two directions, each at a right angle to the other, and an average of the two readings used for determining the Brinell hardness number.
5. Refer to the chart corresponding to the load used in the test, and the Brinell hardness number will be found opposite the diameter of the impression.

PRECAUTIONS

Certain precautions should be observed in operating the Brinell hardness tester:

1. Bring the specimen into contact with the penetrator before applying the load. Damage to the machine and errors in measurement may result if this rule is not followed.
2. After each test, open the relief valve a sufficient length of time to allow the penetrator to return to its original starting position.
3. Keep clean. Dirt in the hydraulic fluid or on the surface of the machine will cause faulty operation and defeat the purpose of the tests.
4. See that all hydraulic fittings are tight.

DEPTH-MEASURING ATTACHMENT

A depth-measuring attachment, by which the depth of the impression can be obtained directly from a dial gage, can be mounted on the Brinell hardness tester as shown in Figure 504. The advantage of this attachment lies in the fact that the measurement of the impression with the microscope is eliminated, thus saving considerable time when making tests involving a large number of identical parts.

The depth of impression, either with the load applied or with the load removed, can be obtained directly in millimeters on the dial gage and referred to a chart upon which is tabulated the Brinell hardness number corresponding to the depth of impression. The standard Brinell hardness table for use with the depth measuring attachment is shown in Figure 505. The smallest division on the gage represents .01-mm depth. This table was prepared on the basis of the standard Brinell formula, which follows:

Formula for determining Brinell hardness number corresponding to the depth of indentation:

$H = K \div 2\pi r d$

H = Brinell hardness number

K = Load in kilograms

r = Radius of ball penetrator

Figure 504. Depth-measuring Attachment (Brinell)

Diameter of Ball = 10 mm. One Mark on Depth Instrument = 1-100 mm.

Depth of Ball Impression	Hardness Number for a Load of KG.		Depth of Ball Impression	Hardness Number for a Load of KG.		Depth of Ball Impression	Hardness Number for a Load of KG.		Depth of Ball Impression	Hardness Number for a Load of KG.		Depth of Ball Impression	Hardness Number for a Load of KG.	
MM.	500	3000	MM.	500	3000	MM.	500	3000	MM.	500	3000	MM.	500	3000
.11	145	868	.36	44.2	265	.61	26.1	157	.86	18.5	111	1.11	14.33	86.0
.12	133	796	.37	43.0	258	.62	25.7	154	.87	18.3	110	1.12	14.22	85.3
.13	123	735	.38	41.9	251	.63	25.3	152	.88	18.1	109	1.13	14.08	84.5
.14	114	682	.39	40.8	245	.64	24.9	149	.89	17.9	107	1.14	13.97	83.8
.15	106	637	.40	39.8	239	.65	24.5	147	.90	17.7	106	1.15	13.83	83.0
.16	99.5	597	.41	38.8	233	.66	24.1	145	.91	17.5	105	1.16	13.72	82.3
.17	93.7	562	.42	37.9	227	.67	23.7	143	.92	17.3	104	1.17	13.60	81.6
.18	88.5	531	.43	37.0	222	.68	23.4	140	.93	17.2	103	1.18	13.48	80.9
.19	83.8	503	.44	36.1	217	.69	23.1	138	.94	17.0	102	1.19	13.37	80.2
.20	79.5	477	.45	35.3	212	.70	22.9	136	.95	16.8	101	1.20	13.27	79.6
.21	75.8	455	.46	34.5	208	.71	22.5	134	.96	16.62	99.7	1.21	13.15	78.9
.22	72.4	434	.47	33.8	203	.72	22.1	132	.97	16.42	98.5	1.22	13.05	78.3
.23	69.2	415	.48	33.2	199	.73	21.8	130	.98	16.23	97.4	1.23	12.93	77.6
.24	66.3	398	.49	32.5	195	.74	21.5	129	.99	16.07	96.5	1.24	12.83	77.0
.25	63.7	382	.50	31.8	191	.75	21.2	127	1.00	15.92	95.5	1.25	12.73	76.4
.26	61.3	367	.51	31.2	187	.76	20.9	126	1.01	15.75	94.5	1.26	12.63	75.8
.27	59.0	354	.52	30.6	184	.77	20.6	124	1.02	15.60	93.6	1.27	12.53	75.2
.28	56.8	341	.53	30.0	180	.78	20.3	122	1.03	15.45	92.7	1.28	12.43	74.6
.29	54.8	329	.54	29.5	177	.79	20.1	121	1.04	15.30	91.8	1.29	12.34	74.0
.30	53.0	318	.55	29.0	174	.80	19.9	119	1.05	15.17	91.0	1.30	12.25	73.5
.31	51.3	308	.56	28.5	171	.81	19.7	118	1.06	15.02	90.1	1.31	12.15	72.9
.32	49.7	298	.57	28.0	168	.82	19.5	117	1.07	14.87	89.2	1.32	12.05	72.3
.33	48.2	289	.58	27.5	165	.83	19.3	115	1.08	14.73	88.4	1.33	11.96	71.8
.34	46.8	281	.59	27.0	162	.84	19.0	114	1.09	14.60	87.6	1.34	11.86	71.2
.35	45.5	273	.60	26.5	159	.85	18.7	112	1.10	14.47	86.8	1.35	11.77	70.7

Figure 505. Brinell Hardness Numbers for Depth-measuring Attachment

d = Depth of impression

π = 3.1416

Because of variation in the flow of the material under test, the hardness values obtained with the depth-measuring attachment do not correspond exactly with those obtained by measuring the diameter of the impression with the microscope. The depth-measuring attachment, however, can be used to advantage when testing a large number of identical parts, whose hardness can vary five or more hardness numbers.

BRINELL MOTOR-DRIVEN HARDNESS TESTER

With the motor-driven hardness tester (Figure 506), tests can be made with accuracy and reliability on a fast production basis and with a minimum of effort on the part of the operator.

As on the hand-operated type, the specimen is brought into contact with the penetrator by turning handwheel (A). The load is then applied with the control lever (B), which automatically closes the relief valve and imposes the required load upon the penetrator. The check valve (C) controls the automatic cutoff of the load. It is usually set to take the standard load of 3,000 kg; however, it can be adjusted for various loads between 500 and 3,000 kg by turning the knurled knob.

The load applied to the penetrator is in-

Figure 506. Brinell Motor-driven Hardness Tester

dicated on the pressure gage at (D). By referring to the pressure gage, the operator can check the setting of the check valve and be assured that the load is the same for successive tests on identical parts. The load is removed by reversing the control lever (B), and the penetrator automatically assumes its normal position. One to two minutes are required to fix the impression in soft metals, and thirty seconds for hard materials. The length of time the load is applied depends upon the nature of the material under test.

The Brinell hardness number is obtained by measuring the diameter of the impression and referring to a table of Brinell hardness numbers or by calculating on the basis of load and diameter of indentation.

PRECAUTIONS

1. Do not operate the machine without a full supply of oil in the reservoir.

2. The machine should be allowed to run for several minutes to warm up before making tests.

3. Changes in temperature affect the flow of the oil, and the pressure may vary during long tests. Adjust pressure with the check valve.

4. Inspect the penetrator periodically; replace it if it becomes worn.

5. Machines used continuously should have the oil changed once a year; those used only periodically should have the oil changed every two or three years.

BRINELL HIGH-SPEED HARDNESS TESTER

The high-speed production Brinell hardness tester (Figure 507) was designed to meet requirements for fast, accurate, 100-percent inspection. Dials indicate both the load applied to the work and the depth of hardness impression. The upper dial indicates in kilograms the load applied to the ball indentor. The load can be set or adjusted as required. The lower dial indicates the depth of hardness impression.

Before the tester is set up for a production run on any one material, it is necessary to make a test on that material and check whether the desired hardness characteristics are present. Depth values are secured for the hardness tolerance required by readings taken from the Brinell microscope. From these readings a tolerance range is determined, and the limit hands on the lower dial are set. The tester is then ready for "go" and "not-go" testing.

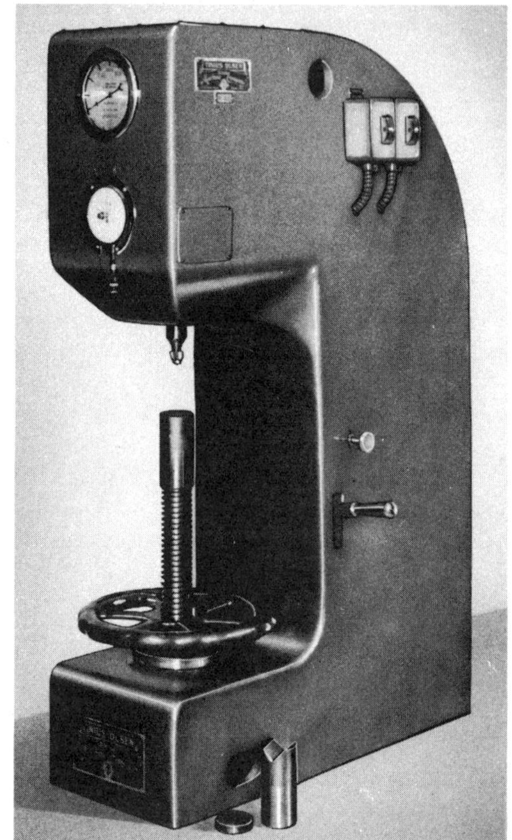

Figure 507. High-speed Brinell Hardness Tester

These machines are designed primarily for testing steel products using 3000-kg load. Adaptations for other Brinell loads, such as 500 or 1000 kg, are available for testing non-ferrous alloys.

SCLEROSCOPE

The Scleroscope differs from the Rockwell and Brinell hardness testers in that the hardness number, or value, is based upon the amount of rebound of a small hammer or "tup" dropped upon the surface of the work when making a hardness test.

DIRECT-READING SCLEROSCOPE

In the direct-reading Scleroscope hardness tester (Figure 508), the rebound of the "tup"

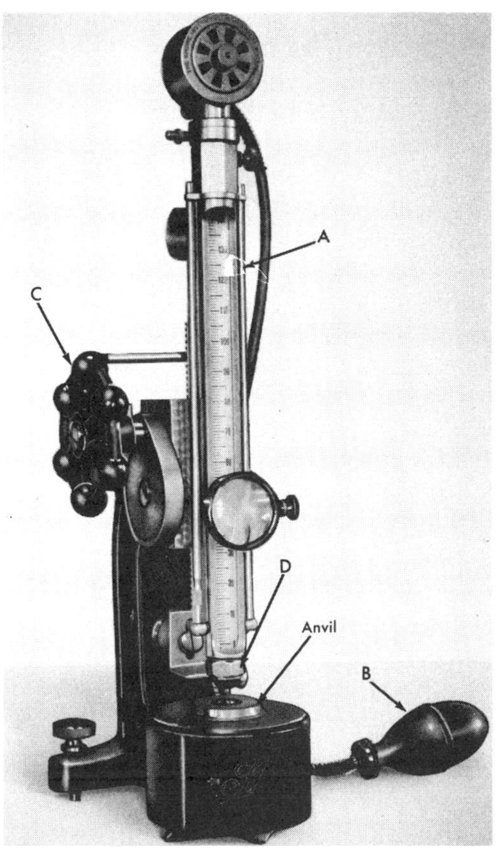

Figure 508. Direct-reading Scleroscope

is noted, visually, against a vertical scale in a glass tube (A). The "tup" is drawn to a predetermined height by a rubber bulb (B) and held there by a spring-operated catch. Upon release by the same rubber bulb, the "tup" falls, striking the surface of the work, and the rebound is noted and recorded. The reading of the scale at the extreme height of rebound is the Scleroscope hardness value of the material.

The Scleroscope is raised and lowered by means of a handwheel (C), and clamped in position with a clamp at (D). The work is placed on the anvil, and the Scleroscope is adjusted until the bottom ring makes solid contact with the surface to be tested. Pressing the bulb releases the "tup;" the rebound is read on the scale as the Scleroscope hardness number.

DIAL-RECORDING SCLEROSCOPE

The dial-recording Scleroscope (Figure 509) records the height of rebound of the "tup" on a dial scale at (A). The dial hand remains at its extreme position until released preparatory to making another test. Because the operator need not watch the scale during the test, this feature saves time; it also eliminates any error in operator judgment when reading the extreme height of rebound on the vertical scale.

Figure 509. Dial-reading Scleroscope

The Scleroscope (B) is mounted on a rack at (C) and is adjustable vertically by the handwheel (D). Rise and fall of the "tup" are controlled by the knob (E).

One of the most important requirements in the use of the Scleroscope is that it be level. Since the fall of the "tup" must be entirely unrestricted, it is necessary that the Scleroscope be in a perfect vertical position during the test. Thumbscrews in the base at (F) provide a convenient means of leveling the instrument. A plumb rod at (G) swings freely when the instrument is level.

OPERATION

1. The specimen is placed upon the anvil at (H) and securely held in place by adjusting the Scleroscope down upon the specimen with handwheel (D), so that it is held firmly between the anvil (H) and the bottom ring (J) of the Scleroscope. It is important that the surface of the specimen upon which the test is to be taken be clean, flat, and comparatively smooth.

2. The "tup" is raised to its initial height by turning knob (E) counterclockwise.

3. When the specimen is in place, the "tup" is released by turning the knob (E) clockwise. At the same time, a ball clutch is automatically moved into position so that when the "tup" rebounds it will be caught at the extreme height of rebound and held there until released preparatory to another test.

4. Thus far there will be no indication on the gage as to the hardness of the material. If the knob (E) is now turned counterclockwise again, the top of the "tup" will contact the spindle of the dial gage and will register the hardness of the material on the scale. A spring catch in the knob holds the "tup" and the pointer of the dial gage in this position until released.

5. In preparing for another test, the knob (E) is turned clockwise, and the "tup" and pointer return to their normal positions.

MINIMUM THICKNESS OF SPECIMEN

When testing hardened steels with the Scleroscope, the minimum thickness that can be tested satisfactorily is about .005". For cold-rolled steel and unannealed brass and steel, the minimum thickness is .010"; for annealed brass or steel, the minimum thickness is .015".

PRECAUTIONS

1. The instrument must be leveled until the plumb rod swings freely.

2. The specimen must be clamped tightly upon the anvil and held there during the test.

3. Surface of the specimen must be clean, flat, and perpendicular to the axis of the Scleroscope.

4. When the specimen is properly mounted, a dull thud can be heard when the "tup" strikes it. If the sound is hollow, or a jingle is heard, the specimen is not properly located.

PORTABLE HARDNESS TESTERS

Portable hardness testers have the advantage of being able to take hardness tests where the non-portable type cannot be used. Time and effort can be saved by taking them to the work, in testing bars and sheets without cutting off samples; in testing large parts, cylindrical parts, odd-shaped parts, assembled units and work difficult to test by other means.

AMES PORTABLE HARDNESS TESTER

The portable hardness tester shown in Figure 510 uses the Rockwell penetration method of testing the hardness of metals by applying pressure to the penetrators by screw action instead of by weights and levers such as are used in large bench-type hardness testers. Tests are made directly in the Rockwell scales, with the penetrators and pressure loads specified in the Rockwell Conversion Chart (Tables XIV-A and

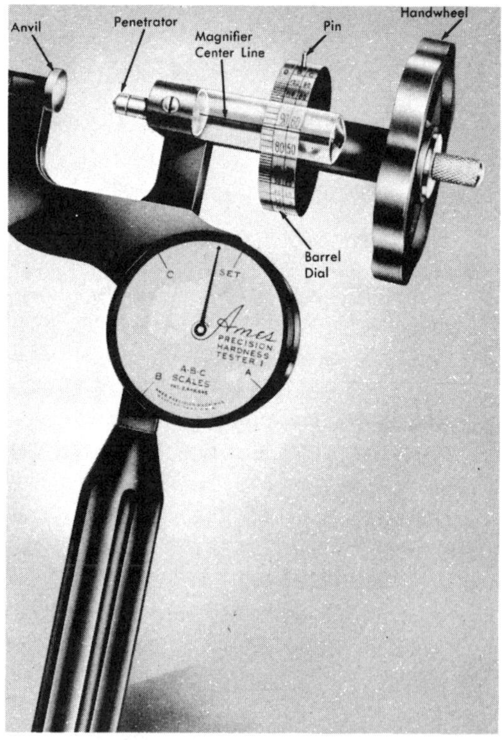

Figure 510. Ames Portable Hardness Tester

XIV-B). Brinell equivalents can be figured.

As the large handwheel is turned to increase pressure on the work being tested, the tester frame is forced open, the lever on the front of the frame lifts, causing the indicator hand to move around the dial.

OPERATION

1. Select the penetrator and test block for the Rockwell scale in which tests are to be made. The indicator hand should rest above the dot at the top of the dial before pressure is applied to the penetrator. Turn the dial by means of the bezel to locate the dot. The handwheel is turned to bring the indicator hand to the line on the dial marked "set;" this applies the minor pressure load to the penetrator. The barrel dial is then turned on the spindle to bring the pin against the top side of the magnifier and the zero line on the barrel beneath the center line of the magnifier.

2. Pressure is then applied to the penetrator by turning the handwheel until the indicator pointer rests above the A, B, or C line on the dial. Line A indicates that 60 kg has been applied; line B, 100 kg; and line C, 150 kg. (On a model for lighter loads, the lines on the dial are marked 15, 30, and 45 for kilogram pressure in the N and T scales.)

3. The handwheel is then turned back to bring the indicator pointer back to the line "set," and readings are taken on the barrel dial through the magnifier (Figure 511). Take a few readings on the test block to be certain that the readings on the tester barrel agree with the markings on the test block.

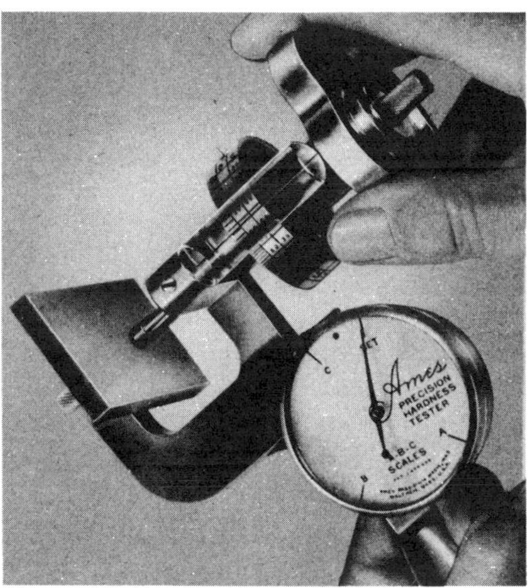

Figure 511. Application of Ames Portable Hardness Tester

This tester is available in different sizes for testing round and flat metals up to four inches. Supplied as standard equipment are diamond and ball penetrators, flat and V-anvils, adaptors and extensions, and test blocks for steel and brass.

ERNST PORTABLE HARDNESS TESTER

The portable hardness tester shown in Figure 512 is a direct-reading instrument designed to test flat, curved, round, and various other shapes of metals. Hardness readings are given directly on a graduated dial. No reference to conversion tables or calculations is necessary.

This tester is available in Rockwell A, B, and C scales, and Brinell scales in medium and low ranges. Either a hardened-steel conical indentor for testing soft and medium-hard metals, or a diamond penetrator for testing extremely hard metals is supplied with the tester.

OPERATION

In use, the tester is placed on the work, and the hand-grips are pressed downward to lower the indentor so that its point (under the

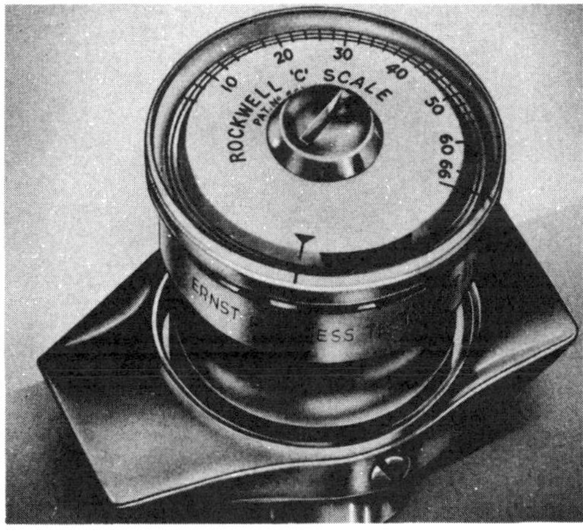

Figure 512. Ernst Portable Hardness Tester

action of a calibrated coil spring exerting a constant load of 15-1/2 lbs.) penetrates into the surface of the material being tested. The movement of the indentor into the material is magnified about 3000 times by displacing a green-colored liquid from a container under the scale, into a capillary tube. The tube, supported in a recess, encircles the scale so that the final position of the liquid indicates the hardness value directly.

Accurate readings are obtained on sheet metal as thin as .020" and round bars above 3/16" diameter. Thin material should be placed on a firm support for testing.

Because of the small size of the instrument (3-1/2" high and 2" in diameter) tests can be made in restricted areas as well as in the center of large sheets or strips. Hardness tests can also be made on work in process without removing parts from machines

THE DUROMETER

The Durometer (Figure 513) is a small portable instrument for measuring hardness and density of materials such as rubber, plastics, leather, and wood. The instrument consists of a quadrant-shaped case, a graduated scale, and a pointer. At the bottom of the case, near the square corner, is an impresser pin projecting downward through an abutment plate. This pin is free to move back into the instrument when pressure is applied. The case contains a miniature beam-weigher mechanism which acts on a standardized resistor spring to measure hardness, in terms of kilograms per square centimeter, by forcing the impressor pin into the material being tested. The Durometer hardness number is indicated by the pointer on the graduated scale.

In use, the Durometer is pressed downward squarely upon the surface of the work being tested so that the abutment stop plate, which is concentric with the impresser point, will rest flat on the work. This is important, whether the instrument is used freehand or mounted in its operating stand. If the abutment plate is not flat against the work, the impresser pin (which operates the indicator hand) will not receive its full stroke, and comparatively low readings will be shown.

The Durometer can be used either freehand as shown in Figure 514 or mounted in a specially constructed operating stand (Figure 515). The operating stand positions the instrument perpendicularly to the work and facilitates its application with constant pressure. The Durometer is screwed into a sliding boss in the operating stand pinion block. After locking the pinion block and the test specimen platform in place with the two knobs on the left, the Durometer can be lowered into firm contact with the work by turning the knob at the right.

The Durometer is available in several types:

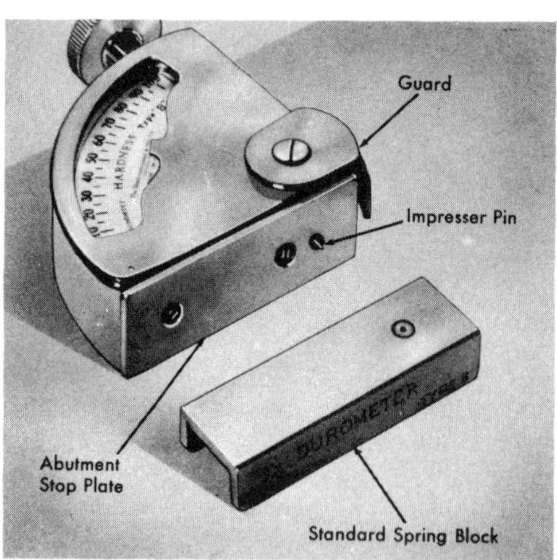

Figure 513. Durometer

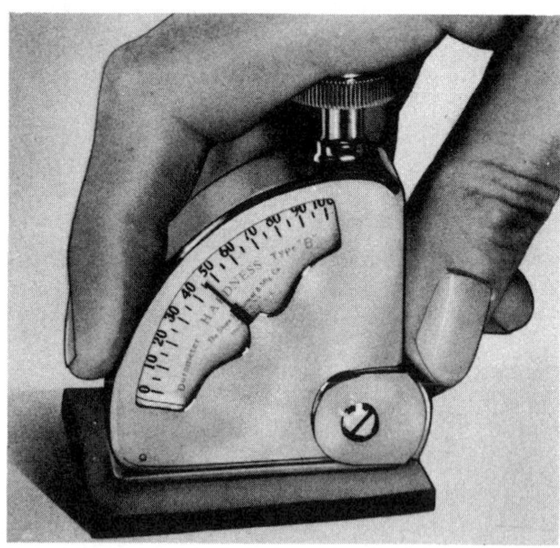

Figure 514. Application of Durometer

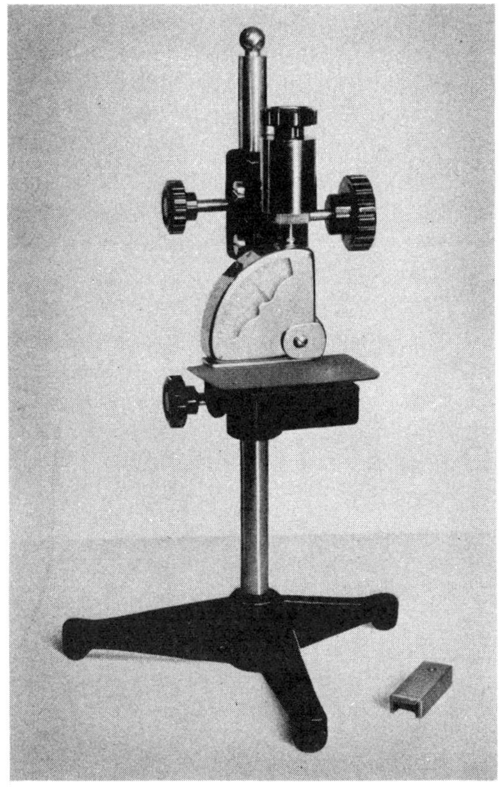

Figure 515. Durometer in Operating Stand

Type 00 for sponge rubber.

Type 0 for printer's rollers, etc.

Type A and A-2 for all elastic rubbers and the softer plastics.

Type B for harder plastics, typewriter rolls, etc.

Type C for medium hard plastics, golf balls, etc.

Type D (sharp point) for materials up to a hardness approaching brittleness.

The Type A-2 Durometer (Figure 516) is an improvement over the quadrant-shaped Durometer shown in Figure 514. The Type A-2 is calibrated in units rather than in multiples of five; the dial is larger and provides more than twice the sweep for the dial hand, resulting in finer reading and improved visibility. It is used the same as the Type A.

A standard spring block is provided with each Durometer for testing accuracy of the instrument. The hardness number of an intermediate part of the scale, usually 60, is stamped

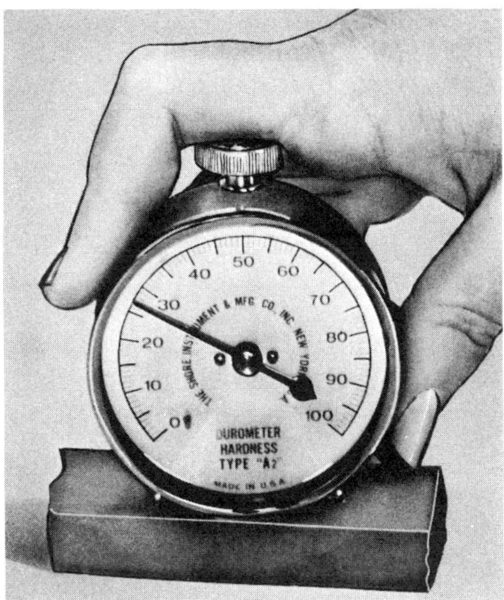

Figure 516. Dial-type Durometer

on the side of the block. In use, the impressor pin is placed in a recess in a button on the spring block and the instrument pressed downward squarely until the bottom plate is in full contact with the flat surface of the test block. If the instrument is in good working order, the pointer should indicate the number stamped on the test block, within a tolerance of plus or minus one point.

QUESTIONS AND PROBLEMS

1. Define hardness.
2. Why do we heat-treat steel?
3. What is the principal hardening agent in steel?
4. Define tempering; normalizing.
5. Give the advantages of casehardening.
6. How is hardness determined with a test file? Is it an accurate method?
7. When is the "brale" penetrator used on the Rockwell hardness tester? The ball penetrator?
8. Why is the Rockwell hardness test made in two stages?
9. How is the minor load indicated on the dial of a Rockwell tester?
10. Is the Rockwell hardness number based upon

the diameter or the depth of the impression made in the material?

11. What limits the minimum thickness of a specimen that can be checked on the standard Rockwell hardness tester?
12. For what special purpose was the superficial hardness tester designed?
13. Specify the various major loads used on the standard Rockwell tester; the superficial tester.
14. Describe the penetrators used on the superficial hardness tester.
15. Interpret the following hardness numbers:
    RC 63
    RB 57
    RA 72
    45 N 63
    15 T 45
16. Explain the principle oı tne Brinell hardness tester.
17. State the range of the applied loads used on the Brinell hardness tester and how they are applied when making a test.
18. State the precautions which should be observed when using the Brinell tester.
19. Explain the principle of the Scleroscope.
20. Give the advantages of the Ames portable hardness tester.
21. How is the load applied to the penetrator for making A, B, or C Rockwell readings on the Ames hardness tester?
22. How is the load applied when using the Ernst hardness tester? How is hardness indicated?
23. Name three materials that can be checked for hardness and density with the Durometer.
24. How is the accuracy of the Durometer checked?

Chapter XV

NON-DESTRUCTIVE TESTING

EXTERNAL INSPECTION with the instruments previously described in this book has as its purpose the development of parts, units, and finished products that meet the specifications of engineers. But what about internal flaws? Little is to be gained by producing a part that meets specifications if the internal structure is such that it breaks down in actual performance. Engineers have long been concerned with the problems of internal flaws in critical parts. Until the techniques of internal testing were developed, attempts to avoid breakdowns were largely confined to specifications for heavier sections at critical places or the use of materials with greater strength. Internal inspection is concerned with methods of determining the presence of hidden flaws, those defects beneath the surface that will cause trouble when the product is placed in actual operation.

Until recently the only means of internal testing involved destruction of the part. In many plants the usual practice was to select parts either at random or at stated intervals, saw them, and examine the internal structure. At best, destructive testing gives only a few of the answers. While the piece destroyed may be satisfactory, the one on either side of it may be faulty. Even if a part appears satisfactory, a saw blade may miss a defect by a few thousandths of an inch or burr over porous metal in a shrink. Destructive testing is truly destructive and on large parts becomes expensive.

A satisfactory solution to the problem of detecting internal flaws is provided by non-destructive testing. Certain types of tests give information about the properties of materials and methods used to produce them which is valuable to research workers and engineers. Non-destructive testing is also important in quality control and inspection. These latter are the functions considered here. The various methods used to detect internal flaws without destroying the part may be considered in four groups: magnetic, fluorescent penetrant, ultasonic, and radiographic. For a basic understanding of magnetic action, some of the fundamentals of magnetism are given.

MAGNETISM

Magnetism is the property of certain substances, usually iron or steel, by which a force (called magnetic force) is created within them and in an area surrounding them. This force either repels or attracts other like substances, depending upon the relation of their respective poles.

The force seems to leave one end of a piece of steel and enter the other end. The end from which it seems to leave is called the north pole. The end at which it seems to enter is called the south pole (Figure 517).

Magnetic force has the property of magnetizing other pieces of steel or iron placed within its field; these other pieces of steel will be magnetized in various directions, depending upon their position when placed within the magnetic field. When placed in the same magnetic field, two pieces of steel, each in the form of a T, will be magnetized in a direction corresponding to their positions (Figure 518).

If a magnetized bar of steel is cut into a number of parts as shown in Figure 519, each

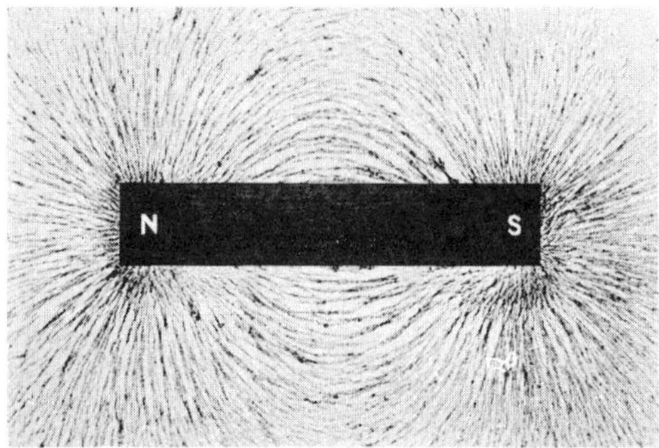

Figure 517. Magnetic Field of Bar Magnet

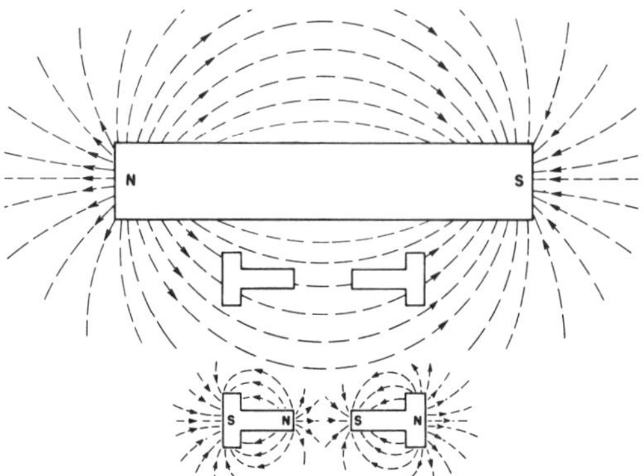

Figure 518. Induced Magnetic Flux

part will become an individual magnet with a field in the direction of the field of the original bar of steel.

If magnetic particles such as iron filings are placed between the ends of these small magnets, each minute particle becomes a magnet and will tend to join together in an effort to bridge the space between the sections of the bar.

MAGNETIC PARTICLE INSPECTION

MAGNAFLUX

The efficiency and life of a machine are continually menaced by the presence of non-visible flaws on or near the surface of the material from which the individual parts are made. Flaws such as cracks, inclusions, laps and folds may, if undetected, eventually reduce the efficiency of a machine and possibly cause a complete breakdown, with attendant loss of production and with hazard to safety.

These defects are sometimes obviously critical, as in jet engine blades, aircraft engines, or automotive steering gear. Less obvious--but well worth detecting for saving money--are defects in tools and dies, in castings and rough forgings before expensive machining, and even cracks in ice skates, sewing-machine parts, or refrigerator compressors. All these are inspected regularly by industry.

A means of detecting the presence of flaws in magnetic materials, such as iron and steel, has been developed in the form of an electrical process with equipment and materials called Magnaflux. In this process, the parts to be inspected are first magnetized. When magnetized, flaws in the metal become leakage fields which are revealed by attraction of finely-divided magnetic particles in powder or paste form, which may be applied to the magnetized parts, either dry or suspended in liquid. Magnaflux powders and paste are made in several colors so as to contrast with the parts being inspected.

Basically, use of Magnaflux employs three units: (1) the first unit magnetizes the work; (2) the second unit applies the magnetic powder; (3) the third unit demagnetizes the work.

In addition to the foregoing, accessory units in the form of inspection tables, viewers, automatic conveyors, and agitators, are available to complete the setup.

MAGNETIZATION

The work is magnetized by inducing a magnetic flux in the work either by passing an electrical current through the work or placing it in the path of magnetic flux. When an electric current is passed through the work, a circular flux is induced which is in a plane at a right angle to the direction in which the current is flowing. For example (Figure 520): if an electrical current passes lengthwise through a conductor, the resulting lines of force will take the form shown in Figure 520B.

When the work is placed in a magnetic field, the induced flux assumes the same direction as that of the field in which it is placed (Figure 518). Such fields are most significant when they can lie along a major axis of the work.

The most important requirement in the magnetization of the work is that the magnetic

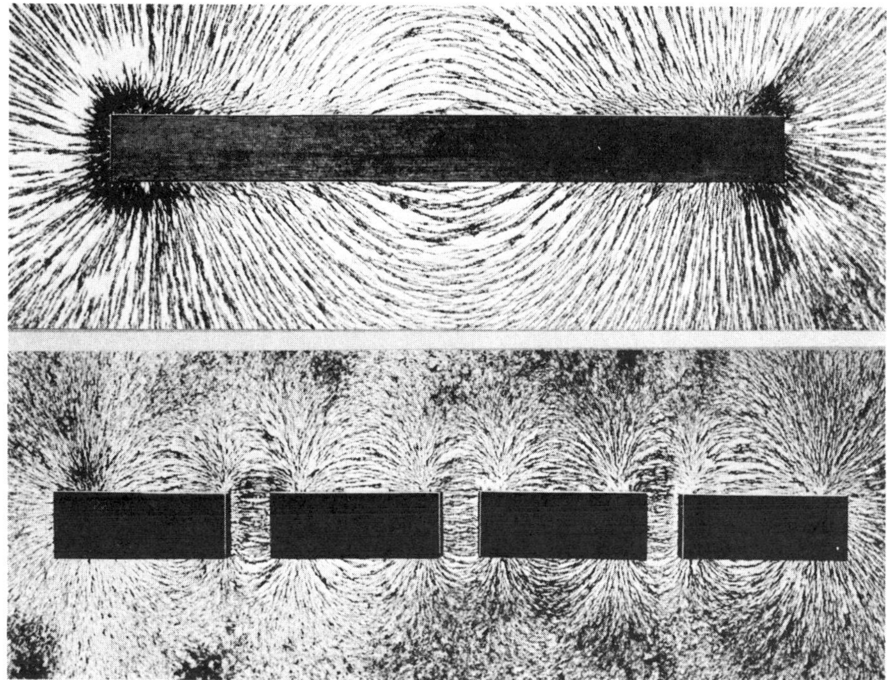

Figure 519. Broken Bar Magnet

flux be induced in a direction at a right angle to the defect. In case of a miniature crack in the material, the flux must pass across the crack and not parallel to it. Therefore, the work must be magnetized in a direction at a right angle to that in which the flaws are assumed to lie. If the direction of flaws cannot be estimated, the work must be magnetized first in one direction, tested with the powder, then the whole process repeated for other directions of magnetization, followed by demagnetization.

If the work is a shaft which the operator desires to check for flaws parallel to its axis, a circular flux should be induced in the work as shown in Figure 520. If it were required to check the shaft for flaws at a right angle to the axis, the flux would be induced by a method similar to that shown in Figure 518. In both cases, the flux would pass at a right angle to the flaw.

APPLICATION OF POWDER

Two methods of applying the powder are available.

The first (called the dry method) consists in dusting dry powder over the magnetized part, then blowing gently to remove excess and to aid in the concentration of the powder over the flaws. Dry powder can be applied by hand shaker, screens, blowing the powder with an air stream, or by tumbling or rolling parts in the powder. When the dry method is used, the part must be perfectly clean and free from grease and dust, although presence of thin paint films will not affect results. Inspection with dry powder is usually carried out with portable magnetizing units on large objects such as refinery vessels, tank or steam turbine castings, heavy weldments, and heavy castings.

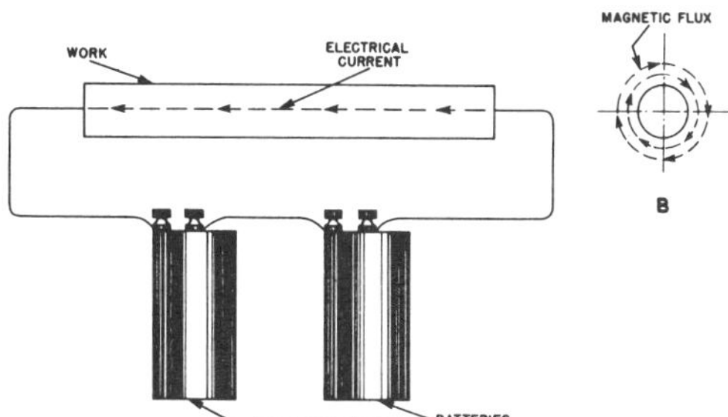

Figure 520. Circular Flux

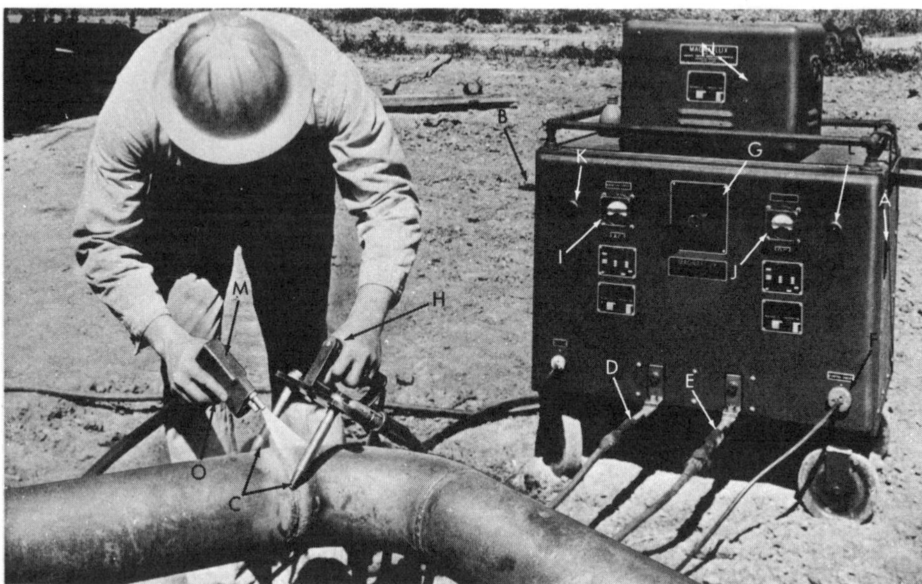

Figure 521. Magnaflux Portable Unit for Use with Dry Powder

Inspection with one of the most typical and widely used portable Magnaflux units is shown in Figure 521. This unit furnishes both AC magnetizing current, and DC (half-wave rectified AC) magnetizing current. This was developed for the maximum sensitivity for sub-surface defects available where surface defects only are to be located, and where freedom from confusion with sub-surface defects is desired. The unit in Figure 521 is used in the field or anywhere in the plant connected to either 220-volt, or 440-volt electric supply.

To operate, supply current switch (A) is turned on. Selector switch (B) is drawn to either the AC or DC position, as desired. Portable hand-held copper prods (C) are attached through two cables to the unit at (D and E). The control cable from the prod is plugged into socket (F). Current is regulated by setting knob (G). Prods are placed in firm contact with the weld or casting, at 6 to 8 inches apart, and the finger switch in the prod handle (H) is depressed; this closes the circuit and closes magnetizing current flow as long as depressed. A bell rings to indicate flow, and the appropriate meter for AC (I) or for DC (J) indicates the magnetizing amperes flowing through the part. At the same time the appropriate light, (K or L) will light to show that current is flowing, visible at a distance.

Magnetic powder is applied while current is flowing through the work, for maximum sensitivity, and the excess is blown off with a gentle air stream while the current is flowing. In Figure 521 this is being done with the hand-held powder gun (M) which is attached to the powder hopper and powder blower unit (N). Powder is blown gently from the gun whenever a finger contact switch is depressed on the gun (O).

After the current actuating switch in the prod handle (H) is released, the part surface is viewed for sharp lines of powder marking surface defects, or diffuse lines of powder marking sub-surface defects. The dry powder is by far most effective in locating such sub-surface defects.

In the second method (called the wet method) the powder is suspended in a thin fluid (low viscosity). The fluid is usually a select high grade of kerosene, or a similar liquid, in which the powder will be suspended without being dissolved, and in which it will not gather together in small lumps.

Either the work is immersed in the liquid, or the liquid is poured over the part. In either method, the powder will build up over the flaw in the form of distinctive patterns which will indicate the nature and size of the flaw.

Figure 522. Magnaflux Unit (Horizontal Wet Type)

As the powder is magnetic, it will tend to bridge the interrupted magnetic path in a manner similar to that shown in Figure 519. The accumulated powder is held magnetically so that the excess powder may be removed without disturbing the pattern.

MAGNETIZING UNIT

Figure 522 shows one of the most widely used types of general-purpose magnetizing units, which supplies both circular and longitudinal magnetization, combined with a means of applying the wet method to the work. The work is held between two contacts at (A) and (B). Contact (A) is crank adjustable to the work by releasing the clamp at (C). Contact (B) is actuated by depressing switch (D) to clamp the part under pressure so that proper electrical contact is made between contact plates and work.

For circular magnetization, the transfer switch (E) is thrown to the "direct contact" position, as shown. A current regulated by the knob at (F) and turned on with the switch at (G) is passed through the work. The strength of the current and length of time it is applied is governed by the size and nature of the part. As a general rule, the amount of current will be approximately 500 to 1,200 amperes per inch-diameter of the part.

For longitudinal magnetization, the transfer switch at (E) is changed to the coil position. In this case, depressing the switch (G) causes current to flow through the magnetizing coil around the part, to magnetize the part longitudinally. The current is registered in the meter at (H). A light at (J) indicates when the magnetizing current is flowing, and when the part is being magnetized. Liquid suspension of magnetic particles is normally flowed over the part by the hand hose and nozzle (K), while the part is being magnetized. The pump is turned on in advance with the switch at (L), and the liquid is agitated in the tank and circulated, but does not flow over the work until the valve in nozzle (K) is opened.

Duration of current shots should be short, and must be governed by the nature of the work. Usually current shots of from 1/10 to 1/2 second are sufficient to magnetize the work. Longer shots may heat the work and burn it at the contact points. If one shot does not magnetize the work sufficiently, successive short-period shots should be applied rather than one long shot.

Shafts, bars, etc., are magnetized by placing them between the contact plates. Hollow cylindrical parts such as bearings, rings, bushings, and nuts can be placed on a standard copper bar as shown in Figure 523. Both these applications will result in a circular flux.

CONTINUOUS OR RESIDUAL METHODS

Inspection can be by either the continuous or residual methods, depending upon the magnetic retentivity of the parts, and the types of defects to be located.

The continuous method consists in applying the magnetic particles while the magnetizing current is being applied. This is the procedure that has just been described in connection with operation of the unit in Figure 522. With the continuous method, inspection for indications of defects can be done right on the unit immediately after the magnetizing and liquid application steps, or parts can be passed to an inspection table, as shown in Figure 524.

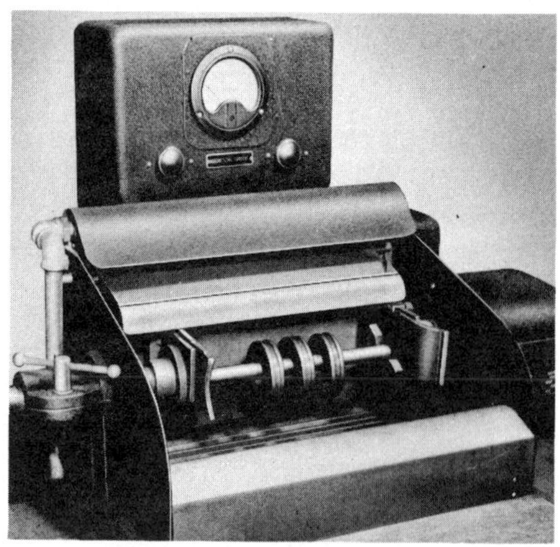

Figure 523. Magnetizing Circular Part

The inspection table (Figure 524) contains a drip board for draining the cleansing liquid from the work, and an overhead lighting system for minimizing annoying high lights.

The viewer being used in Figure 524 contains its own source of illumination and incorporates a lens which magnifies the work so that the patterns are more easily detected.

The residual method can be used on parts which have high magnetic retentivity, such as hardened steel parts that will retain a magnetic field after magnetizing current has been turned off. These parts can have the liquid flowed over them on the unit after the magnetizing current is applied, or can be loaded in baskets and immersed in an agitated liquid suspension of magnetic particles in a separate tank, as shown in Figure 525. This is more convenient in some production situations, or may somewhat affect the types of indications produced in a way that may be desirable on certain types of parts.

The combination immersion and cleaning tank (Figure 525) is divided into two sections: one section contains the immersion liquid in which the pattern is formed after magnetization when the residual method is being used; the other section contains a cleansing liquid in which the excess magnetic material is removed so that any patterns resulting from flaws can be readily identified.

A motor-driven pump enclosed in the unit provides: (1) a means for agitating the liquid so that the magnetic material will be kept in proper suspension; (2) a source of pressure for applying the liquid with a hose.

Alternating current or direct current magnetization may be used. AC is used only where location of surface defects is desired, and where confusion with subtle sub-surface defects is to be avoided. DC magnetization is used wherever

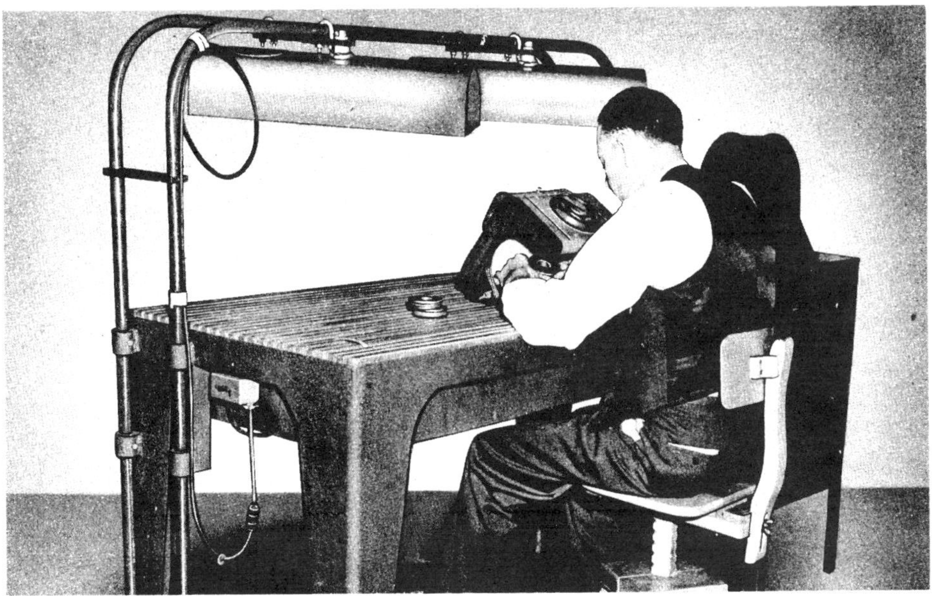

Figure 524. Inspection Table

there is interest in locating sub-surface defects, such as inclusions or internal seams in steel.

To accomplish most efficiently any of these varied combinations of circular or longitudinal magnetization, or use of the continuous or residual method, and for large lots of certain pieces or wide varieties of different sized pieces, a large assortment of units have been designed and are available. Any one of them will be most efficient on a certain inspection problem for certain volumes and varieties of parts.

DEMAGNETIZATION

After the work has been inspected it is usually desirable to demagnetize it. For all practical purposes, demagnetization is not necessary when the material is such that it does not retain the magnetic charge. Iron and soft steels usually lose the charge immediately upon being removed from the magnetic field.

When the work is magnetized first in one direction, then in another, the second charge usually nullifies the first charge so that the work need not be demagnetized between charges.

Certain materials, such as hardened steels, retain the magnetism and must be demagnetized after test. The method most frequently used for demagnetization is to pass the work through the field of a solenoid (coil wound on an iron core) through which an alternating current with a frequency of 25 or 60 cycles is flowing. The 25-cycle current is the more effective; however, as 60 cycles is the prevailing frequency in most localities, it is the frequency most commonly used.

Figure 526 shows a typical demagnetizer used

Figure 525. Combination Immersion and Cleaning Tank

Figure 526. Demagnetizer

in conjunction with a Magnaflux setup. A current controlled by switch (A) is impressed upon the solenoid at (B). A pulsating magnetic field is created in the space within the coil, which has the property of neutralizing or removing the magnetism of parts passed through the field. If the part remains in this pulsating field when the current is shut off, the part will be magnetized; if it is passed in and out of the field, it will be demagnetized. A light at (C) indicates when the current is turned on.

Figure 527 is an example of the pattern created by subjecting a piece of work to the Magnaflux treatment.

MAGNAGLO

The term Magnaglo is applied to auxiliary material for the detection of flaws and irregularities in magnetic materials. It is similar to Magnaflux and uses the same equipment, except that a fluorescent paste is substituted for the regular paste used in Magnaflux, and provisions are made for viewing the work in semi-darkness under "black light."

Figure 527. Magnaflux Patterns

Black light is a term popularly applied to the invisible radiant energy in that portion of the ultra-violet spectrum just beyond the violet of the visible spectrum. It ranges between 3,200 and 4,000 Angstroms in wave length. An Angstrom = .000,000,004 (four billionths of an inch)--a range invisible to the human eye and harmless to the human eye and skin.

Fluorescence is a term used to describe the effect produced by certain chemicals which emit visible light from within when activated by the black light. These chemicals absorb the invisible energy, alter its wave length, and emit the energy in the form of visible light.

Some of the advantages of Magnaglo are:

1. Absence of highlights on the surface

inspected.

2. Increased sensitivity, particularly for defects on the inner surface of hollow parts, and for very fine surface cracks.

3. Time saving in that the indication of the flaws is detected immediately because of the ease with which the eye focuses upon the emitted light.

4. Detection of defects on obscure surfaces like the surfaces of coil springs and threads.

5. Fluorescent indications can be easily seen on rough surfaces. They are independent of surface finish, color, or shape.

Using Magnaglo, a special paste must be used to make a fluorescent wet bath. The visual examination is carried out in semi-darkness under black light. Suitable booths designed for this purpose are provided for inspection with Magnaglo, Figure 528 shows an automatic conveyorized Magnaflux-Magnaglo unit. Automotive steering forgings of a wide range of size and shape are placed on the conveyor at the left, and processing is automatic. After magnetization and application of Magnaglo bath, parts are raised by heads and rotated before the inspector under black light. When he sees a fluorescent line marking a defect he presses a button to cause ejection. Good parts pass out separately through the demagnetizer to be carried to the next processing.

Typical examples of the appearance of Magnaglo treated parts viewed under black light are shown in Figure 529.

DUOVEC

A new technique has been developed for magnetic-particle inspection, called Duovec. This uses the magnetic fields and magnetic particles as does the Magnaflux, and of course, is confined in use to metals that can be magnetized.

The chief advantage claimed for Duovec is that processing and inspection time is about halved. This saving of time is the result of using a magnetic field that magnetizes the part in many directions during one shot, while Magnaflux or Magnaglo bath is applied, and makes possible total inspection in one operation.

Duovec is primarily intended for high-volume inspection in production of particular parts or groups of similar parts and is most effective for parts in cylindrical or bar shape such as bolts, wrist pins, small gears and cam shafts.

FLUORESCENT PENETRATION METHOD

Non-magnetic materials such as aluminum,

Figure 528. Automatic Magnaflux -- Magnaglo Unit

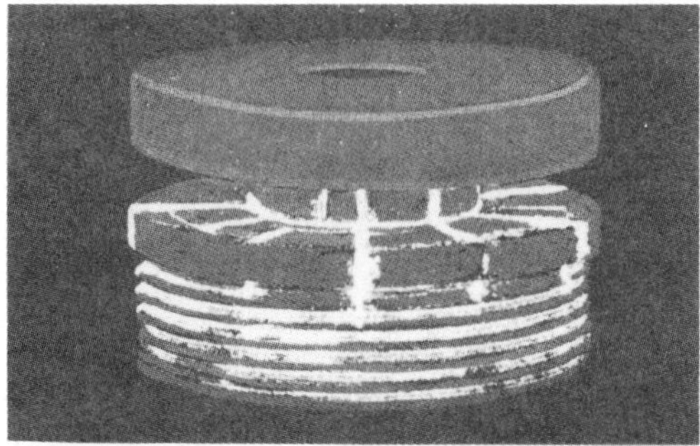

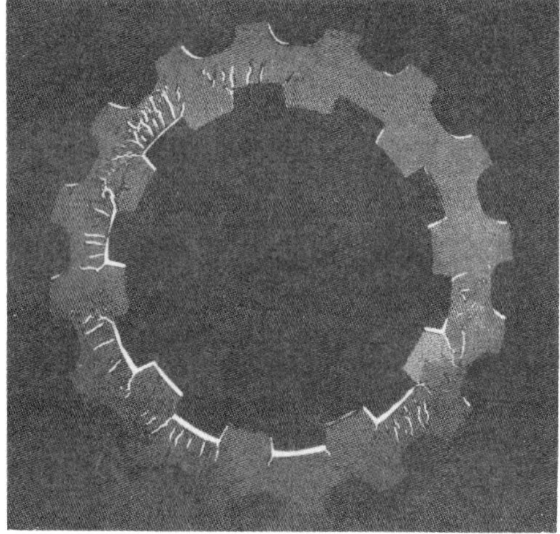

Figure 529. Magnaglo Patterns

brass, bronze, carbides, plastics and some types of stainless steel may be inspected with Zyglo. Also some small magnetic parts difficult to handle, such as ball bearings, are inspected by this method.

As with Magnaglo, Zyglo involves the use of a penetrant containing a highly fluorescent material, but no magnetism is involved. One of the smaller general-purpose Zyglo units is shown in Figure 530.

APPLICATION

The Zyglo oil-base penetrant is applied by dipping, brushing, or spraying. Dipping is the most convenient for small and medium-sized parts. After dipping, the excess liquid is drained from the parts during a time interval required for penetration of the flaws. This interval varies from a minute or so to several hours, depending on the nature of the defects to be located. It is usually from five to fifteen minutes.

Figure 530. Zyglo Unit

RINSE AND DRY

After draining, the parts are rinsed and dried. Rinsing is done most satisfactorily by water sprayed over the parts in a rinse tank. Use of water instead of rinsing by solvents makes this step much less critical, because water does not remove the penetrant from true defects. Parts may be dried after the rinse by wiping, use of a hot-air oven, or an air blower.

DEVELOPING

There are two methods of developing--dry and wet. In the dry method, parts are placed in a tank containing a developing powder and the powder is dusted over them. The developing-powder film on the part helps draw the penetrant out of the flaws by capillary action, thereby intensifying the indications, making them more brillant. In the wet method, after rinsing, parts are dipped into a water suspension of a special developer and then placed in a dryer. A uniform layer of developer is deposited on the surface of the part, resulting in increased brillance (and contrast) of the indication. The developer can be washed off with water.

The operation of actual inspection is conducted under black-light units in a darkened booth, a dark room or a dark area in the plant. Figure 531 shows how Zyglo fluorescent indications glow under black light to mark patterns of cracks in an aluminum casting.

Figure 531. Zyglo Indication on an Aluminum Casting

ELECTRIFIED PARTICLE INSPECTION

A non-destructive means of testing called Statiflux is now available for detection of defects in non-conducting materials, such as ceramics, porcelain enamel, glass and plastics.

Statiflux materials and equipment are used with the electrostatic-particle inspection method to develop indications of cracks and through-pores, or surface scratches, when desired. Whether discontinuities are visible or non-visible, they can be rapidly located with a large visible powder indication, marking each defect so that its extent and nature can be determined immediately.

METHOD

Electrified-particle inspection using Statiflux consists of:

1. Dipping or otherwise coating the part with a hot liquid-penetrant conductor which enters the discontinuities, however fine. Penetrant is not needed on metal-backed objects, such as porcelain enamel.
2. Drying the surface with cloth, air blast, hot air, or other means.
3. Blowing a cloud of fine electrostatically charged particles on the non-conducting surface to be inspected. The powder particles are held electrostatically at the defect and build up a highly visible powder indication.

The penetrant used is usually a water-base material of special preparation which is non-injurious to the non-metal and to the operator, and which develops the best possible indications. The powder is a special non-injurious material available in colors to contrast with various background colors.

Figure 532. Statiflux Indications

Specially developed Statiflux powder gun and nozzles are used to blow the powder in a dispersed cloud. Each powder particle is dynamically charged by its passage through the gun nozzle to be emitted over the surface of the part.

The inspection procedures can be carried out by hand on individual small or large objects, or parts can be carried on conveyors and processed automatically with rapid viewing by the inspector.

Figure 532 shows typical Statiflux powder indications marking invisible cracks in porcelain enamel on a small pan. The star-shaped cracks were caused by impact on the inside, and the large crack pattern was due to heating and quenching.

PARTEK

A non-destructive test for porous materials, called Partek, is a recently developed method for locating cracks in unfired clay bodies such as ceramic insulators, dinnerware, and sanitary ware. Partek quickly locates checks and cracks in any porous bodies, such as powdered-metal compacts, showing errors causing defects in early stages of manufacture, and enables correction of these errors before they become obvious after long firing cycles.

APPLICATION

The porous body is flooded with a special liquid suspension of particles by dipping in an agitated bath, or by spray application to the surface. Differential absorption between a crack and the unmarred surface causes the particles to line up along the crack. A crack indication is clearly marked by a highly visible line of colored particles which build up along the defect and can easily be seen.

No manual operation need be added; all Partek materials burn off completely in the furnace or kiln. Visibly colored Partek materials are generally used in dinnerware inspection to contrast with clayware colors. Inspection under white light permits rapid withdrawal of cracked parts from good products before firing.

To increase speed and visibility fo: some inspection needs, Partek indications can be fluorescent, glowing under black light, rather than visibly colored.

Inspection for cracks in formed or machined clay insulator bodies before firing, or in finished insulators which are porous, is easily carried out with Partek. Figure 533 shows Partek indications of invisible cracks in an unfired ceramic insulator.

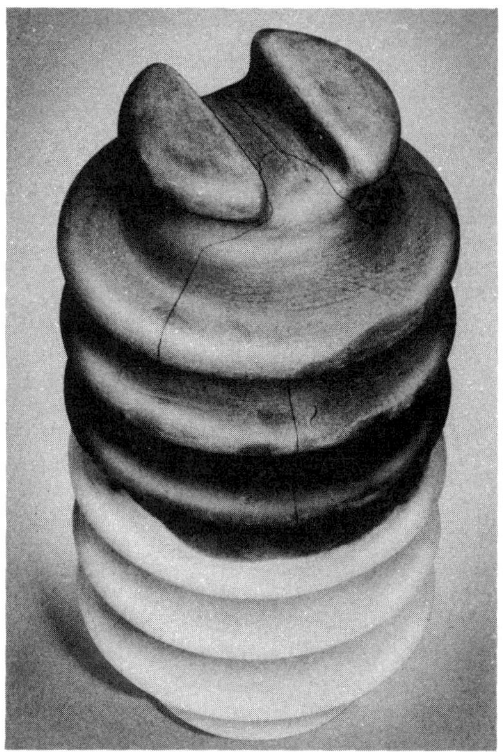

Figure 533. Partek Indications on Ceramic Insulator

ULTRASONIC FLAW DETECTION

One of the newer non-destructive methods is ultrasonic testing, which uses sound waves or vibrations to detect internal flaws in a wide variety of metals and plastics.

REFLECTOSCOPE

Figure 534 shows the reflectoscope, an ultrasonic instrument widely used in non-destructive testing. This instrument can be used to detect flaws in products of fairly uniform thickness and at depths up to thirty feet, depending on the type of material being tested.

The Reflectoscope develops high-frequency, pulsating sound waves which are sent into the material to be tested at right angles to the testing surface by means of a searching unit which contains a quartz crystal. The crystal at the end of the cord is passed over the material. The waves pass from the crystal through the material and are reflected back or echo from the other side. The elapsed time between sending and return is registered and shown on the screen of a cathode-ray tube as vertical indication along a line

Figure 534. Ultrasonic Reflectoscope

of square wave markers. The markers represent units of distance. Internal defects cause a difference in the elapsed time between sending and returning waves and result in deviations from the line. Thus the presence of flaws is shown. Supersonic waves travel in a straight line, or have rectalinear propagation, since they are of a very high frequency. This makes possible the determination of the direction of a defect.

Large objects of relatively simple designs with a minimum thickness of two inches can be tested by using a straight crystal search unit. This is called the "through" method and will detect cracks, bursts, flakes, voids, inclusions and segregations. By using an angle crystal searching unit, fabricated parts, thin metal, and curved sections may be tested for internal flaws.

This method differs from the straight-beam testing only in the manner in which the ultrasonic waves pass through the material under test. Using the angle crystal, the beam is projected into the material at an acute angle to the surface.

According to the manufacturer, sheet materials of .025" thickness have been tested with angle crystals. The crystal is placed near one edge of the sheet, pointing to the opposite edge, and moved sideways to scan the sheet. Scanning along two edges of a sheet will make a complete determination. Scratches .002" deep have been picked up by the Reflectoscope in a sheet .064" thick.

Tubing one inch or larger in outside diameter may be tested by the Reflectoscope using angle crystals. Welds, other than spot welds, can be tested for soundness. Defects such as poor fusion, inclusions and cracks will cause abrupt breaks on the line and are called "indications." While indications show the presence of flaws, there is some difficulty in distinguishing between the various types of defects.

The manufacturers of the Reflectoscope state that the operation of the instrument requires considerable skill in tuning, searching for flaws, and the interpretation of the results, However, a skillful operator can determine the presence and location of defects with assurance.

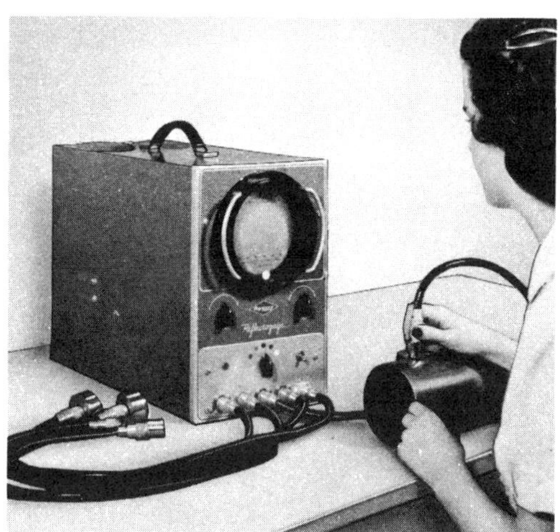

Figure 535. Reflectogage

SONIGAGE AND REFLECTOGAGE

Another instrument that makes use of ultrasonic mechanical waves is the Sonigage. Several firms have been licensed by the developers of the Sonigage to manufacture the instrument under their own trade names, one of which is the Reflectogage, shown in Figure 535. This instrument was developed primarily for measuring thickness where only one side of the object to be tested is available: for example, measuring the wall thickness of hollow propeller blades; it can also be used for detecting defects in materials.

The Reflectogage transmits a mechanical wave of continually varying frequency by means of a quartz crystal. The thickness of the material being tested can be determined by calculating the wave length at the frequency indicated by the peak on the screen. Another application of the Reflectogage is to check bonds in brazed joints. It measures the thickness of the metal at the bond.

SONIZON

The Sonizon, another ultrasonic instrument manufactured under license from the developers of the Sonigage, is shown in Figure 536. The illustration shows the instrument being used to measure wall thickness from one side only. Maximum accuracy is obtained through close contact between the crystal and the material being tested. Flat crystals operate well on flat

Figure 536. Sonizon

surfaces and may also be used on large pipe, or other convex surfaces with greater than four to six-inch radius, when surface conditions are good. For smaller radius convex surfaces, such as one-inch tubing, curved crystals are ground for best match with the surface.

The manufacturer of the Sonizon states that the instrument is effective for any of the following general types of measurements: determining wall thickness of propeller blades, pipe, fabricated metal shapes, and irregularly shaped products, with wall thickness from .010 to 4.00"; locating lack of bond such as bond failures between silver and metal-backed bearings, soldered, brazed, or welded surface layers and back material, laminar built-up materials with sound conducting materials on the outside; detecting defects such as laminations in steel or brass sheets or rolled stock, and thin spots in cast, formed, welded, or ground products.

As in the Reflectogage, the Sonizon scale is marked off to read directly in thickness across the face of the cathode-ray tube. When the crystal searching unit is held against a part, a sharp peak appears on the line across the tube. Location of this peak gives a direct reading of either the thickness of the part or the depth of any defect. Some skill is required to read and interpret the results. Operator training time compares favorably with that required to train a Rockwell hardness tester.

MEASUREMENT OF SURFACE COATING

There are four main reasons for coating surfaces:

1. Protection--to protect the surface against corrosive action of air, water, or other agents with which the surface may come in contact.

2. Decoration--to improve the appearance. Painting a house involves both protection and decoration.

3. Manufacturing Purposes--intermediate, manufacturing operations sometimes require the covering of the surface temporarily to limit the action of treatments given the material. An illustration of covering for manufacturing purposes is the copper plating of part of the surface of the work so that it will not be affected by the carburizing treatment prior to case-hardening.

4. Reconstruction--building up of worn parts to bring them back to their original size.

OXIDATION

Oxidation may be defined as the act of a substance uniting with oxygen. Most metals oxidize when brought into contact with air or water. A common example of oxidation is the rust that appears on iron or steel when they are exposed to the atmosphere.

Oxidation is corrosive because it eats away the substance which is oxidized. Therefore, it is necessary to preserve with a protective coating the metals ordinarily used in the production of machinery and other metallic products. This coating may be oil, grease, paint, or metal upon which the action of oxygen has only a minor effect.

ELECTROPLATING

Metal coating is usually applied by a method known as electroplating, by which the metal is deposited upon the surface by an electrical process. The action is similar to that which takes place in a storage battery. The work and a bar of the metal to be deposited on the work are immersed in a solution. The negative side of the plating current is attached to the work, and it becomes the cathode. The positive side of the current is attached to the metal bar, and it becomes the anode.

When current is passed through the solution, some of the metal leaves the anode and is deposited upon the work. The amount of metal deposited upon the work depends upon the strength of the current and the length of time the work is subjected to the plating action.

Thickness of plating is usually measured because:

1. If metal coating is too thin, it is porous and will not afford full protection to the surface beneath.

2. If too thick: (a) an excess of metal will be used; plating metal is expensive and should not be wasted; (b) extra time will be consumed, which will increase the final cost of the work; (c) work will not be within certain dimensional tolerances required for assembly; (d) it becomes spongy and so will not have the wearing quality of a thinner covering.

THE MAGNE-GAGE

The Magne-Gage (Figure 537) is designed to measure the thickness of surface coating. The principle of the instrument is based upon the magnetic properties of the work and the covering. Measurements can be made of: (1) magnetic coating on non-magnetic base material--for example, nickel coating on metals such as copper, brass and zinc-base die castings; (2) non-magnetic coatings on magnetic materials--for example, zinc, copper and brass coating on iron or steel; (3) magnetic coatings on magnetic materials--for example, polished nickel coating on iron and steel.

The Magne-Gage consists of a base (A) with a column (B) upon which the measuring head (C) is raised and lowered through a rack and pinion by knob (D). When in use, the instrument is placed upon a surface plate (E) having sponge-rubber feet to absorb any vibration, Vibration has a marked effect upon the accuracy of the measurements and should be entirely eliminated if possible.

MEASURING HEAD

The measuring head (Figure 538) contains at (F) a means of suspending a small permanent bar magnet from a horizontal lever arm actuated by a spiral spring (G) so that the attraction between the magnet and the work is indicated on a dial at (H) (Figure 537).

Figure 537. The Magne-Gage

Figure 538. Measuring Head of Magne-Gage

Readings on the dial are converted into thickness of the coating in inches by means of calibration curves supplied with the instrument. Figure 539 shows a calibration curve used in measuring the thickness of non-magnetic plating on steel.

Two dials, one white and the other black, are mounted at (H) (Figure 537) and are turned with knob (J). The scales of the two dials bear no relation to each other, because they are used separately on individual tests; however, the two dials are connected by a gear mechanism having a ratio of 5 to 1. This gear ratio provides a means of rotating the white dial very slowly to determine the reading of the scale at the exact moment the magnet breaks away from the work.

The Magne-Gage must be calibrated with coatings of known thicknesses before it can be used for measurements. This calibration is made at the National Bureau of Standards. Four samples of each of the coatings used in the calibration are furnished with the instrument for the purpose of checking the calibration periodically.

MAGNETS

Four magnets are supplied with the instrument. They are numbered to correspond with the numbers designated on the thickness samples and the calibration curves. Tips of the magnets have been carefully shaped and polished. The tips should be carefully cleaned with soft cloth, lens paper, or clean compressed air before being used.

A magnet holder in which to store the magnets is provided. The magnets should be returned to the holder when not in use. Never allow the magnets to approach one another, or other magnetized material, closer than one inch. The calibration of the instrument is based on the strength of the magnets as supplied. If the magnetism were changed even slightly, the accuracy of the instrument would be impaired.

OPERATION

No attempt to operate the Magne-Gage should be made without first reading carefully the instructions provided with the instrument, and becoming thoroughly familiar with its construction and operation. The following outline is given here for the purpose of illustrating the principle of operation.

Seven different calibration curves are furnished with the Magne-Gage. They are designated alphabetically as A, B, C, D, E, F, and H. The range of each calibration and the magnet used are shown in Figure 540.

MEASURING EXAMPLE

OBJECTIVE: To measure the thickness of zinc plating on steel.

INFORMATION: Thickness of plating less than .002".

PROCEDURE: As zinc is non-magnetic, according to the table (Figure 540), use magnet 1 and calibration chart (A) as shown in Figure 539.

1. Raise measuring head about one inch above the base. Select magnet 1 from the magnet holder. Insert the lower end of the magnet into the magnet guard (A) (Figure 541); then slip the loop of the magnet over both extension pins at (B) on the balance arm so that the loop locates in the small groove in the upper extension pin. The lower end of the magnet should hang freely inside the magnet guard.

2. Place the work on the base, directly under the magnet guard; then lower the measuring head until the magnet guard rests, gently, upon the work.

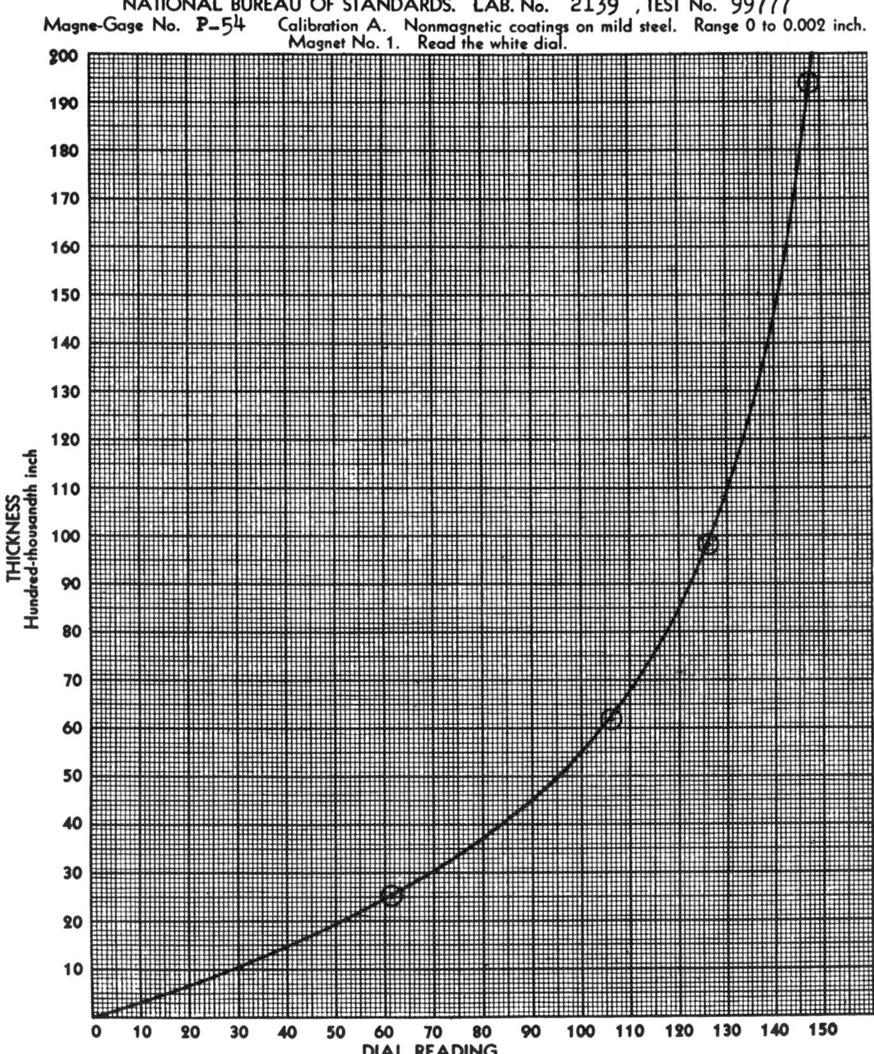

Figure 539. Calibration Curve

3. Slowly rotate the large dial knob counter-clockwise until the magnet makes and retains contact with the work. Then slowly rotate the knob in the opposite direction and note the white dial reading where the magnet leaves the work. This will be indicated by a sudden movement of the pointer (C) (Figure 541). The reading of the dial at this point is then referred to the curve (Figure 539) and the thickness of plating obtained. The reading on the white dial at the moment the magnet left the part is 68. Referring to the chart, we find that the thickness of the zinc plating is approximately .0003".

MAGNET NO.	NON-MAGNETIC COATINGS ON STEEL		NICKEL COATINGS ON STEEL		NICKEL COATINGS ON NON-MAGNETIC METALS	
	CALI-BRATION	RANGE, IN.	CALI-BRATION	RANGE, IN.	CALI-BRATION	RANGE, IN.
1	A	0 TO 0.002	B	0 TO 0.00075		
2	C	0.002 TO 0.007	E	0.0005 TO 0.002		
3	D	0.007 TO 0.025			F	0 TO 0.001
4	H	0.025 TO 0.080				

Figure 540. Range of Magne-Gage Calibrations

CARE OF MAGNE-GAGE

Certain precautions must be observed in operating the Magne-Gage. Some of the most

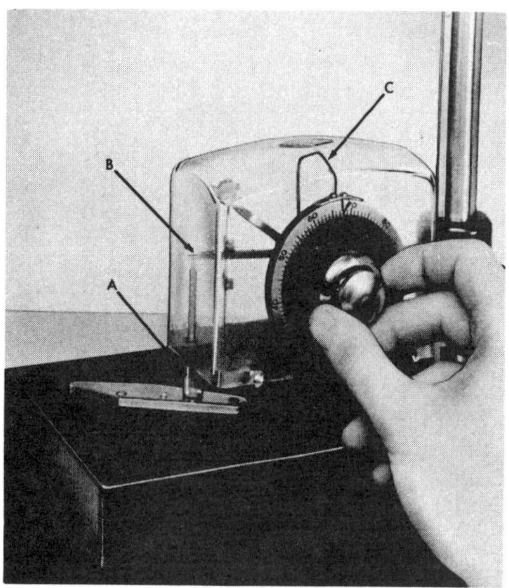

Figure 541. Measuring Thickness of Zinc Plating on Steel

important of these are:

1. Handle carefully.

2. Do not drop the magnets, or allow the tip of the magnet to rub on any material other than the cloth or tissue used when cleaning it.

3. The work must be stationary and the surface of the work flat, smooth, and square with the axis of magnet.

4. Cover the measuring unit at all times except when replacing magnets.

5. Do not lubricate any part of the Magne-Gage.

6. Do not attempt to repair or adjust the Magne-Gage. Any adjustment or change in the mechanism will change the calibration. If magnets or any part of the instrument have been impaired, the whole instrument including magnets and calibration samples must be returned to the manufacturer for repair and recalibration.

METALS COMPARATOR

A fast, non-destructive test of the quality of all magnetic and non-magnetic metals is provided by an electronic device called the metals comparator. By comparing with a standard, metal parts and stock--appearing alike but differing either in composition, heat-treatment,

Figure 542. Testing Part in Metals-Comparator Coil

hardness or plating thickness--can be separated into acceptable and non-acceptable groups.

The metals comparator consists of an electronic unit in a steel cabinet, and a test unit which may be either a test coil or a test head. The electronic unit consists of a balancing network, an oscillator, and an indicator. The control knobs, for setting to the desired sensitivity and frequencies, and the indicator are mounted on the front panel.

When testing small metal parts or parts with small diameter, test coils are used as shown in Figure 542. The test head is used on surfaces of large equipment and parts which cannot be inserted in a test coil, as shown in Figure 543.

In operation a reference specimen, as determined by laboratory or operating tests, is placed in the test coil or against the test head to secure the initial balance for the group of

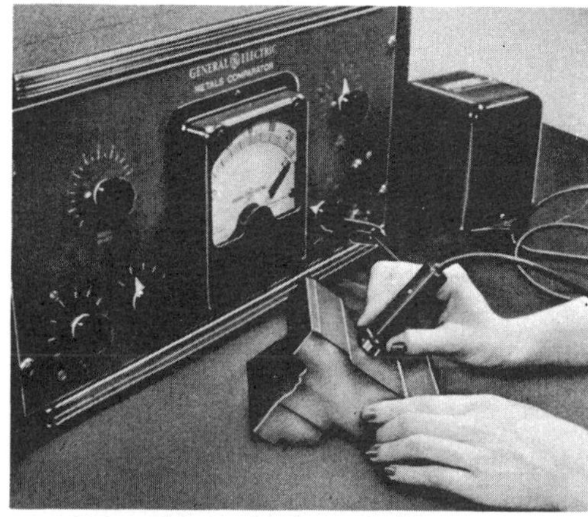

Figure 543. Using Test Head with Metals Comparator

specimens being compared. This balance is indicated by a zero reading on the indicator. After the reference specimen has been removed, the parts to be tested are held briefly against the test unit, just long enough to allow the indicating pointer to come to rest. If the parts to be tested have different properties than the reference specimen, there is a change of test-unit impedance which upsets the balance and will show on the indicator scale.

After tolerances have been established, a specimen can be accepted or rejected on the basis of the scale reading.

Typical applications of the metals comparator are:

1. Separation of metals such as: steel differing by one point of carbon, annealed from unannealed steels, nickel-chrome steel from nickel-chrome molybdenum steel, hard brass from soft brass, aluminum alloys.

2. Indicating changes in plating thickness to .0005" on nickel on non-magnetic or magnetic metals.

3. Measuring depth of casehardening for materials that have been casehardened by induction heating.

The manufacturers of the metals comparator state that, unless the operator of the instrument is experienced, the preliminary work of setting up the test should be done by the supervisor. This inculdes determining and selecting the proper test coil or test head, and determining the sensitivity, frequency, and test limits of acceptance. When these have been established, the regular operator can proceed with the routine tests. However, the regular operator of this equipment should be under the direct supervision of someone with technical experience in metallurgy or electrical engineering, or one who has acquainted himself with the principles and operation of the comparator.

RADIOGRAPHIC FLAW DETECTION

X-ray, long recognized as a valuable aid in the medical profession, has now made a place for itself in inspection procedures in the metal-working industry.

Inspection specifications designed to secure satisfactory performance by determining the soundness of internal structure are largely responsible for the use of radiography in industry. Radiography provides a permanent, visible record of internal condition that can be used as a basis for positive decisions as to soundness or unsoundness of a product or any of its parts. In addition to assuring the customer a sound product, the manufacturer often achieves considerable saving by the elimination of unsatisfactory parts before expensive machining is done on them or valuable cutting tools are damaged.

THE X-RAY IN INDUSTRY

Industrial radiography is finding many new uses, and spectacular reports have been made of its advantages. Its use by industry extends from basic operation to final assembly. Its job is to assure satisfactory performance, improve quality, and lower costs. It is proving itself indispensible to designers, engineers, production men, and inspectors.

X-rays are invisible but, like visible light, are a form of radiant energy. Their distinguishing feature is their extremely short wave length--approximately 1/10,000 the length of visible light, or even less. The shortness of X-rays permits them to penetrate materials which absorb or reflect ordinary light.

X-rays exhibit all of the properties of visible light, but in such a different degree as to modify greatly their practical behavior. For example, light is refracted by glass and, consequently, is capable of being focused by a lens in such instruments as cameras, telescopes, and spectacles. X-rays are also refracted, but to such a slight degree that the most exact experiments are required to detect this phenomenon. Therefore, it is impossible to focus X-rays.

Basically, the device in which X-rays are generated is an X-ray tube in which electrons, traveling at great speed in a high vacuum, under stress of high voltage, collide with a tungsten target. The sudden stopping of these electrons in the surface of the target results in the generation of X-rays. Figure 544 shows a diagram of a high-vacuum X-ray tube with tungsten target.

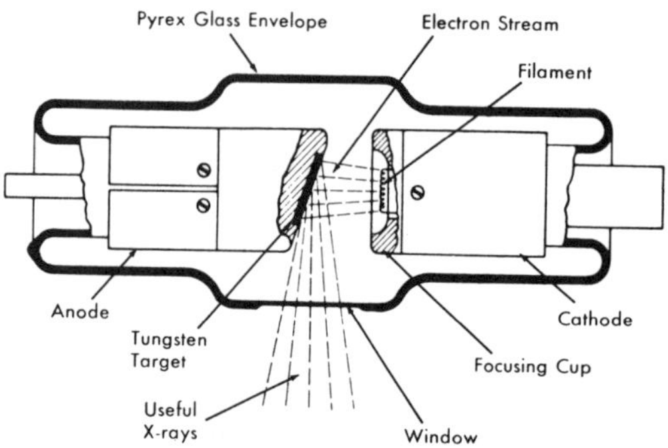

Figure 544. High-Vacuum X-Ray Tube

An X-ray film, or radiograph, of any object is similar to a photograph, except that it is fundamentally a shadow picture rather than a picture produced by reflected light. The source of "light" is an X-ray tube instead of an ordinary lighting appartus, and the "light and darks" of the film depend on the differences in the density of the object situated in the path of the X-rays. The denser or the thicker the object, the more X-radiation will be absorbed by it, and less will reach the film. The amount of X-radiation which reaches the film, in a light-tight holder behind the object, determines the amount of exposure which is expressed in terms of "black and white." The greater the density or thickness of the object radiographed, the whiter will be that particular area in the film. So the radiograph is not simply a picture but a record of shadows of different parts according to their respective densities or resistance to X-ray.

A radiograph, such as the one in Figure 545 of a pump housing showing shrinkage, is a shadow picture. The picture is secured by placing the object in the path of an X-ray beam. When properly developed, the resulting picture may be viewed to determine whether the object has internal flaws that will cause it to break down in operation.

Because of the great need for ammunition that will not fail in use by the armed forces, large X-ray installations such as the one shown in Figure 546 are used in ordnance plants. Shells filled with liquid TNT are placed in the continuously-moving circular rack with film behind them. As they pass into the thick-walled room the radiographs are made. In some instances, radiographs showed that sufficient shrinkage or other defects existed to make the shells unsound.

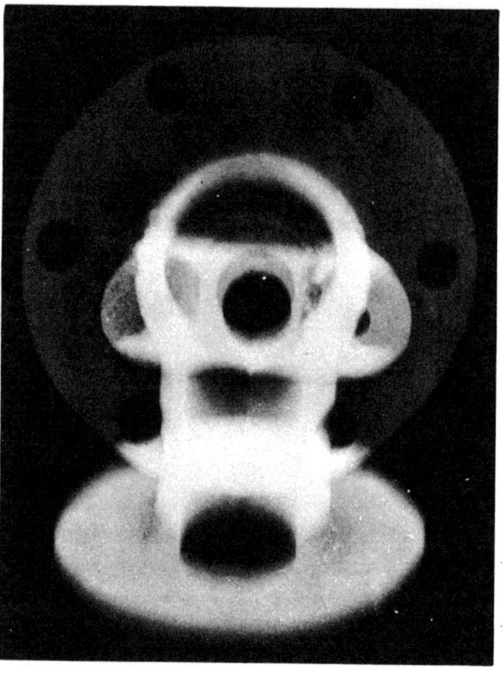

Figure 545. Radiograph of Pump Housing Showing Shrinkage

Figure 547 shows a large crankshaft being made ready for X-ray inspection. The film has been placed under the area to be X-rayed by the 1,000,000-volt unit.

Defects in airplane parts are critical because of the potential loss of life and property in case of failure. Radiographs have prevented failure of many such parts. Figure 548 is a radiograph of an airplane crankcase in which shrinkage has occurred. No other method of inspection shows the internal structure of such a part so clearly or forms such a valuable basis for judgment as to the soundness of the part.

Radiographic inspection is not only applicable to large single parts but to assemblies as well. Figure 549 is a radiograph of a pump and valve. Inspection by radiographs of either parts or assemblies after a period of service gives information on wear and fatigue that is valuable in developing a better product.

Radiography of welds for defects has assumed a definite place in routine tests of boilers and

Figure 546. Two-Million-Volt Industrial X-Ray

pressure vessels and its practicability and usefulness are no longer questioned. A radiograph gives assurance of the homogeneity of the deposited weld material. Figure 550 shows a radiograph of a series of welds.

Various X-ray machines commercially available may be roughly classified according to their maximum voltage. The choice of a machine depends on the type of work to be done. Machines may be either fixed or mobile depending on the specific uses for which they are intended. When the material to be radiographed is portable, the X-ray machine is usually permanently located in a room so that nearby personnel are protected against the escape of X-radiation.

GAMMA RAYS

Gamma rays, also used extensively in making radiographs, are the product of the disintegration of radium and are similar to X-rays but are confined to the shorter wave-length area. However, the intensity of radiation is generally less than that obtained by electrically produced X-rays; so longer exposure times are required in making radiographs.

Capsules containing radium salts are used for making radiographs. Gamma rays emerge from a radium capsule with equal intensity in all directions. It is therefore possible to take radiographs in all directions from the radium at the same time, if the areas to be tested can be arranged on all sides of the radium. For example, in radiographing a cylindrical casting, it is possible to place the radium on the centerline of the cylinder and place a band of films entirely around the circumference of the cylinder for simultaneous exposure. This can also be done with 1,000,000 and 2,000,000 volt X-ray units now in use.

Some of the advantages claimed for gamma-ray radiography are that no expensive equipment is needed, no electrical power is consumed, and the necessary equipment is highly portable. The

Figure 547. X-Ray Inspection of Large Crankshaft

purchase price of the radium is the only cost. While the initial cost is high, the half-life period is about 1700 years. Radium may be leased by those who have short-term or infrequent use for it.

As with any other type of inspection, radiography has its advantages and disadvantages. Used properly, radiography supplies information not available to the inspector in any other way.

The thickness of material which can be penetrated by X-ray varies with the voltage of the equipment, the distance of the film from the X-ray tube, the density of the material, and the time of exposure. For example, a 250,000-volt X-ray unit will penetrate two inches of steel in about 12 minutes from a 36" distance. With a 400,000-volt unit the same exposure requires only two minutes. Under the same conditions, a 1,000,000-volt unit will penetrate six inches of steel in 60 minutes, while a 2,000,000-volt unit will do the job in less than 1-1/2 minutes. Working at a 48" distance, a 10,000,000-volt betatron-type X-ray machine, will penetrate 11" of steel in 10 minutes, or 12 inches of steel in 17 minutes. By contrast, 200 mg of radium will penetrate 3-1/2" of steel in 24 hours.

Operation of the X-ray equipment requires careful instruction, but is not difficult to learn. Specially-trained technicians are not needed

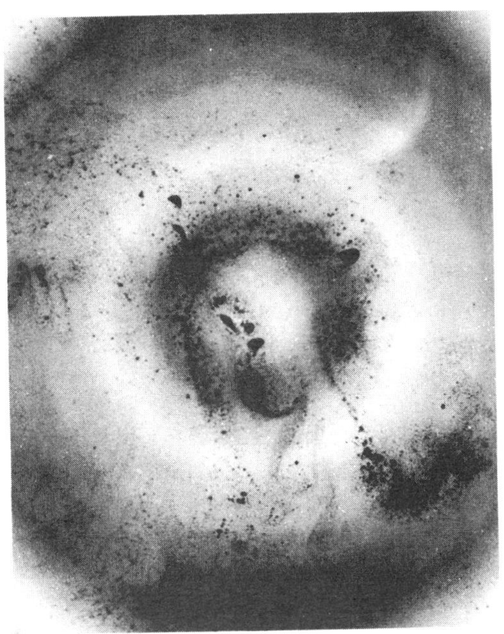

Figure 548. Radiograph of Airplane Crankcase Showing Shrinkage

for making radiographs using either X-rays or radium. A few days' instruction is all the operator needs. However, interpretation of the information that can be derived from an examination of the radiographs requires considerable practice and skill.

SAFETY PRECAUTIONS

One of the most important considerations in the X-ray or gamma-ray laboratory is the provision and use of adequate safe-guards for the personnel. The American Standards Association has developed safety standards for X-ray or gamma-ray installations.

Body tissues may be injured by over-exposure to X-rays or gamma rays, the blood, skin, and some internal organs being particularly sensitive to these rays. For this reason, persons who work with X-ray may be exposed regularly to small quantities of X-rays or gamma rays and should wear film badges or radiation monitoring devices, and have periodic blood counts and physical examinations under the direction of a physican.

SUMMARY

A combination of radiographic and ultrasonic or magnetic and non-magnetic particle inspection gives a complete analysis of internal flaws. Deep-lying flaws are best detected by radiographs, but not all X-ray equipment is suitable for revealing the location of fine cracks or laminations in or near the surface. Inspection for such flaws by magnetic-particle tests supplements radiography, and the combination of the two techniques provides what may be considered a complete inspection.

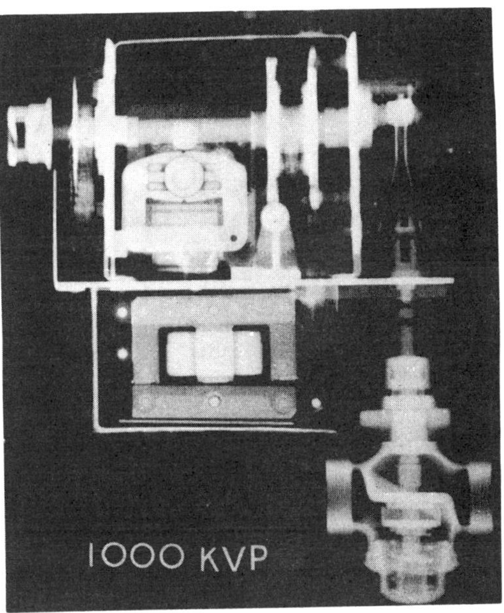

Figure 549. Radiograph of Pump and Valve

In addition to aiding the search for hidden flaws that make parts break down in service, non-destructive inspection has other advantages.

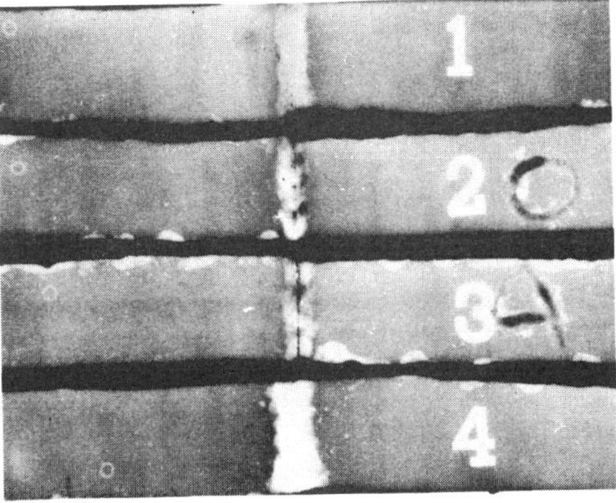

Figure 550. Radiograph of a Series of Welds

The raw materials coming into a fabricating plant can be inspected and defective parts can be eliminated before expensive machining is done or valuable tools damaged. Non-destructive testing provides information upon which judgments can be based as to the weight of the metal that is necessary to provide for certain safety factors. Non-destructive testing is also valuable in maintenance because it is a means of measuring fatigue cracks that may finally result in failure of the equipment.

NOTE: Materials and illustrations for this chapter were supplied by the following firms: American Instrument Company; Ansco; Canadian Radium and Uranium Corporation; Eastman Kodak Company, X-Ray Division; General Electric Equipment Division; General Electric X-Ray Corporation; Magnaflux Corporation; Shore Instrument Company, and Sperry Products, Inc.

QUESTIONS AND PROBLEMS

1. What factors have been responsible for the recent emphasis upon non-destructive inspection?
2. What type of defects are detected by magnetic particle inspection?
3. Describe the basic steps in the Magnaflux method of inspection.
4. If the direction of flaws cannot be estimated on a part, in what direction would you magnetize the part? Why?
5. Explain how the powder used in the Magnaflux method helps to detect flaws in iron or steel parts.
6. How may the powder be applied when checking a part with Magnaflux?
7. What determines the strength of the current and the length of time it is applied when Magnafluxing a part?
8. How is a part demagnetized after inspection by the Magnaflux method?
9. Explain briefly the Magnaglo method of inspection and give some of its advantages.
10. In the Zyglo method, why is a fluorescent liquid penetrant used?
11. Describe the Statiflux method of inspection.
12. For what type of materials is the Partek process particularly adapted.
13. Describe the Reflectoscope method of thickness measurement.
14. List three uses of the Magnaflux Sonizon other than measuring material thickness.
15. Give four chief reasons for coating surfaces.
16. Why is it necessary to measure the thickness of metal plating?
17. Name three different types of plated surfaces that can be checked with the Magne-Gage.
18. Explain the principle of the Magne-Gage.
19. What would the plating thickness on a part be if the dial reading on the Magne-Gage were 90? Use chart in Figure 539.
20. Give three typical applications of the metals comparator.
21. What valuable information can be obtained from use of X-ray in the metal working industry?
22. What are gamma rays?
23. What are some of the advantages of gamma-ray radiography over X-ray radiography?
24. How is the density or thickness of a part expressed on an X-ray film?
25. What factors determine the thickness of materials that can be penetrated by X-ray?

INDEX